Springer Series in Synergetics

Editor: Hermann Haken

Synergetics, an interdisciplinary field of research, is concerned with the cooperation of individual parts of a system that produces macroscopic spatial, temporal or functional structures. It deals with deterministic as well as stochastic processes.

The Physics of
Structure Formation

Theory and Simulation

Proceedings of an International Symposium,
Tübingen, Fed. Rep. of Germany,
October 27 – November 2, 1986

Editors: W. Güttinger and G. Dangelmayr

With 207 Figures

Springer-Verlag Berlin Heidelberg New York
London Paris Tokyo

Professor Dr. Werner Güttinger
Dr. Gerhard Dangelmayr
Institut für Informationsverarbeitung, Universität Tübingen
D-7400 Tübingen, Fed. Rep. of Germany

Series Editor:

Professor Dr. Dr. h. c. Hermann Haken
Institut für Theoretische Physik der Universität Stuttgart, Pfaffenwaldring 57/IV,
D-7000 Stuttgart 80, Fed. Rep. of Germany and
Center for Complex Systems, Florida Atlantic University,
Boca Raton, FL 33431, USA

ISBN-13: 978-3-642-73003-3   e-ISBN-13: 978-3-642-73001-6
DOI: 10.1007/978-3-642-73001-6

© Springer-Verlag Berlin Heidelberg 1987
Softcover reprint of the hardcover 1st edition 1987

2153/3150-543210

Preface

The formation and evolution of complex dynamical structures is one of the most exciting areas of nonlinear physics. Such pattern formation problems are common in practically all systems involving a large number of interacting components. Here, the basic problem is to understand how competing physical forces can shape stable geometries and to explain why nature prefers just these.

Motivation for the intensive study of pattern formation phenomena during the past few years derives from an increasing appreciation of the remarkable diversity of behaviour encountered in nonlinear systems and of universal features shared by entire classes of nonlinear processes. As physics copes with ever more ambitious problems in pattern formation, summarizing our present state of knowledge becomes a pressing issue. This volume presents an overview of selected topics in this field of current interest. It deals with theoretical models of pattern formation and with simulations that bridge the gap between theory and experiment. The book is a product of the International Symposium on the Physics of Structure Formation, held from October 27 through November 2, 1986, at the Institute for Information Sciences of the University of Tübingen. The symposium brought together a group of distinguished scientists from various disciplines to exchange ideas about recent advances in pattern formation in the physical sciences, and also to introduce young scientists to the field.

Our sincere thanks are due to many. First and foremost to the lecturers for their stimulating presentations and the extra effort they took to make them into authoritative reports. The generous financial support for the symposium was provided by the Stiftung Volkswagenwerk of the Federal Republic of Germany. We are particularly indebted to Dr. H. Plate of the Foundation for his interest and help with our projects. Furthermore, we are pleased to thank the Vice-President of the University, Dr. E. Lindner, and the Chancellor, Dr. G. Sandberger, for their efficient cooperation.

It is a most pleasant task to express our gratitude to Mrs. E. Allmendinger, G. Ambros, C. Ott, and V. Müller for their invaluable assistance in the organization of the symposium and to Mrs. A. Kluwe, M. Resch, and C. Trost for their help in preparing the manuscripts. Finally, we thank our colleagues D. Armbruster, C. Geiger, P. Haug, D. Lang, and M. Neveling for their help throughout and also Dr. H. Lotsch of Springer-Verlag for his patient cooperation.

Tübingen
June, 1987

W. Güttinger
G. Dangelmayr

Contents

Part III   Interfacial Patterns

Part IV   Diffusion Limited Aggregation and Fractals

Part V   Chaos and Turbulence

Part VI   Bifurcations and Symmetry

Introduction

The universe we see is a ceaseless creation, evolution and annihilation of structures or patterns which are endowed with a degree of stability: They take up some part of space and last for some period of time. It is a central problem of the natural sciences to explain this change of structure and, if possible, to predict it.

Nature surprises us with the fact that the tremendous amount of physical data and results can be condensed into a few simple laws and equations that summarize our knowledge. These laws are essentially qualitative and the question arises how nature generates the enormous variety and novelty of structures by starting from such simple principles. Unaware of the scope of simple equations, man has often concluded that more than mere equations is required to explain the complexities of the world. Today, we are not so sure, for already a simple random walk described by the Laplace equation provides a prototypical situation with behaviours ranging from orderly to chaotic. The formation of spatio-temporal patterns is, first of all, a geometrical problem. Thus, in our quest to understand how structures are generated by nature, the relationship between physical forces and the geometries they can shape is of fundamental importance. This relationship and the ensuing universal character of nonlinear pattern-forming processes are the central themes underlying this book.

The mechanisms that lead to pattern formation are best understood in continuous nonlinear systems, which to external probes appear to be governed by a few essential degrees of freedom. In such systems, structural changes are characterized by the appearance of singularities and bifurcations which occur if a balance existing between competing system-immanent effects breaks down. As a consequence, an initially quiescent system becomes unstable and, in a sequence of bifurcations, restabilizes successively in ever more complex space- and time-dependent configurations. Within particular symmetry constraints, the bifurcations occur in certain definite and classifiable ways if the system is structurally stable. For large continuous systems, in which whole bands of modes become unstable, such a geometrical formalism is presently under development. New types of pattern formation problems occur in complex discrete systems composed of a very large but finite number of interacting elementary units, exemplified by neural networks and cellular automata. They possess self-organizing properties and the capacity to store and retrieve information. The following parts of this volume present reports on these classes of systems and the structure formation processes displayed by them on various scales.

The contributions of the first part deal with DISORDERED SYSTEMS and NETWORKS. Their scientific interest derives from the desire to design and anlyze systems which, by means of sensory and perceptual mechanisms, can construct internal representations of physical structures in the outside world. The typical problems one encounters here include dynamical computing, learning of patterns and adaptation. D.J. Amit begins with a survey of models of neural networks, analyzes their pattern-forming and retrieving capabilities and discusses future prospects in this field. J.L. van Hemmen examines the statistical mechanics and bifurcation properties of nonlinear neural networks. D.W. Heermann and K. Binder then discuss the formation of ordered structures in quenching experiments. G. Weisbuch shows how random Boolean nets develop patterns ranging from orderly to chaotic. J.J. Koenderink guides us through the structure of the visual field and R.D. Levine concludes the part with a discussion of the role of the maximum entropy formalism in pattern formation.

The book then turns, in the part on STRUCTURE FORMATION in NON-LINEAR SYSTEMS, to pattern formation problems in continuum physics, with mode interaction and symmetry constraints, which are amenable to a bifurcation analysis. F.H. Busse begins with a discussion of pattern formation in systems with spherical symmetry. M. Lücke, M. Mihelcic, B. Kowalski and K. Wingerath then investigate the propagation of patterns into an unstable homogeneous region. E. Knobloch, A.E. Deane and J. Toomre show how travelling, standing and modulated waves interact in doubly diffusive convection and compare experimental results with theoretical predictions. L. Kramer, E. Bodenschatz, W. Pesch and W. Zimmermann then discuss pattern selection and defects in anisotropic systems related to electrohydrodynamic convection, while J. Rehberg and H. Riecke investigate experimentally and theoretically the induction of moving convection patterns in a cell with a ramp. M. Neveling, D. Lang, P. Haug, W. Güttinger and G. Dangelmayr develop a theory of interacting modes in systems with two and three spatial dimensions and apply it to directional solidification and thermohaline convection. R. Meyer-Spasche and M. Wagner close the section with an analysis of mode interaction in Taylor vortex flows.

The papers in the part on INTERFACIAL PATTERNS survey mechanisms by which patterns are generated by the motion of an interface between different phases, a subject that is also of technological interest. M.J. Bennet, R.A. Brown and L.H. Ungar begin with a discussion of the formation of cellular patterns and their interaction in directional solidification. B. Caroli, C. Caroli and B. Roulet then survey recent developments in our understanding of the problem of velocity selection in dendritic growth processes. D.A. Kessler and H. Levine conclude the part by reviewing the dynamics of side-branching and tip-splitting in solidifying systems and in viscous flows.

The next part, on DIFFUSION LIMITED AGGREGATION and FRACTALS, is devoted to the study of the behaviour of aggregating particles and structures formed by random growth. This part starts with H.E. Stanley's survey of the principles of diffusion limited aggregation (DLA) and of its role in fluid dynamics and in dendritic solidification. J. Nittmann explores the implications of anisotropy in DLA for the growth of fractal and nonfractal forms. L.M. Sander then discusses radial and dendritic fractal growth in electrochemical deposition. M. Kolb reviews different variants of DLA and shows how the self-similar structure of fractal forms is influenced by anisotropy and by fluctuations. R. Röhricht, W. Metzler, J. Parisi, J. Peinke, W. Beau and O.E. Rössler present a topological classification of fractals and M.A.V. Axelox, D. Tchoubar and R. Jullien conclude the chapter with a comparison between experiments and simulations of cluster aggregation.

The contributions to the part on CHAOS and TURBULENCE discuss theoretical approaches to spatial and temporal chaos and to turbulence in extended systems. P. Coullet, C. Elphick and D. Repaux begin with a discussion of spatial disorder, dislocations and defect dynamics in terms of Silnikov bifurcations. K. Kirchgässner shows how resonance between gravity and pressure waves leads to spatial chaotic patterns. A. Arneodo and O. Thual then compare normal form theory with simulations of temporal chaos in thermohaline convection. R. Friedrich and H. Haken show how two-mode interaction leads to chaotic patterns in systems with spherical geometry. H. Effinger and S. Großmann conclude the chapter by demonstrating how the Navier-Stokes equations give rise to self-similar solutions.

The final part on BIFURCATIONS and SYMMETRY describes mathematical methods related to problems of pattern formation. J. Montaldi, M. Roberts and I. Stewart begin with an anlysis of oscillatory behaviour in Hamiltonian systems. D. Chillingworth describes pattern-forming processes in terms of singularity theory. G. Dangelmayr and E. Knobloch investigate interactions of standing and travelling waves in systems with broken rotational symmetry. C. Geiger and W. Güttinger discuss symmetry-breaking bifurcations in the evolution of the early universe. F.J. Wright and R.G. Cowell introduce the concept of model-mapping and D. Armbruster closes the section by applying computer algebra to dynamical systems.

The articles in this volume describe various classes of physical structure formation problems. Thus, one wonders whether there are unifying principles linking them. For the phenomena described in Parts II, III and V bifurcation geometry provides a common basis which might possibly be extended to include some of the phenomena discussed in Part IV. Whether one can conceive a geometrical theory for the dynamics of the discrete systems of Part I remains an open question.

The Editors

Part I

Disordered Systems,
Networks and Patterns

Neural Networks – Achievements, Prospects, Difficulties

D.J. Amit

Racah Institute of Physics, The Hebrew University,
Jerusalem 91904, Israel

1. The Background

a. Quasi history

There seems to be an incessant wave of attempts to construct models which capture
essential features of the human central nervous system. The types of features one
has been after are: pattern recognition; classification; learning; extraction of
concepts or rules from instances; adaptive computation, etc. The perceptron has
been the most influential idea so far. It ruled for about 12 years, between 1957-
1969, from the proposal by Rosenblatt [1], to its fascinating demise in the hands
of Minsky and Pappert [2]. This phase did not touch the world of physics.

In the sixties and early seventies there were two disjoint attempts to model
McCulloch-Pitts neural networks within the physics community [3,4]. They were both
very short-lived and produced no significant results, though the work of Little is
having much influence on present research in the field.

Another large research effort has been started by Grossberg [5] around 1968 and
is still in progress [6]. This modelling project, like the physicists' attempts,
has engaged in studying networks of elements which are intended to map more or less
faithfully cortical neurons. Despite the wealth of papers and books and the announce-
ment of ever escalating collection of psychophysical results, this effort has had
so far no noticeable impact on either physics, artificial intelligence or biology.
It is my view that the reason is that the description is too general to provide
meaningful clues for any of the problems that are on the immediate agenda of these
three communities.

Finally, in an overview like this one should mention at least one more major
trend, closely associated with the names of Kohonen [7], Cooper [8] and Palm [9],
that of the *correlation matrix*. This approach seems to have had an influence on
neurophysiologists [10,11], mainly, I believe, due to its simplicity. This simplicity is directly related to the fact that correlation matrix models *are* perceptrons.
As such they partake in the advantages of the simplicity of perceptrons (see e.g.
the quote from Minsky and Pappert's book *Perceptrons* at the end of this section),
but also in the deficiencies of perceptrons, to which the book *Perceptrons* is devoted. Moreover, as a model for higher functions of the human central nervous system
they have a much more basic shortcoming: they fail to generate *meaning*, in a sense

in which I shall try to elaborate following the description of the Hopfield model in Sect. 2d. This they share with all networks whose connections feed exclusively forward.

b. The current program

In the present phase, starting in 1982, the world of physics is obsessed by Hopfield's research program into neural networks [12,13]. One may view it as a mere social phenomenon, but it has the makings of a major one. To give a few indications I should mention that the *disease* has spread to essentially every major physics center in the world - from California to Paris, Rome, Moscow, Edinburgh etc. etc. It has crossed all the relevant disciplines: biologists at MIT, or at the Institut Pasteur are as interested as computer scientists and engineers in Paris, Murray Hill or Yorktown Heights.

The conjuncture which led to the success of Hopfield's approach has been complex. The essential feature has been the outrageous simplifications that were introduced into the formulation of the questions, as well as into the description of the system. The question of "mind" has been reduced to:

Can a set of *synaptic efficacies*, coupling *formal neurons* be so chosen as to be able to memorize an arbitrary, random selection of patterns? (A memorized pattern being one which acts as an *attractor* in the dynamical behavior of the network).

All the glorious questions about learning, computation, abstraction, intention, invariances, conditioning have been suspended, pending the clarification of the feasibility test for the internal "Carnot cycle" of a neural network. All issues of coupling to the input and output have also been set aside.

At the level of the construction of the model the massacre has been even more severe. All the internal dynamics of neurons (relative refractory period, time dependent summation of potentials etc.) have been replaced by a linear, instantaneous threshold automaton. This has already been done by Caianiello and Little, but Hopfield has gone further and chosen the matrix of synaptic coefficients to be <u>symmetric</u>. Biologically this is untenable, but it turned out to be extremely productive and was later found not to be too restrictive [14,15].

To paraphrase Toulouse, all these backward steps taken by Hopfield are of the rare type that only a very highly evolved organism can take in order to make a large evolutionary leap forward. The conceptual strength of a truncated simplified model has been beautifully captured by Minsky and Pappert [2]:

Although we do not have an equally elaborated theory of "learning", we can at least demonstrate that in cases where "learning" or "adaptation" or "self-organization" does occur, its occurrence can be thoroughly elucidated and carries no suggestion of mysterious little-understood principles of complex systems. Whether there are such principles we cannot know. But the perceptron provides no evidence; and our success in analysing it adds another piece of circumstantial evidence for the thesis that cybernetic processes that work can be understood, and those that cannot be understood are suspect.

But there has been another triple historical conjuncture that has made success possible for Hopfield's program: Experience has increased familiarity with simple stochastic processes of the Monte-Carlo-Glauber [16] type. One has learnt that ergodicity is broken when a phase transition occurs to a phase of broken symmetry - such as ferro-magnetism. This was combined with the new vistas opened by the study of spin-glasses, which led to the realization that a large, frustrated thermodynamic system may have a very diverse set of attractors [17].

2. The Basic Model

a. The machine

Let me now move to a brief description of the model: there are two types of variables:

i) two-valued neuronal variable - $S_i = \pm 1$ (action potential or refractory, respectively), i = 1,...,N lists the neurons.

ii) synaptic variables - $(+J_{ij})$ (synaptic efficacy) is the amount of post-synaptic potential (PSP) on neuron no. i if neuron j fired; $(-J_{ij})$ is the PSP on i if neuron j refractory. $J_{ij}S_j$ is the PSP on i due to state of S_j.

In this model the S_i are *dynamic* variables, the J_{ij} are *quenched* - there is no learning.

The network is schematically described in the Figure 1 below. It is fully connected and the feedback is complete.

b. The operation

It operates in the following way:

i) Its states are the 2^N possible configurations of N ±1's, depicting the activity states of all the neurons at a given instant.

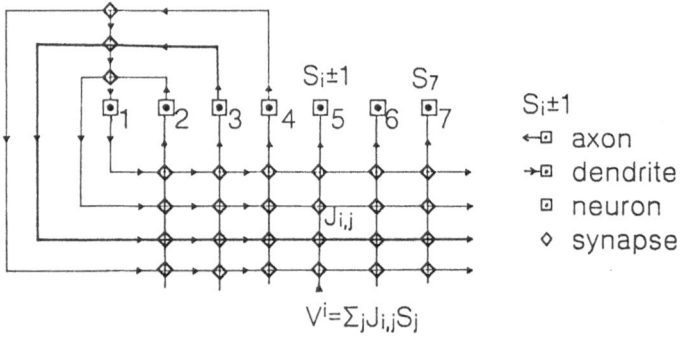

$$V^i = \Sigma_j J_{i,j} S_j$$

Fig. 1. Schematic representation of a neural network

ii) At every moment each neuron sums the PSP arriving to it from all others, due to their activity states one instant (1-2 ms) earlier:

$$(PSP)_i = V_i = \sum_j J_{ij} S_j .$$

iii) Each neuron has a threshold T_i and if

$$V_i > T_i \rightarrow S_i = +1 \text{ (will fire action potential)},$$
$$V_i < T_i \rightarrow S_i = -1 \text{ (will not fire action potential)}.$$

Given any initial state and the collection of J_{ij}'s the system marches according to this rule in the space of 2^N possible states.

The rule (iii) may be generalized to include synaptic noise (as suggested by Little [4]). It can be formulated in a probabilistic way (see e.g. Fig. 2); where $\bar{S}_i$ is the state of neuron i at time t+Δt. The parameter $T \equiv 1/\beta$, the width in Fig. 2, plays a role analogous to temperature in statistical mechanics. The Hopfield process is the limit $\beta \rightarrow \infty$, or $T \rightarrow 0$.

This is a stochastic process of a Monte-Carlo type. If the J_{ij} are symmetric, it satisfies detailed balance, and hence the march of the system is down into valleys of an "energy" surface, over the space of states.

c. The dynamics and its interpretation

There are basically two types of "marches"

a) the network, starting from some initial state may quickly reach a state which reproduces itself under the dynamics.

b) The network may wander for a relatively long time before reaching such a re-curring state.

The time scale separating (a) from (b) is a typical time separating consecutive inputs - tens of milliseconds.

Interpretation:

i) An initial state leading to a recurrent state will be said to have retrieved the memory, which is the N-bit word given by the activity list in the recurrent state. This is associative recall. The memory has been addressed by its content.

$$P(\bar{S}_i) = \frac{1}{1+\exp[-\beta(V_i-T_i)\bar{S}_i]}$$

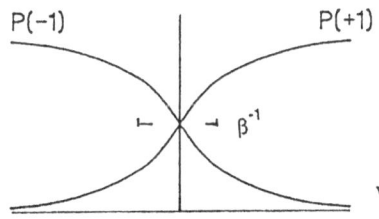

P(−1) P(+1)

Fig. 2. The probability for an action potential in a noisy neuron.

ii) An initial state which took too long to retrieve is simply wiped out by the
   next input.
(Another possibility that has been suggested recently by Shinomoto [18], whereby
unidentified objects have their own attractor).

   Hence the question is reduced to the control we can exercise over the dynamics of
the networks. In other words, given a set of p N-bit words, can the dynamics of the
system provide for valleys at, or very near, those memories.

   As a statistical physicist one knows that the dynamics of the prescribed stochastic
process is determined by the J_{ij}, and thus the J_{ij}'s determine the attractors as well.
In fact, if all $J_{ij} > 0$ and T is low enough, there will be two attractors, one all
$S_i = +1$ the other all $S_i = -1$ - this is the ferromagnetic state. It is not very
exciting as a memory - the first may be considered as epilepsy, the second as coma.
Boring as it may be, it shows the promise of the approach - ergodicity is broken
and all initial states are classified by one or the other of the two FM states.
This takes place rather rapidly. Ferromagnetism insures us also that due to the
collective nature of the phenomenon, it should be robust.

   Next one had to find the origin of the diversity. Biology provides the clue, the
J_{ij}'s can have both positive and negative signs - synapses can be both excitatory
and inhibitory. The neural network will be frustrated and thus can have a multiplicity of attractors, unrelated by simple symmetry transformations.

   At T=0 the dynamics draws the system to the local minima of the energy:
$$E = -\textstyle\sum_{ij} J_{ij} S_i S_j \quad .$$

   Figure 3 indicates how frustration gives rise to a multiplicity of low energy
states, in a 3-neuron system.

   Diversity is therefore assured. But can it be controlled?

   Hopfield has shown that it can provided the J_{ij}'s are cleverly selected. If one
wants the network to memorize p patterns $\{\xi_i^\mu\}$ ($\mu=1,...,p$; $i=1,...,N$), each an N-bit
word, then if· they are random and uncorrelated, choosing
$$J_{ij} = \frac{1}{N} \textstyle\sum_{\mu=1}^{p} \xi_i^\mu \xi_j^\mu$$

insures that $S_i=\xi_i^\mu$ is an attractor of the dynamics. Since the S_i are the activities
in the network this choice of J_{ij} has a number of attractive features:
   i) it is reminiscent of Hebb's rule [19], which says that one learns by modi-
      fying synapses according to the activities of neurons on both sides of a
      synapse.

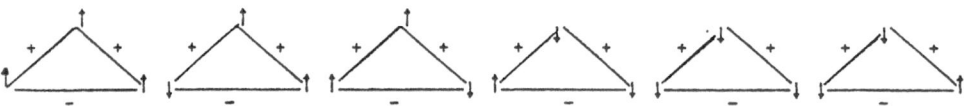

E=-2 (six ground states)

Fig. 3. A three-spin example of diversity from frustration

ii) it identifies memory with the synaptic efficacies, and thus <u>learning</u> can be
   related to <u>self-organization.</u>
iii) the behavior of the system can be fully analyzed [20,21,22].

d. <u>Self-generation of meaning [23]</u>

In subsection 1b we have mentioned that feed-forward networks do not generate meaning. Now that the metaphor captured by the Hopfield network has been explained, we
will try to elaborate on the meaning of *meaning*. In this context it is appropriate
to adopt a biological point of view. In biology information is not merely structure
[24,23]. It is that part of structure which leads to biological function. And different biological functions give different *meaning* to different parts of the structure.

   Any cognitive process must separate the meaningful from the wealth of impinging
stimuli entering the system through the senses organs. This has been perceived by
Shinomoto who chose the title "A Cognitive Associative Memory" [18]. Before the
mind can compute, classify or abstract it must choose which input to operate on and
which one to ignore, or perhaps try to learn. These may be conceived as the different
biological functions by which meaning is assigned.

   The underlying concepts in the Hopfield description are the states - the snapshots
of the activity states of the network at a given moment, and the dynamics - the
prescription by which the network moves from state to state. The Hopfield network
assigns *meaning* by its dynamics, in the sense that it provides two classifications.
The first, between initial states which do not lead to an attractor (or lead to a
very special, easily identifiable one, à *la* Shinomoto); the second, within the diversity of attractors. This is a candidate for cognitive classification in the absence of a knowledgeable arbiter.

   It is not difficult to envisage a different biological reaction of the system
to a course of activity in which the same group of neurons in the network fires
repeatedly and the same neurons remain silent for a while, or to a course in which
the state wanders around. All one needs is a set of probing neurons, which sum their
potential on a longer time scale. Those will be activated by a recurring pattern,
but not by a wandering one. Among the different recurring patterns a finer classification of meaning can take place by an identification of the particular attractor,
via different groups of probing neurons, each leading to a different biological
reaction again.

   In contrast, feed-forward networks, which are potentially very useful devices,
[25,26,27], provide an output for every input. Thus, they can serve as classifiers,
which recall associatively. But such a network either identifies every stimulus or
has special outputs for stimuli that are not "familiar". In the first case it assigns
no meaning, in the sense sketched above. In the second it requires an arbiter to decide which output was meaningful and which was not.

   These observations would put Shinomoto's network in the class of the feed-forward
systems, since it provides an attractor for every stimulus. It does this by a deci-

mation of synapses which leaves behind a net ferromagnetic interaction alongside the memorized patterns. Stimuli which are too remote from all memorized patterns reach one of the two ferromagnetic ground states, which are themselves attractors. Yet, they are attractors with a distinguishable feature - the mean level of activity, which is either extremely high or extremely low.

This may seem like a way out for the feed-forward networks. The special, macro-scopically different, mean level of activity may serve to classify the familiar separately from the unfamiliar, without detailed pre-knowledge, and lead to a dif-ferent biological function. But to be effective the unfamiliar must have an attrac-tor with a level of activity very significantly different from the typical level - "epilepsy" or "coma" in Shinomoto's case, and this seems problematic from the bio-logical point of view.

On the other hand, the idea that the gross classification of cognition - between meaningful and meaningless - is affected by two differing types of dynamics, one being attractor dynamics, the second wandering dynamics, with very low temporal mean activity per neuron seems to me more acceptable. It can be implemented in a variety of ways. The simplest is the one mentioned above, that of the usual Hopfield model, in which the length of time required for reaching an attractor separates the two types of dynamics. Parisi [28] has put forward another alternative. If the net-work is asymmetrically connected it may have quasi-attractors, cycles as well as chaotic trajectories. These may serve for the desired cognitive classification, as suggested by Parisi.

Finally, it should be pointed out that to the extent that *types* of dynamics are the tool for the assignment of meaning, one is not forced to resort to fixed points, cyclic attractors can be just as useful.

3. Properties of the Network

a. Finite number of memories in a large network

If p is held fixed as $N \to \infty$ [20]

   i) For T > 1 the system is ergodic (no retrieval possible).

  ii) For T < 1 $S_i = \pm \xi_i^\mu m(T)$ are 2p attractors with very large basins of attraction.

 iii) At T=0 these attractors have m(T) = 1, and associative retrieval is perfect.

  iv) Since the process is highly non-linear, spurious attractors appear as well. They are mixtures of the memorized patterns. Their basins of attraction are relatively small, but as p increases their number grows faster than 3^p.

   v) Fast noise, or temperature, eliminates spurious states. As T is raised higher mixtures are destabilized first. Then, above T = 0.46, only the pure memories are attractors. This is a rather fascinating role played by noise.

  vi) Sequential up-dating of neurons has very similar behavior to parallel dynamics - same attractors.

These features are exemplified in Figures 4 and 5 below. The first shows the rate of retrieval as a function of temperature, starting with an initial stimulus which has an overlap of 0.4 with a memory (70% of spins in correct direction). The retrieval quality is measured by the overlap of the attractor with one of the memories:

$$m^\mu = \frac{1}{N}\sum_i \xi_i^\mu <S_i>$$

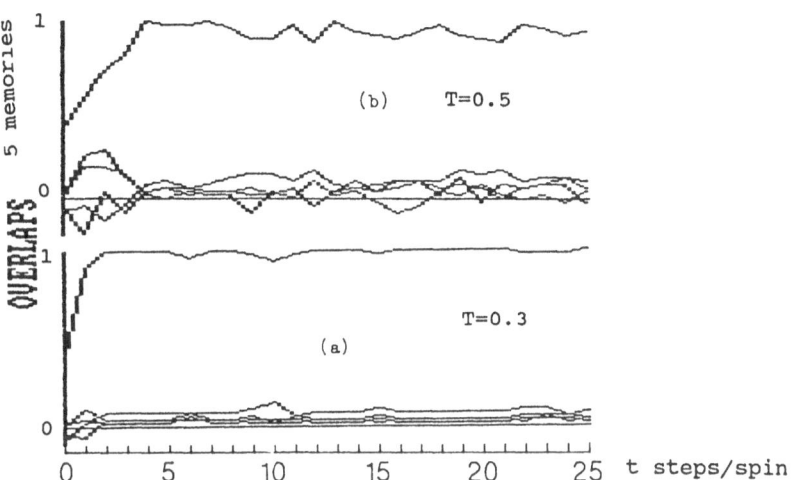

Fig. 4. Retrieval at two different temperatures

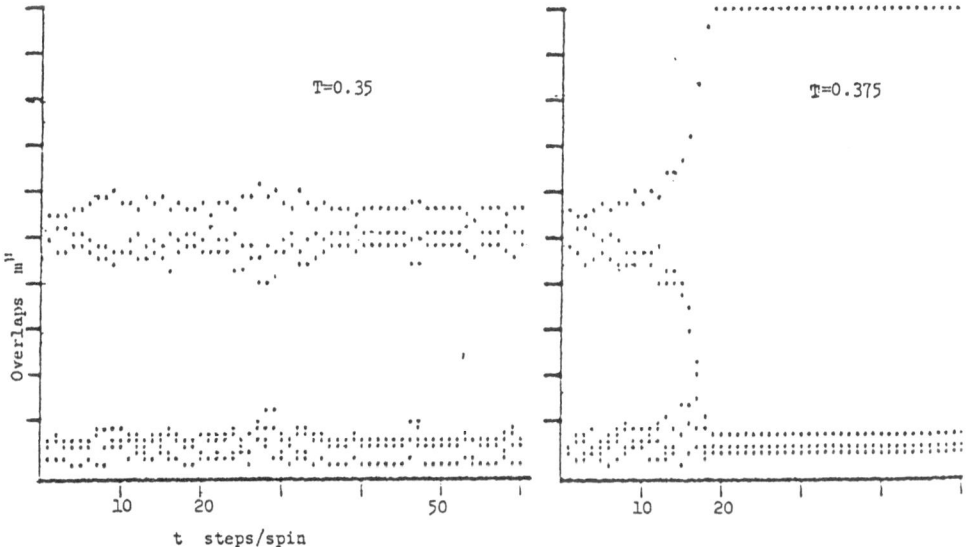

Fig. 5. Elimination of the 3-mixture spurious state by raising temperature

in terms of which one can express the Hamming distance of the attractor from the memory as $d=(1-m^{\mu})N/2$. The second shows the stability of a 3-mixture at low T and its destabilization at higher T.

b. Storage of a macroscopic number of patterns

In trying to determine the storage capacity of the neural network its properties have been studied for $p \to \infty$ as $N \to \infty$ [21,29,31]. If memories are chosen at random - $\xi_i^{\mu}=\pm 1$ with equal probability, as p increases the random <u>correlations between memories create noise</u>, which acts to destabilize the memories as attractors. This is the main factor which sets a limit on the storage capacity of the network.

There are two possible definitions of the capacity:

   i) The largest p for which the <u>exact</u> memories $\{\xi^{\mu}\}$ are attractors. In this case

$$p_c \approx \frac{N}{2\ell n N} \ [29].$$

   ii) The largest p for which the network still possesses attractors which are <u>very near</u> the memories. In this case $p_c=.145N$ [30].

If $\alpha (\equiv p/N)$ becomes $> \alpha_c (\equiv p_c/N)=.145$, no memory can be retrieved. The noise level is too high (an effect similar to that of temperature), but instead of a paramagnetic ergodic phase, one has a spin-glass (SG) phase. At saturation $(\alpha=\alpha_c)$ there is a discontinuous change in the behavior of the network. Very precise retrieval of the memorized patterns becomes suddenly possible. The amount of deviating spins, at α_c, is <1% of the N-bit word, and decreases like $exp(-1/2\alpha)$ as α decreases.

In the spin-glass phase overlaps are small but still macroscopic. They are not detected by our computations because the barriers between these attractors are not of O(N). But they are still quite effective in a single-spin dynamics [21].

The sizes of the basins of attraction of the memories, at high levels of storage, for large networks are described in Figure 6.

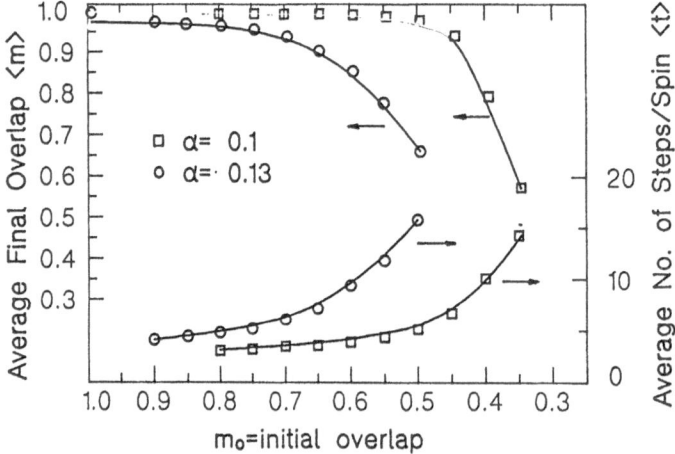

Fig. 6. Basins of attraction and retrieval times at high storage levels

The situation at α_c has provoked two objections.

i) Toulouse *et al* [32] have criticized the sudden blackout which overwhelms the network as the amount of storage increases through α_c. They, as well as Parisi [33], have proposed a different scheme for the construction of the J_{ij}'s, which they have called "learning within bounds", or "a memory that forgets", respectively. These extensions are described in somewhat greater detail below. They have very attractive features, but the simple scenario above has the property that its forgetfulness is reversible.

ii) The second objection has been raised by Parisi [29] and is concerned with the assignment of *meaning*, as discussed in Sec. 2d. The existence of a multitude of attractors, which are not closely associated with the memorized patterns, both below and above α_c (see e.g. Fig. 6), implies that the random groups of stimuli will be classified associatively, and spurious meaning will be assigned. Parisi's cure is asymmetry in the synaptic connections, which is expected to eliminate the spin-glass state, as was also suggested by Hertz *et al* [34]. Another option, suggested by the properties of the network with finite p, is that the spurious attractors be eliminated by temperature, since they are protected by relatively low barriers, a possibility that has not been tested yet.

4. Robustness

The robustness of these results has been tested under the relaxation of many of the unrealistic simplifications introduced into the model. In fact, robustness is one of the outstanding virtues of this class of models of neural networks. In terms of device building it allows for tolerance of a very high level of faults in components.

a. Dilution of synapses.

If synapses are zeroed at random, keeping J_{ij} symmetric, the system remains a very effective associative (CAM) memory [31]. The variation of storage capacity - $\alpha_c (=p_c/N)$ and retrieval quality - m_c, with the level of dilution is shown in Figure 7.

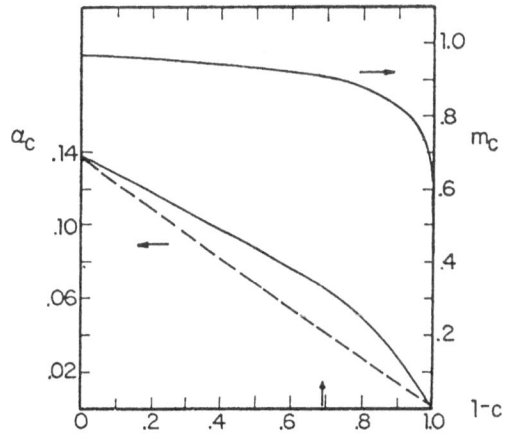

Fig. 7. Storage capacity and retrieval quality *vs.* dilution level. Position of arrow is dilution equivalent of clipping.

b. Clipping of synapses

This connects to the question of clipping. If synapses cannot have the analog depth implied by our formula for J_{ij} but can keep the sign only, i.e.

$$J_{ij} = sgn(\Sigma_\mu \xi_i^\mu \xi_j^\mu),$$

what is the effect on the memory? It has been shown [31] that clipping is equivalent to a symmetric dilution down to a concentration of $2/\pi$. This leaves $\alpha_c \approx .1$ and $m_c \approx .95$.

c. Asymmetric dilution

The biological community is strongly opposed to networks with symmetric synaptic connections, and probably for a good reason. But asymmetric networks are much harder to analyze, though significant progress is being made in developing tools for such analysis [34,35]. Here we describe two tests, based on extensive simulations:

  i) Random asymmetric dilution:

  The procedure is to generate a J_{ij} as in Hopfield and then to zero couplings, but only one out of J_{ij} and J_{ji}. Large numbers of initial configurations, all of which are memorized patterns, are then run at T=0 and the system is tested, at a given level of asymmetric dilution, for the presence of attractors which are very close to the memories. The results are shown in the Figure 8 [36].

  At 50% dilution, i.e. 50% of the bonds are symmetric, 25% are one-sided, the network retrieves with $m_c > .9$ until $\alpha_c \approx .09$. At 100% dilution - all bonds unidirectional - α_c goes down to 0.05. The main difference is that at 50% dilution, above α_c we find the SG (attractors with low overlaps). At 100%, above α_c, the dynamics is dominated by wandering trajectories, essentially no SG [29,34].

  ii) Neuron specific synapses

  Another constraint which biologists are rather adamant about is that all synapses emanating from the axons of the same neuron are of the same type - they are

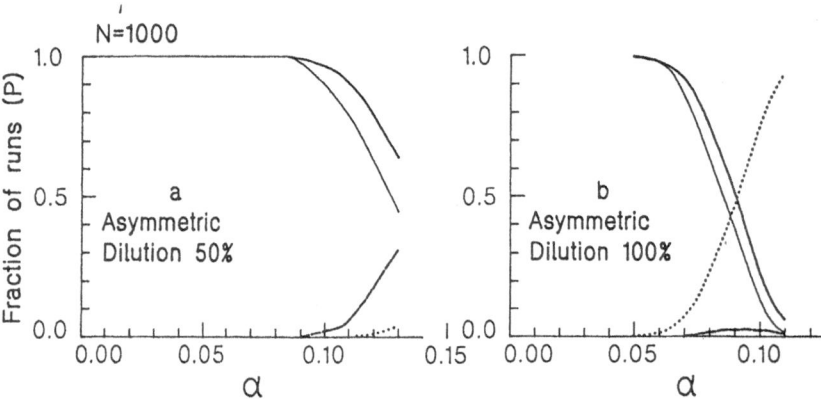

Fig. 8. Simulation results for mean retrieval quality *vs.* storage level α for two levels of asymmetric dilution

all either excitatory or all inhibitory. In terms of the variables of the model
the sign of J_{ij} is determined by j only, leading to a network with asymmetric connections.

This problem has been recently attacked by Shinomoto [18]. He modifies the connection matrix to read

$$J_{ij} \rightarrow K_{ij} = 2J_{ij}\theta(\zeta_j J_{ij}) ,$$

where ζ_j is a random variable taking on the values +1 and -1 with a given distribution, and J_{ij} is the original connection. Thus for every i connected to neuron j the
connection will either have the type prescribed by ζ_j, or be zero. This again is a
random dilution, where the randomness is in the choice of type assignment for a given
neuron and once the type has been assigned all synapses conflicting with the assignment are eliminated. The resulting system appears to be a very effective CAM, for
a wide range of ratios of excitatory to inhibitory <u>neurons</u>. The Shinomoto variant
has an additional ferromagnetic or anti-ferromagnetic attractor, mentioned in Sec. 1,
above.

d. <u>Low levels of activity</u>
Another serious simplification in the original Hopfield model has been the excessive
neural activity level of about 50%. Since the bits of a memory are chosen as ±1 with
equal probability, about one half of the neurons fire when the network retrieves.
Apart from being biologically unrealistic, this also leads to the artificial feature
that the network stores with each memory also its opposite: $\{S_i\} \rightarrow \{-S_i\}$. It should
be pointed out that because the "magnetization" in each pattern is essentially zero,
a uniform magnetic field is not effective in eliminating this doubling feature.

Again, this simplification can be lifted and one can construct and analyze networks with any given ratio of active to passive neurons. But the patterns stored
by such networks automatically have large, though simple correlations. In lifting
this restriction one had to learn to deal with correlated patterns without going far
beyond the simple locality of the J_{ij}'s [22].

5. <u>Further Developments</u>

a. <u>Retrieval without errors</u>
It was already mentioned that random overlaps undermine retrieval. In particular,
even before blacking out the network it causes retrieval with errors. A very useful
remedy was proposed by the ESPCI group [37], extending the correlation matrix idea
of Kohonen [7], whereby the synaptic coefficients are so constructed from the stored
memories as to <u>eliminate correlations</u>. They propose

$$J_{ij} = \sum_{\mu\nu} \xi_i^\mu A_{\mu\nu} \xi_j^\nu$$

and

$$(A^{-1})_{\mu\nu} = \sum_i \xi_i^\mu \xi_i^\nu \; .$$

With this coupling matrix retrieval is <u>without errors</u> until p = N (or α=1). For large α, while noise is avoided, the limitation on the level of storage comes from the shrinking size of the basins of attraction of the memories [38]. At α = 1, J_{ij} is the unit matrix which gives a completely flat energy landscape.

This approach, while very popular among device designers, has met with less interest among biologically motivated students. The main reason being the non-locality of the J_{ij}'s in terms of the neural activities ξ_i^μ's, which seems implausible from a Hebbian learning point of view.

It is my view that in the absence of any <u>dynamic</u> learning model the many attractive features of the <u>projectionist</u> network should not be abandoned too quickly. It may be essential for solving some of the difficulties involved in a more direct adaptation of the model to biological reality.

b. <u>Learning without bounds</u>

We have mentioned above that when memory is stored à la Hopfield, whereby all patterns enter with <u>equal weights</u>, all patterns become uniformly irretrievable, beyond a certain level of storage. This unpleasant feature has been shown by Nadal et al [39] and Parisi [33] to have its remedy in a modification of the "learning rule". If the original coupling matrix is assumed to evolve by adding linearly terms like $\xi_i^\mu \xi_j^\mu$ to J_{ij}, as μ goes from 1 to p, they suggest that J_{ij} be modified by the next pattern only if the new value of J_{ij} remains below a certain absolute value, i.e.

$$J_{ij}^{(p+1)} = J_{ij}^{(p)} + \xi_i^{(p+1)} \xi_j^{(p+1)} \quad \text{if} \quad |J_{ij}^{(p+1)}| < K \; ,$$

$$J_{ij}^{(p+1)} = J_{ij}^{(p)} \qquad\qquad \text{otherwise} \; .$$

Thus, a given new pattern will modify only synapses which are modifiable. This was called "learning within bounds", but it provides "learning without bounds". As additional patterns are entered old ones are erased, "forgotten", in an indefinite process. Thus, in addition to <u>indefinite "learning"</u>, it provides a process of forgetting due to learning. The maximum storage capacity of such a network (the number of patterns which can be simultaneously retrieved) is $p_c \approx .05N$.

c. <u>Retrieval of tunes</u>

It is a common experience to recall an entire ordered time sequence following a brief stimulus. This is the way we recall tunes, count, or recite the alphabet. Sompolinsky and Kanter [40] have shown that the Hopfield network can be extended to account for such phenomena. The idea is to superpose on top of the original connections, which are related to instantaneously operating synapses, a second type of synapses which are weaker but act with a time delay and an accumulating effect. Their role is to bring about an orderly transition from one memorized pattern to another <u>only after</u> <u>each of the patterns in the sequence has been clearly identified</u> (recalled).

Thus, the network operates in the following way: given a stimulus it associatively retrieves a sequence by drifting to the first pattern in the sequence and staying in it for a few cycle times. This takes place under the influence of the instantaneous couplings which, like in the Hopfield case, are symmetric and provide energy basins of attraction. The delayed couplings are asymmetric and time dependent. They have the general form

$$J_{ij}^L = \lambda \sum_{\mu} \xi_i^{\mu} \xi_j^{\mu+1}$$

and, effectively, λ carries the time dependence. The attractors are only quasi-stable, and after a while, determined by the ratio of asymmetric to symmetric connections and the accumulation time, the network moves out of the attractor and drifts to the next one.

d. Memorizing hierarchies

The desire to have neural networks which store hierarchically organized information has affected the field rather early [32,41,42]. The motivations have ranged from the synaptic pruning paradigm of Changeux [32] to the attempt to capture the apparent psychophysical phenomenon in which one seems to identify classes before individuals [41], to the search for networks with storage capacity significantly higher than linear in the number of neurons [42].

Recently two new efforts in this direction have produced manageable results:

i) In the first [35] it is a single network which stores and retrieves the entire hierarchy. A two-level hierarchy is stored as follows:

There are k_1 primary random patterns $\{\xi_i^{\mu}\}$ each one carries a bunch of k_2 secondary patterns given by

$$\xi_i^{\mu,\rho} = \xi_i^{\mu}\gamma_i^{\mu,\rho} \qquad \mu=1,\ldots,k_1; \ \rho=1,\ldots,k_2$$

with γ's distributed according to

$$p(\beta)=c\delta(\gamma+1)+(1-c)\delta(\gamma-1).$$

The couplings are then constructed as

$$J_{ij} = \frac{1}{N}\sum_{\mu}\xi_i^{\mu}\xi_j^{\mu} [1+\frac{1}{4\Delta}\sum_{\rho}(\gamma_i^{\mu,\rho}-\bar{\gamma}) \ (\gamma_j^{\mu,\rho}-\bar{\gamma})]$$

with $\bar{\gamma}=2c-1$. The prescription can be extended to any number of levels.

For this network it has been shown that there are two transition temperatures for every level of storage and every value of the parameter $\Delta>\Delta$ $(=c1-c))$. At the higher transition temperature the primary states become attractors. At the lower temperature the secondary satellites become retrievable. This allows for a scenario in which a rough recognition takes place quickly, leading to a lowering of temperature and a finer retrieval.

The level of storage is, of course, $p=k_1k_2$. It is found to be limited by $a_c(\approx.14)$, of the basic model. No gain in storage is found.

The double temperature scenario implies that while this model for the storing of hierarchical memories is elegant and efficient, it misses the main objective of hierarchical storage, which is the possibility of early retrieval of rough categorization, and a slower followup of the detailed picture. To do this in the Feigelman-Ioffe scheme one has to vary the temperature, which introduces additional architectural elements. This is allowed for *ab initio* in the second approach.

ii) Gutfreund [43] has proposed a different neural organization for storing hierarchies. The tree of patterns is constructed by first choosing a primary set of k_1 patterns $\{\xi_i^\mu\}$, which may have an arbitrary mean level of activity $(1+a)/2$ $(-1<a<1)$, as in Ref. [34]. These patterns serve as ancestors for a set of secondary satellite patterns $\{\xi_i^{\mu,\rho}\}$ - k_2 satellites per ancestor. The bits in a detailed satellite are chosen from the distribution

$$p\{\xi_i^{\mu,\rho}\} = 1/2(1+\xi_i^\mu b)\delta(\xi_i^{\mu,\rho}-1) + 1/2(1-\xi_i^\mu b)\delta(\xi_i^{\mu,\rho}+1).$$

In this set of patterns one has the following correlations:

$$<<\xi_i^{\mu,\rho}\xi_i^{\mu,\sigma}>>=b^2; \qquad <<\xi_i^{\mu,\rho}\xi_i^\nu>>=a^2b^2;$$

the first is the intra-bunch correlation and the second is inter-bunch. The correlations between levels are

$$<<\xi_i^{\mu,\rho}\xi_i^\mu>>=b; \qquad <<\xi_i^{\mu,\rho}\xi_i^\nu>>=a^2b.$$

The mean level of activity in the detailed patterns is given in terms of the mean *bias* (magnetization) of these patterns:

$$<<\xi_i^{\mu,\rho}>>=ab$$

to be $1/2(1+ab)$. The tree of patterns can be generated to any desired level.

The proposal is then that the full class of patterns at a given level in the hierarchy is stored in a separate N-neuron network. Thus, a hierarchy of l levels is stored in l interconnected networks. The storage capacity is determined by the number of patterns at the extreme branches of the tree, i.e. $k_1k_2...\approx.14N$, where N is the number of neurons per network. Again no gain in storage.

The top network operates with the dynamics

$$J_{ij} = \frac{1}{N}\sum_\mu(\xi_i^\mu-a)(\xi_j^\mu-a)$$

as in Ref. [34]. The second network down the hierarchy has the dynamics

$$J_{ij} = \frac{1}{N}\sum_{\mu,\rho}(\xi_i^{\mu,\rho}-b\xi_i^\mu)(\xi_j^{\mu,\rho}-b\xi_j^\mu).$$

Retrieval of a pattern, which is associated with $\xi_i^{\mu,\rho}$, proceeds as follows: The pattern is presented to the top network and, if b is not too small, it retrieves

16

very rapidly one of the ξ_i^μ. This is a common ancestor of bunch μ. The top network is connected to the second one by N neurons, each producing contact between one neuron in the higher level with one in the lower one (see e.g. Fig. 9). This is the fast retrieval of the "category".

Once retrieval is achieved at the higher level the communication neurons produce local fields (PSP) $h_i = h\xi_i^\mu$ on each neuron in the lower network. This implies a single synaptic connection, of strength h, between neurons number i in the two networks. Gutfreund shows that there is a range of values of h for which the local field, projected from the higher to the lower network, acts to constrain the dynamics of the lower one to states $\{S_i\}$ which satisfy the constraint

$$\frac{1}{N}\sum_i \xi_i^\mu S_i = b.$$

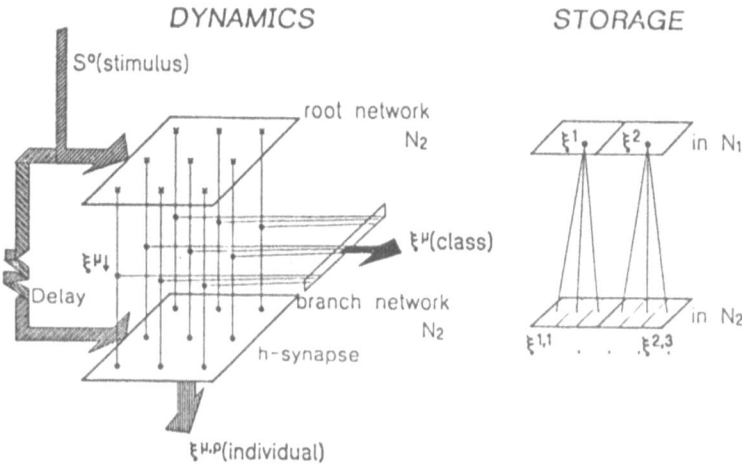

Fig. 9. Schematic representation of a neural arrangement for the retrieval of a two-level hierarchy

This constraint, together with the energy function of the lower network, ensures effective retrieval of the detailed patterns out of the bunch μ, provided that the original stimulus, which retrieved the class representative at the higher level, has been propagated with some delay as a stimulus for the lower network, a nontrivial biological requirement.

The new phase of research into neural networks has not yet found its limitations. Most questions have found their answers within the paradigm, if the latter is interpreted broadly enough. One may react that the questions have been chosen cautiously. This may indeed be the case, but if one is to judge from the history of physics, it is a very promising methodology.

6. Difficulties

Major issues are still to be attended to even at this elementary component level. Some of them are:

a. The basic time scale

On the most basic level is the question of the correspondence of the model to a biological neural network. The simplest attitude is that $S_i = +1$ at any moment represents a single "spike" - an action potential. Retrieval then implies a burst of spikes at the fastest possible rate - the same neurons fire at every cycle, the same neurons remain quiescent. The natural single neuron cycle time is 1-2ms. Yet, the fastest bursts one observes in the cortex have a spike every 7-8 ms.

This gap may have a number of possible implications. The first may be that one must include more involved internal dynamics in the model description of single neurons - their threshold must increase after firing (relative refractoriness) and then decay to normal (to prevent the fast cycle) and their PSP must not disappear following a cycle in which they have not fired (to allow for rekindling). This type of modification has been implemented [44], but the question of relevance to us has not yet been answered. Can such dynamics have enough coherence to emerge with an effective Hopfield model with bursts of a longer time cycle?

Another implication may be that meaning and retrieval are not signalled by bursts but rather by some special types of cyclic time behavior. In fact some neurophysiologists have been eliminating bursts altogether from their data [45]. In either case one must reconcile the amount of time required for retrieval in the model with the accumulating experimental data on the times needed for changes in cortical neural activity due to perceived inputs [10].

b. Neural sampling procedure

This question connects naturally with another - the type of sampling done by a neuron. Both in Hopfield's and in Little's dynamics every neuron knows the firing state of all others. This is clearly unrealistic. Neurons sample different groups of other neurons at different moments in time and can make their decision before the sampling is complete, or after some of the PSP has diffused away. Again, the question is: can such random time-dependent sampling achieve coherent behavior on a macroscopic scale? And if so, how robust can such behavior be?

c. "Hebbian" learning

Essentially all models within the Hopfield program have had static (quenched) memory. Learning has been used as a euphemism for an interpretation of the algebraic structure of the quenched coupling constants. There are some notable exceptions such as the Rumelhart-Ackley-Hinton-Sejnowsky project [46,26] or that of von der Malsburg [47]. I am not sure that either of them would consider itself within the Hopfield paradigm. But, in any case, both indicate the extent of the difficulty rather than a solution to the problem.

First, there is a biological difficulty. Hebb's learning mechanism is universally interpreted as one of synaptic modification due to correlated activity of pre and post synaptic neurons. Yet it is rather difficult to propose a scheme by which the post synaptic activity can be registered at the synapse and still correlate with the pre synaptic activity.

Another major lacuna is the absence of combined dynamics of neurons and synapses, which changes appropriately under external stimulation, to provide stable learning separable from retrieval. It is my impression that wherever both dynamics have been implemented and collective non-ergodicity looked for, the networks operated much too slowly [48]. It appears as if the innate component must be much more prominent than has been considered so far, though it may suffice that the innate part be random, as had been suggested by Changeux and Toulouse [32].

The major difficulty is that as a network learns synapses are modified and consequently "valleys" are dug in the energy landscape. Thus, a subsequent pattern, presented for learning, drifts to previously formed attractors, rather than digs its own. One way to avoid this difficulty and remain with an autonomous network is to employ Parisi's [28].proposal. That is, to have synapses form asymmetrically in learning. Then, apart from quasi-attractors - learnt memories - there are chaotic trajectories, which will not hamper the learning of new patterns. This scenario still remains to be tested.

It may be the case that a single autonomous network is not adequate for the task, and a more elaborate architecture is called for, with some networks serving as attention raisers for others which are supposed to learn. One simple scenario of this type would be that when the system becomes aware that some part of it is to learn a new pattern the temperature of this part is raised, so the previously stored memories do not effect the dynamics of the pattern being learnt. This approach has also not been implemented.

Another possibility, suggested by the Edinburgh group [49], has the network stop its normal dynamics altogether and modify its synapses on the basis of a set of patterns, persistently presented to it until they have become stable. The algorithm is as follows:

Given any initial set of couplings J_{ij}^{o} and a set of p patterns $\{\xi_i^{\mu}\}$, each pattern in turn is tested for stability, namely whether the local field at each site is in the direction of the "spin". The couplings are modified only at synapses connecting spins which are in the wrong direction, i.e. would have wanted to flip. Thus, if $h_i \xi_i^{\nu} < 0$, then

$$J_{ij} \rightarrow J_{ij}^{!} = J_{ij} + k\xi_i^{\nu}\xi_j^{\nu}, \quad \text{for i fixed and all } j \neq i$$

This is repeated for the same pattern until it is stabilized with no errors. Then the next pattern is treated in the same way and so on until the last pattern in the group is embedded. By that time the previous ones might have been destabilized, so the whole procedure is repeated. The perceptron learning theorem [2] ensures that this process will converge in a finite number of steps per spin, provided there exists a matrix J_{ij} for which the whole group of patterns is stable.

There is a number of variants in this "learning" algorithm. The one described above leads to a non-symmetric synaptic matrix. Others [48] lead to symmetric ones. Yet, this is not a satisfactory solution for the learning problem. First, because it is not biologically plausible. But even as far as device learning is concerned, construction of a pseudo-inverse, à la Kohonen, leads to a better memory matrix [38]- the resulting matrix in the algorithm described above seems to have tiny basins of attraction for the memories.

Finally, every time a new pattern is to be memorized, one must go back to refresh old ones. In the pseudo-inverse coupling one can embed new patterns without recourse to the old ones.

The list of difficulties comes only to show how much is still to be done in this very dynamic and exciting area. No doubt, the tools developed in physics for the study of collective emergent properties of strongly interacting many-particle systems have still much to contribute.

Acknowledgements

I have been constantly benefitting from discussions with Haim Sompolinsky. I am also indebted to Dr. D. Lehmann for discussions concerning his learning experiments, and to Henri Atlan for bringing the issue of self-generation of meaning to my attention. This work has been supported in part by a grant from the Fund for Basic Research of the Israel Academy for Science and Humanities.

References

1. F. Rosenblatt, *Principles of Neurodynamics*, (Washington DC: Spartan, 1961).
2. M. Minsky and S. Pappert, *Perceptrons*, (Cambridge, Mass: MIT Press, 1969)
3. E.R. Caianiello, J. Theor. Biol., 2 204 (1961)
4. W.A. Little, Math. Biosci., 19 101 (1974)
5. S. Grossberg, Proc. Nat. Acad. Sci., 60 758 (1968)
6. M.A. Cohen and S. Grossberg, IEEE Trans. SMC, 13 816 (1983)
7. T. Kohonen and M. Ruohonen, IEEE Trans. Comput., C22 701 (1973)
8. L.N. Cooper, F. Liberman and E. Oja, Biol. Cybern., 33 9 (1979)
9. G. Palm, Biol. Cybern., 36 19 (1980)
10. E.T. Rolls, Information Representation, Processing and Storage in the Brain: Analysis at the single Neuron Level, In: J.P. Changeux and M. Konishi (eds), *Learning* (Berlin, Springer Verlag, 1986)
11. A. Aertsen, T. Bonhoeffer and J. Kruger, Coherent Activity in Neuronal Populations: Analysis and Interpretation, In: E.R. Caianiello (ed): *Physics of Cognitive Processes* (Singapore: World Scientific, 1987).
12. J.J. Hopfield, Proc. Natl. Acad. Sci. USA, 79 2554 (1982)
13. J.J. Hopfield, Collective Processing and Neural States, In: C. Nicollini (ed): *Modelling and Analysis in Biomedicine* (NY: World Scientific, 1984)
14. D.J. Amit, The Properties of Models of Simple Neural Networks, In: I. Morgenstern and L. Van Hemmen (eds): *Heidelberg Symposium On Glassy Dynamics* (Berlin: Springer-Verlag, 1987)

15. H. Sompolinsky, The Theory of Neural Networks: The Hebb Rules and Beyond, In: I. Morgenstern and J.L. Van Hemmen (eds): *Heidelberg Symposium on Glassy Dynamics* 1986 (Berlin: Springer-Verlag, 1987)
16. K. Binder, *The Monte-Carlo Method in Statistical Physics* (New York: Springer-Verlag, 1978)
17. G. Toulouse, In: I. Morgenstern and J.L. Van Hemmen (eds): *Heidelberg Colloquium on Spin Glasses* (NY: Springer, 1983)
18. S. Shinomoto, Biol. Cybern., to be published (1987)
19. D.O. Hebb, *The Organization of Behavior*, (NY: Wiley, 1949)
20. D.J. Amit, H. Gutfreund and H. Sompolinsky, Phys. Rev., A32 1007 (1985)
21. D.J. Amit, H. Gutfreund and H. Sompolinsky, Annals of Physics, 173 30 (1987)
22. D.J. Amit, H. Gutfreund and H. Sompolinsky, Phys. Rev., A in press (1987)
23. H. Atlan, Physica Scripta, to be published (1986)
24. A. Lwoff, *Biological Order*, (Cambridge, Mass.: MIT Press, 1962)
25. T. Kohonen, *Self-Organization and Associative Memory*, (Berlin: Springer-Verlag, 1984)
26. P.K. Kienker, T.J. Sejnowski, G.E. Hinton and L.E. Schumacher, Perception, 15 197 (1986)
27. E.B. Baum, J. Moody and F. Wilczek, ITP-SB preprint (1987)
28. G. Parisi, J. Phys., A19 L675 (1986)
29. G. Weisbuch and F. Fogelman-Soulie, J. Physique Lett., 46 L623 (1985)
30. A. Crisanti, D.J. Amit and H. Gutfreund, Europhys. Lett., 2 337 (1986)
31. H. Sompolinsky, Phys. Rev., A34 2571 (1986)
32. G. Toulouse, S. Dehaene and J.P. Changeux, Proc. Nat. Acad. Sci. (USA), 83 1695 (1986)
33. G. Parisi, J. Phys., A19 L617 (1986)
34. J. Hertz, G. Grinstein and S. Solla, Neural Nets with Asymmetric Bonds, In: L. Van Hemmen and I. Morgenstern (eds): *Glassy Dynamics* (Berlin: Springer-Verlag, 1987)
35. M.V. Feigelman and L.B. Ioffe, Int. J. Mod. Phys., to be published (1986)
36. H. Gutfreund and Y. Stein, to be published (1987)
37. L. Personnaz, I. Guyon and G. Dreyfus, J. Physique Lett., 46 L359 (1985)
38. I. Kanter and H. Sompolinsky, Phys. Rev., A35 380 (1987)
39. J.P. Nadal, G. Toulouse, J.P. Changeux and S. Dehaene, Europhys. Lett., 1 535 (1986)
40. H. Sompolinsky and I. Kanter, Phys. Rev. Lett., 57 2861 (1986)
41. M. Virasoro, Ultrametricity, Hopfield Model and all That, In: E. Bienenstock, F. Fogelman-Soulie, and G. Weisbuch (eds): *Disordered Systems and Biological Organization* (Berlin: Springer-Verlag, 1985)
42. V.S. Dotsenko, J. Phys., C18 L1017 (1985)
43. H. Gutfreund, ITP-SB preprint (1987)
44. J.W. Clark, J. Rafelski and J.V. Winston, Phys. Reports, 123 215 (1985)
45. M. Abelles, *Local Cortical Circuits*, (New York: Springer-Verlag, 1982)
46. J.L. McClelland and D.E. Rumelhart, *Parallel Distributive Processing: Explorations in the Microstructure of Cognition: Foundations*, (Cambridge, Mass: The MIT Press, 1986)
47. C. von der Malsburg and E. Bienenstock, Statistical Coding and Short-Term Synaptic Plasticity: a Scheme for Knowledge Representation in the Brain, In: E. Bienenstock, F. Fogelman-Soulie, and G. Weisbuch (eds): *Disordered Systems and Biological Organization* (Berlin: Springer-Verlag, 1986)
48. D. Lehmann, HU preprint (1987)
49. A.D. Bruce, A. Canning, B. Forrest, E. Gardner and D.J. Wallace, Edinburgh preprint, 86/366 (1986)

Pattern Recognition in Nonlinear Neural Networks

J.L. van Hemmen

Sonderforschungsbereich 123, Universität Heidelberg,
D-6900 Heidelberg, Fed. Rep. of Germany

Abstract : In this paper some new techniques to analyze nonlinear neural networks are reviewed. A neural network is called nonlinear if the introduction of new data into the synaptic efficacies has to be performed through a *non*-linear operation. The original Hopfield model is linear whereas, for instance, clipped synapses constitute a nonlinear model. We examine the statistical mechanics of a nonlinear neural network with finitely many patterns and *arbitrary* synaptic kernel, study the information retrieval, and show how the abundantly present spurious states which are a consequence of the nonlinearity can be eliminated.

1. Introduction

Associative memory is a characteristic property of the brain. Its power and efficiency are hard to surpass - and hard to model. Recently, however, Hopfield [1] has proposed a rather attractive picture [2]. Information which has been stored is related to certain attractive sets (equilibrium states) in the phase space of an Ising spin glass with Hamiltonian

$$H_N = -\frac{1}{2} \sum_{i,j} J_{ij} \, S(i) \, S(j). \tag{1.1}$$

The neurons, say there are N of them, are described by Ising spin [3] variables $S(i)$, $1 \leq i \leq N$, which can assume the values $+1$ (firing) and -1 (quiescent), and the dynamics of the network is a downhill motion in the energy landscape associated with H_N. In other words, it is a zero-temperature Monte Carlo dynamics [1]. To take care of the internal noise of the system one may postulate [4] a positive temperature $T > 0$ and let the system operate with a Monte Carlo dynamics at this temperature.

The information is provided in the form of patterns. These N - bit words are specific configurations in the phase space of the Ising spin system which is used to describe the neural network. For instance, I could introduce you to the crook of Fig. 1.

Since he escaped from the computer some time ago you will understand that he is very dangerous. To portray him, we have used a 20 x 20 Ising spin system. A cross means $S(i) = +1$ and an empty site

Fig. 1: Portrait of the crook referred to in the main text.

stands for S (i) = - 1. So the portrait represents a specific configuration of N = 400 Ising spins. If you meet this person again, you would like to remember him but this is usually nontrivial. You see either a noisy pattern (you see him in a hurry) or an incomplete one (for instance, during a holdup in your bank) and the burning question is: How can we efficiently store the information of Fig. 1 amidst the many other patterns (there are certainly more crooks to be stored) and how can we retrieve it fast and completely ? We will return to this problem later on, in Sec. 3.

For a given Hamiltonian, such as H_N in (1.1), the basins of attraction of the Monte Carlo dynamics are the free-energy valleys or *ergodic components* [5,6] associated with H_N. The system is able to recall a memory if all the stored patterns can be associated with certain specific ergodic components, to be called retrieval states, and if the initial state, e.g., a noisy or incomplete pattern, is in the ergodic component it belongs to. It will turn out that nonlinear neural network models can function in a way as satisfying as the original Hopfield model. However, in addition to the retrieval states there exist many other ergodic components, which are unwanted. For the moment we call them *spurious states* . They deteriorate the memory function because they also give rise to basins of attraction and, hence, may occupy a considerable part of the phase space. In view of this, it is of paramount importance to obtain a detailed knowledge of the ergodic components and, therefore, to exactly solve the equilibrium statistical mechanics of the underlying Ising spin glass. To do this, we have to specify the coupling constants J_{ij} in (1.1).

For suitable couplings J_{ij}, the network with Hamiltonian (1.1) operates as a fault-tolerant, content-addressable (associative) memory [1,2]. That is, the J_{ij} can be chosen in such a way that the stored patterns are free-energy *minima* of the Hamilton function H_N. Additional patterns may be learned by appropriately modifying the J_{ij}. To facilitate the modeling, the patterns $\{\xi_{i\alpha}; 1 \leq i \leq N \}$, say with $1 \leq \alpha \leq q$, are taken to be random and unbiased. Then the $\xi_{i\alpha}$ are independent, identically distributed random variables which assume the values ± 1 with equal probability.

23

Following Hebb [7], one stores the data in the synaptic efficacies

$$T_{ij} = \sum_{\alpha=1}^{q} \xi_{i\alpha} \, \xi_{j\alpha} \equiv \xi_i \cdot \xi_j \qquad (1.2)$$

while taking [1]

$$J_{ij} = J N^{-1} T_{ij}. \qquad (1.3)$$

Note that $\xi_i = (\xi_{i\alpha} ; 1 \leq \alpha \leq q)$ is a vector in R^q, representing the information available to neuron i, and that ξ_i and ξ_j determine J_{ij} completely. So *local* information fixes the synaptic efficacies and, thereby, the coupling constants J_{ij}.

More generally, it is desirable to study models with [8,9]

$$J_{ij} = J N^{-1} \phi (T_{ij}), \qquad (1.4)$$

the synaptic function ϕ being arbitrary. If $\phi(x) = x$, then (1.4) reduces to (1.3), which may be called a linear neural network since (1.3) is *linear* in the T_{ij}. The linearity greatly simplifies the ensuing analysis [2]. A typical example of a nonlinear neural network is provided by the case of clipped synapses,

$$J_{ij} = J N^{-1} \operatorname{sgn} (\xi_i \cdot \xi_j), \qquad (1.5)$$

where $\phi(x) = \operatorname{sgn}(x)$ denotes the sign of x, vanishing at x = 0. There are several reasons to concentrate on (1.5). First, with clipped synapses, a *minimal* amount of information is stored for only the sign is saved. Furthermore, the inner product $\xi_i \cdot \xi_j$ may assume $2q+1$ values, varying between $-q$ and $+q$. As q becomes large, this does not seem plausible physiologically nor is it efficient from a computational point of view. The spins themselves allow digital processing but $-q \leq \xi_i \cdot \xi_j \leq q$ does not. If, however, one takes $\phi(x) = \operatorname{sgn}(x)$, then analog - to - digital conversion in storing the data is not needed any more. Finally, this type of function simulates a flip-flop and, therefore, is far easier to implement in silicon versions than the original, linear synapses (1.3).

As to the number of patters q, there are two distinctive cases. We may require that q be finite or that q be extensive, i. e., $q = \alpha N$ for some fixed $\alpha > 0$. The first case can be treated exactly [8,10], and this is what I will review in the present paper, whereas the second case needs certain approximations, which are not that easy to control [9,11]. Furthermore for finitely many q, nonlinear neural networks present a surprising richness of structure which, as we will see shortly, does not show up in the Hopfield case [1].

In Sec. 2, we describe how the model (1.1) with $J_{ij} = N^{-1} Q (\xi_i ; \xi_j)$, i. e., with the most general symmetric, *local* description of a synapse, can be solved. Q is called the *synaptic kernel* ; it is taken to be arbitrary. We specify the set of order parameters (for a continuous probability distribution of the ξ's, the order parameter function) and identify the significant ergodic components. Information retrieval in nonlinear neural networks is discussed in Sec. 3 (see in particular Figs. 2 - 5) and the problem of eliminating the inherent spurious states is analyzed in Sec. 4. Finally, we discuss our results in Sec. 5. Full details may be found in an extensive study [10] which will be published elsewhere.

2. Statistical mechanics

The asymptotics of a Monte Carlo dynamics is determined by the free energy and the structure of the ergodic components of the Hamilton function associated with the neural network. So we start by evaluating the free energy,

$$- \beta f(\beta) = \lim_{N \to \infty} N^{-1} \ln \ \mathrm{tr} \ \exp \ (- \beta H_N) \tag{2.1}$$

of the Hamiltonian (1.1) with

$$J_{ij} = J N^{-1} Q (\xi_i; \xi_j) \tag{2.2}$$

for some synaptic kernel $Q(x;y) = Q(y;x)$ on $R^q \times R^q$. For the moment we absorb J into Q. The ξ_i are independent random vectors in R^q (q fixed), whose components need not necessarily be ± 1. The $\xi_{i\alpha}$ have *fixed* values, randomly chosen according to their distribution (usually, by a random number generator).

The free energy (2.1) will be obtained in two steps. We first specify the relevant order parameters and then use a simple large-deviations argument [8, 11 – 13].

I. *The order parameters.* Let us suppose first that the ξ's have a discrete probability distribution. Say, the vector ξ may assume, with probability p_γ, n different positions γ, where γ denotes a q - vector. For instance, taking the neural network model (1.5), we see that in this case γ labels the $n = 2^q$ corners of the q-dimensional hypercube $[-1, 1]^q$. Each of them has probability $p_\gamma = 2^{-q}$. At each site i the ξ_i can assume only one out of n possible positions, which we have denoted by γ (to facilitate the labeling). So the index set $\{1 \le i \le N\}$ may be divided into n disjoint subsets

$$I_\gamma = \{ i : \xi_i = \gamma \} \tag{2.3}$$

whose sizes $|I_\gamma|$ become *deterministic* [14] as $N \to \infty$,

$$N^{-1} |I_\gamma| = p_\gamma . \tag{2.4}$$

With each I_γ we associate [8, 13, 15] a sublattice magnetization

$$m_\gamma = |I_\gamma|^{-1} \sum_{i \in I_\gamma} S(i), \tag{2.5}$$

25

which will play the role of order parameter. In passing we note that, if $\gamma \neq \gamma'$, then I_γ and $I_{\gamma'}$ are disjoint while m_γ and $m_{\gamma'}$ are not directly correlated.

Using (1.3) - (2.5) we rewrite the Hamiltonian,

$$- \beta H_N = - \frac{1}{2} \beta \sum S(i) N^{-1} Q(\xi_i ; \xi_j) S(j)$$

$$= \frac{\beta}{2N} \sum_{i \in I_\gamma ; j \in I_{\gamma'}} S(i) \ Q(\gamma ; \gamma') S(j)$$

$$= \frac{\beta}{2} N \sum_{\gamma, \gamma'} m_\gamma \ [p_\gamma \ Q(\gamma ; \gamma') p_{\gamma'}] \ m_{\gamma'}$$

$$\equiv N Q(m), \tag{2.6}$$

where m is vector with n components m_γ. In the second line of (2.6) we have used (2.3) together with the fact that the index set may be written as a disjoint union of the I_γ. In the third line we have taken advantage of (2.4) and (2.5).

II. *Large deviations.* We may write the normalized trace in (2.1)

$$\text{tr ex} \ (- \beta H_N) = 2^{-N} \sum_{\{S(i) = \pm 1\}} \exp (- \beta H_N)$$

$$= 2^{-|I_1|} \sum_{\{ S(i) : i \in I_1 \}} \dots .2^{-|I_n|} \sum_{\{ S(i) : i \in I_n \}} \exp (-\beta H_N). \tag{2.7}$$

By virtue of (2.6) we can express $-\beta \ H_N$ in terms of the m_γ. So we are left with the task of determining the asymptotic distribution of the m_γ. This is done by using the theory of large deviations [12, 16]. In the present case, where we have (identically distributed) Ising spins, there is a simple combinatorial argument [8, 13] to show that asymptotically, as $| I_\gamma |$ becomes large,

$$2^{-|I_\gamma|} \sum_{\{S(i) : i \in I_\gamma\}} \psi \ (|I_\gamma|^{-1} \sum_{i \in I_\gamma} S(i)) \sim \int_{-\infty}^{+\infty} dm_\gamma \ D(m_\gamma) \ \psi \ (m_\gamma) \tag{2.8}$$

where

$$D(m_\gamma) = \exp \{- | I_\gamma | c^* (m_\gamma) \} \tag{2.9}$$

is a density and

$$c^* (m) = \frac{1}{2} \left[(1+m) \ln (1+m) + (1-m) \ln (1-m) \right] \tag{2.10}$$

if $|m| \leq 1$, and $+\infty$ elswhere. Returning to (2.7) we now see that asymptotically ($N \to \infty$) the multiple sum becomes a product integral,

$$\text{tr} \exp(-\beta H_N) \sim \int_{-\infty}^{\infty} dm\, D(m)\, e^{NQ(m)} \tag{2.11}$$

with

$$D(m) = \prod_\gamma D(m_\gamma) = \exp\left\{ -N \sum_\gamma p_\gamma c^*(m_\gamma) \right\}. \tag{2.12}$$

Here we also used (2.9) and (2.4). Combining (2.1) and (2.11) we then find, using a Laplace argument [17],

$$-\beta f(\beta) = \lim_{N \to \infty} N^{-1} \ln \int dm \exp\left[N\left\{ Q(m) - \sum_\gamma p_\gamma c^*(m_\gamma) \right\} \right]$$

$$= \sup_m \left\{ Q(m) - \sum_\gamma p_\gamma c^*(m_\gamma) \right\}. \tag{2.13}$$

The supremum in (2.13) is realized among the m that satisfy the fixed-point equation

$$m_\gamma = \tanh\left\{ \beta \sum_{\gamma'} Q(\gamma, \gamma')\, p_{\gamma'} m_{\gamma'} \right\} \equiv \tanh(x_\gamma). \tag{2.14}$$

To get (2.14) we took advantage of (2.6) and the fact that $d/dm\, c^*(m) = \tan h^{-1}(m)$.

A fixed point m is stable, i.e. gives rise to a (local) maximum, if the second derivative of the free-energy functional (2.13) is negative-definite or, in other words, if the matrix with elements

$$\beta p_\gamma Q(\gamma, \gamma')\, p_{\gamma'} - p_\gamma \delta_{\gamma\gamma'} (1 - m_\gamma^2)^{-1} \tag{2.15}$$

has negative eigenvalues only.

Using (2.14) we can simplify (2.13); cf. Ref. 18, Sec. III A. Let

$$c(t) = \ln \text{tr}\, e^{tS} = \ln\left[\cosh(t) \right]. \tag{2.16}$$

One readily verifies that if m satisfies (2.14), then $c^*(m_\gamma) = m_\gamma x_\gamma - c(x_\gamma)$ and upon substituting this into (2.13) one gets

$$-\beta f(\beta) = -\frac{1}{2}\beta \sum_{\gamma,\gamma'} m_\gamma p_\gamma Q(\gamma, \gamma')\, p_{\gamma'} m_{\gamma'} + \sum_\gamma p_\gamma c(x_\gamma). \tag{2.17}$$

One has to take the solution(s) m of the fixed-point equation (2.14) which maximize(s) (2.17). A global maximum corresponds to a stable phase and a local maximum to a metastable phase.

For n-component or soft spins the very same expression (2.17) holds, provided one replaces (2.16) by an appropriate - and rather evident - modification [12,13,18]. Combinatorial arguments to get the analog of (2.9) do not work anymore.

The above results also require some minor changes if we take a *continuous* probability distribution μ for the $\xi_{i\alpha}$, e.g., a Gaussian one. We now have to reinterpret m_γ as a function m (γ) or, more explicitly, m (x) on the probability space. Instead of (2.14) we then find [13], under mild restrictions on Q,

$$m (x) = \tanh \{ \beta \int d\mu (y) Q (x; y) m (y) \} \qquad (2.18)$$

while

$$-\beta f (\beta) = - \frac{1}{2} \beta \iint d\mu (x) d\mu (y) m (x) Q (x; y) m (y) +$$

$$\int d\mu (x) c (\beta \int d\mu (y) Q (x;y) m (y)) \qquad (2.19)$$

replaces (2.17). In fact, (2.18) and (2.19) hold for a general probability distribution μ, whether discrete or continuous. Throughout what follows, however, we will restrict our attention to the bimodal distribution where $\xi_{i\alpha} = \pm 1$ with equal probability.

As $N \rightarrow \infty$, the ergodic components (free-energy valleys) associated with the Hamiltonian H_N are parametrized by the order parameters m (x) where, for the bimodal distribution, x ranges through the $n = 2^q$ corners of the hypercube $[-1, 1]^q \subseteq R^q$. As we already noted, an ergodic component is (meta)stable when the m (x) give rise to a (local) maximum of the free-energy functional (2.19) , *i.e.*, when the matrix with elements

$$\beta 2^{-q} Q (x; y) - [1 - m^2 (x)]^{-1} \delta_{x,y} \qquad (2.20).$$

has negative eigenvalues only ; cf. (2.15). This criterion together with (2.18) singles out the relvant ergodic components.

3. Information retrieval

A necessary condition for m (x), $x \in \{ -1, 1 \}^q$, to specify an ergodic component is that it satisfy the fixed-point equation

$$m (x) = \tanh \{ \beta 2^{-q} \sum_y Q (x; y) m (y) \}, \qquad (3.1)$$

which is (2.14) or (2.18) for a bimodal distribution. For high enough temperature T, i.e. for small enough inverse temperature β, the only solution to (3.1) is m (x) $\equiv 0$. Let us denote by λ_1 the largest (positive) eigenvalue of the real-symmetric matrix Q (x; y) and by $\widetilde{\lambda}$ the second-largest eigenvalue [19]. Then (a) nonzero solution(s) branch(es) off [20] into the direction of the eigenvector(s) corresponding to λ_1 and a phase transition occurs as T reaches $T_c = 2^{-q} \lambda_1$, i.e., when $\beta_c 2^{-q} \lambda_1 = 1$. Physically, it is a reasonable requirement that the ergodic components or, for short, the *states* corresponding to the stored patterns bifurcate first and remain stable down to T = 0. It, therefore, is of prime importance to solve the spectral problem for Q. The solution will be sketched in subsection 3.1. In subsection 3.2 we show that, as compared to the Hopfield model [1,2], a multitude of new states appear as T passes through $2^{-q} \widetilde{\lambda}$,........ All these states are associated with positive eigenvalues of Q. The Hopfield model is distinguished by the fact that λ_1 is positive and q - fold degenerate while $\widetilde{\lambda} = 0$. Information retrieval is hampered by too many metastable states. In spite of that, a nonlinear neural network may function well as is illustrated in subsection 3.3 for the case of clipped synapses.

3.1 Spectral theory

Let $(x)_i$ denote the component x_i of the vector x in R^q and let g_α be the inversion with respect to the coordinate axis α. That is, $g_\alpha x$ has the same coordinates as x except for $(g_\alpha x)_\alpha = - x_\alpha$. A complete spectral theory can be obtained if Q satisfies the invariance condition [10]

$$Q (g_\alpha x ; g_\alpha y) = Q (x; y), \qquad 1 \le \alpha \le q, \qquad (3.2)$$

for all x and y in $\{-1, 1\}^q$. Nearly all known neural network models, including the forgetful ones [1, 21 – 23], satisfy (3.2). We will see that all such Q have a *common* set of eigenvectors though the eigenvalues may and, in general, will be different.

Let ρ be one of the 2^q subsets of $\{1,......, q\}$ and let

$$v_\rho (x) = \prod_{i \in \rho} x_i. \qquad (3.3)$$

The empty product ($\rho = \emptyset$) is one. It is not hard to prove that the v_ρ 's are orthogonal,

$$\sum_x v_\rho(x) \, v_{\rho'} (x) = 2^q \, \delta_{\rho,\rho'}. \qquad (3.4)$$

Endowing $\{-1, 1\}^q$, i.e., the corners of the hypercube $[-1, 1]^q$, with a group-theoretical structure one can show [10] that the v_ρ 's are eigenvectors of Q with eigenvalue

$$\lambda_\rho = \sum_x Q (e; x) \, v_\rho(x) \qquad (3.5)$$

and e = (1, 1, ..., 1).

If Q is odd (even) in the sense that

$$Q (e; - x) = \mp Q (e; x) \qquad (3.6)$$

where - stands for odd (and + for even), then $\lambda_\rho = 0$ for $|\rho|$, the number of elements in ρ, being even (odd). This directly follows from (3.3), (3.5) and (3.6). In practical work Q always has a definite parity, usually odd.

3.2 Bifurcation of pure (and other) states

For the Hopfield model, $Q(x; y) = x \cdot y$. Using (3.5) one easily verifies that $2^{-q}\lambda_\rho = 1$ if $|\rho| = 1$, i.e., if ρ contains only one element, while $\lambda_\rho = 0$ for all other ρ with $|\rho| \neq 1$. Hence $\lambda_\rho = \lambda_{|\rho|}$ and the *only* nonzero eigenvalue λ_1, is q - fold degenerate. As we have seen in Sec. 2, at $T_c = 2^{-q}\lambda_1$ we have a bifurcation into the direction of the q eigenvectors belonging to λ_1. This reminds us of the q stored patterns. Indeed, making the Ansatz $m(x) = \alpha_\rho v_\rho(x)$ in (3.1) we find, since $|v_\rho(x)| = 1$,

$$\alpha_\rho = \tanh\{(\beta 2^{-q}\lambda_\rho)\alpha_\rho\} \tag{3.7}$$

and a nonzero solution to (3.7) exists if $\lambda_\rho > 0$ and $T < T_\rho = 2^{-q}\lambda_\rho$. As $T \to 0$, α_ρ approaches one at an exponental rate. For the Hopfield model we only have to consider $|\rho| = 1$. Is it then true that $m(x) = \alpha_\rho v_\rho(x)$, $|\rho| = 1$, corresponds to a bifurcation into the direction of one of the stored patterns? To answer this question we return to (2.5) and substitute $S(i) := \xi_{i\alpha}$. Then, by the very construction (2.3) of the I_y, m_y or $m(x)$ is nothing but $v_\rho(x) = x_i$ with $\rho = \{i\}$ for some i between 1 and q. So in the Hopfield case the *retrieval states* bifurcate first.

The above considerations show more, however. To be precise, the Hopfield model belongs to the class of "inner product" models where

$$Q(x; y) = \phi(x \cdot y) \tag{3.8}$$

for some synaptic function ϕ. One can prove that for this type of model [10]

$$\lambda_\rho = \sum_{k=0}^{|\rho|}\sum_{l=0}^{q-|\rho|}(-1)^k \binom{|\rho|}{k}\binom{q-|\rho|}{l}\phi(q - 2(k+l)) \tag{3.9}$$

so that, given q, $\lambda_\rho = \lambda_{|\rho|}$ only depends on the *size* $|\rho|$ of the set ρ. Moreover, the multiplicity of $\lambda_{|\rho|}$ is $\binom{q}{|\rho|}$,

except for possible accidental degeneracies .

A second observation is the following. The retrieval states bifurcate first if λ_ρ for $|\rho| = 1$ is the largest eigenvalue. This we assume throughout what follows. [Hence the name λ_1, for the largest eigenvalue.] If so, the retrieval states are stable down to $T = 0$. Eq. (3.7) shows that other, new states bifurcate from $m(x) \equiv 0$ at $T_\rho = 2^{-q}\lambda_\rho < T_c$ for any ρ with $\lambda_\rho \neq \lambda_1$. Except for the case $|\rho| = 1$, these states, which we call *pure states* [10], do not occur in the Hopfield model. In spite of

30

that, they all become metastable as T → 0. This directly follows from (2.20) if one notes that m (x) = α$_\rho$ v$_\rho$ (x) implies m^2 (x) = α$_\rho^2$ and that α$_\rho$ approaches one at an exponential rate as T → 0. [This conclusion is independent of the assumption (3.8).] So a *non* linear neural network model has many more metastable states than the Hopfield model. These states all correspond to products [10] of the stored patterns; cf. (3.3). As we will see in Sec. 4, this is not a real drawback, however.

One might wonder, though, whether the pure states are the only ones that bifurcate from zero at T$_c$. The answer is no. One can show [10] that in addition to the pure states only *symmetric states* bifurcate from zero. These are elaborate linear combinations of the v$_\rho$ (x), i.e.,

$$m (x) = \sum_{\rho \subseteq \mathcal{N}} \alpha_\rho \, v_\rho (x) \qquad (3.10)$$

where ρ ranges through all the subsets of $\mathcal{N} \subseteq \{ 1,...., q \}$ and

$$\alpha_\rho = \left\{ \begin{array}{ll} \alpha_{|\rho|} & , \quad \rho \subseteq \mathcal{N}, \\ 0 & , \quad \text{otherwise.} \end{array} \right. \qquad (3.11)$$

The states (3.10) are called symmetric because the right-hand side of (3.10) is invariant under all permutations of the elements of $\mathcal{N}$. If $|\mathcal{N}| > 1$ they are not (meta)stable as they bifurcate from zero but they may become so at lower temperatures. In fact, for clipped synapses with ϕ (x) = sgn (x), it turns out [10] that the bifurcation structure directly below T$_c$ reduces to that of the Hopfield model with the same number of patterns [2, 24].

3.3 Clipped synapses

The case of clipped synapses corresponds to storage of a minimal amount of information. Instead of the full synaptic efficacy $\xi_i \cdot \xi_j$ we only keep its sign. I can imagine that the reader may wonder whether information retrieval is possible at all - despite the above, rigorous mathematics. We, therefore, return to Fig. 1. The portrait of our crook (no. 1) was stored by using an array of 20 x 20 Ising spins. Together with this picture, the portraits of twenty-four other (random) crooks were stored and the picture of Fig. 1 was purposely deteriorated either by flipping the spins with a finite probability (noise level = 30 %) or by deleting most of the picture; cf. Figs. 2 - 5. Given the noisy or incomplete pattern, the system had to retrieve the original pattern by using a zero-temperature Monte Carlo dynamics. The original and final Hamming distances HD0 and HD1, i.e., the number of sites where the patterns differ, have been stored and are indicated in the figures. We see from Figs. 2, 3 and 4 that the network can function pretty well but, as is exemplified by Fig. 5, there is no guarantee of perfect retrieval. Note, however, that in case of convergence (Figs. 2, 3 and 4) the system typically needs two or three Monte Carlo steps per spin (MCSS). So the procedure is fast, much faster than the conventional one, which compares the noisy / incomplete pattern with all the twenty five stored patterns and picks the one that gives the minimal Hamming distance (if any). The latter procedure needs at least 25 steps per spin.

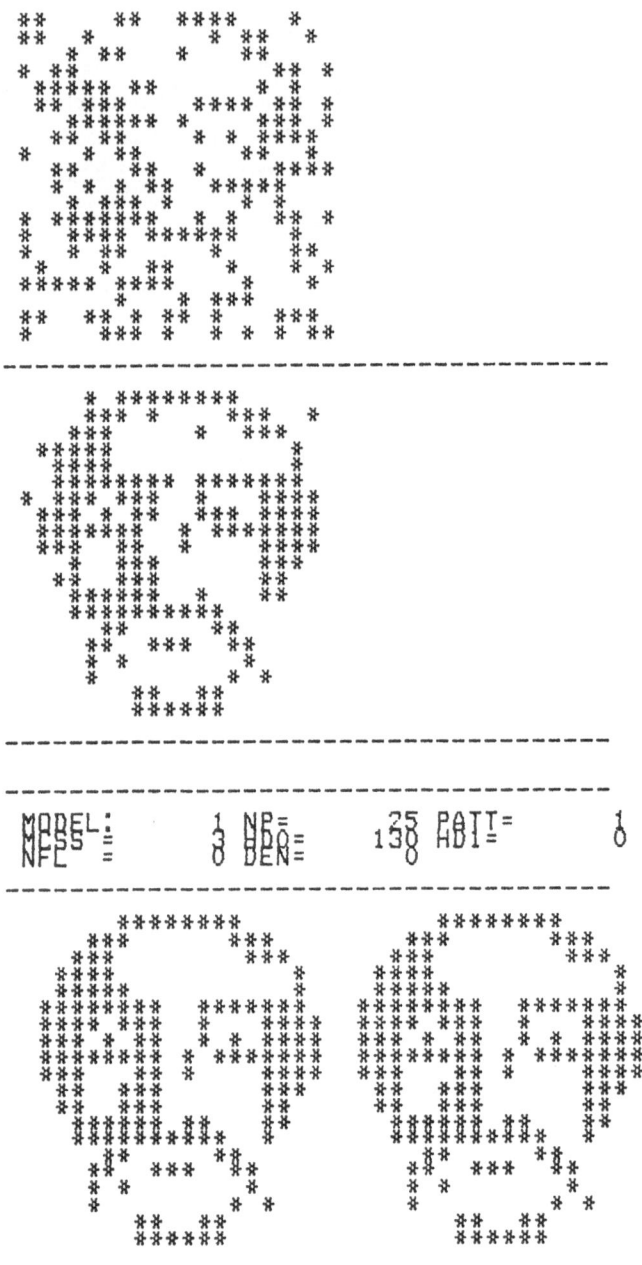

Fig. 2. Starting with a noisy pattern (the *upper* one, with 30 % noise; HD0 = 130), retrieval usually works extremely well, as is exemplified by the two patterns below (HD1 = 0). The one on the left is the original pattern. The one on the right was obtained after 2 Monte Carlo steps per spin (MCSS). The intermediate picture after 1 MCSS is in between. After one more MCSS the procedure automatically stopped if the picture did not change any more. This then gave 3 MCSS in the table.

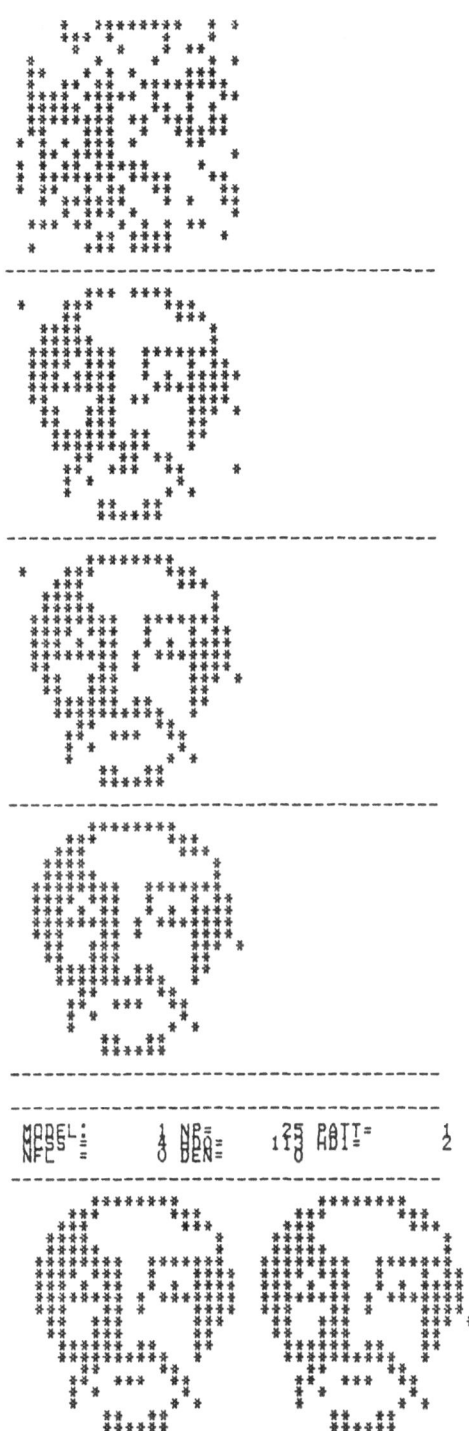

Fig. 3. However, some errors may still occur in the final pattern (HD1 = 2).

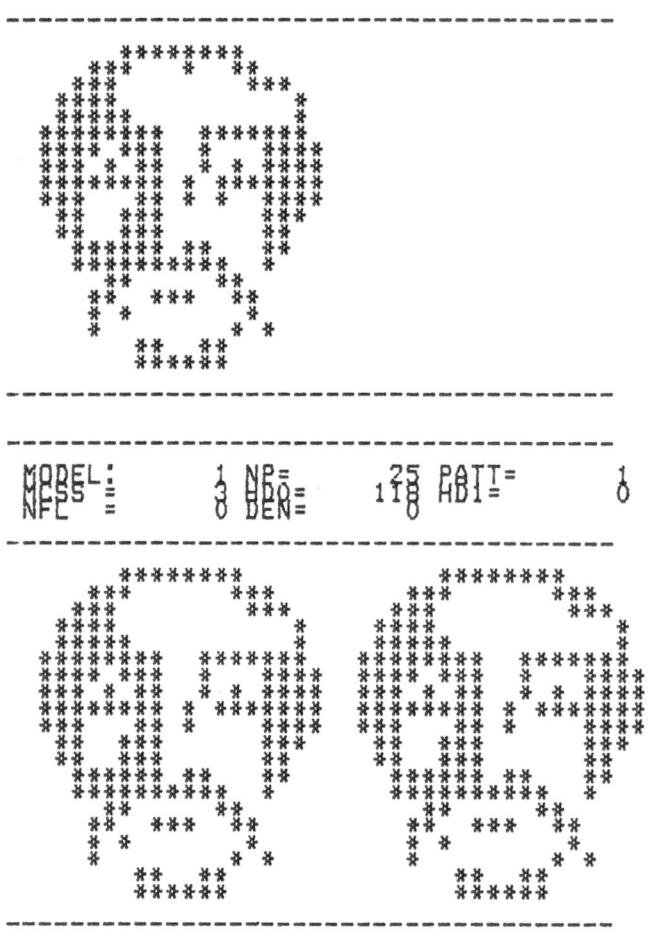

Fig. 4. If the starting configuration is incomplete instead of noisy, the system recovers the original pattern equally well.

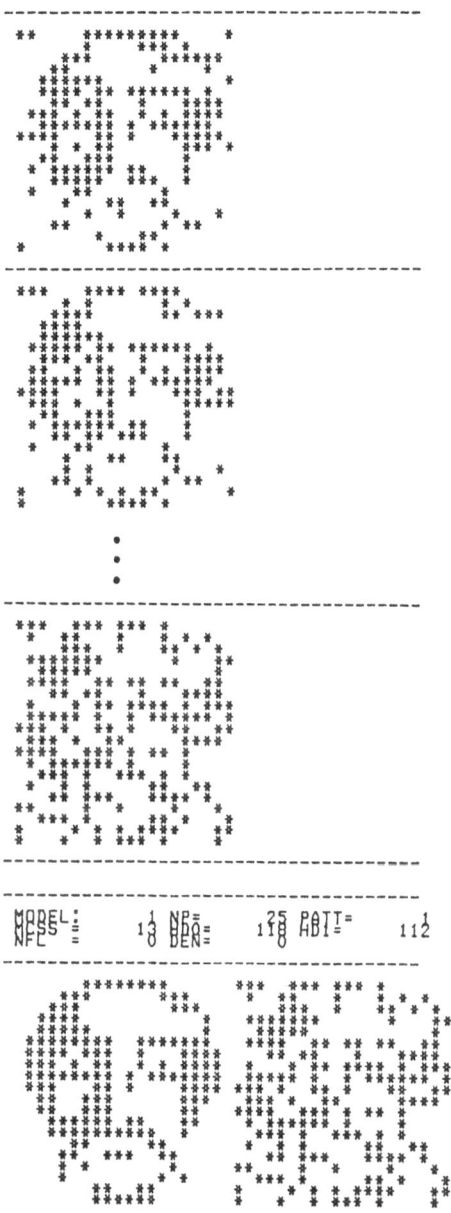

Fig. 5. The system does not always recall a memory correctly. Surprisingly, after one MCSS it looks as if the dynamics drives the system in the right direction but after 12 more steps (the missing pictures have been indicated by three dots) we land in a wrong valley (HD1 = 112). The frequency of such an event is rather low, though.

For clipped synapses, the retrieval states bifurcate first. The maximal eigenvalue λ_1 is given by

$$2^{-q}\,\lambda_1 = 2^{1-q}\,\binom{q-1}{(q-1)/2} \quad \text{or} \quad 2^{1-q}\,\binom{q-1}{(q-2)/2} \tag{3.12}$$

according to whether q is odd or even. For large q it follows by Stirling's formula [25] that

$$T_c = 2^{-q}\,\lambda_1 = \sqrt{\frac{2}{\pi\,(q-1)}} \;. \tag{3.13}$$

To fix T_c as $q \to \infty$ one has to rescale J in (1.5) by putting $J \to J\,\sqrt{q}$ so that we are left with (J=1)

$$T_c = \sqrt{\frac{2}{\pi}} \approx 0.80 \;. \tag{3.14}$$

Directly below T_c the retrieval states are the only ones which are stable [10]. The symmetric states associated with λ_1 also appear at T_c but they cannot become metastable before the temperature is much lower [10] ($T < 0.5\,T_c$), as in the Hopfield case [2, 24] . We have seen that at $T = 2^{-q}\lambda$ new states appear which are associated with products of the original patterns. These spurious states are definitely not wanted but at low enough temperatures they cannot be dispensed with. In fact, including multiplicities we have about $2^{\,q-2}$ positive eigenvalues [26] . At $T = 0$, the corresponding pure states are all stable.

4. Spurious states and how to get rid of them

Throughout this section we suppose that at $T_c = 2^{-q}\lambda_1$ the retrieval states bifurcate first. At a lower temperature $\widetilde{T} = 2^{-q}\widetilde{\lambda}$, where $\widetilde{\lambda}$ is the second-largest eigenvalue of Q, other pure states appear and many more follow at still lower temperatures - as many as there are positive eigenvalues.

As we lower the temperature, the thermal noise is reduced and the retrieval quality α of the stored patterns,

$$\alpha = \tanh \{ (T_c/T)\,\alpha \}, \tag{4.1}$$

approaches one at an exponential rate. Phrased differently, the fraction of errors, which is given by $1/2\,(1 - \alpha)$, goes to zero at the very same rate. For example, if $T_c/T = 5$, then $\alpha \approx 1 - 0.9 \times 10^{-4}$ and the error percentage is already negligible: One out of every 20 000 spins has the wrong sign.

The question now is : Can we choose the temperature T so low that $T_c/T \gtrsim 5$, i.e., $T \lesssim T_c/5$, and at the same time $T > \widetilde{T}$? If so, the retrieval quality produced by the Monte Carlo dynamics is fine and the new, unwanted states, which are inherent to the nonlinearity of the synapses, do not awake yet. To see whether we can realize this, we take a typical case, clipped synapses. If the $\xi_{i\alpha}$ assume the values ± 1 with equal probability, then one can show [27] that for every other q

$$\frac{T_c}{\widetilde{T}} = \frac{\lambda_1}{\widetilde{\lambda}} \propto q^{-2} .$$ (4.2)

For Gaussian $\xi_{i\alpha}$ the same result holds for all q. This easily allows a value between 10 and 50 for T_c/T in (4.1) - and solves the problem.

5. Discussion

We now want to discuss some specific issues. A natural one [28] is the apparent synaptic sign change that may ocur in (1.5) as q varies. We will see that this is not really the case. One also may dilute the bonds in model (1.5). This gives rise to some interesting physics. Biased patterns [2] present physiologically relevant generalization of the Hopfield model [1]. Here the large-deviations philosophy turns out to be extremely natural and helpful. Finally, we turn to the question why a model such as (1.4) or (2.2) can retrieve information. A summary of the main results of this paper can be found at the end of this section.

Synaptic sign changes and external noise. The model (1.5), which refers to the important case of clipped synapses, can be extended easily so as to include static external noise and eliminate synaptic sign changes altogether:

$$J_{ij} = -J N^{-1} + a_{ij} \Theta (\xi_i \cdot \xi_j) + \varepsilon\, b_{ij}.$$ (5.1)

The constant term $-J N^{-1}$ provides an *anti* ferromagnetic background and favors configurations with zero magnetization. The a_{ij} and b_{ij} are independent Gaussians with suitable mean and variance (say N^{-1}) while $\Theta (x) = 1/2\ [\ sgn\ (x) + 1\ ] \geq 0$ is the Heaviside function. The ε determines the strength of the external noise and is still at our disposal. The model is surprisingly robust [29]. At $T = 0$, there exists a critical ε_c so that for $\varepsilon > \varepsilon_0$ no pattern can be stored. But even at $\varepsilon = \varepsilon_c/2$ the error percentage is below 1 %. If ε vanishes and $a_{ij} = 2 J N^{-1}$, we recover (1.5).

Now a "typical" pattern always has vanishing magnetization, i.e., $N^{-1} \sum_i \xi_{i\alpha} \approx 0$. This is consistent with the antiferromagnetic background, which we therefore hypothesize [8] to be an *intrinsic* element of the system. Imagine a being is offered patterns with magnetization zero only. By the selection principle it must take care of this in order to survive. This being so, the second term in (5.1) which represents the synaptic strength, will *never* change sign - as it should.

Dilution: One may, for instance, dilute the bonds of model (1.5),

$$J_{ij} = J N^{-1}\ sgn\ (\xi_i \cdot \xi_j)\ \Theta\ (|\xi_i \cdot \xi_j|\ - x)$$ (5.2)

by taking $x > 0$. As before, $\Theta(x)$ is the Heaviside function. Eq. (5.2) is of the form (1.4), so all the considerations of Secs. 2 - 4 apply and the problem is in principle solved.See Morgenstern´s

37

contribution to these Proceedings for some interesting numerical results. There is an amusing special case, however, which can be analyzed in a completely elementary way. We choose x to be a number between q - 1 and q (i.e., q - 1 < x < q). Then Θ $(\,|\,\xi_i \cdot \xi_j\,|\,- x\,)$ is nonzero only if $\xi_i = \xi_j$ or $\xi_i = - \xi_j$. In terms of the sets I_γ of (2.3), the intragroup coupling (inside I_γ and $I_{-\gamma}$) is ferromagnetic whereas the intergroup coupling (between I_γ and $I_{-\gamma}$) is *anti* ferromagnetic. That is, for this special choice of x , Eq. (5.2) gives an extension of the chopper model [30] in that we now have gotten a partition of the index set { 1,...., N } into *pairs* of sets $\{I_\gamma , I_{-\gamma}\}$. If we perform the gauge transformation $S(i) \rightarrow - S(i)$ for all $i \in I_{-\gamma}$, then the $N/2^{\,q-1}$ spins in $I_\gamma \cup I_{-\gamma}$ interact ferromagnetically. Hence all the $2^{\,q-1}$ groups $I_\gamma \cup I_{-\gamma}$, which are to be extensive (q << ln N), perform a phase transition at the same temperature $T_c = 2^{\,1-q} J$ and below T_c we obtain $2^{\,2^{q-1}}$ ergodic components [6], which are *all* at our disposal. The noise level may be rather high , up to 50 %, but the price one has to pay is that q << ln N. This is precisely the condition under which the considerations of Sec. 2 hold; cf. Ref. 14.

We can easily recover the above results from Secs. 2 - 4. Combining (5.2), (3.6) and (3.9) one finds that all nonzero eigenvalues are identical: $\lambda_\rho = 2$ for all ρ with $|\rho|$ odd. So $\lambda_1 = 2$, the retrieval states bifurcate from zero at $T_c = 2^{-q} \times 2 J = 2^{\,1-q} J$, and they are stable down to T = 0. However, all the other pure states belonging to λ_ρ with $|\rho|$ odd also branch off at T_c and they also are stable down to T = 0. There are $2^{\,q-1}$ of these pure states and for each of them there is a component with opposite sign of α_ρ and, hence, of the order parameter m (x) = α_ρ v_ρ (x); cf. Eq. (3.7). So we get $2^{\,2^{q-1}}$ free energy valleys, which all appear at the same $T_c = 2^{\,1-q} J$ and states which are all *true* valleys, *i.e.*, stable and metastable states, down to T = 0 - as advertized.

Biased patterns. Until now the $\xi_{i\alpha}$ were taken to be ± 1 with equal probability. This would imply that about half of the neurons would be firing (cf. Sec. 1), an activity we would not survive that long. A much more realistic model is obtained by assuming that neurons are biased [2]. That is, S (i) = + 1 with probability p < 1/2, say p = 0.10. Then only a small minority is active. If so, what has to be modified?

An important aspect of large-deviations theory is that the Ising spins are considered as *stochastic* variables themselves [12, 16, 31] . More precisely, they assume the values ± 1 with equal probability. [That´ s why we took a normalized trace in (2.7).] So it is very natural to embed a random pattern [1] into the phase space of an Ising spin system. Both assume the values ± 1 with equal probability. If, however, we suppose that $\xi_{i\alpha} = \pm 1$ with probability p $\neq$ 1/2, then we also have to modify the *phase space*. Each spin now gets an a priori measure, e.g. through an external field, so that S (i) = + 1 has weight (probability) p and, hence, S (i) = -1 is left with 1-p.

One also has to consider the Hamiltonian itself. The Hopfield model assigns energy zero to a "typical" spin configuration S = { S (i); $1 \leq i \leq N$ } . Only configurations related to the stored patterns have an energy less than zero. However, if we modify the a priori measure of the phase space, the energy $N^{-1} H_N$ (**S**) of *almost any* [14] spin configuration **S** turns out to be less than zero. So one has to modify the J_{ij} in such a way that *only* configurations which are associated with the stored patterns get an energy less than zero. Details of this construction will be given elsewhere.

Why does the system recall a memory ? Though physiologically objectionable, it is not harmful to the stability of the patterns if the J_{ij} change sign. As we have already pointed out at the beginning of this section, one cannot take (1.4) or (1.5) as such (and complain that it changes sign) because if one did and took, for instance, the absolute value of sgn ($\xi_i \cdot \xi_j$) , then the retrieval states would never bifurcate from zero since the parity argument related to (3.6) implies that λ_ρ vanishes if $|\rho| = 1$ [32]. It cannot be overly stressed that it is the subtle *dependence* among the coupling constants J_{ij} with i fixed, say, and $1 \leq j \leq N$ which is responsible for the retrieval function. This dependence is seen most clearly in the construction (2.3) of the L_γ

Summary. Through a novel method we can analyze neural network models à la (2.2) with arbitrary nonlinearity. The equilibrium statistical mechanics is obtained as a variational problem where we optimize the free-energy functional (2.19) with respect to the solutions of the fixed-point equation (2.18), or (3.1) in the special case of unbiased patterns. If the synaptic kernel Q satisfies the rather weak invariance condition (3.2) , then a complete spectral theory can be given, the bifurcation analysis of the fixed-point equation can be performed, and all solutions which branch off from the high-temperature phase m (x) ≡ 0 can be obtained. For specific models, such as the clipped synapses (1.5), it turns out that the retrieval states bifurcate first - at T_c - and that additional states, inherent to the nonlinearity, appear at a temperature $\widetilde{T}$ much below T_c and such that $\widetilde{T} / T_c \rightarrow$ o as q →∞. Furthermore, in this limit the bifurcation structure converges to that of the Hopfield model with the same number of patterns. This suggests that for extensively many patterns the two models have nearly the same behavior. The dependence among the coupling constants is responsible for the retrieval function.

Acknowledgments

I thank D. Grensing, A. Huber, R. Kühn, I. Morgenstern and E. Pehlke for stimulating discussions and helpful advice. Furthermore, I gratefully acknowledge the invaluable assistance of R. Kühn and D. Grensing in preparing Figs. 1 - 5.

References

1.   J. J. Hopfield, Proc. Natl. Acad. Sci. U.S.A. **79** (1982) 2554

2.   See D. J. Amit, these proceedings, for details

3.   W. C. McCulloch and W. Pitts, Bull. Math. Biophys. **5** (1943) 115

4.   P. Peretto, Biol. Cybernet. **50** (1984) 51

5. R. G. Palmer, Adv. Phys. **31** (1982) 669

6. A. C. D. van Enter and J. L. van Hemmen, Phys. Rev. A **29** (1984) 355

7. D. O. Hebb, *The Organization of Behavior* (Wiley, New York, 1949)

8. J. L. van Hemmen and R. Kühn, Phys. Rev. Lett. **57** (1986) 913

9. H. Sompolinsky, Phys. Rev. A **34** (1986) 2571

10. J. L. van Hemmen, D. Grensing, A. Huber and R. Kühn, *Nonlinear neural networks: I. General theory, II. Information processing,* to be published.

11. D. Grensing, R. Kühn and J. L. van Hemmen, J. Phys. A.: Math. Gen. **20** (1987)

12. J. L. van Hemmen, Phys. Rev. Lett. **49** (1982) 409

13. J. L. van Hemmen, D. Grensing, A. Huber and R. Kühn, Z. Phys. B **65** (1986) 53

14. J. Lamperti, *Probability* (Benjamin, New York, 1966) Sec. 7. More generally, one needs ergodicity.

15. D. Grensing and R. Kühn, J. Phys. A: Math. Gen **19** (1986) L 1153

16. J. L. van Hemmen, in : *Heidelberg Colloquium on Spin Glasses*, edited by J. L. van Hemmen and I. Morgenstern, Lecture Notes in Physics **192** (Springer, Berlin, 1983) pp. 203 - 233, in particular the Appendix.

17. The intuitive justification of the Laplace argument is clear. For a proof, see: N. G. de Bruyn, *Asymptotic Methods in Analysis*, 2nd Ed. (North-Holland, Amsterdam, 1961) Sec. 4.2 ; E. T. Copson, *Asymptotic Expansions* (Cambridge University Press, Cambridge, 1965) Chap. 5

18. J. L. van Hemmen, Phys. Rev. A **34** (1986) 3435

19. This convention differs from the one in Ref. 8. There λ denotes an eigenvalue of QP where P = diag (p_γ).

20. G. Iooss and D. D. Joseph, *Elementary Stability and Bifurcation Theory* (Springer, Berlin, 1980)

21. J. J. Hopfield, in : *Modelling and Analysis in Biomedicine*, edited by C. Nicolini (World Scientific, Singapore, 1984) pp. 369 - 389, in particular p. 381

22. G. Parisi, J. Phys. A: Math. Gen. **19** (1986) L 617

23. J. P. Nadal, G. Toulouse, J.-P. Changeux, and S. Dehaene, Europhys. Lett. *1* (1986) 535

24. D. J. Amit, H. Gutfreund, and H. Sompolinsky, Phys. Rev. A **32** (1985) 1007

25. K. L. Chung, *Elementary Probability Theory and Stochastic Processes*, 2nd Ed. (Springer, Berlin 1975) Eqs. (7.3.5) - (7.3.7) on pp. 211 - 212.

26. Since sgn (x) is odd, λ_ρ vanishes for $|\rho|$ even ; cf. (3.6). About half of the remaining eigenvalues is positive.

27. See Ref. 10. An upper bound is provided by q^{-1} .

28. J.-P. Changeux, S. Dehaene and G. Toulouse, Proc. Natl. Acad. Sci. U.S.A. *83* (1986) 1695

29. J. L. van Hemmen and K. Rzążewski, to be published

30. J. L. van Hemmen and A. C. D. van Enter, Phys. Rev. A **34** (1986) 2509

31. J. L. van Hemmen, A. C. D. van Enter, and J. Canisius, Z. Phys. B **50** (1983) 311

32. Interestingly, states which are *products* of *two* patterns bifurcate first. This allows logical operations such as EQUIVALENCE.

Formation of Ordered Structures in Quenching Experiments: Scaling Theory and Simulations

D.W. Heermann and K. Binder

Johannes-Gutenberg-Universität Mainz, Postfach 3980
D-6500 Mainz, Fed.Rep.of Germany

I. INTRODUCTION

In this note we want to address the particular problem of the formation of ordered structures resulting from "quenching experiments". The generic experimental situation is depicted in Figure 1. Initially the system is in an unordered random state in the one-phase region. Then the temperature is lowered (for some systems like polymers the coexistence curve is inverted so that the temperature must be raised) until the system is in the two phase region. The system is now in a non-equilibrium situation and evolves toward equilibrium. It is during the evolution toward equilibrium that the system develops ordered structures /1,2/.

There are two generic models which we want to consider. In both models the interaction is described by a Ginzburg Landau Wilson Hamiltonian /3/ on a lattice (d dimensional)

$$H = \alpha \sum_i [\tfrac{1}{2}\psi^2(i) + \tfrac{1}{4}\psi^4(i)] - \beta \sum_{\langle i,j \rangle} \psi(i)\psi(j) \qquad (1.1)$$

$$-\infty < \psi(i) < \infty$$

with the order parameter ψ. The two models are

Model 1: The interaction between the lattice sites is governed by (1.1) and the order parameter is not conserved

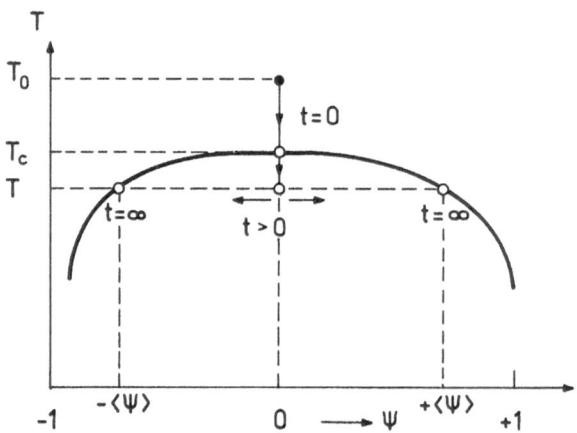

Fig. 1:   Shown is a schematic phase diagram with order parameter ψ and temperature T. We consider in this paper a quench (marked by the arrow) from a disordered state to a state in the two—phase region

Model 2: The interaction is governed by (1.1) and the order
         parameter is conserved

Model 1 is the prototype for a system showing a order-disorder
transition /4-13/. Model 2 is the prototype showing the unmixing of
two components /14-20/. In both cases the system develops patterns
after a quench from the one-phase region into the two-phase region.
Due to the conservation law we expect different time developments
for the patterns. The central questions which we address in the next
sections are: How do the patterns evolve? Are there universal
properties?

Let us look at the structures forming after a quench. In Figure 2
is shown a typical structure for the first model. From an initial
random structure the system coarsens into ordered regions of either
+ψ or -ψ. This coarsening reduces the walls between domains and
lowers the free energy. After a short time essentially two large
domains, separated by a domain wall, have formed. The evolution for
the conserved case is shown in Figure 3. Again the pictures show
snapshots of the evolution starting from a disorderd state for a
symmetrical quench. Here we see the coarsening of interpenetrating
networks.

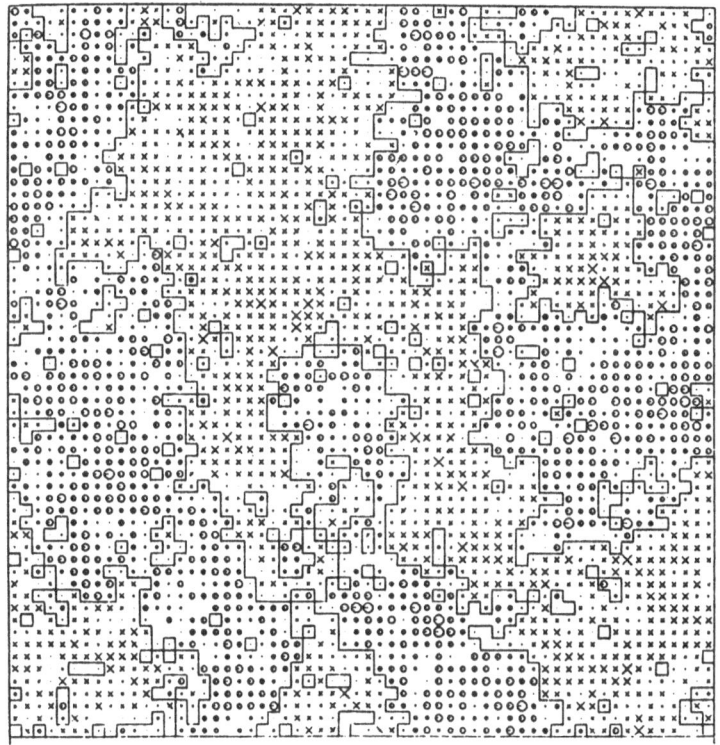

Fig. 2:   Shown is a snapshot picture of the evolution of domains
           after a quench of model 1 after 150 time steps. The two
           symbols denote the possible signs of the order
           parameter.

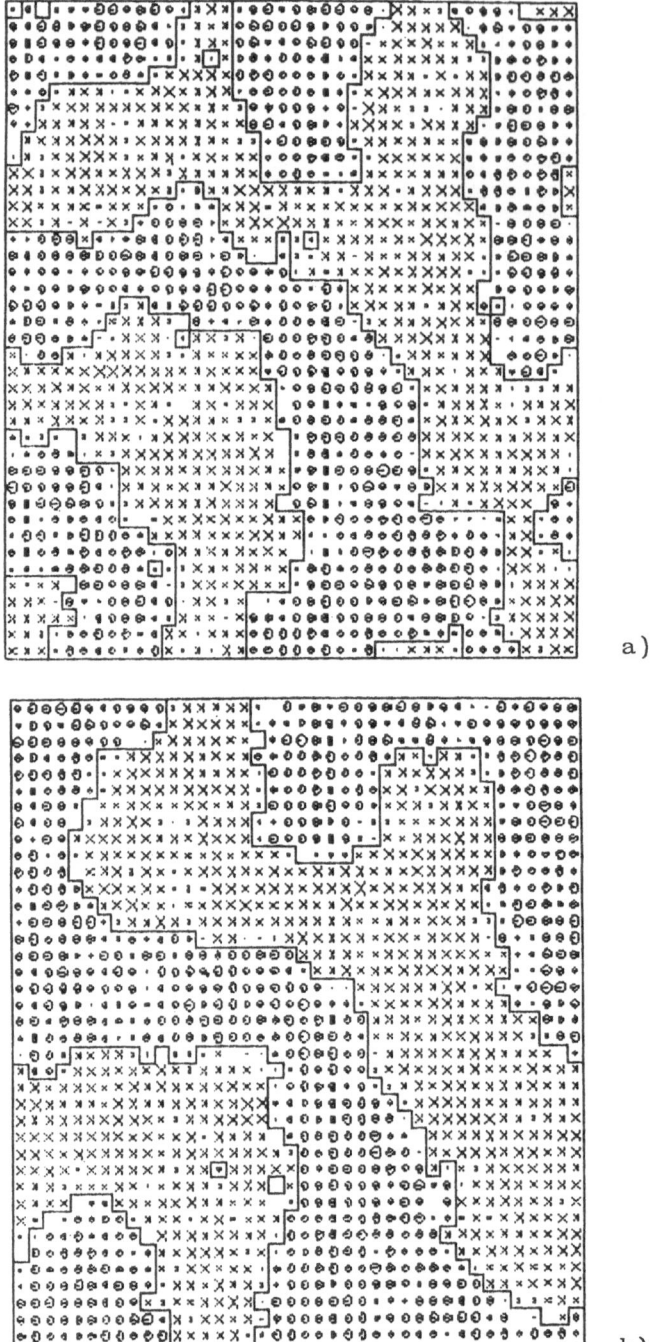

a)

b)

Fig. 3:   Shown are snapshot pictures of the evolution of model 2
          after a quench. Figure a) is a snapshot after 10000 time
          steps and b) after 30000 time steps

44

Common to both problems is the development of a characteristic length with time. In the case of model 1 this length (the typical domain size) is defined as /4/

$$l(t) \quad \alpha \quad \langle \psi^2 \rangle^{1/d} \qquad (1.2)$$

and for the second model it is convenient to define a characteristic by the inverse of the wave length /21/

$$k_m(t) \quad = \quad \max_{\{k\}} S(k,t) \qquad (1.3)$$

where $S(k,t)$ is the equal-time structure factor.

In much the same way as in critical phenomena /22/ one observes in simulations of the models 1 and 2 /23-26/ and in experiments /27/ a power law behaviour:

$$\text{Model 1:} \quad l(t) \quad \alpha \quad t^x, \qquad (1.4)$$

$$\text{Model 2:} \quad k_m(t) \quad \alpha \quad t^{-a}. \qquad (1.5)$$

Of course, there are transient power law behaviours and we expect to find true power law behaviour only in the late stages of the domain growth and unmixing. In these late stages the characteristic length will be larger than any other length (e.g. lattice spacing or correlation length). From the analogy with critical phenomena we then also expect to find scaling behaviour of for example the structure factor /20,21/. In the sequel we investigate the power law behaviour analytically and numerically with the Monte Carlo method.

II. Model 1: Domain Growth

In simulations, using such methods as Monte Carlo or Molecular Dynamics /28,29/, of a model system and in real experiments one always has to deal with effects due to the finite size of the sample. In a simulation the system has a volume L^d (computational cell). Similar to the finite size scaling theory of critical phenomena /30/ we develop in the following a scaling theory for the process of domain growth. This exposition will be limited to the case where the symmetrical quench is performed to a temperature T below T_c but T still in the critical region. A full treatment has been published elsewhere /13/.

Above we emphasized the analogy with critical phenomena which we will now draw upon. We start out by considering the static property since eventually (t → ∞) the system must end up in the static situation. In zero field the order parameter probability distribution can be expressed in terms of scaling /31/

$$P_L(\psi) = L^{\beta/v} \tilde{P} (\psi L^{\beta/v}, \xi/L). \qquad (2.1)$$

Here β is the order parameter critical exponent and v the correlation length exponent /30/ and of course P_L is normalized to unity.

As the system evolves in a quenching experiment, we expect to find two lengths which compete. The first length is the correlation length ξ which is large but finite since $T < T_c$. The second length is the typical domain size $l(t)$. In addition to the correlation length, we also need the typical domain size in the scaling relation for the order parameter. If we were right at T_c, we would have to replace ξ by $l(t)$. At T_c the system relaxes to equilibrium with the dynamical exponent z so that

$$l(t) \quad \alpha \quad t^{1/z} . \qquad (2.2)$$

This suggests the relation

$$P_L (\psi, t) = L^{\beta/\upsilon} \, \overline{P} \, (\psi \, L^{\beta/\upsilon}, \, t^{1/z}/L, \, \xi/L) . \qquad (2.3)$$

We now want to analyze the limits $t \to \infty$, $L \to \infty$ and ξ finite. In this limit $\chi'' = \xi/L$ tends to zero. Anticipating a possible singular behaviour of $\overline{P}$ we allow for a scaling of this variable. The exponents are determined from the condition that critical exponents must cancel out from all scaling powers of L:

$$P_L (\psi, t) \;=\; \xi^{\beta/\upsilon} \; \overline{P}_< (\psi \, \xi^{\beta/\upsilon}, \, L \, \xi^{zx-1} \, t^{-x}), \quad T < T_c . \; (2.4)$$

Having arrived at a scaling relation for the probability distribution of the order parameter we can introduce the characteristic length /4/ via the moments

$$[l(t)]^d \;\equiv\; \frac{\langle \, \psi^2 \, \rangle_t \; L^d}{\langle \, \psi \, \rangle_T^2} \quad \underset{L\to\infty}{\alpha} \quad [t^x \, \xi^{1-zx}]^d \qquad (2.5)$$

where

$$\langle \, \psi^2 \, \rangle_t \;=\; \int_{-\infty}^{+\infty} \psi^2 \, P_L(\psi, t) \; d\psi \qquad (2.6)$$

and

$$\langle \, \psi \, \rangle_T \quad \alpha \quad (1 - T/T_c)^{\beta} . \qquad (2.7)$$

Hence our scaling theory predicts a critical behaviour of the rate R

$$l(t) = (Rt)^x, \quad R \; \alpha \; \xi^{(\frac{1}{x} - z)} \quad \alpha \quad (1-T/T_c)^{\upsilon(z-1/x)} . \qquad (2.8)$$

How does this compare to the standard theory /10/? In the standard theory the rate is constant as one approaches the critical point and the exponent x is predicted to be $x = 1/2$. Since the dynamical exponent z is close to 2 ($z = \gamma/\upsilon = 2 - \eta$) the non-trivial temperature dependence of the rate on T is very hard to observe.

In Figure 4 we show simulation results for the characteristic length for model 1 and for an Ising model. Model 1 differs from the Ising model, which can be obtained from Model 1 in the limit $\alpha \to \infty$, in that it has "soft walls" /6/ between domains. These walls can be seen in Figure 2 from the gradual transition between the domains (the size of the symbols reflect the value of the order parameter ψ_i at the particular lattice site). For systems with soft walls, it was predicted that $x = 1/4$ /6/. In contrast our simulations of both systems show a growth exponent x of one-half.

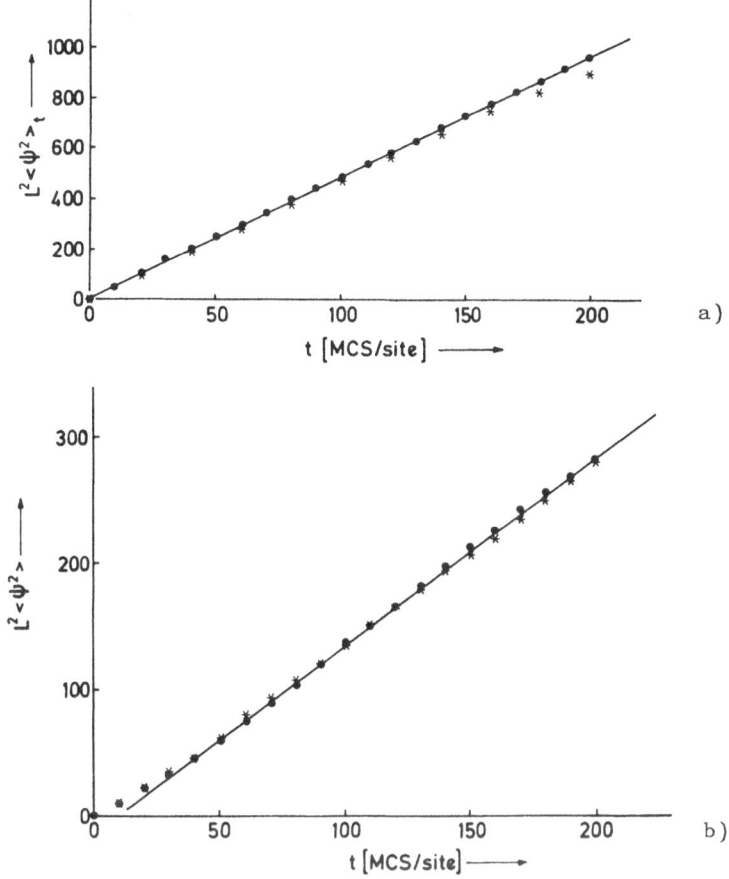

Fig. 4:   Plot of the characteristic length in the domain growth
           problem for the Ising model a) and model 1 b).

Hence at least for the growth exponent the "softness" of the wall
does not play any role in our model.

So far we have derived a prediction for the growth of domains
from a finite size scaling theory. Since our system is finite, and
hence fluctuations due to the finiteness of the system are present,
we need to consider the errors involved in a numerical determination
of quantities.

These considerations will also hold, of course, for real
experiments. Our question is

Does the error Δ of a quantity A reduce to zero as $L \to \infty$. If this
does not happen, A exhibits "lack of self-averaging".

As an example consider the error of $l(t)$. The error $\Delta(n, L)$ of any
quantity A is

$$\Delta(n,L) = (\langle A^2 \rangle_L - \langle A \rangle_L^2)^{1/2} \, n^{-1/2}, \quad n \gg 1 \qquad (2.9)$$

47

where n is the number of statistically independent samples. Calculating the error for l(t) we find

$$\frac{\Delta(\psi^2)}{\langle\psi^2\rangle_t} \xrightarrow[L \to \infty]{} n^{-1/2}\sqrt{2 - 3U^*}$$ (2.10)

where U^* is a characteristic constant (which for domain growth is even believed to vanish) /13/. This result shows that the typical length l(t) exhibits a lack of self-averaging!

III. Model 2: Unmixing

In the previous section we discussed the growth of domains after an instantaneous quench from a disordered state to a temperature T below T_c. Here we study model 2 under the same conditions. Due to the conservation law we expect a different power law behavior and a somewhat different coarsening pattern. For the symmetrical quench there are two interpenetrating networks. The networks coarsen by diffusion of material again reducing the free energy.

Figure 5 shows the time evolution of the characteristic wave vector as observed in a Monte Carlo simulation of Model 2 /32/. The observed behaviour does suggest a power law. However, one must be careful. The data does show the power-law behaviour already in the very early stages of the unmixing process where there is no reason that a power law holds.

In this note we stressed at several points the analogy between the power law and scaling in critical phenomena and the power law

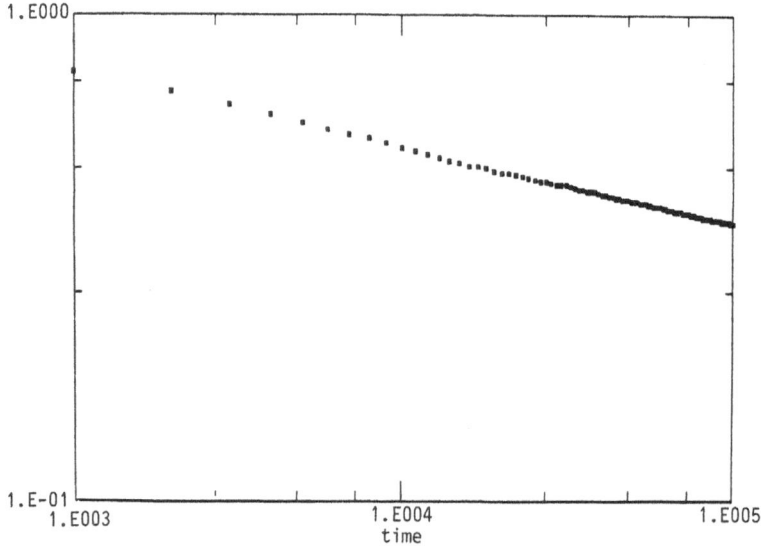

Fig. 5:   Shown on this log-log plot is the characteristic wave
              vector after a quench of model 2 as a function of time
              measured in units of Monte Carlo steps per lattice site.

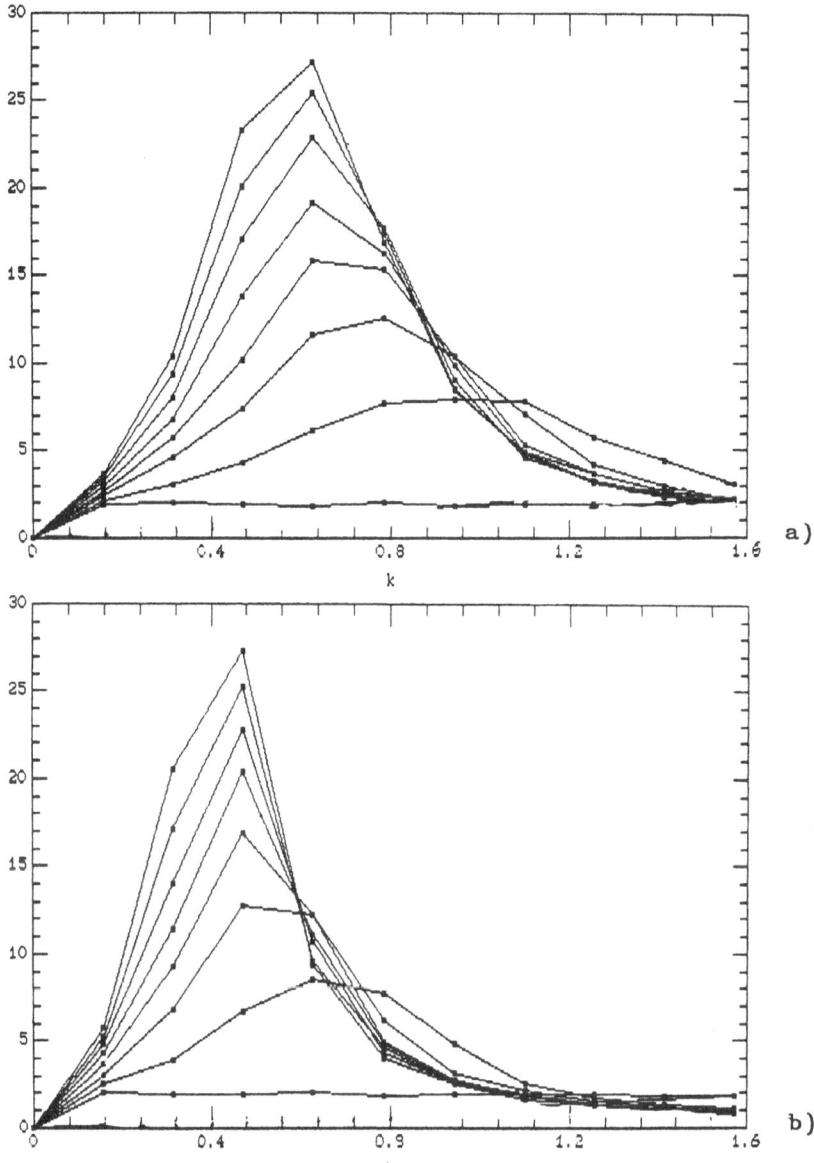

Fig. 6:   Evolution of the structure factor after a quench of model 2 into the two – phase region. Shown are the structure factors for the (0,1) direction (a) and (1,1) direction (b) from time t=0 to t=300

and scaling in the growth of patterns. There is, however, a difference in the structure at T_c and the structures which evolve below T_c. At T_c the critical correlations are isotropic leading to a self-similar structure. The underlying geometry of the lattice is irrelevant. This is not the case in the process of unmixing, as was observed experimentally /33/. The structure factor must then depend on the direction of $\underline{k}$. Then, of course also the scaling function must be anisotropic.

That the structure factor indeed reflects the underlying geometry is shown in the Figure 6. Shown are the results /32/ of a quench from $\beta = 0$ (disordered state) to $\beta = 0.3$ ($\alpha = 0.1175$) of model 2. There is a clear difference between the (1,1) direction and the (0,1) [(1,0)] direction of the underlying two-dimensional square lattice. It is interesting to observe that the anisotropy has manifested itself quite early. Already after about 30 time steps the two major directions differ in their peak position of the structure factor by one unit. This does not mean that there will be two growth exponents. Rather the amplitudes are different for the two major directions.

Summary

In this paper we have studied the problem of the formation of structures after a quench. We have developed a finite size scaling theory for the domain growth problem and our principle result of that theory is the temperature dependence of the rate constant. In addition we found that certain quantities are not self-averaging. In the case of unmixing we have presented Monte Carlo simulations showing that the structure factor reflects the underlying lattice geometry.

Acknowledgment: We would like to thank A. Milchev for stimulating discussions and his valuable cooperation in some of the original work (Refs. 13,32) on which this talk was based.

References

1. J.D. Gunton, M. San Miguel, and P.S. Sahni, in Phase Transitions and Critical Phenomena, vol 8, eds C. Domb and J.L. Lebowitz, Academic Press, New York, 1983
2. K. Binder and D.W. Heermann, in Scaling Phenomena in Disorderd Systems, eds. R.Pynn and A. Skjeltorp, p. 207, Plenum Press, New York, 1985
3. A. Milchev, D.W. Heermann, and K. Binder, J. Statist. Phys. 44, 749, (1986)
4. A. Sadiq and K. Binder, J. Statist. Phys. 35, 617 (1984)
5. E.T. Gawlinski, M. Grant, J.D. Gunton, and K. Kaski, Phys. Rev. B 31, 281 (1985)
6. O.G. Mouritsen, Phys. Rev. B 28, 3150 (1983); B 31, 2613 (1985); B 32, 1632 (1985); Phys. Rev. Lett. 56, 850 (1986)
7. G.S. Grest, D.J. Srolovitz, and M.P. Anderson, Phys. Rev. Lett. 52, 1321 (1984); G.S. Grest and D.J. Srolovitz, Phys. Rev. B 32, 3014 (1985)
8. I.M. Lifshitz, Sov. Phys. JETP 15, 939 (1962)
9. I.M. Lifshitz and V.V. Slyozov, J. Chem. Solids 19, 35 (1961)
10. S.W. Allen and J.W. Cahn, Acta Metall 27, 1085 (1979)
11. T. Ohta, D. Jasnow and K. Kawasaki, Phys. Rev. Lett. 49, 1223 (1982)
12. G.F. Mazenko and M. Zannetti, Phys. Rev. Lett. 53, 2106 (1984)
13. A. Milchev, K. Binder and D.W. Heermann, Z. Phys. B 63, 521 (1986)

14. J.W. Cahn, Trans. Metall. Soc. AIME 242, 166'(1968); Acta Met. 9, 795 (1961)
15. J.W. Cahn and J.E. Hilliard, J. Chem. Phys. 28, 258 (1958)
16. H.E. Cook, Acta Met. 18, 297 (1970)
17. J.S. Langer, Ann. Phys. 65, 53 (1971)
18. J. Marro, A.B. Bortz, M.H. Kalos, J.L. Lebowitz, Phys. Rev. B 12, 2000 (1975); A.B. Bortz, M.H. Kalos, J.L. Lebowitz and M.A. Zendejas, Phys. Rev. B 10, 535 (1974)
19. K. Binder, in Condensed Matter Research using Neutrons, eds. S.W. Lovesey and R. Scherm, p. 1, Plenum Press, New York, 1984
20. D.W. Heermann, Phys. Rev. Lett. 52, 1126 (1984), Z. Phys. B 61, 311 (1985)
21. J.L. Lebowitz, J. Marro and M.H. Kalos, Acta Met. 30, 297 (1982)
22. H.E. Stanley, Introduction to Phase Transitions and Critical Phenomena, Oxford University Press, Oxford, 1971
23. K. Binder and D. Stauffer, Phys. Rev. Lett. 33, 1006 (1974)
24. K. Binder, Phys. Rev. B 15, 4425 (1977)
25. K. Binder and C. Billotet and P. Mirold, Z. Phys. B 30, 183 (1978)
26. H. Furukawa, Progr. Theor. 59, 1072 (1978); Phys. Rev. Lett. 66, 60 (1978), Phys. Rev. A 28, 1717 (1983), Adv. Phys. 34, 703 (1985)
27. See for example the review by W.I. Goldburg, in Scattering Techniques Applied to Supramolecular and Nonequilibrium Systems, ed. Sow-Hsin Chen et al, Plenum Press, New Nork, 1981
28. K. Binder (ed), Monte Carlo Methods in Statistical Physics, Springer Verlag, Berlin, 1979
29. D.W. Heermann, Computer Simulation Methods in Theoretical Physics, Springer Verlag, Berlin, 1986
30. For a recent review, see N.M. Barber in Phase Transitions and Critical Phenomena, eds. C. Domb and J.L. Lebowitz, vol 8, Academic Press, New York, 1983
31. K. Binder, Z. Phys. B 43, 119 (1981)
32. A. Milchev, D.W. Heermann, and K. Binder, Acta Met., in press
33. J.R. Simon, P. Guyot and A. Filarducci de Salva, Phil. Mag. A 49, 151 (1984)

Patterns in Random Boolean Nets

G. Weisbuch

Laboratoire de physique de L'E.N.S., 24 rue Lhomond,
F-75231 Paris Cedex 5, France

In some respects my talk could be considered as a possible answer to Professor Güttinger's question:

"What is to be done about systems made up of a large number of variables?"

Fortunately, in a number of cases, these variables take on only a small set of values: a neuron in the brain is either firing or it is in a quiescent state - a chemical has a vanishingly small concentration in a cell or is present at saturation. In such cases the variables can be represented by Boolean variables, which take on only values 0 and 1. Like any variables in a dynamical system, they change their values in relation with some other variables. But the system, instead of being driven by differential equations, is completely discretized. Time is now an integer, and at each time step, each variable is updated according to a transition function whose arguments are the other variables. Each variable is called an automaton, and their set a network (or a net) of automata.

We shall first introduce a few definitions concerning the structural and dynamical properties of networks. Chaotic and frozen (or organized) behavior will then be described by their temporal and spatial properties. A number of "dynamical quantities", some in analogy with magnetism, will then be introduced, and their relation to the frozen/chaotic transition established. The greatest part of this presentation will be concerned with computer simulation results, but I shall mention the few cases for which a theoretical approach is possible.

1.1 Definitions

For the sake of simplicity, we shall use here a more restricted definition of automata than in Computer Science.

An **automaton** is a binary device which computes at finite time steps its own binary state as a function of binary input signals coming from interacting automata. This function is called the **transition function**. The number of interconnected automata, k, is called the **connectivity** of the automaton. The most general way to specify a transition function is to specify the function for each of the 2^k input configurations. The total number of different transition functions for Boolean automata with k inputs is thus 2 to the 2^k. Logical functions AND, OR, NAND, XOR are examples of Boolean functions.

A **network** (or a net) is obtained by the connection of several automata. The inputs which update the state of one automaton at time t are the states of the connected automata at time $t-1$. Depending upon the application, automata might model for instance genes, nerve cells or chemicals, while the network models a living cell, the central nervous system or a primeval soup at the origin of life.

The **connection graph** is the set of connections established between the automata. This graph might be complete (all automata connected), random or regular. The latter case is that of cellular automata, more frequently used in physics or in technological applications.

Finally, for a given network, several modes of operation can be used to apply the transition rules: either all automata apply simultaneously the transition rule, which is called **parallel** iteration, or this process is done sequentially, one automaton at a time, in a **sequential** iteration mode. We shall only discuss in this paper the parallel iteration mode.

1.2 Dynamical Properties

Once a network is defined by its connections among the automata, by the transition rule for each automaton and by the choice of an iteration process (parallel iteration in our case), one is interested in its dynamical properties, i.e. the description of the configurations reached by the network when the transition rules of the automata are applied at regular time intervals. For networks made up of a small number of automata, say 10, the complete **iteration graph** (the set of all the configurations of the net with arrows pointing to the successor of each state) can be drawn. From the example given in figure 1, several concepts can be defined:

Attractors. If starting from an initial configuration, a configuration is reached a second time, the net indefinitely cycles through the subset of configurations between these two occurrences. This subset is an **attractor** of the dynamics. Attractors composed of only one state are called **limit points,** the others being called **limit cycles.** The number of states of an attractor is its **period.** It is the time separating reoccurrences of the same configuration.

Transients. The states not belonging to any attractor are called **transients.** The subset of all transients which evolve towards the same attractor plus the attractor, is called an **attraction basin.**

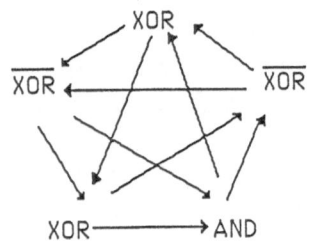

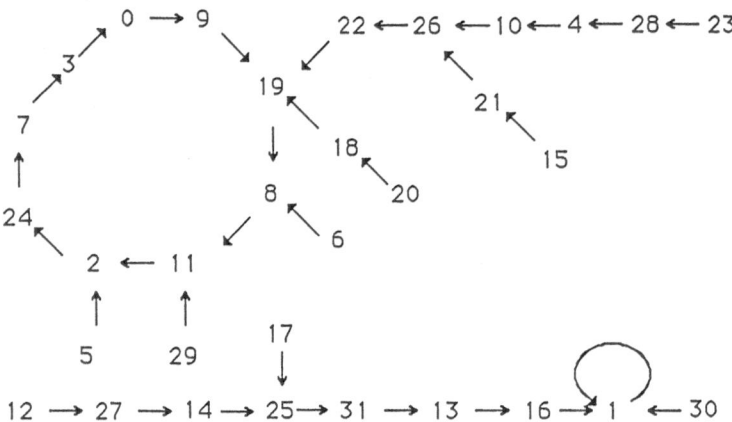

Fig. 1. A network of 5 Boolean automata (represented above), and its iteration graph (below). The arrows indicate the connections, and the name of each automaton refers to the logical function of the automaton. Function AND yields 1 if both inputs are 1, XOR yields 1 if one and only one input is 1, $\overline{\text{XOR}}$ yields 0 if one and only one input is 1.

These definitions are of course valid for nets of any size. Since the number of configurations of a net increases exponentially with its size (2^N for size N), it is in general impossible to know the iteration graph for large N. One then characterizes the dynamics of a net by its attractors: their number, their periods, or the size of their attraction basins. Other quantities will be later defined.

The lower graph is the iteration graph. The numbers varying from 0 to 31 refer to the decimal representations of the binary configurations of the net - the least significant bit is the state of the upper XOR automaton and the significance increases clockwise. Arrows show the succession of the configurations. Two independent subgraphs appear, leading to two different attractors. The loop is a limit cycle of period 9. Configuration 1 is a fixed point.

2.1 Random Boolean Nets

As you might imagine, the choice of the net depends upon the problem under study and I certainly do not intend to describe the "zoology" of all possible nets in this talk. Let me restrict my attention to Random Boolean nets proposed by Stuart Kauffman [1] to model cell differentiation. The biological problem is the following:

"All our cells use the same genetic material, coded in the chromosomes, but there exist different cellular types: liver cells, epithelial cells, neural cells... How come?"

It is generally agreed upon that cellular activity is determined by a choice among all the proteins that the cell can synthesize. The verb 'can' refers to the fact that each gene codes for a protein. But the presence of the protein with a finite concentration interacts with the actual synthesis of other proteins by activating or depressing the corresponding genes. The gene coding for these proteins is said to be expressed when the protein is actually synthesized. The fact that a gene is expressed or not can be represented by a binary state, the interactions among genes via the proteins by Boolean automata with their connection structure, and the genome by the Boolean network. A large problem still remains: very little is really known about the complete set of interactions among the genes of even the simplest organisms. The idea of S. Kauffman was then to look at the properties of randomly constructed nets with the hope that they would exhibit generic dynamical properties.

The nets are composed of Boolean automata with transition functions randomly chosen among Boolean functions with k inputs (k, the connectivity being constant), and with random connections. The parallel iteration mode is selected. The question is to determine whether there exist properties which are **generic** on the set of random nets: generic in the sense that they are exhibited by almost all nets with, possibly, a few exceptions corresponding to special designs.

Among the properties that were first exhibited by computer simulations are the following:

There exists a difference in behavior between nets with connectivity 1 or 2 and those with larger connectivity. When one increases the number of automata, for low connectivities small periods and small numbers of attractors are observed (**organized** or **frozen behavior**), whereas exponential growth of periods and linear growth in the number of attractors is observed in the case of large connectivities (**chaotic behavior**).

For $k = 2$, both the period and the number of attractors vary roughly as the square root of the number of automata. For large k, the period varies as $exp(\frac{N}{2})$.

Within Kauffman's formalism, the different cell types are interpreted as the different attractors of the net. This interpretation is supported by the fact that the cell division time and the number of different cell types scale as the square root of the DNA mass of the genome, the same law as for the corresponding quantities for Boolean nets with connectivity 2.

2.2 Patterns

The first cellular implementation of Kauffman nets on a cellular lattice is due to Atlan et al. [2]. It consists in placing Boolean automata with connectivity 2 at the nodes of a square lattice with a connectivity matrix described in figure 2. Since connectivity is 2, these random nets only have a frozen behavior.

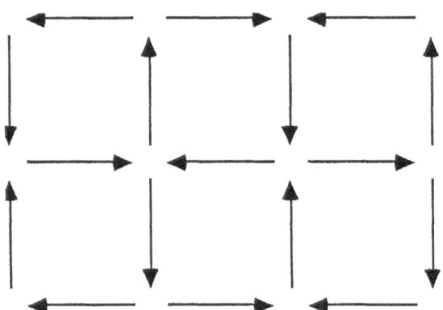

Fig. 2. Pattern of connectivity for 2-input cellular automata on a square lattice.

One then observes that during the limit cycles some automata remain stable while others oscillate (cf. figure 3). The set of stable automata is often connected and isolates subnets of oscillating automata. The details of the **patterns** depend upon the initial conditions and are specific for the attractor.

A possible analysis of this dependence on initial conditions, is to summarize, as in figure 4, how many times each automaton is oscillating for a given number of initial conditions (see Atlan et al. [3]). Figure 4 shows that a large proportion of nodes remain always stable, the **stable core**.

Those which oscillate are grouped in clusters, with contiguous but discontinuous probability. Each probability step corresponds to a different limit cycle and the interval between the nearest probabilities to the width of the attraction basin.

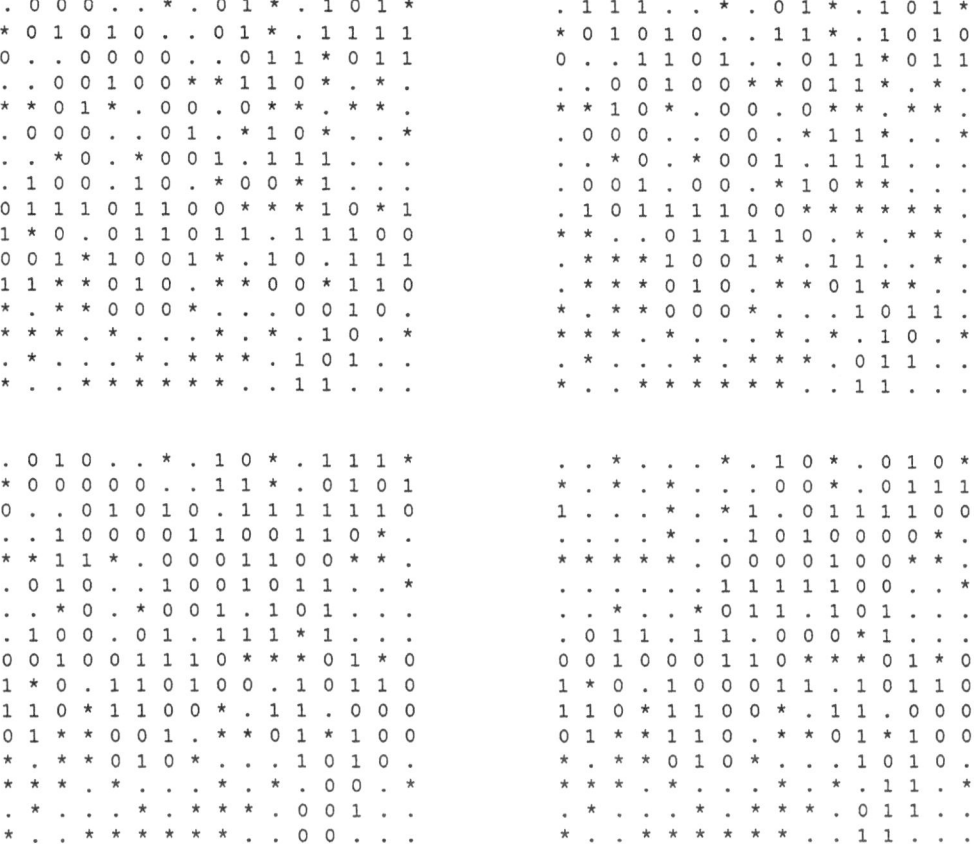

Fig. 3. Patterns of activity of the same 16 × 16 net during the limit cycles reached from 4 different initial conditions: The 0's and 1's correspond to oscillating automata, while . and * correspond to automata that remain fixed.

Kauffman's simulation results have never been rigorously derived in spite of numerous attempts. However, two extreme cases can be treated by analytical methods.

For $k = N$, the iteration graph is random: every new configuration is randomly selected from the previous one. The mean period varies as the square root of the number of configurations and is equal to $exp(\frac{N}{2})$ [4], [5].

For $k = 1$, the net is divided into independent subnets, which we call **crabs** [6], [7]. There are only 4 Boolean functions with one input. Two are constant and we shall not use them. One simply transmits its input, and we denote it +. The other one, which inverts it, is denoted −. Each crab is made up of a central loop which sends signals into branches. Signals can nowhere enter a loop since the input automaton would have two inputs. For

31	533	533	533	0	0	0	0	999	999	0	0	999	999	999	31
31	533	533	533	533	533	0	0	999	999	0	0	905	999	999	999
999	0	0	533	533	564	564	749	0	999	936	936	717	936	999	999
0	0	564	564	533	564	564	749	749	999	936	936	655	655	0	0
0	0	564	564	0	0	999	999	749	999	718	655	655	0	0	0
0	564	564	564	0	0	999	999	749	749	936	936	655	0	0	0
0	0	0	564	0	0	999	999	999	62	936	936	936	0	0	0
0	999	999	999	0	999	999	62	780	936	936	62	407	0	0	0
438	999	999	999	999	999	999	999	999	62	62	62	438	438	0	438
438	0	438	0	999	999	999	999	999	999	62	438	438	438	438	438
438	438	438	0	999	999	999	999	63	63	782	782	0	438	438	438
438	438	0	0	999	999	999	0	63	63	782	782	0	438	438	438
31	155	155	0	999	999	999	0	63	63	0	782	782	937	937	31
31	155	155	155	0	0	0	0	63	63	0	0	999	999	0	31
31	0	0	0	0	0	0	0	0	0	0	875	999	999	31	31
0	0	0	0	0	0	0	0	0	0	0	875	875	31	31	31

Fig. 4. Statistics of those initial conditions (out of 999) which lead each automaton to oscillate during the limit cycle.

the same reason a crab has only one loop. Crabs dynamics can be reduced to loop dynamics. When a loop contains an odd number of − functions, it cannot be stable since the signal cannot be identical to itself after having gone through the loop. Such a loop is said to be frustrated and the period of its limit cycle is twice its length, or an odd divider of this figure. On the other hand a non-frustrated loop can be stable.

The generalization of the latter approach to 2 input nets is fully described in F. Fogelman-Soulié's thesis [8]. It is based on the distinction between forcing functions such as AND, OR, NAND, NOR, whose state can be determined by only one of their inputs, and non-forcing functions such as XOR or EQU, whose state needs the knowledge of the two inputs in any case. Forcing functions have a strong tendency to evolve towards simple attractors while non-forcing functions are difficult to stabilize. Even connectivity 2 random nets made of only OR and XOR functions have a chaotic behavior. The increase in the proportion of non-forcing functions with k explains the explosion of the periods for large k.

3.1 Four-Input Square Lattices. (See figure 5)

A nice way to study the frozen/chaotic transition is to use a continuous parameter. Derrida and Stauffer [9] have proposed to work with 4-input square lattices. If the transition functions of the automata are chosen symmetrically with respect to 0 and 1 the net's behavior is chaotic. But one can choose to bias the probability so that the transition function yields 1 for any input configuration. If this probability is 0 for instance, the net has only 1 stable attractor: in one iteration step any initial configuration evolves towards 0. When p varies from 0 to 0.5 the transition should occur somewhere (the region from 0.5 to 1 is symmetrical). Computer simulations in [10] show that for low values of p the periods vary slowly with N, which indicates a frozen regime, while they grow exponentially with N at larger p (chaotic regime).

The **local periods** are also quantities of interest [10]. The state of each single automaton evolves periodically with its own local period which divides the period of the whole net. Figures 6 and 7 show these periods

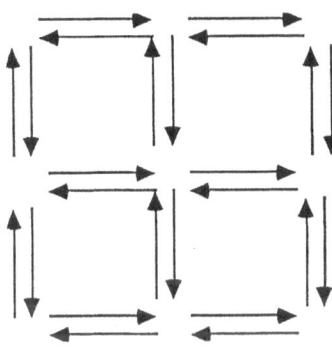

Fig. 5. Pattern of connectivity for 4-input cellular automata on a square lattice.

1	1	4	4	8	8	8	1	1	4	4	1	1	1	1	1
1	1	1	1	8	8	4	1	1	1	1	1	1	4	1	1
4	1	1	1	8	72	4	1	1	1	1	1	1	4	1	4
4	1	1	72	72	36	4	1	2	1	1	1	1	1	1	1
4	4	1	18	1	18	36	2	2	2	1	1	1	1	6	1
4	1	1	18	18	18	36	1	1	1	1	1	1	12	6	12
12	1	18	18	18	18	18	1	1	1	1	1	12	12	1	12
1	1	18	18	18	1	18	1	1	1	1	1	1	12	12	12
1	1	1	1	1	1	1	1	1	4	4	1	12	12	1	1
1	1	1	1	1	1	1	1	1	1	4	4	4	12	1	1
1	4	4	1	1	1	1	1	1	1	4	4	1	4	1	4
1	1	1	1	1	1	1	1	1	1	1	4	1	4	4	4
1	1	1	1	2	2	1	1	1	1	1	1	1	1	4	1
1	1	1	1	1	1	1	1	1	1	1	1	1	1	4	1
1	1	1	8	8	1	1	1	1	1	1	4	4	1	1	1
1	1	8	8	8	8	1	1	1	1	4	1	1	4	4	1

Fig. 6. Local periods in the frozen regime, $p = 0.22$

```
1********  1  1  1   1  1  1   1  1   1  6   6   1
 1*****************   1   1  1  1***************
**************************   1   1   1   1******      1
*****************   1   1***   1   1   1   1******      1
 ****************   1***************    1*********
**************   2   2   2   2***   4***   1*********
************  1   1   1   2   1   1*****************
********  1   1   1   1   1   1   1*****************
********  1   1   1   1   1   1******************
**************  1   1   1********************    1***
***********  1   1   1   1************************
***   1   1***************************    1*********
***   1   1************************  1***   1   1
 1   1   1*************************** 12***  12   1
 1   1********************  1   1 12   6   6   1
 1   1******  1   1******  1   1   1   1   1   2   6   1
```

Fig. 7. Local periods in the chaotic regime, $p = 0.30$ (the stars correspond to periods larger than 1000).

for both regimes. In the frozen regime oscillating automata are grouped into small clusters with medium periods. On the other hand, in the chaotic regime, automata seem to be oscillating with either a very large period or a very small one.

3.2 "Magnetization"

Figures 8 and 9 are histograms of the time average of the states of the automata [11]. In the frozen state, the periods are small, and these averages are fractions with small integer numerators and denominators. The his-

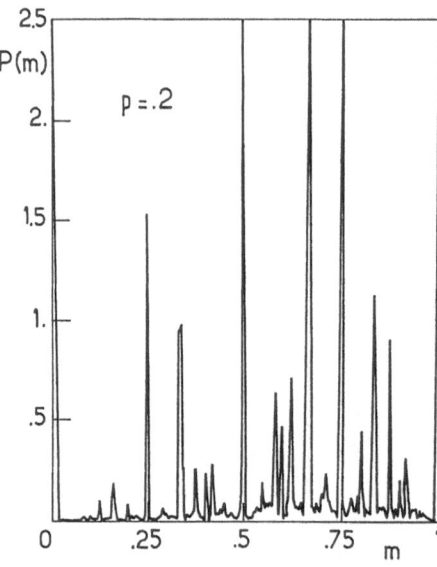

Fig. 8. Histograms of the time average of the states of the automata during the limit cycle. Frozen regime.

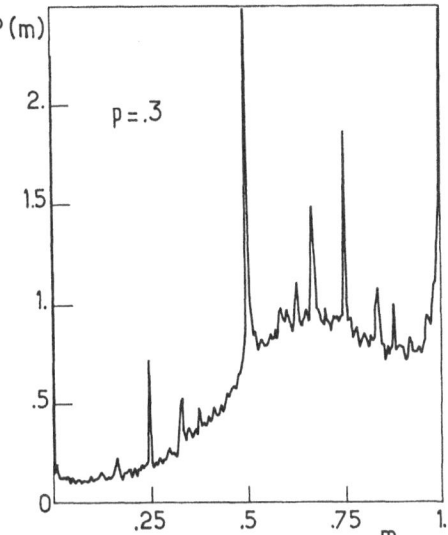

Fig. 9. Histogram of the time average of the states of the automata during the limit cycle. Chaotic regime.

togram for $p = 0.2$ looks like a series of peaks corresponding to $\frac{1}{4}, \frac{1}{3}, \frac{1}{2}, \frac{2}{3} \ldots$ while for the chaotic regime the histogram has a strong continuous part corresponding to the automata oscillating with a large period. This reminds us of the velocity power spectra observed in fluid dynamics in the vortex regimes or in the turbulent regime.

3.3. Percolation

Figures 10 and 11 are histograms of oscillations equivalent to figure 4. They show how many times out of 9 initial conditions each automaton is oscillating during the limit cycle. For small values of p, the oscillating regions are small clusters separated by the stable core. For larger p, the oscillating clusters percolate through the sample. Computer simulations show that the percolation threshold is $p = 0.26 \pm .02$ which is the same value for the frozen/chaotic transition as determined by other methods (see later).

3.4 Evolution of the Overlaps

As in the case of continuous systems we expect some sort of strong sensitivity to the initial conditions in the chaotic behavior. In order to compare trajectories in phase space, one computes the overlap between successive configurations which is defined as the number of automata that are in the same state, divided by N. If starting from some initial condition a few automata are flipped, the evolution of the overlap between the two configu-

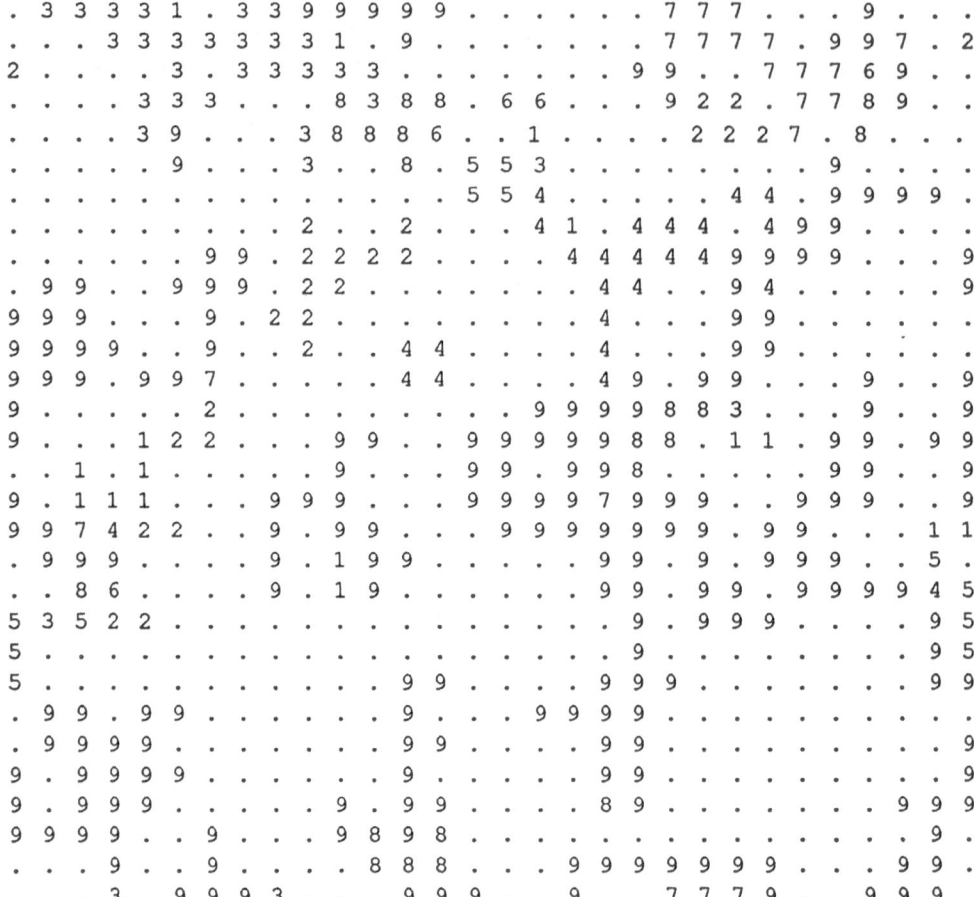

Fig. 10. Number of initial conditions for which an automaton oscillates
during the limit cycle in the frozen regime, $p = 0.21$

rations (the perturbed and unperturbed configurations) indicates whether
they converge to the same attractor, whether they remain at some distance proportional to their initial distance (which would be analogous to
Hamiltonian dynamics) or whether they diverge as in continuous chaotic
dynamics. Figure 12, from [9], compares $d\infty$, the distances at "infinite
time", as a function of the initial distance for 4-input cellular nets. In the
chaotic regime the relative distance evolves towards a finite value of the
order of 0.1, for however small the initial perturbation is. In the frozen
regime $d\infty$ is proportional to $d0$: because the frozen behavior corresponds
to independently oscillating subnets, for a small value of $d0$, only a few subnets are perturbed, in proportion to $d0$. So $d\infty$ varies. Derrida and Stauffer
obtained 0.26 for the transition threshold by plotting $d\infty$ as a function of
p for fixed $d0$.

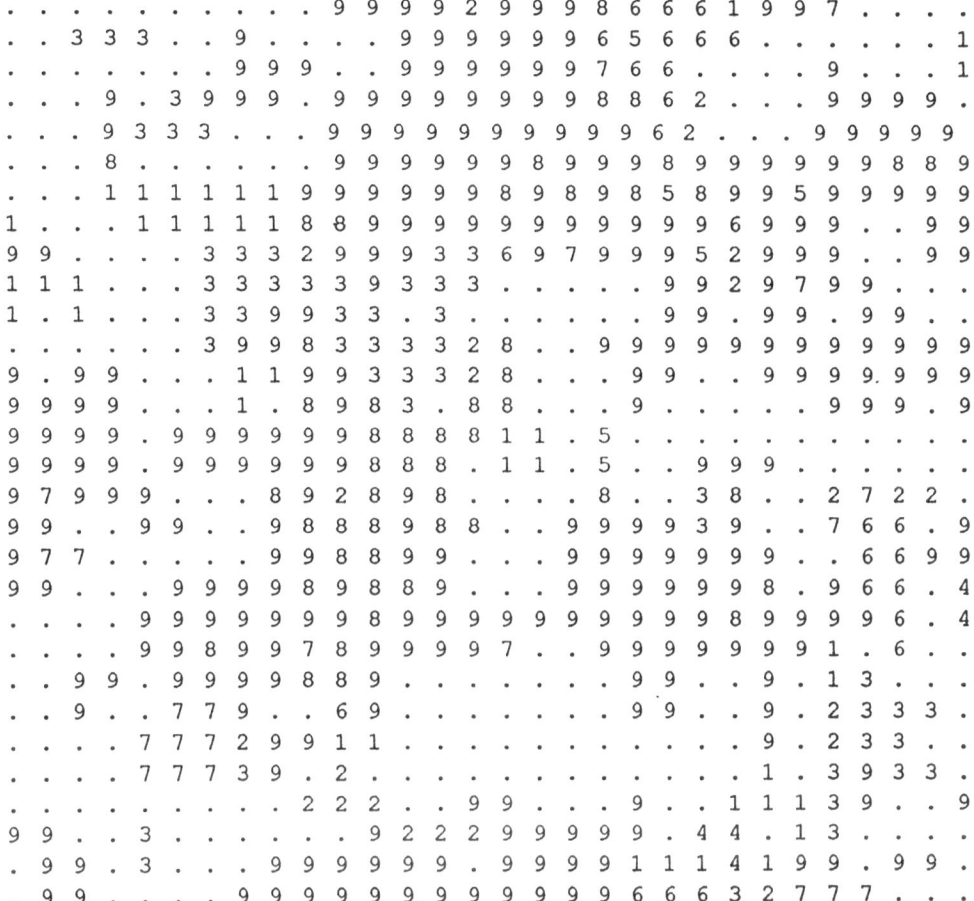

Fig. 11. Number of initial conditions for which an automaton oscillates during the limit cycle in the chaotic regime, $p = 0.28$

4. Annealed Nets

In the case of Kauffman nets of random connectivity k, the evolution in one time step of the overlap $x(t)$ of two random configurations can be predicted. x^k is the proportion of automata whose k inputs are in the same state for both configurations. These automata will be in the same state at the following time step and all the other automata will be in the same state with probability $\frac{1}{2}$. x then varies as:

$$x(t+1) = \frac{1 + x^k(t)}{2} \tag{1}$$

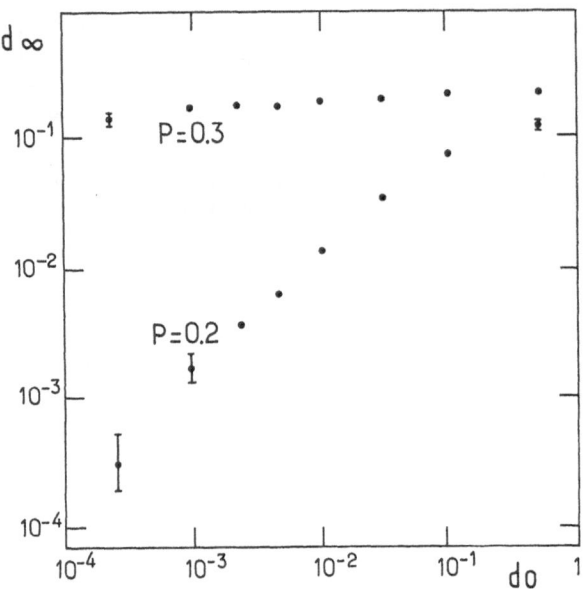

Fig. 12. Distances at 'infinite time' versus initial distances, in the frozen $(p = 0.2)$ and the chaotic regimes $(p = 0.3)$.

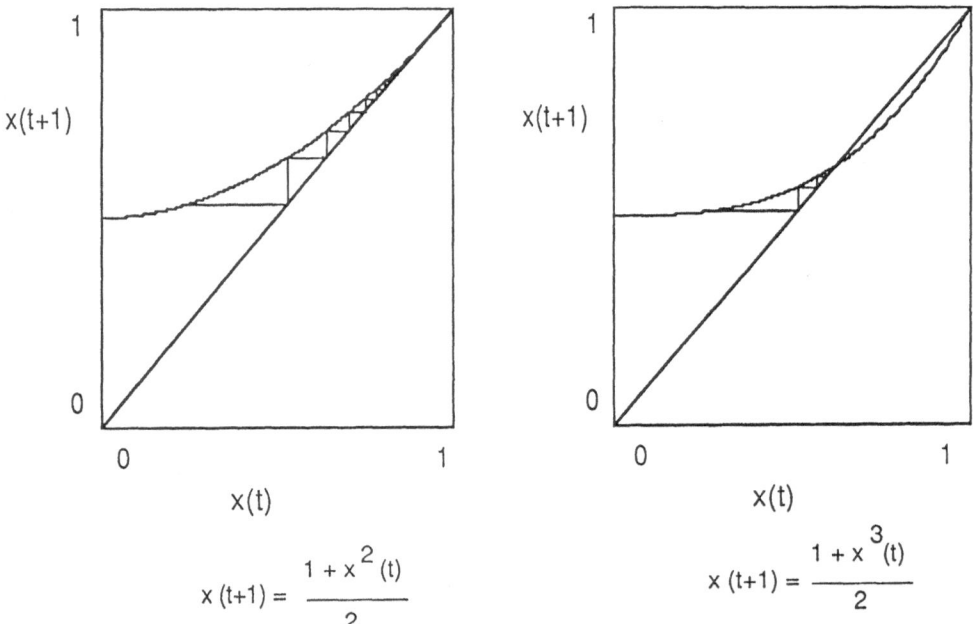

$$x(t+1) = \frac{1 + x^2(t)}{2}$$

$$x(t+1) = \frac{1 + x^3(t)}{2}$$

Fig. 13. Iteration graph of the relative overlap for $k = 2$ and $k = 3$

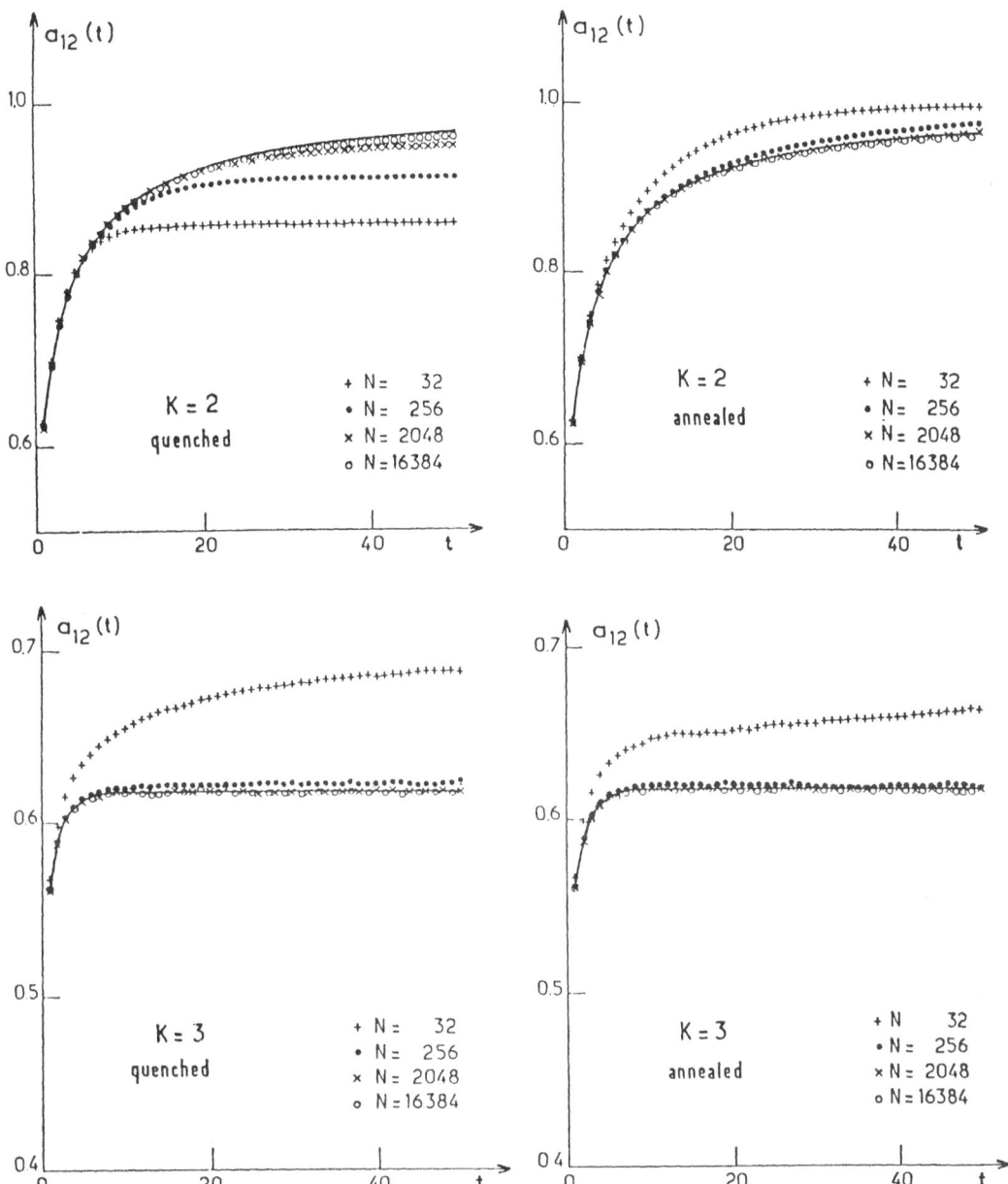

Fig. 14. Comparison of the time evolutions of the overlaps between two configurations for quenched and annealed nets, for connectivity $k = 2$ and 3. The continuous line is obtained by iterating equation (1).

Such an expression is only valid for random configurations and, in principle, cannot be indefinitely iterated, which would give us interesting indications about the infinite time behavior of the system.

One way to get rid of this difficulty is to invent a new type of automaton whose function is randomly changed at each time step: these "annealed" nets have been proposed by Derrida and Pomeau [12]. In this case expression (1) can be iterated. Two types of iterated maps exist: for k less than or equal to 2, x goes to 1 (identical configurations) at infinite time for infinitely large nets. This implies that the volume of the configuration space available to the system goes to 0 and that an organized behavior has been reached. This is not the case for k larger than 2 (see figure 13).

We have done computer simulations both for annealed nets and for normal Kauffman nets [13]. Surprisingly enough annealed and deterministic ("quenched") nets exhibit the same behavior (see figure 14), except at large time for $k = 2$. In this latter case scaling effects are observed which show that the overlap saturates at intermediate times. This similarity in behavior can be explained. Modeling a deterministic net by an annealed net on several time steps is valid as long as the ancestors of an automaton are all different - by ancestor we mean all the automata which influence the state of an automaton after t time steps. In such a case there are no correlations between the inputs of a given automaton and expression (1) can be iterated. This approximation is thus valid for time intervals which increase with the size of the net.

Conclusions

Pattern formation is definitely responsible for the existence of a frozen regime. The isolation of small patches of oscillating automata is the origin of the small periods of the attractors. It is also responsible for the robustness of the dynamics with respect to small changes in initial conditions or in the transition rules of a few automata. It is the basis of the interpretation of the attractors of the dynamics as an "organized behavior".

Acknowledgments.

The review presented here is the result of common efforts involving Henri Atlan, Bernard Derrida, Françoise Fogelman-Soulié, Eric Goles, Stuart Kauffman, Jean Salomon and Dietrich Stauffer.

Bibliography

1 Kauffman S.A.: (1969), *J. Theor. Biol.*, **22**, pp. 437-467

2 Atlan H., Fogelman-Soulié F., Salomon J. and Weisbuch G.: (1982), *Cybernetics and Systems*, **12**, p. 103

3 Atlan H., Ben-Ezra E., Fogelman-Soulié F., Pellegrin D. and Weisbuch G.: (1986), *J. Theor. Biol.*, **120**, pp. 371-380

4 Kauffman S.A.: (1986), pp 339-360 in Bienenstock E., Fogelman-Soulié F., Weisbuch G., (Eds): *Disordered Systems and Biological Organization* NATO ASI Series in Computer and Systems Sciences, Springer-Verlag, Berlin, Heidelberg

5 Coste J. and Henon H.: (1986), pp. 361-366 in *Bienenstock* et al, ref. 4

6 Holland J.H.: (1960), *J. Franklin Inst.*, **270**, pp. 202- 226

7 Fogelman-Soulié F., Goles-Chacc E. and Weisbuch G.: (1982) *Bull. Math. Biol.*, **44**, pp. 715-730

8 Fogelman-Soulié F.: (1985) *Contribution á une théorie du calcul sur réseau*, Thesis, Grenoble University

9 Derrida B. and Stauffer D.: (1986), *Europhysics Letters*, **2**, 739

10 Weisbuch G. and Stauffer D.: (1987), *J. de Physique*, **48**, 11 - 18

11 Derrida B.: in *Chance and Matter*, Les Houches Summer School, July 1986

12 Derrida B. and Pomeau Y.: (1986), *Europhysics Letters*, **1**, pp. 45-49

13 Derrida B. and Weisbuch G.: (1986), *J. Physique*, **47**, pp. 1297-1303

The Structure of the Visual Field

J.J. Koenderink

Physics Laboratory, Utrecht University, P.O. Box 80000,
NL-3508 TA Utrecht, The Netherlands

1. Introduction

The "visual world" is our awareness of the physical world through optical means (eyes open). It is a three-dimensional construct that extends beyond our backs. It is invariant against head and movements. We observe "things" and "relations" that are coloured through experience (e.g. causal relationships) and other modalities such as touch (e.g. fur looks "warm" and "soft"). We measure distances in terms of motor programs (e.g. "ten paces", "graspable", etc.).

The "visual field" is what you get with monocular vision (one eye closed), without head movements and with normal fixation (i.e. the optical axis roams about within a solid angle of about a square degree or less). The visual field is a two-dimensional entity. It is cut off by the nose and part of the face (Fig. 1). It has a graded resolution. If we switch to an "impressionistic mood" we may appproach the state where we see a two-dimensional system of coloured patches, like a complicated abstract painting.

The visual field is amenable to experimental studies in laboratory settings. Psychophysics has revealed the spatiotemporal structure of the field.

2. The Empirical Evidence

2.1. The Fine Grained Structure

If you inquire after the graininess of the visual field you get answers that depend on the operational definition you use. If you want to know whether an observer can discriminate some structure from a uniform illumination of the retina you get simple results. At daylight values of the illuminance you can resolve about 60 Hz of temporal variations (it depends on the target size) and - at the optical axis - about a minute of arc of visual angle. Far from the optical axis you resolve about the eccentricity (distance from the axis) divided by 120, i.e. 10 minutes of arc at 20 degrees, etc. Thus the detector array can be expected to look like that shown in Fig. 2.

Fig. 1: The visual field of one eye of Ernst Mach [1]

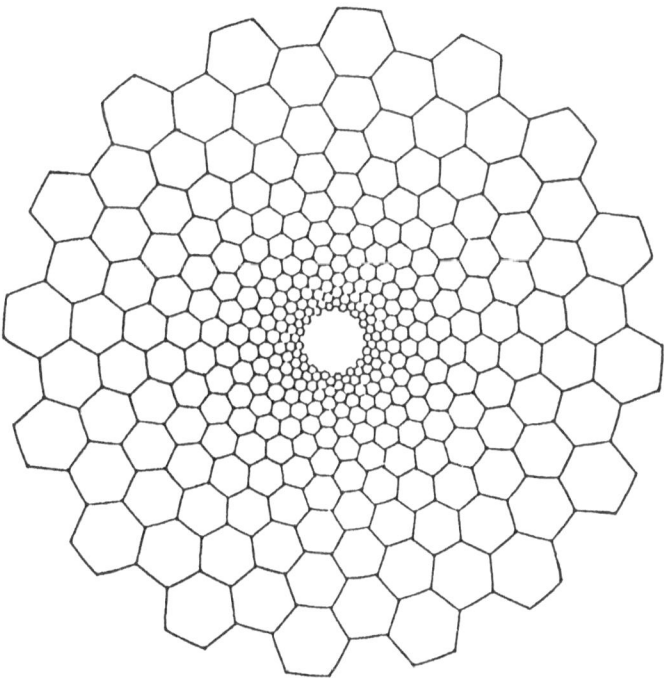

Fig. 2: " Sunflower model" of the fine-grain detector system of human vision [2]

An interesting complication is that there is also a cut-off at low spatio-temporal frequencies (about 0.1 Hz and about 0.1 of a cycle per degree for central fixation). Thus the system is "AC" rather than "DC-coupled".

A 50 Hz temporal variation is perceived as "flicker": you can't even guess at the frequency! If you aim at temporal resolution in a real sense (being able to report on the temporal structure) the limit is more like 4 Hz. In the spatial domain we fare a lot better. The total visual field is very large, the optics support a retinal image up to 105 degrees from the optical axis; a solid angle of 0.63 times 4π steradians. Thus the retinal image has several million degrees of freedom (which agrees with a number of optic nerve fibers of $2 - 3 \times 10^6$) and a temporal bandwidth of 4 Hz, i.e. about 10^7 independent samples per second.

2.2 The "Deep Structure"

The human observer is able to "focus" the attention in the resolution domain, i.e. switch between the perception of the leaves to that of the tree. Many visual objects do not exist on the fine grained scale at all! Evidence for such processes can also be found psychophysically in the so-called "size invariances".

If you measure the contrast detection threshold for Gaussian blobs, i.e. the retinal illuminance varies as

$$I(1 + m \exp(-\|r\|^2/2\sigma^2)),$$

you find a surprising result: the threshold modulation depth (smallest m at which the blob is visible) is constant (about 5%) over a very large range - more than two decades - of sizes (σ). One would never expect such a result for a system with sampling elements of a single fixed size. We have to assume that the visual system has a large range of sample apertures and switches between them. Then the system effectively samples the retinal illumination pattern at many resolutions simultaneously. Several other experimental results corroborate this conclusion.

In some cases the full range of sampling apertures may not be available. For instance, at low luminances the photon shot noise makes the small apertures effectively unusable. Then the system has to switch to coarser sampling and visual acuity drops in a predictable manner. When you show the pattern to more eccentric locations (farther from the optical axis) you get the same effect but this time because the small sampling apertures just aren't there. Empirically one has nice scaling laws: if I scale a pattern to the smallest apertures at a given location and at the same time scale luminance such as to keep the quantum catch per smallest aperture at a constant value then visual performance will be identical. Such scaling laws hold over many decades of parameter values.

3. Speculations on Mechanisms

3.1. The Front-end Visual System as a Self-similar Detector Array

The basic sampling units of the visual system do have a lot of internal structure (e.g. they are not just photocells but compute something like the temporal derivative of the Laplacean of the logarithm of the retinal illuminance). However, for the moment we consider only a single parameter, namely their size δ . (You may think of δ as roughly the diameter of a disclike integration area.) In order to cover the visual field without gaps or excessive overlap one needs about δ^{-2} units per unit area.

In order to cover the resolution dimension one needs a range of δ's . In a scale-invariant system one needs a uniform density on a logarithmic axis. Such considerations lead to the following hypothetical density of sampling units: there are $\mu dA\, d\delta$ units per area element dA in the size range $[\delta, \delta+d\delta]$. The density μ must be proportional with the third power of δ. Then one has a scale-free, selfsimilar structure. In practice μ will depend on the eccentricity E (the distance from the optical axis) and we have:

$$\mu(E,\delta) = \begin{cases} 0 & \text{for } \delta < d^*(E + E_0) \text{ or } \delta > D \\ \frac{A}{\delta^3} & \text{otherwise,} \end{cases}$$

where $E_0 \sim 2$ degrees; $d^* \sim 1/120$ and D about $20 - 100$ degrees. The constant A determines the amount of overlap and can be guestimated from the total number of optic nerve fibers or from electrophysiology. (Each point of the retina is covered by about 30 "receptive fields".) Conceptually, we may subdivide the dimension in (logarithmically equal) intervals such that the units with sizes in the interval just cover their part of the visual field without gaps or overlap. Then one obtains a series of similar sampling arrays of different sizes (Fig. 3).

This density can be used in theories of the visual threshold in order to predict performance for all kinds of pattern. The results are very good indeed: such models are unique in predicting reasonable values over large parameter ranges.

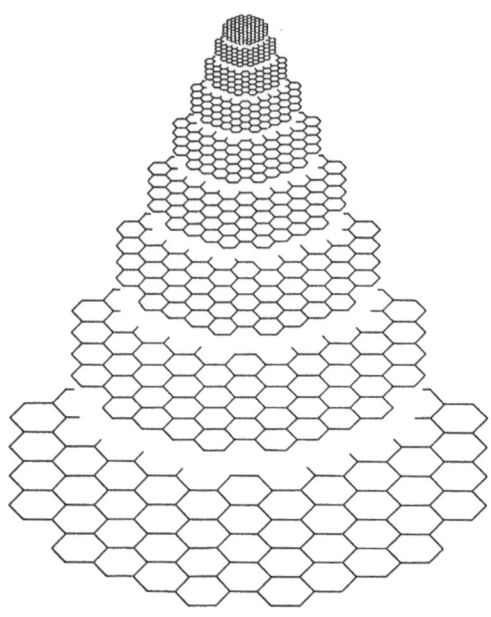

Fig. 3: "Stack model" of detector system of human vision [3]

3.2. The Forest and the Trees

How does one use a self-similar optical sampling array? The system described above acts like a "zoom-lense": one obtains wide angle to tele-views when one runs through the resolution dimension. However, one has all these views simultaneously. The situation is like that with a geographical atlas where one can "zoom" from a view of the world to continents, countries, city plans, etc. Note that one doesn't get the structure of the city from the world map (no use putting it under the microscope!) for there the city is represented by an arbitrary symbol. When one flips pages to more and more detailed maps one really gets a sequence of singularities being "unfolded".

A simple model is obtained in the following way: suppose we describe the retinal illuminance pattern as a one parameter family of images L(x,y;t). Here (x,y) denote Cartesian coordinates in the image plane, t a resolution parameter. Now ask the question: when do points (x,y) and (x^*, y^*) at resolution t and t^* correspond to the same feature? You may require that for this to be the case one should have $L(x, y; t) = L(x^*, y^*; t^*)$, whereas $(x - x^*)^2 + (y - y^*)^2$ is at a minimum. This problem is easily solved, and we find that the integral curves of the vector field

$$\vec{s} = (-L_x \Delta L, -L_y \Delta L, L_x^2 + L_y^2)$$

connect similar features in (x,y;t)-space ("scale space"). At stationary points of the images ($\vec{\nabla}L = 0$) the $\vec{s}$ field is singular and one may expect bifurcations. Now we require that scale-space has a natural "causal" structure: only when you zoom in to higher resolution does new structure occur, i.e. the bifurcation is only one-way. This yields the condition

$$\Delta L.L_t > 0 \quad whenever \quad \vec{\nabla}L = 0.$$

A neat structure is obtained when one runs the diffusion equation $\Delta L = L_t$ on the finest scale image: then "causality" is guaranteed. (One can show that this is also necessary.)

The typical singularity in scale space looks like

$$L(x + \delta x, y + \delta y; t + \delta t) = L(x, y; t) + \frac{\delta x^3}{6} + \frac{\delta y^2}{2} + (1 + \delta x)\delta t + ...$$

at $(x, y; t)$.

An extremum meets with a saddle point and both disappear on blurring, or a pair of critical points is created on "deblurring". A point on a low resolution image explodes into a region on a high resolution image, just like the point representing a city on the world map explodes into the city plan when we "zoom" through a geographical atlas (Fig. 4).

The discrete structure of singularities in scale-space can be coded as a data structure (a tree) that supposedly represents the multiresolution or deep structure. We may hypothise that the visual system deals with the "forest-and-the-tree-problem" in this way.

3.3 The Problem of "Shape"

What do we really mean when we say that the cypress is conical, the oak-tree hemispherical? Surely they have extremely intricate shape on many scales (and we are no doubt also able to appreciate those).

Many aspects of "shape" are of a differential geometric nature (e.g. "boundary curvature") and are easily expressed in terms of algebraic combinations of partial derivatives of the retinal illuminance pattern. Consider boundary curvature for instance. If the illuminance is L(x,y), then the curvature of an isophote ($L(x, y) = L_0$ say) is simply:

$$K = \frac{L_{\xi\xi}}{L_\eta}$$

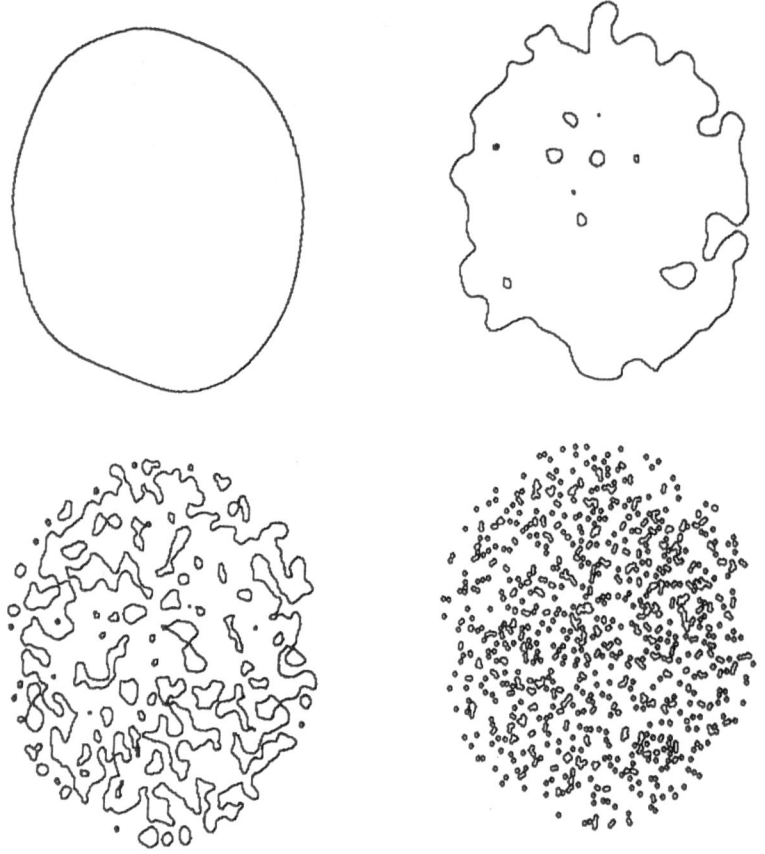

Fig. 4: The same object at different levels of resolution. Note the changes in topology and boundary curvature [4].

where the coordinates (ξ, η) are locally chosen such that η runs along the gradient $\vec{\nabla}L$ and η is tangential to the isophote. Thus the descriptions of shape are simple algebraic combinations of partial derivatives. Of course the result depends on the level of resolution! We may introduce resolution dependent "fuzzy derivatives" as follows:

$$\frac{\partial^{n+m}}{\partial x^n \partial y^m} \leftrightarrow \Psi_{nm}(x,y;t) = \frac{\delta^{n+m}}{\delta x^n \delta y^m} \frac{\exp(-\frac{x^2+y^2}{4t})}{4\pi t} \quad .$$

Thus the (n+m)th order mixed derivative (n times with respect to x, m times with respect to y) of the blurred image L(x,y;t) equals the convolution of the retinal illuminance (L(x,y;0)) with the kernel Ψ_{nm} .

The functions $\Psi_{nm}(x, y; t)$ have a striking likeness to the "receptive fields" that the electrophysiologists have described in the visual system of mammals. One has found at least the following fuzzy derivatives:

$$\frac{\partial}{\partial x}, \quad \Delta \left(= \frac{\partial^2}{\partial x^2} + \frac{\partial^2}{\partial y^2}\right), \quad \frac{\partial^2}{\partial x^2}, \quad \frac{\partial^3}{\partial x^2 \partial y}.$$

In the primary visual cortex one finds primarily second and third order derivatives. The visual system does not use Cartesian coordinates, but instead represents e.g. $\frac{\partial^2}{\partial \xi^2}$ for a great many directions $\vec{e}_\xi$ (Fig.5).

It seems not unlikely that the basic unit ("pixel value") is a local jet (or truncated Taylor series) of at least the third order. Then the visual cortex represents a jet-extension of the scale-space based on the retinal illuminance.

The problem of "shape" then boils down to a study of the generic singularities in this jet-extended scale-space. One finds such phenomena

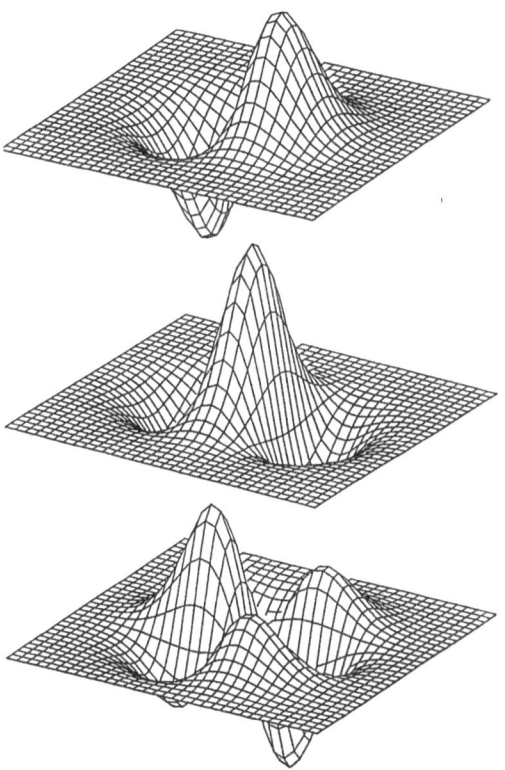

Fig. 5: The "fuzzy derivatives" $\frac{\partial}{\partial x}$, $\frac{\partial^2}{\partial x^2}$ and $\frac{\partial^3}{\partial x^2 \partial y}$ [4]

as merges and splits of blobs (light or dark blotches) or the creation or annihilation of blobs, the creation or destruction of pairs of inflexions of a boundary, etc. Such studies also have an importance in the design of artificial vision systems.

4. Some Further Research Problems

Empirically one finds certain pathologies ("amblyopic" or "lazy" eyes) in which the patients have a completely intact visual system by all counts and normal resolution as far as discriminating structure from uniform fields is concerned but are yet unable to read the headlines of the newspaper! Such patients have "all the pieces of jigsaw-puzzle" but seem to be unable to piece the visual field together. We may assume that they have the jet representation at the local level but miss the ability to connect pixels, i.e. they miss a global topology. These patients have the same problem with simple figures as normals with slightly more complicated ones (Fig. 6).

Fig. 6: Of these two black blobs one actually consists of two blobs, the other of one. Even normal human vision has difficulties in assessing such simple topological features as the connectedness of simple blobs [5].

Thus it becomes of some interest to investigate which shape descriptions are "point-properties" and for which ones a geometrical expertise is required (i.e. a connection such as the ability to compare orientations of line segments in different jets). This depends on the order of the jet and moreover, some properties can be computed on the local as well as on the multilocal level.

Here lies a rich field of enquiry of which we are largely ignorant at the present time.

Literature

1. E. Mach: *Analysis of sensations* (English translation 1962, originally published in 1886. Dover, New York 1962)
2. A.J. van Doorn, J.J. Koenderink, M.A. Bouman: *The influence of the retinal inhomogeneity on the perception of spatial patterns.* Kybernetik **10**, 223-230 (1972)
3. J.J. Koenderink, A.J. van Doorn: *Visual detection of spatial contrast; influence of location in the visual field, target extent and illuminance level.* Biol. Cybernetics **30** , 157-167 (1978)
4. J.J. Koenderink, A.J. van Doorn: *Representation of local geometry in the visual system.* Biol. Cybernetics, in press (1987)
5. M. Minsky, S. Papert: *Perceptrons* (M.I.T. Press, Cambridge, Mass. 1968)

Patterns of Maximal Entropy

R.D. Levine

The Fritz Haber Research Center for Molecular Dynamics,
The Hebrew University, Jerusalem 91904, Israel

1. Introduction

The interplay between energy and entropy governs the equilibrium structure of macroscopic systems. At low temperatures, it is usually energetic considerations that are important while at higher excitation entropic aspects are more evident. One purpose of the maximum entropy formalism [1-4] is to extend the range of applications of this way of analysis. Must the system be at equilibrium?, must it be macroscopic?, must the relevant variable be energy and is there just one such relevant variable are some of the questions worth considering. The advantages of such an extension are clear in that a whole range of systems and phenomena, many of current interest, could be examined by methods analogous to those of equilibrium statistical mechanics [1,2]. The required generalization is that the state of the system is one of maximal entropy subject to constraints. When the value of the constraints (or 'relevant variables') already very much restrict the possible states, the state of the system is primarily specified by them. When many states are consistent with the given constraints, maximizing the entropy amongst that subset of states does serve to specify the one. The generalization does therefore subsume equilibrium statistical mechanics and can indeed be used as a unifying principle for it [1,2,5]. In this paper we review the reasons for considering the principle to have a wider range of applicability and choose as examples the simple structures ('patterns') that are possible in two dimensions, [6-12].

The common feature in all applications of the maximum entropy formalism is that the constraints in themselves do not specify a unique state of the system. In outline, one is given m values a_r, $r=1,...,m$ which the unknown probability distribution p_i, $i=1,...n$, $n>m+1$ satisfies

$$\sum_{i=1}^{n} p_i a_{ri} = a_r , \qquad r=1,...,m . \tag{1}$$

Here a_{ri} is the value of the r'th constraint in the i'th state and the probabilities p_i are mutually exclusive and collectively exhaustive. Since the number m of constraints is smaller than $n-1$ (and typically much smaller), (1) cannot be uniquely inverted without some additional input. The maximum entropy formalism is equivalent to providing a unique inversion [13,14]. Amongst the subset of distributions which satisfy (1), it selects the one whose entropy is maximal. The original motivation, due to BOLTZMANN is that this selects the most probable distribution, i.e. the one which will be overwhelmingly more probable when the experiment is repeated a very large number of times. This justification is sufficient when one in fact is dealing with an experiment which can be so repeated. The familiar example is when the state i is the (quantum) state of a molecule in a dilute gas phase, and p_i is the fraction of molecules (out of Avogadro's number) which are in that state. It is not sufficient for the present problem and other strongly interacting systems where i is the state of the entire system. One can introduce the notion of an ensemble of replicas of the system of interest, but in the final analysis that is an ad-hoc device as there are, in reality, no such very large ensembles. The justification championed by JAYNES [2,4,5] is based on SHANNON's axiomatic introduction of entropy as the measure of uncertainty. The distribution of maximal entropy is consistent with what is known and is otherwise most uncertain. Indeed one can derive the procedure itself on axiomatic grounds [15] as the distribution which does not bring in any correlations not warranted by the constraints. This point of view, where the procedure is regarded as a 'least biased' or 'maximally non-committal' method of inference has many proponents and opponents. The primary objection of the latter is that the subjective character of the procedure is not appropriate for the natural sciences.

For the problem at hand, an important consideration is that the resulting distribution is the 'most random' or 'most uniform' one amongst the subject which is consistent with the constraint. Another relevant argument applies to 'reproducibility': If experiments or simulations show that the given constraints are sufficient to reproduce the distribution, then it is one of maximal entropy [14]. The range of arguments (and additional ones by JAYNES in [4]) indicates to us that the 'physics' enters the solution via the choice of constraints. It is indeed possible to show that one can identify a set of constraints equivalent to an exact solution for time-evolving mechanical (quantal or classical) systems [16]. For the problem at hand, an *a-priori* specification of the constraints is unrealistic. Progress is however being made [17] on the specification of the exact constraints for non-equilibrium dissipative time evolution [18].

To conclude. There is no reason to limit the application of the maximum entropy formalism to systems at equilibrium or to systems with widely different time scales. Of course, if the system is at equilibrium or if there are slow variables then that strongly influences the choice and number of the relevant constraints. It is intuitively reasonable (and remains so under examination [17]) that the state of the system is as random as possible subject to the instantaneous values of the slow variables. To use the phrase of GIBBS: 'the state of the systems is one of *constrained equilibrium*'. For the problems of interest here, this is all that is required. It is however of interest to note that one can introduce constants of the motion such that the system is rigorously in a state of constrained equilibrium even without a separation of time scales. So much so that we are using this point of view to numerically solve quantum mechanical scattering problems.

2. The Maximum Entropy Formalism

The formalism itself has a deceptive simplicity and is well reviewed [1-5]. We outline it here so as to emphasize key points which are sometimes taken as obvious. First, the expression for the entropy

$$S = -\sum_{i=1}^{n} p_i \ln p_i .$$

(2)

This is such a familiar expression that one can easily overlook the important physical assumption made by writing the entropy as in (2). As we shall shortly point out, (2) implies that in the absence of any constraints on the distribution beyond normalization, the distribution of maximal entropy is uniform. Now that may or may not be physically reasonable. For example, if we are dealing with a particle with a spin and i is the spin label then in the absence of any constraints we expect $p_i \propto (2i+1)$. For distributions in phase space, Planck and later von Neumann determined the proper rule: (2) is correct if i enumerates quantum states. For patterns, we do not have a corresponding backlog of experience to fall on. By analogy to Ising models (and the lattice gas, etc.) we shall build the pattern on a lattice and, in the absence of other constraints take all possible patterns to be equally probable. It is important to recognize that this is an assumption and one that is made by analogy only.

Specifically, we take a square lattice of N sites which can be occupied or not. There are $n = 2^N$ different possible patterns and the entropy is written as in (2) in terms of the probabilities of these 2^N patterns. It cannot be overemphasized that this is an assumption that will eventually have either to be justified on the basis of an invariance argument or to be replaced. An alternative could be, for example, that we consider the distribution p_s of patterns that cover s squares and write the entropy as in (2) over the N different possibilities. One can even imagine conditions for which the latter choice is more compelling. One cannot simply write down (2) for the entropy. In principle however one could always use (2) and bring in the symmetry considerations that govern the specification of the most uniform distribution through a suitable choice of additional constraints. A very definite physical implication is made when (2) is written down and it needs justification.

To maximize the entropy subject to the m constraints (1) and the ever present normalization one introduced $m+1$ Lagrange undetermined multipliers λ_r, $r=0,1,..,m$ (with $r=0$ as the normalization constraint and $a_{oi}=1$) and seeks the unconstrained maximum of the Lagrangian

$$L = S - \sum_{r=0}^{m}(\lambda_r - \delta_{r,0})(\sum_{i=1}^{n} a_{ri} p_i - a_r) . \tag{3}$$

The (unique) result is the canonical exponential distribution

$$p_i = \exp(-\sum_{r=0}^{m} \lambda_r a_{ri}) . \tag{4}$$

If there are no other constraints beyond normalization (i.e. $m=0$) we have $p_i = \exp(-\lambda_0)$, i.e. a uniform distribution.

It is necessary here to reiterate that the exponential form (4) is only valid for that distribution which, in the absence of other constraints, is uniform. That an observed (or computed) distribution is not of the simple form (4) is not, in itself, a proof that it cannot be described by the procedure. After all, many familiar equilibrium distributions are not of exponential type. Indeed, the only equilibrium distribution which is exactly of the form (4) is the one over the quantum states of the system. Another point worth noting is that the functional form of the distribution (4) is prescribed by the *nature* of the constraints. At this point we have not yet used the *value* of the a_r's in (1). It is enough to know that the value of a_r is a constraint to be able to write (4) down. What this value is remains, so far, unspecified.

The values of the constraints are however required for the determination of the value of the Lagrange multipliers. It is customary to determine λ_0 first since typically $a_0 \equiv 1$. Imposing normalization

$$\exp(\lambda_0) = \sum_{i=1}^{n}\exp(-\sum_{r=1}^{m}\lambda_r a_{ri}) \tag{5}$$

makes λ_0 a function of the other m Lagrange multipliers. Using (5) and (4) in (1) leads to an implicit equation

$$a_r = -(\partial \lambda_0 / \partial \lambda_r)_{\hat{\lambda}_r} \tag{6}$$

for the value $\hat{\lambda}_r$ of the Lagrange multiplier conjugate to the value a_r for the constraint. For certain values a_r one may find that (6) will return the result $\hat{\lambda}_r = 0$. Such a constraint is not relevant since it does not constrain the distribution. If one knows this beforehand, an irrelevant constraint will not be imposed from the very beginning. For example, we could have imposed the n very obvious constraints $p_i \geq 0$, $i=1,...,n.$, but did not, for unless the constraint is $p_j \equiv 0$ for one or more states j, it is irrelevant and automatically satisfied, as is evident from (4).

As long as the number of states is finite and the constraints are linear in the probabilities, (6) will have a unique solution which can be obtained by minimizing a potential function with a single global minimum [19]. It should also be noted that one can use either the m constraints as given or the m Lagrange multipliers (as discussed, e.g. by LEVINE in [4]) or any combination of m variables from the two sets using the Legendre transform [18] to change variables. If all the variables are changed so that it is the Lagrange multipliers are given, (6) is replaced by

$$\hat{\lambda}_r = (\partial S (\{a_r\})/\partial a_r)_{a_r} \tag{7}$$

as the implicit equation which determines the value of the constraints. In (7), $S (\{a_r\})$ is the entropy at its maximal value (over all distributions) for the specified value of the constraints.

3. Patterns

A pattern is here specified on the square lattice of N cells where each cell is occupied or not. Hence a pattern can be represented by a binary number of length N, which we denote by $\mathbf{N}$. A priori, all patterns are taken to be equally probable so that the entropy is

$$S = -\sum_{\mathbf{N}} P(\mathbf{N}) \ln P(\mathbf{N}) . \tag{8}$$

Here $P(\mathbf{N})$ is the probability of a particular pattern. The sum in (8) is over the 2^N possible patterns. As already discussed, in the absence of additional constraints, each one of these patterns is equally probable. The number of patterns where n cells are occupied is $N!/n!(N-n)!$. (A simple proof of this well-known result is by expanding $(1+1)^N$ in a binomial series). In general, the mean number of occupied cells, $<n>$, can be written as an expectation value

$$<n> = \sum_{\mathbf{N}} g(\mathbf{N}) P(\mathbf{N}) \tag{9}$$

where $g(\mathbf{N})$ is the decimal representation of the binary number $\mathbf{N}$ (using 1 to denote an occupied cell). For the uniform distribution $<n>/N=1/2$. Hence the constraint $<n>$ is not relevant for the value $<n>=N/2$ as the distribution $P^o(\mathbf{N})=1/2^N$ is of maximal entropy whether this constraint is imposed or not. For other values of $<n>$ this will no longer be the case.

An occupied cell can have (four) near neighbors, some of which may also be occupied. Let there be n_j clusters consisting of j occupied cells,

$$n = \sum_j n_j . \tag{10}$$

For a given value of $<n>$, the distribution of maximal entropy is

$$P(\mathbf{N}) = \exp(-\lambda_0 - \lambda n) \tag{11}$$

$$= \exp(-\lambda_0 - \lambda \sum_j j n_j) .$$

The uniform distribution $P^o(\mathbf{N}) = 1/2^N$ corresponds to $\lambda = 0$ in (11). The use of the identity

$$\sum_{\mathbf{N}} (P(\mathbf{N}) - P^o(\mathbf{N})) \ln(P(\mathbf{N})/P^o(\mathbf{N})) \geq 0 \tag{12}$$

for $P(\mathbf{N})$ given by (11) implies that

$$\lambda((N/2) - <n>) \geq 0 \tag{13}$$

with equality if and only if $\lambda = 0$. For higher mean occupancy than $N/2$, λ must therefore be negative, which makes the distribution (11) favor larger than average clusters. The opposite behavior is found for $<n> < N/2$ when $\lambda > 0$ and smaller clusters are more probable. The connection with the (bond) percolation threshold on a square lattice which is at $<n>/N = 1/2$, [6], is clear.

When only $<n>$ is imposed as a constraint, all the patterns where n cells are occupied are equally probable. This, irrespective of whether these n cells are present as one cluster or dispersed as n clusters or anywhere in between. To differentiate amongst those possibilities one can introduce a second constraint, the mean number, $<m>$, of clusters

$$m = \sum_j n_j . \tag{14}$$

With this constraint, the distribution of maximum entropy

$$P(\mathbf{N}) = \exp(-\lambda_0 - \lambda n - \mu m) \tag{15}$$

$$= \exp(-\lambda_0 - \sum_j (j\lambda + \mu) n_j)$$

will depend on the individual cluster sizes and not only on the overall number, n, of occupied cells. We have used μ to designate the Lagrange parameter conjugate to $<m>$ since it acts as a chemical potential. For $\mu > 0$, the distribution (15) favors the low m patterns, i.e. occupied cells tend to aggregate forming as few clusters as possible. The results are opposite for $\mu < 0$. One can use the same route that led to (13) except that we are now comparing the distributions (11) and (15) at the same $<n>$ value. Then

$$\mu(<m>_0 - <m>) \geq 0 . \qquad (16)$$

In (16), $<m>_0$ is the irrelevant value of $<m>$, i.e. the value of $<m>$ computed for the distribution (11). We reiterate that (16) has been derived at a given value of $<n>$. $<m>_0$ and hence the value of $<m>$ for which μ changes sign can therefore depend on $<n>$.

The use of the pair of constraints derived from (10) and (14) to describe 'patterns' goes way back [20]. By regarding $<n>$ and $<m>$ as the (reduced) density and pressure respectively one can see the similarity with the problem of condensation. In this connection, the discussion by HILL [21] is particularly thorough and FOWLER [22] remains a useful reference. Note however that here no Hamiltonian has been invoked nor an equilibrium assumption. Imposing $<n>$ and $<m>$ as constraints and regarding the Lagrange multipliers λ and μ as the variables one can generate a wide gamut of patterns in which occupied cells tend to aggregate ($\mu > 0$) or repel ($\mu < 0$) one another whether at low ($\lambda > 0$) or at high ($\lambda < 0$) density.

In performing averages over $P(N)$ it simplifies the work to note that all patterns of a given n and cluster size distribution $\{n_j\}$, are equally probable. There are $g(\{n_j\})$ such patterns,

$$g(\{n_j\}) = \frac{n!}{\prod_j n_j! \prod_j (j!)^{n_j}} \qquad (17)$$

where $j=1,...,n$, [23]. Also

$$\sum_{(\{n_j\}|\Sigma n_j = m)} g(\{n_j\}) = S_n^{(m)} \qquad (18)$$

where $S_n^{(m)}$ is the number of ways of partitioning a set of n elements into m non-empty subsets, also known as the Stirling number of the second kind [23]. In particular, for $N \to \infty$ we obtain

$$j! <n_j> = \exp(-\lambda j - \mu) . \qquad (19)$$

The factor of $j!$ on the left-hand side of (19) is for the $j!$ equivalent clusters of size j. Using square brackets to denote concentrations as chemists do, we see from (19) that

$$\frac{[n_j][n_k]}{[n_{j+k}]} = \exp(-\mu) \tag{20}$$

which is just what one would expect from the quasichemical [6,21] approximation. Yet, we have not imposed any such approximation. Rather, we have imposed two very reasonable constraints, maximized the entropy and followed the consequences, without any further assumptions. That these two constraints subsume, *inter-alia*, the quasichemical approximation explains why the patterns generated using (15) appear so familiar. In particular, the role of μ in shifting the distribution from low m values (at $\mu<0$) to high ones (at $\mu>0$) is nicely demonstrated by (20).

The exponential size distribution

$$[n_j] = \exp(-\lambda j - \mu) \tag{21}$$

is due to the assumptions implicit in (17), i.e. that the clusters are interchangable. A different distribution would obtain otherwise. Roughly [13] ,$[n_j] \approx 1/(\lambda j + \mu)$, for small λ.

Concluding Remarks

The purpose of this brief overview is to call attention to the advantages inherent to the maximum entropy formalism. With one or two relevant constraints one can already span a wide variety of patterns. Not mentioned are more elaborate lattices where a topological constraint (on the mean number of sides of a cluster) will be operative [8], nor any energetic constraints. The latter however are not necessarily independent and are definitely not relevant in the mean field or in the quasichemical approximations of statistical mechanics [6,21] where $<n>$ and $<m>$ suffice to reproduce many familiar results. We have not discussed growth processes. There is no reason however why we cannot use the formalism with $<m>$ and $<n>$ being time dependent, [18]. When this is the case, the Lagrange multipliers and the patterns will change with time. A special case is where $<n>$ is constant but $<m>$ is changing, corresponding to 'aging' of a pattern.

Acknowledgment

This work was supported by the Stiftung Volkswagenwerk. The Fritz Haber Research Center is supported by the Minerva Gesellschaft für die Forschung, mbH, München, BRD.

Literature

1. M. Tribus: *Thermostatics and Thermodynamics* (Van Nostrand, Princeton 1961).

2. A. Katz: *Principles of Statistical Mechanics* (Freeman, San Francisco 1967).

3. R.D. Levine and M. Tribus, eds.: *The Maximum Entropy Formalism.* (M.I.T. Press, Cambridge 1979).

4. J.H. Justice, ed.: *Maximum Entropy and Bayesian Methods in Applied Statistics* (Cambridge Univ. Press 1986).

5. E.T. Jaynes: Phys. Rev. **106**, 620 (1957).

6. J.M. Ziman: *Models of Disorder* (Cambridge Univ. Press, 1979).

7. F. Yonezawa and T. Ninomiya, eds.: *Topological Disorder in Condensed Matter.* Springer Ser. Solid-State Sci., Vol. 46 (Springer, Berlin 1983).

8. D. Weare and N. Rivier: Contemp. Phys. **25**, 59 (1984).

9. B. Gruenbaum and G.C. Shephard: *Tillings and Patterns* (Freeman, San Francisco 1985).

10. H.E. Stanley and N. Ostrowsky, eds.: *On Growth and Form* (Martinus Nijhoff, Boston 1986).

11. S.A. Trugman: Phys. Rev. Lett. **57**, 607 (1986).

12. M. Silverberg, A. Ben-Shaul and F. Rebentrost: J. Chem. Phys. **83**, 6501 (1985).

13. R.D. Levine: J. Phys. A. **13**, 91 (1980).

14. Y. Tikochinsky, N.Z. Tishby and R.D. Levine: Phys. Rev. A**30**, 2638 (1984).

15. J.E. Shore and R.W. Johnson: Trans. IEEE IT **26**, 26 (1980).

16. R.D. Levine: Adv. Chem. Phys. **47**, 239 (1981).

17. R.D. Levine and C.E. Wulfman: Physica (1987).

18. R.D. Levine: J. Chem. Phys. **65**, 3302 (1976).

19. N. Agmon, Y. Alhassid and R.D. Levine: J. Comput. Phys. **30**, 250 (1979).

20. W.H. Stockmayer: J. Chem. Phys. **11**, 45 (1943).

21. T.L. Hill: *Statistical Mechanics* (McGraw Hill, New York 1956), Ch. 7.

22. R.H. Fowler: *Statistical Mechanics* (Cambridge University Press 1936), Ch. 5.

Part II

Structure Formation in
Nonlinear Systems

Patterns of Bifurcation
from Spherically Symmetric States

F.H. Busse

Physikalisches Institut, Universität Bayreuth,
Postfach 101251, D-8580 Bayreuth, Fed. Rep. of Germany

Summary

Patterns associated with bifurcations from spherically symmetric states
are determined to a considerable extent independently of the structure
of the physical problem. The solvability conditions induced by the nonlinear properties of the bifurcating solution select a small number of
possible distinct patterns. The final selection of the physically realized pattern is obtained through a stability analysis. Even in the
stability analysis a number of results can be stated which do not depend on the particular physical problem.

1. Introduction

Most states which exhibit the process of a symmetry-breaking bifurcation do not possess the property of isotropy with respect to two or
more spatial dimensions. When the basic state is approximately homogeneous in the dimension with respect to which the symmetry-breaking
occurs, the bifurcating state usually exhibits a characteristic wavelength in that dimension which characterizes the physical process
leading to the instability. Taylor vortices between two long coaxial
cylinders are a familiar example. Another example which is more closely
related to the subject of this review are the convection columns in a
cylindrical fluid annulus cooled from within and heated from the outside and rotating about its vertical axis. The centrifugal force gives
rise to the onset of convection rolls in this case which are aligned
with axis of rotation [1,2]. The azimuthal wavenumber m of the rolls
or columns is well determined and indicates the influence of physical
parameters such as the ratio of conductivity between the boundaries and
the fluid or the inclination of the conical end surfaces of the annular
region. Only when the preferred value of m become sufficiently large,
can a certain amount of variability be expected depending on the initial conditions from which the onset of convection is reached.

The situation is quite different when the basic state is characterized by an isotropy with respect to two dimensions. If instead of
the axially symmetric annulus problem the problem of onset of convection in a self-gravitating spherical fluid shell heated from within is
considered, the bifurcation problem becomes degenerate. Only the degree
ℓ of the spherical harmonics describing the horizontal structure of
convection is determined by the physical conditions. All superpositions
of the $2\ell+1$ orthogonal spherical harmonics are equally likely patterns
according to linear theory. The partial removal of this degeneracy by
the nonlinear theory represents an interesting and challenging mathematical problem. Physical aspects enter only when a further selection
takes place among the remaining pattern on the basis of their stability
with respect to arbitrary infinitesimal disturbances. But some
general results about the stability of certain patterns can be obtained without reference to the physical properties of the system.

An even richer structure must be expected for the problem of a bifurcation from a basic state which exhibits the symmetry of a three-dimensional spherical surface in a four-dimensional space. The approximately homogeneous basic state of the universe as described by the Robertson-Walker metric represents a physical example. Deviations from isotropy of the universe could thus be governed by the general mathematical theory of patterns of bifurcations from spherically symmetric states [3].

The general mathematical results which can be derived without specification of the physical problem are limited to a neighborhood to the point of bifurcation. Whenever the physical problem permits secondary bifurcations, the patterns of the solutions will be changed considerably in general. On the other hand, the neighborhood in which the mathematical results remain applicable may extend over a large domain of the control parameter in some problems as long as only quantitative, but not qualitative aspects of the physical conditions are changed.

In the following we shall describe the method of analysis in section 2 and then outline in sections 3 and 4 the main results obtained so far for bifurcations from spherically symmetric states in 3- and 4-dimensional spaces. Some special cases and extensions will be briefly mentioned in the concluding discussion.

2. Mathematical Procedure

We assume that the bifurcating solutions are governed by the following homogeneous nonlinear vector equation:

$$(\underset{\sim}{W} - R\underset{\sim}{U}) \cdot \underset{\sim}{X} = \frac{\partial}{\partial t} \underset{\sim}{V} \cdot \underset{\sim}{X} + \underset{\sim}{Q}(X, X) \tag{2.1}$$

with corresponding homogeneous conditions applied at the boundaries of the spherical shell. The vector $\underset{\sim}{X}$ may denote velocity components and the deviation of the temperature field from that of basic state in the case of convection or the displacement vector in the case of a spherically symmetric buckling problem. The linear operators $\underset{\sim}{W}$, $\underset{\sim}{U}$ and $\underset{\sim}{V}$ are spherically symmetric and the same property holds for the nonlinear operator Q. We restrict the attention to stationary solutions of (2.1) and introduce the expansion in powers of the amplitude ε

$$\underset{\sim}{X} = \sum_{n=1}^{\infty} \varepsilon^n \underset{\sim}{X}^{(n)} , \qquad R = \sum_{n=0}^{\infty} \varepsilon^n R^{(n)} \tag{2.2}$$

where ε is defined by a suitable measure of $\underset{\sim}{X}$. For simplicity we shall restrict the attention to steady bifurcating solutions, for which the $\frac{\partial}{\partial t}$-derivative vanishes in (2.1). The linear problem corresponds to the order ε of equation (2.1) and has the general solution

$$\underset{\sim}{X}^{(1)} = \underset{\sim}{f}(r) \sum_{m=-\ell}^{\ell} c_m w_\ell^{(m)} , \qquad R^{(0)} = R_\ell \tag{2.3}$$

where the functions $w_\ell^{(m)}$ represent $2\ell+1$ orthogonal spherical harmonics of degree ℓ. In the case of the hypersphere in the 4-dimensional space, the degree ℓ is replaced by the degree n of hyperspherical harmonics and we obtain a $(n+1)^2$-fold degeneracy instead of the $2\ell+1$-fold degeneracy. Otherwise the structure of the two problems is entirely analogous. Of particular interest is the degree ℓ for which the eigenvalue R_ℓ of the control parameter reaches an extremum, usually a minimum, since the corresponding bifurcating solution is likely to be encountered in the physical applications.

A disadvantage of the representation (2.3) is that orientational degeneracy and pattern degeneracy are not separated. For a given solution the representation (2.3) includes all solutions obtained by arbitrary rotations of the sphere, but from the physical point of view only those solutions are of interest which cannot be transformed into each other by rotations. The manifold of these solutions is referred to by the term pattern degeneracy. The group-theoretical approach avoids the orientational degeneracy and therefore has been preferred by mathematicans [4,5,6,7]. In addition, it is usually possible to determine the full set of possible patterns. On the other hand, the approach based on spherical harmonics is more directly accessible and will be used in the following. The special cases $\ell = 0,1$ and $n = 0,1$ in three and four dimensions, respectively, do not show a pattern degeneracy. The cases $\ell = 0$ or $n = 0$ can be excluded from our considerations since the symmetry of the basic spherically symmetric state is not broken. The cases $\ell = 1$ or $n = 1$ correspond to a single solution since all the remaining degrees of freedom in the representation (2.3) describe rotations of that solution over the sphere. But a pattern degeneracy exists for all values $\ell \geq 2$ and $n \geq 2$.

By collecting terms of the order ε^2 we obtain from equation (2.1)

$$(\underset{\approx}{W}-R^{(0)}\underset{\approx}{U}) \cdot \underset{\sim}{X}^{(2)} = Q(\underset{\sim}{X}^{(1)},\underset{\sim}{X}^{(1)})+R^{(1)}\underset{\approx}{U} \cdot \underset{\sim}{X}^{(1)}. \qquad (2.4)$$

For simplicity we are assuming that Q is a bilinear operator. The solvability condition for the inhomogeneous linear equation (2.4) is obtained by multiplying the left-hand side by a complete set of linearly independent solutions of the adjoint homogeneous problem and averaging the result over the spherical shell. Indicating the latter average by angular brackets we obtain

$$R^{(1)}c_k = M_O(\ell) \sum_{m,p} c_m c_p <w_\ell^{(k)} w_\ell^{(m)} w_\ell^{(p)}> \qquad \text{for} \quad -\ell \leq k \leq \ell \qquad (2.5)$$

where the details of the R-dependence enter through the expression $M_O(\ell)$ which depends on ℓ only. We have used the fact that the solutions of the adjoint homogeneous problem can be represented in the form

$$\underset{\sim k}{X}^* = \underset{\sim}{f}^*(r) \, w_\ell^{(k)} \qquad \text{for} \quad -\ell \leq k \leq \ell$$

and that derivatives tangential to the spherical surface can be eliminated by partial integration. For details we refer to Busse and Riahi [8]. Because spherical harmonics possess the property

$$w_\ell(\theta,\phi) = (-1)^\ell w_\ell(\pi-\theta,\phi+\pi) \qquad (2.6)$$

the average over the product of three spherical harmonics always vanishes when ℓ is an odd integer. The solvability conditions (2.5) are thus satisfied with $R^{(1)} = 0$ for odd ℓ without restrictions on the coefficients c_m. But for even ℓ non-trivial equations for the coefficients c_m are obtained as will be discussed below. The property (2.6) that the values of spherical harmonics at opposite positions on a sphere are the same or differ by a sign depending on whether the degree ℓ is even or odd also holds for the hyper-spherical harmonics in the case of 4 dimensions. Thus no restrictions are obtained in the order ε^2 on the patterns of solutions for odd n while a partial removal of degeneracy occurs for even n.

The solvability conditions in the order ε^3 of the problem are of particular interest in the case of odd ℓ or odd n. It is remarkable that the equations for the coefficients c_m are still independent of the radial structure of the problem, at least for the relatively low values of ℓ that have been investigated so far [8]. We refer to the latter paper for further details.

The equations for the coefficients c_m obtained from the solvability conditions in general admit solutions corresponding to several different patterns. A stability analysis is thus required to select the physically preferred solution. By superimposing infinitesimal disturbances $\tilde{X}$ onto the steady solution X we obtain the following linear stability equation from (2.1),

$$(\underset{\sim}{W}-R\underset{\sim}{U}) \cdot \tilde{\underset{\sim}{X}} = \sigma \underset{\sim}{V} \cdot \tilde{\underset{\sim}{X}} + Q(\tilde{\underset{\sim}{X}},X) + Q(X,\tilde{\underset{\sim}{X}}) \, . \tag{2.7}$$

Without losing generality we have assumed an exponential time-dependence of the vector $\tilde{\underset{\sim}{X}}$. The growthrate σ represents the eigenvalue of the stability problem (2.7). If there exists an eigenvalue σ with a positive real part σ_r, the steady solution is unstable; otherwise it is stable. After inserting expansion (2.2) we can solve equations (2.7) by assuming analogous expansions for $\tilde{X}$ and σ,

$$\tilde{\underset{\sim}{X}} = \sum_{n=1}^{\infty} \varepsilon^{n-1} \tilde{\underset{\sim}{X}}^{(n)} \, , \qquad \sigma = \sum_{n=0}^{\infty} \varepsilon^n \sigma^{(n)} \, . \tag{2.8}$$

The most interesting disturbances are those corresponding to the same degree ℓ as the steady solution. They possess a vanishing contribution $\sigma^{(0)}$ to the growthrate while all other disturbances have negative values $\sigma^{(0)}$ if the minimizing value R_ℓ is chosen for $R^{(0)}$. Because of the orientational degeneracy of the problem there exist in general at least two vanishing eigenvalues σ which correspond to infinitesimal rotations of the steady solution. Another special disturbance is given by the derivative with respect to ε of the steady solution,

$$\tilde{\underset{\sim}{X}} = \frac{d}{d\varepsilon} \underset{\sim}{X} \quad \text{with} \quad M_2 \sigma = -\varepsilon R^{(1)} - 2\varepsilon^2 R^{(2)} + \ldots \tag{2.9}$$

where M_2 is a constant. In the following sections some of the results obtained by the methods outlined above will be reviewed.

3. Preferred Patterns in the 3-Dimensional Case

For arbitrary ℓ solutions of the form (2.3) include the axisymmetric solution as a special case. It has already been mentioned that no pattern degeneracy exists in the case $\ell = 1$ and the solution is usually visualized in its axisymmetric form with two hemispheres of opposite sign. In the case $\ell = 2$ patterns distinct from the axisymmetric one can be described by the general representation (2.3). But only the axisymmetric pattern is selected by the solvability condition (2.5) [9]. Even in the self-adjoint case when $M_0(\ell)$ vanishes and when the coefficients c_m are restricted by higher order solvability conditions the axisymmetric solution is still selected [5]. There are actually always two different solutions corresponding to a given pattern in the case of even ℓ. The solution obtained by a reversal of the signs of the coefficients c_m corresponds to a reversed sign of $R^{(n)}$ whenever n is odd. It thus gives rise to a different relationship between the control parameter and the amplitude ε which is assumed to be a positive quantity. Accordingly the bifurcation for even ℓ is transcritical in general, as shown in figure 1. In the case of odd ℓ a change of the signs of the coefficients c_m corresponds to inversion of the solution on the sphere. No change in the relationship $R(\varepsilon)$ occurs since all coefficients $R^{(n)}$ with odd n vanish because of the symmetry property (2.6). A particular aspect of the general preference for even ℓ is the possibility that solutions with even ℓ may exist at lower values

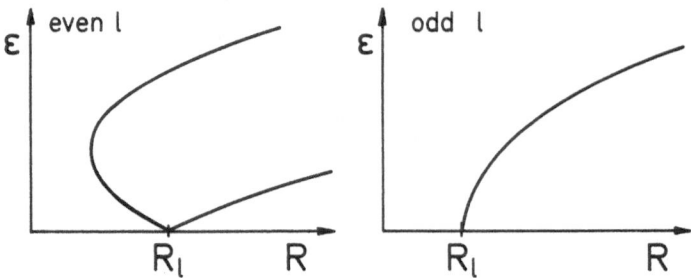

Figure 1: Transcritical and supercritical bifurcations in the
cases of even and odd ℓ, respectively.

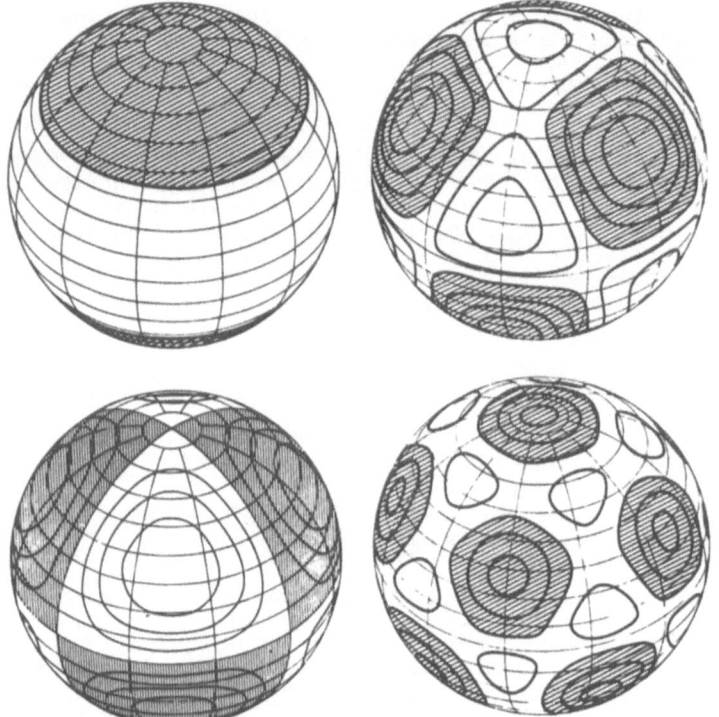

Figure 2: Preferred patterns of bifurcations in cases $\ell = 2,3$
(left) and $\ell = 4,6$ (right) (from references [8,9]).

of the control parameter R than odd ℓ solutions even if the minimum
of $R^{(0)}$ is attained for the latter.

The axisymmetric solution loses its preferred status at higher
values of ℓ. Instead those solutions are preferred which possess the
property that all domains on the sphere where the solution has the
same sign are symmetric, i.e. they can be transformed into each other
by a rotation. The axisymmetric solution has this property for $\ell = 2$,
but loses it for $\ell \geq 3$. The preferred solutions for $\ell = 3,4$ and 6 are
those exhibiting the symmetries of the platonic solids as shown in
figure 2. The preference for the octahedral and the dodecahedral

solutions in cases $\ell = 4$ and 6, respectively, is based on a maximum value $|R^{(1)}|$ which translates into a maximal subcritical range provided $R^{(2)}$ does not vary much for the different solutions as is typical for most problems. In the case $\ell = 3$ the situation is not quite as clear, since there is a competing solution with a sector pattern. The inspection of the various terms contributing to the growthrates suggests that in most cases the tetrahedral solution is stable while the sector solution is unstable. The axisymmetric solution is always unstable independent of the physical situation. This result has been generalized for all odd ℓ up to $\ell = 11$ by Lauterbach [10].

4. Preferred Patterns in the 4-Dimensional Case

We assume that the metric of the hypersphere in the 4-dimensional space is given by

$$d\sigma^2 = a^2[d\chi^2 + \sin^2\chi(d\theta^2 + \sin^2\theta d\phi^2)] \ . \tag{4.1}$$

The solutions $\underset{\sim}{X}^{(1)}$ of the linear problem can then be written in the form

$$\underset{\sim}{X}^{(1)} = F(a)H_n(\chi,\theta,\phi) \equiv F(a) \sum_{\ell=0,m=-\ell}^{n,\ell} c_{\ell m} h_n^{(\ell)}(\eta) w_\ell^{(m)}(\theta,\phi) \tag{4.2}$$

where $\eta = \cos\chi$ has been introduced. The functions $h_m^\ell(\eta)$ are defined by

$$h_n^{(\ell)}(\eta) \propto (-1)^\ell (1-\eta^2)^{\ell/2} \frac{d^\ell}{d\eta^\ell} \frac{\sin(n+1)\chi}{\sin\chi} \tag{4.3}$$

and are normalized by the condition

$$\langle |h_\eta^{(\ell)}|^2 \rangle = \frac{2}{\pi}\int_0^\pi |h_n^{(\ell)}|^2 \sin^2\chi d\chi = 1 \ . \tag{4.4}$$

Explicit expressions for the hyperspherical harmonics $h_m^{(\ell)}(\eta)$ are given in table 1.

Table 1: The functions $h_n^{(\ell)}(\eta)$ for $0 \leq n \leq 4$ with $\xi \equiv (1-\eta^2)^{1/2}$

$h_0^0 = 1$

$h_1^0 = 2\eta,$ $\qquad\qquad h_1^1 = -2\xi/\sqrt{3}$

$h_2^0 = 4\eta^2-1,$ $\qquad h_2^1 = -\sqrt{8}\xi\eta,$ $\qquad\qquad h_2^2 = \xi^2\sqrt{8/5}$

$h_3^0 = 4\eta(2\eta^2-1),$ $\quad h_3^1 = 4\xi(1-6\eta^2)/\sqrt{15},$ $\quad h_3^2 = 8\eta\xi^2/\sqrt{5},$ $\quad h_3^3 = -\xi^3 8/\sqrt{35}$

$h_4^0 = 16\eta^4-12\eta^2+1,$ $\quad h_4^1 = -\xi(8\eta^3-3\eta)\sqrt{8/3},$ $\quad h_4^2 = (8\eta^2-1)\xi^2\sqrt{8/7},$

$\qquad\qquad\qquad\quad h_4^3 = -\xi^3\eta 8\sqrt{2/7},$ $\quad h_4^4 = \xi^4\sqrt{128/63}$

In the following we restrict our attention to the solvability condition in the order ε^2 of the basic equation (2.1), which can be written in a form analogous to (2.5),

$$R^{(1)}c_{ik} = N(n) \sum_{\ell,m,j,p} c_{\ell m}c_{jp} < (h_n^{(i)}w_i^{(k)})(h_n^{(\ell)}w_\ell^{(m)})(h_n^{(j)}w_j^{(p)}) >$$

$$\text{for } 0 \le i \le n, \ -i \le k \le i \ . \tag{4.5}$$

In the case $n = 2$ there exist two physically distinct solutions of this system of equation which are given by expressions (16) of reference [3], but which are repeated here because of misprints in the latter paper,

$$c_{00} = \pm 1, \ c_{\ell m} = 0 \text{ for } \ell + |m| \ne 0, \text{ with } R^{(1)} = \pm N(n), \tag{4.6a}$$

$$c_{00} = \pm 3^{-1/2}, \ c_{20} = \pm(2/3)^{1/2}, \ c_{\ell m} = 0 \text{ otherwise, with } R^{(1)} = 0. \tag{4.6b}$$

Clearly, the solution (4.6a) which is symmetric with respect to two axes is preferred in comparison with the solution (4.6b) which is symmetric with respect to one axis only. Other representations of the solutions (4.6) derived from rotations on the 3-dimensional hyperspherical surface are also solutions of equations (4.5), of course. Because of the properties

$$\frac{2}{\pi}\int_0^\pi \sin^3(n+1)\chi(\sin\chi)^{-1}d\chi = 1, \ \frac{2}{\pi}\int_0^\pi \sin^2(n+1)\chi d\chi = 1 \text{ for } n \ge 0. \tag{4.7}$$

The solution (4.6a) represents the general twice axisymmetric solution for arbitrary n. Using the fact that equations (4.5) represent the Euler-Lagrange equations for a stationary value of the functional

$$R(H_n) \equiv <(H_n)^3><(H_n)^2>^{-3/2} \tag{4.8}$$

it can be stated that the stationary value attained by the twice axis symmetric solution is ± 1, independent of n.

In analogy to the 3-dimensional case it may be expected that the twice axisymmetric solution loses its preferred status for $n = 4$ and higher values of n. Using the symmetries of the analogues in four dimensions of the three-dimensional Platonic solids we search for preferred solutions in the case $n = 4$. Using the symmetry of the 8-cell-polyhedron in the notation of Hilbert and Cohn-Vossen [11] we consider solutions with

$$c_{00}^2 + c_{20}^2 + c_{40}^2 + c_{44}^2 = 1 \tag{4.9a}$$

and with $c_{\ell n} = 0$ for all coefficients not appearing in (4.9a). The result of this search is a solution with

$$c_{00} = \pm 0.5872, \ c_{20} = \pm 0.0541, \ c_{40} = \pm 0.5795, \ R^{(1)} = \pm 0.9035N(4) \tag{4.9b}$$

and several others obtained from solution (4.9b) by suitable rotations on the hypersphere. Obviously the solution with the symmetry of a regular polyhedron is not preferred for $n = 4$. No calculations for the case $n = 6$ have yet been carried out. The fact that the stationary value of the functional (4.8) does not decrease with increasing n for the twice axisymmetric solution appears to be the cause for its preference. Even though there exist domains of unequal shapes in which this solution assumes the same sign, only the polar regions in the neighborhood of the points $\chi = 0$ and $\chi = \pi$ contribute significantly in the triple integral of (4.8). The solution thus retains approximately the property of the solution for n = 2 as n is increased to higher even integers.

For sufficiently large n, however, there exist solutions which exceed the value unity for the absolute value of the functional (4.8). Since for high n the characteristic wavelength of the bifurcating solution becomes small in comparison with the radius of the hypersphere, the metric can be approximated locally by the metric of the three-dimensional Euclidian space. In that case an extremalizing solution for the functional (4.8) can be written in the form

$$\overline{H}_n = \frac{1}{\sqrt{12}} \sum_{j=-6}^{6} \exp\{i\underset{\sim}{k}_j \cdot \underset{\sim}{r}\} \tag{4.10}$$

where the wavevectors k_j have equal length and have the directions of the edges of an octahedron. The corresponding value of $|R^{(1)}/N(n)|$ is $2/\sqrt{3}$ and thus exceeds the value of the twice axisymmetric solution. The solution (4.10) has been obtained by Alexander and McTague [12] in their investigation of the Landau theory of solidification. The solution (4.10) corresponds to a body-centered cubic lattice.

5. Concluding Remarks

It has been the goal of this paper to point out that a large fraction of the properties of bifurcations from spherically symmetric states are independent of the physical mechanism of instability. The special role played by the even degrees ℓ and n of spherical harmonics in three- and four-dimensional spaces, respectively, has been pointed out and the analogies of the problem in both dimensions have been emphasized. Obviously the analogy could be extended to higher-dimensional spaces, but no physical examples of bifurcations from spherically symmetric states are known in that case. Bifurcations from circularly symmetric states (O(2)-symmetry) do not exhibit a pattern degeneracy. For even as well as for odd values of the wave number m the bifurcation assumes the pitchfork form. Only through co-dimension-2 bifurcations involving two different wavenumbers m can more complicated forms of bifurcation be realized.

The complications that can arise in the case when the minimum of $R^{(0)}$ is reached by R_ℓ and by $R_{\ell+1}$ simultaneously have not been addressed in this paper. Some results which have been obtained by Busse and Riahi will be reported elsewhere [13]. Interesting time-dependent and chaotic solutions have been described recently in the case $\ell = 1$ by Friedrich and Haken [14]. They will also be reviewed in this proceedings.

References

1. F.H. Busse: Thermal instabilities in rapidly rotating systems, J. Fluid Mech. 44, 441-460, 1970.
2. F.H. Busse, and C.R. Carrigan: Convection induced by centrifugal buoyancy, J. Fluid Mech. 62, 579-592, 1974.
3. F.H. Busse: Preferred anisotropies of the universe, Phys. Rev. D 28, 1248-1250, 1983.
4. D.H. Sattinger: Group Theoretic Methods in Bifurcation Theory, Lecture Notes in Mathematics, vol. 762, Springer-Verlag, Berlin 1979.

5. P. Chossat: Bifurcation and stability of convective flows in a rotating or not rotating spherical shell, SIAM J. Appl. Math. 37, 624-647, 1979.
6. E. Ihrig, and M. Golubitsky: Pattern Selection with O(3) Symmetry, Physica 13D, 1-33, 1984.
7. R. Lauterbach: An example of symmetry breaking with submaximal isotropy subgroup, IMA preprint 148, University of Minnesota, Minneapolis, 1985.
8. F.H. Busse, and N. Riahi: Patterns of convection in spherical shells II, J. Fluid Mech. 123, 283-302, 1982.
9. F.H. Busse: Patterns of convection in spherical shells, J. Fluid Mech. 72, 67-85, 1975.
10. R. Lauterbach: private communication, 1985.
11. D. Hilbert, and S. Cohn-Vossen: Geometry and Imagination, New York: Chelsea, 1952.
12. S. Alexander, and J. McTague: Should all crystals be bcc? Landau theory of solidification and crystal nucleation, Phys. Rev. Letts. 41, 702-705, 1978.
13. F.H. Busse, and N. Riahi: Patterns of convection in spherical shells, Abstract TERRA cognita 4, 240, 1984.
14. R. Friedrich, and H. Haken: Static, wavelike, and chaotic thermal convection in spherical boundaries, Phys. Rev. A 34, 2100-2120, 1986.

Structure Formation by Propagating Fronts

M. Lücke[1], M. Mihelcic[2], B. Kowalski[2], and K. Wingerath[2]

[1]Institut für Theoretische Physik, Universität des Saarlandes,
 D-6600 Saarbrücken, Fed. Rep. of Germany and
[2]Institut für Festkörperforschung der Kernforschungsanlage Jülich,
 D-5170 Jülich, Fed. Rep. of Germany

1. INTRODUCTION

In many nonlinear dissipative systems spatially periodic structures grow out
of a homogeneous basic state when the driving exceeds a critical value [1].
Taylor vortex flow (TVF) in the annulus between concentric cylinders [2] and convective rolls in a fluid layer heated from below [3] are prototype examples for
such dissipative structures. After increasing the driving from a subcritical to
a supercritical value these structures start growing from imperfections that
break the translational invariance of the system.

Here we investigate numerically a situation where (i) such inhomogeneities are
very small in the bulk of the system and (ii) the structured state is generated
initially, after a step-up in driving, near a boundary. Then the supercritically
stable state expands into the bulk which is, in the absence of sizeable imperfections, still occupied by the now unstable homogeneous state. In the rotating
Couette system and in the Rayleigh-Bénard system this situation gives rise to a
front (c.f. Fig. 1) between the two states that propagates with stationary speed
c and stationary spatial intensity profile f into the homogeneous state.

In Sec. 2 we describe the numerical method to simulate these two systems.
Section 3 deals with the generation of the unstable homogeneous state in the
bulk and how to initiate a front propagating into it. In Sec. 4 we present
our numerical results on propagating convection fronts and Taylor vortex fronts
and compare them with results obtained from an amplitude equation approximation.
Section 5 contains structural analyses of the periodic states growing behind

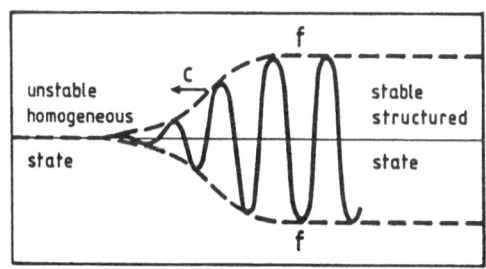

Fig. 1. Propagation of the front intensity profile f (dashed line) of the
stable structured state into the unstable homogeneous state. The structure
itself does not move except for some
phase adjustments.

the propagating fronts. We conclude by summarizing in Sec. 6 some aspects of
structure formation via front propagation.

2. THE SYSTEMS AND THE NUMERICAL METHOD

Front propagation and the associated structure formation was investigated in
numerical detail in the rotating Couette system [4] and in the Rayleigh-Bénard
system [5] in a supercritical driving range of $\varepsilon = R/R_c - 1$ between 0.01 and 0.2.
Here R is either the Reynoldsnumber or the Rayleighnumber, and R_c is the critical
one for growth of the dissipative structure. Figure 2 shows schematically the
systems and properties of the periodic flow states.

2.1 Simulation of the Rayleigh-Bénard system

We numerically solved [5] the full two-dimensional Oberbeck-Boussinesq equations [6] for a fluid of Prandtl number $\sigma = 1$ in a rectangular x-z section of a
box (c.f. Fig. 2a) of height d in z-direction and length L=25d in x-direction.
The planar velocity field with vertical component w and horizontal component u
and the fields of temperature T and pressure p were evaluated as a function of
x,z, and time t subject to realistic rigid boundary conditions. The top and
bottom plates had temperatures T and T+ΔT, respectively. The lateral sidewalls
were heat insulators except for some short time interval during which a lateral
heat pulse was injected into the fluid through the sidewall at x=25d (c.f.
Sec. 3.2). We also made some runs with sidewalls that were thermally attached
to either the top or the bottom plate. The lateral boundary at z=0 was a mirror
plane so that in effect we simulated a system of length 50d with convection
fronts propagating mirror-symmetrically inwards from both ends at x=±L.

To integrate the field equations we used an explicit finite difference
MAC [7] method of first order in time and second order in space. The fields were
evaluated on lattice points that lie on three uniformly spaced staggered grids:
one for u, one for w, and one for T and p. The distance between like points was
$\Delta x = \Delta z = .05d$ and the time step was $5 \cdot 10^{-4} \tau$ where.

$$\tau = d^2/\kappa \qquad (2.1)$$

is the characteristic vertical heat diffusion time. From a comparison with a
control run done with $\Delta x = \Delta z = .025d$ and $\Delta t = 1.25 \cdot 10^{-4} \tau$ we estimate the typical
numerical error of the convective fields to be of the order of 3%.

To determine the critical Rayleigh number, R_c, for onset of convection in our
finite, discretized system we used the fact that close to onset the square of the
supercritical bulk convective flow amplitude of the stationary, final vertical
velocity grows linearly with R-R_c. An extrapolation to zero gave $R_c^\infty = 1691.5$, i.e.,

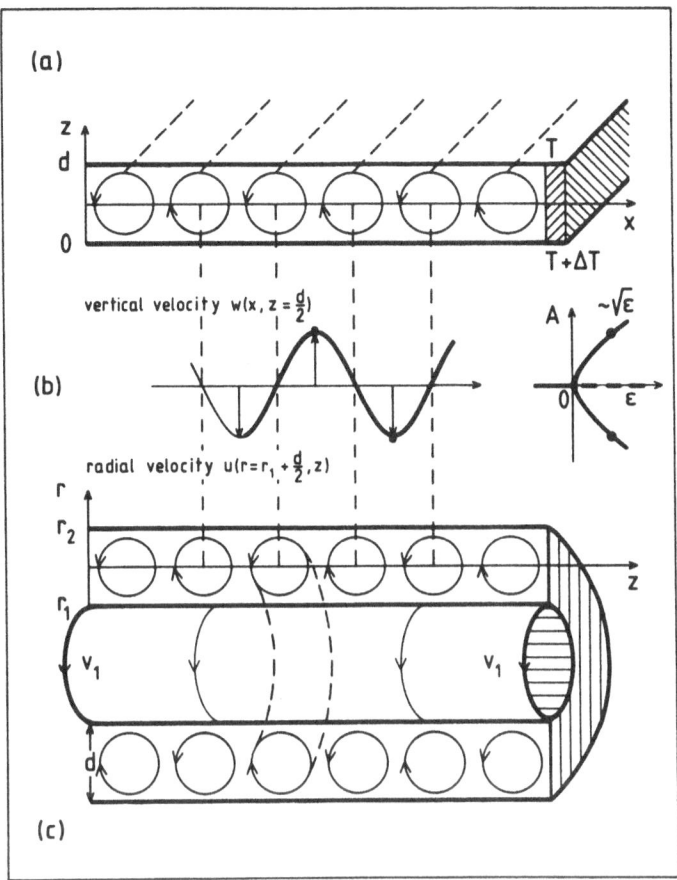

(a)

z
d

T

x

0

T+ΔT

vertical velocity $w(x, z = \frac{d}{2})$

A

$\sim\sqrt{\varepsilon}$

(b)

0

ε

radial velocity $u(r=r_1 + \frac{d}{2}, z)$

r

r_2

r_1

z

V_1

V_1

d

(c)

Fig. 2. Sketch of parallel convective rolls in a fluid layer heated from below (a) and of toroidal Taylor vortices stacked in the annulus between two concentric cylinders around the rotating inner one (c). An order-parameter field associated with these periodic states and its bifurcation behaviour is shown (b) very schematically.

a value which is about 1% below the theoretical one, R_c=1707.76 [6,8], for an infinite system.

2.2 Simulation of the rotating Couette system

Here the time-dependent, axially symmetric Navier-Stokes equations [4] for an incompressible fluid of kinematic viscosity ν were solved in the rectangular r-z-section shown in Fig. 2c. The radial extension of the gap between the concentric cylinders was $d=r_2-r_1$ and the axial extension was L=25d. We used setups with three different radius ratios $\eta=r_1/r_2$=0.5066, 0.75, and 0.893. The last one was chosen for comparison [4] with front-propagation experiments of Ahlers and Cannell [9].

The rotationally symmetric fields of radial velocity u, azimuthal velocity v, axial velocity w, and pressure p were evaluated as a function of r,z, and t subject to the following boundary conditions. The flow field (u,v,w) vanishes all the time at the stationary outer cylinder, $r=r_2$, and at the stationary collar (indicated in Fig. 2c by the vertically hatched part) that bounds the fluid at z=L=25d. For various reasons partly related to questions concerning wavenumber selection [10] we did not impose at the inner cylinder, $r=r_1$, the standard boundary condition, u=w=0, $v=v_1=r_1\Omega_1$, corresponding to rotation with spatially uniform rate Ω_1. We rather used a spatially ramped driving, $v(r=r_1,z)=v_1(z)$, corresponding to a Reynolds number

$$R(z) = \frac{v_1(z)d}{\nu}$$ (2.2)

that varied axially: The driving parameter

$$\varepsilon(z) = \frac{R(z)}{R_c} - 1 = \varepsilon_o \begin{cases} 1-2\left(1 - \frac{z}{12d}\right)^2 & \text{for } z<12d \\ 1 & \text{for } z\geq12d \end{cases}$$ (2.3)

shows in the range 12d≤z≤25d a supercritical plateau of height $\varepsilon_o>0$ and in the range 0≤z≤12d a parabolic ramp that starts with zero slope at z=12d and smoothly falls off to the subcritical value $\varepsilon(z=0)= - \varepsilon_o$ at the left end, z=0, of the cylinder (c.f. Fig. 7a in Sec. 4.3). The boundary condition at z=0 was that of subcritical circular Couette flow (CCF)

$$u(r)=0=w(r) \qquad\qquad \text{at } z=0,$$ (2.4a)

$$v(r)=v(r_1) \left[\frac{r_2}{r} - \frac{r}{r_2}\right] \frac{\eta}{1-\eta^2} \qquad\qquad \text{at } z=0,$$ (2.4b)

with

$$v(r_1)=v(r_1,z=0) = R_c(1-\varepsilon_o) \frac{\nu}{d}$$ (2.4c)

being the tangential velocity at $z=0,r=r_1$ corresponding to the subcritical driving $\varepsilon(z=0) = - \varepsilon_o$ (2.3).

Eqs. (2.2 - 2.4) are the final boundary conditions either after a sudden start from rest, u=v=w=0, or after a step-up in driving from a stationary subcritical state with spatially constant subcritical driving, $\varepsilon = - \varepsilon_o$, $v(r_1)=R_c(1-\varepsilon_o)\nu/d$ and corresponding subcritical CCF boundary condition (2.4) at z=0. This condition eliminates [4] Ekman vortices [2,11] in the subcritical stationary state and strongly suppresses their generation by the step-up to the driving profile (2.2, 2.3) — the latter is still subcritical close to z=0 and increases there gently.

Also here the field equations were integrated [4] with an explicit finite dif-ference MAC [7] method of first order in time and second order in space. The fields were evaluated on lattice points that lie on three uniformly spaced staggered grids: one for u, one for v and p, and one for w. Again the distance

between like points was $\Delta r=\Delta z=.05d$ and the time step was about $5.4 \ 10^{-4} \ d^2/\nu$
where d^2/ν is the characteristic time for momentum diffusion across the gap d.
Also here the typical numerical error was estimated to be of the order of 3%.

3. GENERATING AN UNSTABLE STATE AND A FRONT PROPAGATING INTO IT

In our systems the periodic state was generated after a step-up to a supercritical driving by mechanisms described in Sec. 3.2 such that it was localized initially close to the right boundary and then expanded into the bulk.

3.1 The homogeneous unstable state in the bulk

Let, e.g., prior to the step-up the stationary, homogeneous basic state appropriate to an initial subcritical driving, $\varepsilon_i<0$, be realized in the bulk of the
system. After stepping up the driving to a supercritical value $0<\varepsilon_f\ll1$ the translationally invariant conductive temperature profile of the Rayleigh Bénard system
adjusts homogeneously in a diffusive manner to the new temperature difference
across the fluid layer within the fast time scale of d^2/κ [5]. Similarly the
radial profile of the CCF velocity $v_{CCF}(r)$ adapts diffusively and homogeneously
to the new driving on the time scale of d^2/ν [4].

In addition and simultaneously the periodic state which is stable at ε_f starts
growing from any imperfection that breaks the translational symmetry of the basic
state, i.e., convective flow or vortex flow, respectively, is nucleated by inhomogeneities. It is instructive to keep in mind that in systems without any
symmetry-breaking imperfections whatsoever the symmetry-broken periodic flow
state cannot grow. Here, in our numerically simulated systems both the inhomogeneities in the bulk and the final supercritical driving parameter ε_f are very
small. Furthermore, the time τ_{onset} which it takes the periodic state to grow to
a significant amplitude after nucleation from imperfections is roughly inversely
proportional to their size and the supercritical driving ε_f. Hence, there was no
problem to guarantee

$$\tau_{onset} \gg d^2/\kappa \text{ or } d^2/\nu. \tag{3.1}$$

This condition can also be realized experimentally with well-machined apparatus
and not too large supercritical driving ε_f.

Thus within a time of the order of d^2/κ or d^2/ν after a step-up to $\varepsilon_f>0$ we
have established in the bulk of our system the stationary, unstable, homogeneous
state appropriate to ε_f which, without front propagation, would be destroyed
only after a time τ_{onset} by nucleation-triggered growth of the dissipative
structure. We have checked numerically that this is true irrespective of the
initial state being a state without driving at all, $\varepsilon_i = -1$, or a subcritical
stationary state at finite driving $-1<\varepsilon_i<0$.

3.2 Starting a front

This requires after a step-up to a supercritical driving the initial generation of the dissipative structure in only part of the system,with the rest of it being occupied by the homogeneous state. In both our systems periodic flow was initiated near the right boundary by two different sidewall-forcing mechanisms.

a) Ekman vortices in the rotating Couette system

Here the stationary rigid collar (vertically shaded in Fig. 2.c) which bounds the fluid at z=25d provides a large imperfection that breaks the translational symmetry and thus generates vortexflow, i.e., Ekman vortices. For subcritical driving, $\varepsilon<0$, their stationary intensity falls off (c.f. Fig. 7b) exponentially into the bulk (with a penetration length [4,11] that grows proportional to $1/\sqrt{|\varepsilon|}$ close to $\varepsilon=0$). Such an Ekman vortex configuration can be used as a nucleus for the expansion of the periodic state into the bulk and the front formation after stepping up the driving.

Alternatively one may,in just one sudden start from rest,step up the inner cylinder's rotation rate directly from zero to the final supercritical value. In that case Ekman vortices are generated close to the rigid, stationary plate at z=25d within the fast time scale of about d^2/ν as can be seen in Fig. 3. Thereafter the periodic flow state expands and forms a front propagating into the bulk. We have verified for a particular case, $\varepsilon_f=0.04$, that both procedures — start from the stationary subcritical state at $\varepsilon_i = - 0.04$ and start from rest, $\varepsilon_i = - 1$ — give the same results for the final front propagation. In an annulus of limited length a sudden start from rest has the technical advantage of allowing exploitation of a longer propagation path than that allowed by a step-up from a stationary subcritical state with an extended Ekman vortex system. For this reason we preferred the former procedure for generating TVF fronts.

b) Sidewall heating in the Rayleigh-Bénard system

To initiate convection close to the right side wall after a step-up of the Rayleighnumber we chose a procedure that can easily be reproduced in an experiment: We simulated a heat pulse entering laterally through the wall into the fluid. To do that we applied for a duration of 0.2 d^2/κ a constant, vertically uniform, lateral temperature gradient across the wall at z=25d. The associated lateral heat current turned out to be a very effective force to trigger onset of localized convection. While the thermal nature of the sidewall thereafter influences the local convective flow, e.g., the intensity and flow direction close to the wall [10], it does not affect the properties of the stationary convective front propagating into the bulk.

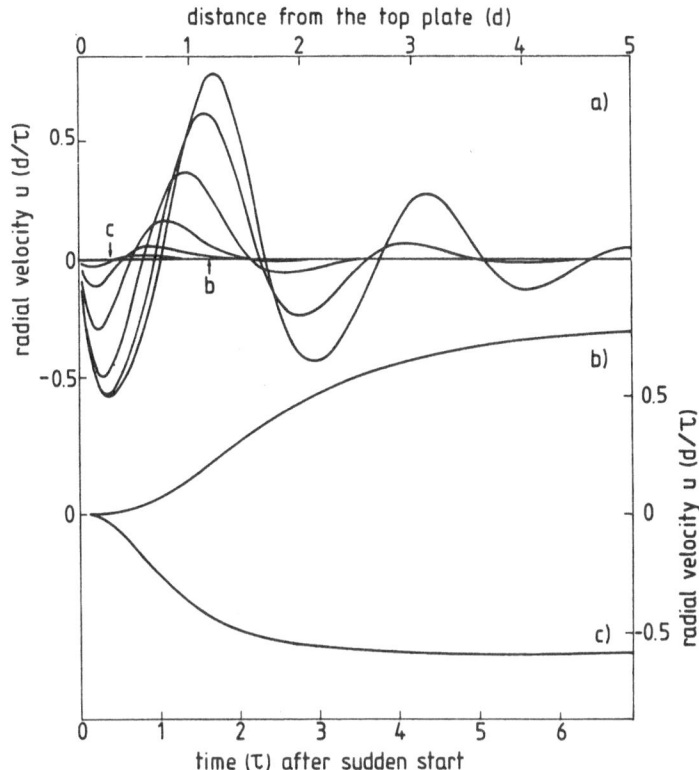

Fig. 3. Growth of Ekman vortices in the rotating Couette system after sudden start from rest to supercritical driving ε_o=0.04. Curves in (a) show radial velocities in the middle of the gap (η=0.893) vs distance from the rigid top plate at z=25d at seven successive times $0.116 \times 2^n \tau$ (n=0,1,...,6). Curves in (b) and (c) represent the radial outflow and inflow velocities at positions marked by arrows b and c, respectively, in (a) as functions of time. Here $\tau = d^2/(2\pi \nu)$.

4. PROPAGATING FRONTS

Before we present our numerical results on the properties of propagating con-vection fronts in Sec. 4.2 and on TVF fronts in Sec. 4.3 we shall briefly recall some analytical results pertaining to propagating fronts obtained [4, 12-14] within the framework of amplitude-equation approximations [15,16].

4.1 Amplitude equation

To lowest order in $\sqrt{|\varepsilon|}$ the vertical velocity field of parallel convective rolls, e.g., at mid height of the heated fluid layer is given by

$$w(x,z=d/2,t) = \text{Re}\left[A(x,t)e^{ik_c x}\right] . \tag{4.1a}$$

Also the radial velocity of TVF in the middle of the gap between the cylinders can be written to order $\sqrt{|\varepsilon|}$ in the form

$$u(r=r_1+d/2,z,t) = \mathrm{Re}\left[A(z,t)e^{ik_c z}\right] . \tag{4.1b}$$

Here k_c denotes the critical wavenumber for onset of the periodic flow structure at R_c in the infinite system. The complex amplitude A being the solution of the amplitude equation

$$\tau_0 \partial_t A = \left(\xi_0^2 \partial^2 + \varepsilon - g|A|^2\right)A \tag{4.2}$$

is for ε close to zero a slowly varying function of time and space $x(\partial=\partial_x)$ or $z(\partial=\partial_z)$, respectively. The length scale ξ_0 is determined by the curvature of the stability boundary of the basic homogeneous state in an infinite system and τ_0 is determined by the growth rate of the critical mode. For the Rayleigh-Bénard system with Prandtl number $\sigma=1$ one finds [8] $\tau_0=(1/13)d^2/\kappa$, $\xi_0=0.385d$. In the rotating Couette system [8] in the range $0.5<\eta<0.9$ explored here τ_0 varies between $0.0372\ d^2/\nu$ and $0.0382\ d^2/\nu$ while ξ_0 varies between $0.276d$ and $0.27d$ such that ξ_0/τ_0 is constant within 1.1%. The coupling constant g follows from a nonlinear analysis.

Equation (4.2) determines, to lowest order in $\sqrt{\varepsilon}$ the time evolution of an initial amplitude profile $A(x,t=0)$ or $A(z,t=0)$, respectively, towards the fully developed periodic state with wavenumber k_c and a homogeneous amplitude (in the infinite system) given by

$$|A_\infty^2| = \varepsilon/g. \tag{4.3}$$

We are interested here in the particular one-parameter family of stationary propagating solutions that can be written in the form

$$A(x,t) = A_\infty\ f(s;\tilde{c}) ; \quad s=\sqrt{\varepsilon}\ \frac{x}{\xi_0} - \tilde{c}\ \varepsilon\ \frac{t}{\tau_0} , \tag{4.4}$$

and similarly for $A(z,t)$ in the TVF case. Here $f(s;\tilde{c})$ is a real valued intensity profile between the unstable state, $f(s=\infty;\tilde{c})=0$, and the stable periodic state, $f(s=-\infty;\tilde{c})=1$, that propagates with velocity $\tilde{c}\ \sqrt{\varepsilon}\ \xi_0/\tau_0$. The parameter $\tilde{c}$ determining the velocity and the shape function f is arbitrary.

Aronson and Weinberger [12] have shown that a very wide class of initial amplitude distributions evolve into a front intensity profile propagating with the speed

$$c_A = 2\ \sqrt{\varepsilon}\ \xi_0/\tau_0 \tag{4.5}$$

corresponding to the borderline [4,13,14] value $\tilde{c}=2$. The associated uniquely determined final stationary front profile f [4] increases monotonously from 0 to 1. The front extension (domain wall thickness) defined as the length ℓ_A [4] over which f increases from 1/4 to 3/4

$$\ell_A = (2.941/\sqrt{\varepsilon})\ \xi_0 \tag{4.6}$$

grows $\sim 1/\sqrt{\varepsilon}$ for small ε. The length over which f increases from 0.1 to 0.5 which we need for comparison in Sec. 4.2 is given by $(2.9/\sqrt{\varepsilon})\ \xi_0$.

We finally mention that the wavenumber of the periodic flow structure growing behind the front is within this solution the critical one, k_c.

4.2 Convection fronts

In Fig. 4 we show as a representative example the time evolution of the vertical velocity field $w(x,z=d/2,t)$ at mid height, $z=d/2$, of the fluid layer at different times after the step-up at $t=0$ to the supercritical Rayleigh number 1.05 R_c. The heat pulse (c.f. Sec. 3.2) was applied to the sidewall at $x=25d$ from $t=0$ to $t=0.2\tau$. Thereafter the sidewall was thermally insulating. The intensity profiles of $w(x)$ at other vertical heights z evolve in a synchronous way so that the convective front propagating inwards is aligned parallel to the sidewall. Note that the convection rolls indicated schematically in the lower part of Fig. 4 practically do not move — the nodes of $w(x)$ marking the centers of the rolls are stationary. Only the envelope profile of the spatial oscillations of the fields of velocity, temperature, and pressure moves.

The final convective state behind the front is stationary and of uniform amplitude except for the weaker convective rolls near the thermally insulating sidewalls. The step-up and sidewall-forcing procedure described in Sec. 3.2

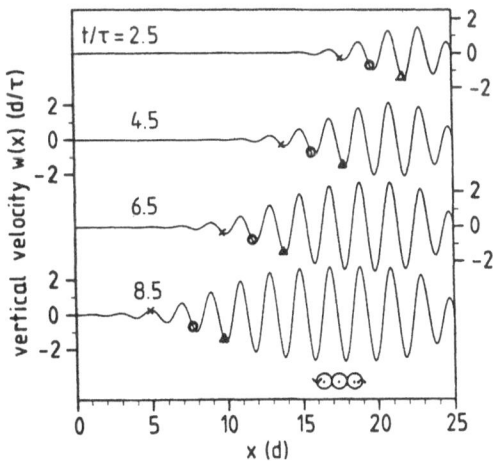

Fig. 4. Vertical convective velocity field $w(x)$ at midheight $z=d/2$ at different times after the step-up to $\varepsilon=0.05$. Zerocrossings mark the centers of the convective rolls, positive (negative) w implies up (down) flow as indicated in the schematic cross sections of the rolls shown in the lower part of the figure (for graphical reasons the flow directions in Fig. 2a are opposite to the above ones). The front positions where the flow has reached the fraction α of the final bulk intensity are denoted by a cross for $\alpha=0.1$, by a circle for $\alpha=0.25$, and by a triangle for $\alpha=0.5$.

always generated upward convective flow near a thermally insulating sidewall with reduced intensity in comparison to the bulk convective flow. Also near sidewalls that were thermally attached to the hot bottom plate the flow was always upwards but of enhanced intensity in comparison to the bulk flow. Near sidewalls that were thermally attached to the upper cold plate, however, the flow was downwards of enhanced intensity [10]. Aside from the turning directions of the rolls the stationary properties of front propagation and of the final periodic flow state in the bulk do not seem to be influenced by the thermal properties of the sidewalls.

To analyse the velocity and the intensity profile of the front we determined at successive times the positions in the front where the convective flow field $w(x,z=d/2)$ has reached given fractions α of the final bulk flow intensity. These front intensity positions marked in Fig. 4 by a cross ($\alpha=0.1$), a circle ($\alpha=0.25$), and a triangle ($\alpha=0.5$) move piecewise continuously. The descontinuity of roughly a roll diameter occurs whenever the next vortex ahead reaches the assigned intensity level. Figure 5 shows the trajectories of the above three front intensity positions as a function of time. After initial transients have died out rather quickly the convection front travels with constant speed c and with a stationary envelope profile: relative distances between different front intensity positions do not change any more. However, when the front arrives at the boundary at x=0 (around 11τ in Fig. 5) the trajectories of the intensity postions level off.

Note that this convective front propagation can be observed in experiments if, after a step-up in driving, the time τ_{onset} for imperfection-triggered onset of convection which competes with sidewall heating initiated convection is longer

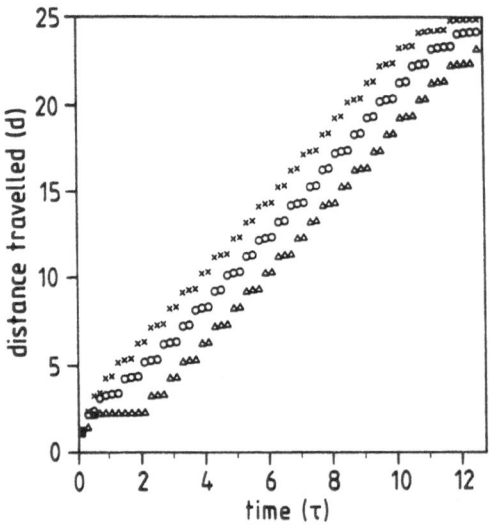

Fig. 5. Front intensity positions (c.f. Fig. 4) vs time. Distance travelled by the front positions at which the flow reached a fraction α of the final bulk intensity are shown for $\alpha=0.1$ (crosses), $\alpha=0.25$ (circles), and $\alpha=0.5$ (triangles).

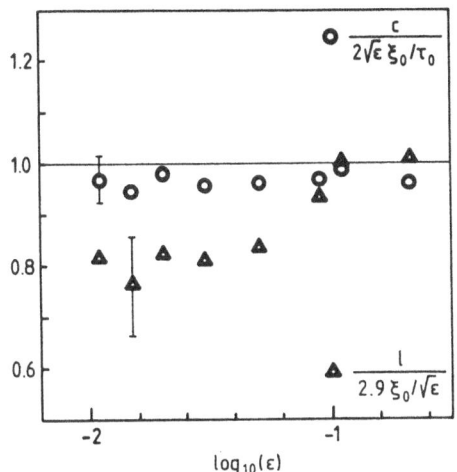

Fig. 6. Stationary velocity c and spatial extension ℓ of propagating convective fronts as a function of driving. Values and error bars were obtained from spacetime diagrams like Fig. 4 with ℓ being the distance between the 0.1 and 0.5 front intensity positions. The line corresponds to the results of the amplitude equation.

than the travelling time of the front. Thus, to observe a front travelling, e.g., over a distance of 15d at $\varepsilon=0.05$ the above onset time should exceed 7 vertical diffusion times which seems to be achievable with well-machined apparatus.

Figure 6 shows the stationary propagation speed of the convective front as a function of driving in the range $0.01 \lesssim \varepsilon \lesssim 0.2$ covered by our numerical experiments. Also shown is the ε dependence of the front extension as measured by the vertical distance between crosses and triangles in Fig. 5, i.e., the distance between the 0.1 and 0.5 intensity level. The propagation speeds determined from our solution of the full Oberbeck Boussinesq equations almost coincide with those obtained from the amplitude equation. The front extensions, $2.9\, \xi_0/\sqrt{\varepsilon}$, predicted by this equation, on the other hand, are somewhat larger than those measured in the numerical experiment (triangles in Fig. 6) but they still agree semiquantitatively.

4.3 Taylor vortex fronts

The formation of the TVF structure in the rotating Couette system via front propagation is shown in Fig. 7. There the radial velocity field u(z) in the middle of the gap, $r=r_1+d/2$, is plotted as a function of z at different times after a step-up of the driving. The intensity profiles of u(z) at other radial positions r evolve in a synchronous way so that the front propagating from the collar at z=25d into the annulus is planar with the front plane being parallel to the collar. Also here the vortices do not move except for minute phase adjustments. Only the envelope profile of the axial oscillations of the flow fields and of the pressure field moves as their intensities grow. The final TVF state (Fig. 7g) is stationary and axially periodic in the range $z \geq 12d$ where the driving $\varepsilon(z)$ (Fig. 7a) is constant. Further details of its structure will be discussed

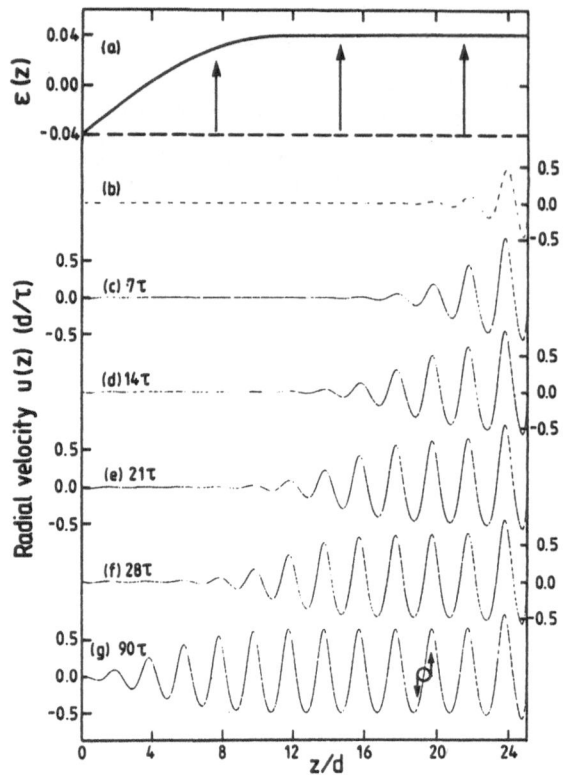

Fig. 7. Time evolution of the radial flow in the middle of the gap of a rotating Couette system (η=0.893) after a step-up to supercritical driving [solid line in (a)]. Radial velocity fields, $u(r=r_1+d/2,z)$, are shown as functions of z in (c)-(g) at different times after the step-up. The initial state at subcritical driving [dashed line in (a)] is stationary CCF in the bulk with localized Ekman vortices near the rigid collar at z=25d. The subcritical CCF boundary condition at z=0 and the spatially ramped driving $\varepsilon(z)$ [solid line in (a)] after the step is discussed in Sec. 2.2. The nodes of u mark the centers of the vortices. In between the radial flow is alternatingly outwards (u>0; away from the rotating inner cylinder and inwards (u<0; towards the inner one). Here $\tau=d^2/(2\pi\nu)$.

later on. Near the rigid, stationary collar at z=25d, however, the radial flow is enhanced [11,17] and directed inwards, which is typical for an Ekman vortex.

a) Propagation velocities

We determined [4] the propagation velocity of the front by monitoring at successive times the positions where the radial flow field reached various fractions of the final bulk radial outflow intensity. The so obtained world lines of the front positions of the 0.1, 0.25, and 0.5 intensity levels show the same behaviour [4] as those of the convective fronts in Fig. 5. The resulting front propagation velocities in our three rotating Couette systems are shown in Fig. 8 by full symbols as a function of the final supercritical driving (ε_0 is the height of the plateau in $\varepsilon(z)$ for z≥12d) in comparison with the prediction (4.5)

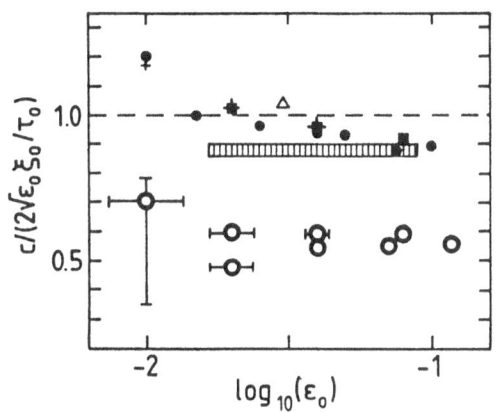

Fig. 8. Propagation velocities of TVF
fronts between cylinders with radius
ratio 0.5066 (dots), 0.75 (squares),
and 0.893 (crosses) as a function of
driving. The dashed line represents
the result of the amplitude equation,
the open triangle denotes another
numerical simulation [18] for $\eta=0.727$,
and open circles show experimental results [9] for $\eta=0.893$. When using the
numerically determined R_c appropriate
for our systems with $\eta=0.75$ and 0.893
all squares and crosses fall into the
shaded area. See text for a full discussion

of the amplitude equation. Included is also, by an open triangle, the only other
numerical result on propagating TVF fronts that we are aware of which was
obtained by Neitzel [18] with a different method than ours. The open circles in
Fig. 8 represent experimental results [9] on propagation velocities deduced from
front-arrival-time measurements. They were obtained by detecting the change in
the scattered light intensity that arises at onset of TVF in a fluid seeded with
Kalliroscope flakes [19]. Further experimental work is needed to investigate the
origin for this substantial discrepancy between these experiments [9] on one
side and the numerical simulations [4,18] and the amplitude equation on the
other side. Note, however, that also the latter two do not agree perfectly.

Before we discuss this disagreement between full numerical simulation and
amplitude equation we have to make a small digression: Some time after our front
propagation work was completed we did — in connection with the problem of
wavenumber selection by a spatially ramped driving — a detailed spatial Fourier
analysis [10] of the radial flow in systems with $\eta=0.75$ and $\eta=0.893$. On this
occasion we found that the critical Reynoldsnumbers for onset of TVF in our
finite, discretized set-ups are 0.68% and 0.7%, respectively, smaller than the
theoretical values [2,8] for the ideal systems (a corresponding analysis for
$\eta=0.5066$ remains to be done). If one rescales ε_0 accordingly then all the squares
($\eta=0.75$) and crosses ($\eta=0.893$) of Fig. 8 come to lie within the shaded area.
While this area is practically parallel to the dashed line it lies about 10-15%
below it.

A comparison with Fig. 6 shows that the disagreement between the front propagation velocity obtained from the amplitude equation and the full numerical
solution of the field equations is larger for TVF fronts than for convection
fronts. This is consistent with our Fourier analysis [10] of the stationary
periodic flow structures which shows that the lowest-order amplitude equation
gives a less quantitative description of Taylor vortices than of parallel con-

vective rolls. Whereas the spatial structure of the order-parameter field $w(x)$ of convective rolls is — in agreement with the result of the lowest-order amplitude equation — practically a pure harmonic one, the radial flow field $u(z)$ of TVF contains substantial higher harmonics. They account for the fact (c.f. Fig. 7g) that the maxima, u_{out}, of the radial outflow velocity are larger than the maxima, u_{in}, of radial inflow and that consequently — by mass conversation — the axial extension of outflow, Δ_{out} (i.e. the node distance of u between which $u>0$), is smaller than the axial inflow range, Δ_{in} (i.e. the node distance of u between which $u<0$).

For example, in the driving range $0.01 \lesssim \varepsilon \lesssim 0.1$ explored here the numerically determined stationary bulk radial flow field in the middle of the gap of our systems can well be represented [10] by

$$u(z) = \sqrt{\varepsilon}\, u_1 \cos k(z-z_0) + \varepsilon\, u_2 \cos 2k(z-z_0) + 0\,(\varepsilon^{3/2}) \qquad (4.7)$$

when z_0 is the axial position of an outflow maximum. For the system with $\eta=0.75$, e.g., we found [10] $u_1=15.1\ v/d$ and $u_2=11.3\ v/d$.

Thus the second harmonic, $\cos 2kz$, which is ignored by the lowest-order amplitude equation contributes with relative weight $0.75\sqrt{\varepsilon}$ such that the ratio

$$u_{in}/u_{out} = (u_1-u_2\sqrt{\varepsilon})/(u_1+u_2\sqrt{\varepsilon}) \qquad (4.8)$$

of the maximal inflow and outflow velocities deviates at $\varepsilon=0.01$ already by 14% from the value 1 predicted by the amplitude equation. The physical origin for the asymmetry $u_{out}>u_{in}$ is the rotating inner cylinder which, by spinning off the fluid radially outwards, causes intensive outflow while the inwards flow returning from the outer nonrotating cylinder is slowed down.

b) Front profiles

Radial outflow and inflow velocities differ not only in the final stationary bulk TVF state. Also the front intensity profiles for the growing TFV state shows a corresponding asymmetry between inflow and outflow which increases as the TVF state grows behind the front. In Fig. 9 we show front intensity profiles of radial outflow and inflow maxima in the middle of the gap between cylinders of ratio $\eta=0.75$ as a function of distance behind the propagating fronts for three different driving rates. The front profiles of TVF between cylinders of radius ratio $\eta=0.5066$ and 0.893 are similar. The origin of the abscissa in Fig. 9 is the position at which the radial flow intensity has reached 10% of the outflow velocity u_{out} of bulk TVF. We determined, about every 2τ, the size of the radial inflow and outflow maxima and their distance from the moving front position of the 10% intensity level. The vertical bars in Fig. 9 reflect the scatter of the so obtained data. Solid lines are hand-drawn, smooth interpolations.

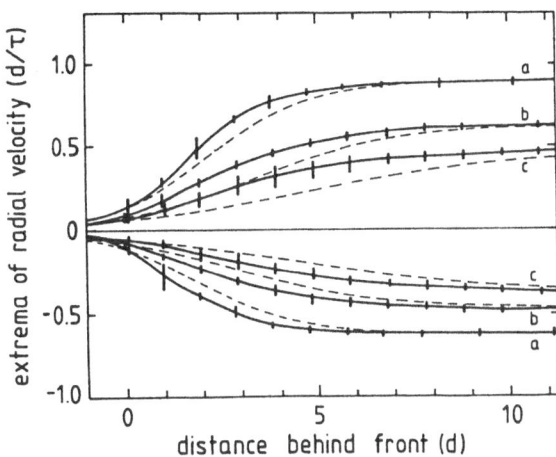

Fig. 9. Intensity profiles of propagating TVF fronts. Upper (lower) solid curves denote smooth interpolations through averaged intensity maxima of radial outflow (inflow) in the moving front in the middle of a gap of $\eta=0.75$ for (a) $\varepsilon_0=0.08$, (b) 0.04, and (c) 0.02. The origin of the co-moving frame is the axial position where $|u(z,t)|$ reaches 10% of the bulk outflow velocity u_{out}. Vertical bars indicate the scatter of intensity extrema monitored about every 2τ. Dashed lines are the universal profiles [4] resulting from the amplitude equation for c=2: they are normalized to our bulk outflow and inflow intensities u_{out} and u_{in}, respectively, and shifted along the abscissa until they coincide at its origin with the numerical results. Here $\tau=d^2/(2\pi\nu)$.

Note that the asymmetry between radial outflow and inflow decreases when the flow intensity becomes small. That happens in the limit of small driving, $\varepsilon \to 0$, and in addition in the extreme precursing head of the fronts. In both cases u_{in}/u_{out} approaches 1.

The front extensions grow $\sim 1/\sqrt{\varepsilon}$ as ε decreases in a way [4] very similar to the convection fronts. However, in the driving range explored by us TVF fronts are consistently, by about 20%, steeper than those predicted by the lowest-order amplitude equation. Since this equation has been derived under the assumption that the TVF intensity distribution A(z) is slowly varying on the scale of a wavelength, $\lambda \approx 2d$, one cannot expect a fully quantitative description of the fronts shown in Fig. 9. In fact, it is surprising how good the predictions of the amplitude equation are, even in such extreme situations as in Fig. 9a, e.g., where the flow intensity varies on the scale of λ.

5. PATTERN SELECTION BY PROPAGATING FRONTS

Here we investigate structural properties of the state growing under the propagating front. At first sight one might expect that, e.g., the wavelength of the growing periodic structure that is selected out of the Eckhaus band [20] of possible stable wavelengths depends on some details of the initial condition and the particular history of the driving until the final supercritical value $0 < \varepsilon \ll 1$. That, however, does not seem to be the case. Instead, all work [4,5,9,13,14] indicates that a particular, uniquely selected, periodic state grows under the fronts which themselves propagate with uniquely selected velocity and intensity profile.

5.1 Convective roll structure

The final stationary convective roll structure is practically a harmonic one and, furthermore, the node distances of the order-parameter field $w(x,z=d/2)$ under and far behind the front are almost the same. Thus the growing convective structure is well characterized, e.g., by its wavenumber. Figure 10 shows that the wavenumber selected by the propagating fronts is very close — in particular for small ε — to the wavenumber $k_{max}(\varepsilon)$ of that infinitesimal perturbation of the homogeneous conductive state that grows fastest according to a linear stability analysis. The full line, $k_{max}(\varepsilon)$, in Fig. 10 was obtained [21] with a Galerkin approximation. It reproduces the exact, numerically obtained [8] result for the initial slope, $\partial k_{max}/\partial\varepsilon$ at $\varepsilon=0$, within 1.2% and extends $k_{max}(\varepsilon)$ to finite $\varepsilon > 0$.

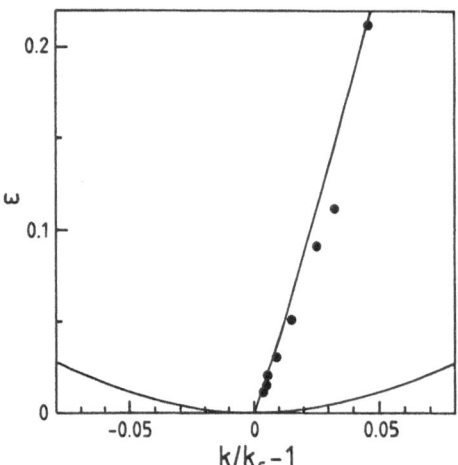

Fig. 10. Wavenumber selected by convection fronts as a function of driving. The full circles denote wavenumbers measured in the stationary propagation mode at the 0.1 front intensity positions (cross in Fig. 4). The line represents wavenumbers $k_{max}(\varepsilon)$ with fastest growth obtained [21] from a linear stability analysis of the conductive state at R_c. The parabola is the Eckhaus stability boundary [20] of the convective roll state.

5.2 Taylor vortex structure

The structure of Taylor vortices growing under the propagating front is more complicated than that of the convective rolls, mainly because with increasing flow intensity the anharmonicity of the TVF structure grows (c.f. Sec. 4.3). Therefore the asymmetry of the node distances of the order-parameter field

$u(r=r_1+d/2,z)$ for radial inflow, Δ_{in}, and radial outflow, Δ_{out}, and the asymmetry between the intensities of inflow, u_{in}, and outflow, u_{out}, increases with growing distance under the front: Δ_{in}/Δ_{out} increases and u_{in}/u_{out} decreases.

Figure 11 shows the axial variation of the distance between successive nodes of the radial velocity field $u(r=r_1+d/2,z)$ along the moving front. The inflow domains, Δ_{in}, increase and the outflow domains, Δ_{out}, decrease monotonously from their common value $\lambda/2$ in the extreme precursing head of the front towards their bulk value far behind the front as the TVF intensifies. Simultaneously the ratio of outflow and inflow intensity, u_{out}/u_{in}, grows monotonously from 1 to a bulk value >1. The local wavelength $\lambda=\Delta_{in}+\Delta_{out}$, however, does not vary much under the front.

In Fig. 12 we show the node distances Δ_{in} and Δ_{out} and the wavelength λ of the TVF structure well behind the moving front as a function of supercritical driving ε_0. Again, λ approaches for small driving, as in the case of convection fronts,

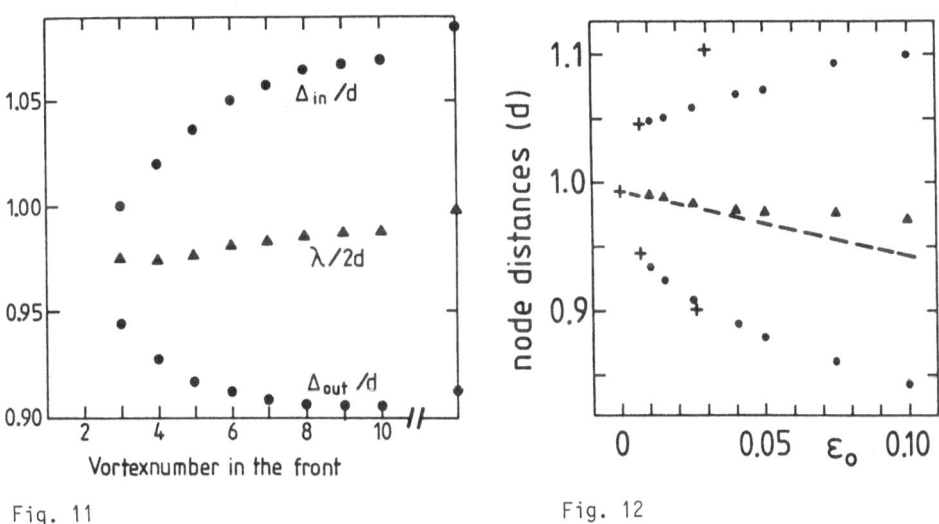

Fig. 11   Fig. 12

Fig. 11. Structure of the TVF state ($\eta=0.893$, $\varepsilon_0=0.04$) unfolded under the propagating front. Intervals for radial inflow, Δ_{in}, and outflow, Δ_{out}, are determined by the axial distance between successive nodes of the radial velocity field between which $u<0$ and $u>0$, respectively. Numbering starts from the vortex near the front position of the 0.1 intensity level. Triangles denote half the local wavelength, $\lambda=\Delta_{in}+\Delta_{out}$. Bulk values behind the front are shown on the right boundary of the figure.

Fig. 12. Structure of the TVF state ($\eta=0.5066$) well behind the propagating front as a function of driving. Upper (lower) dots denote axial extension of radial inflow, Δ_{in}, (outflow, Δ_{out},) as measured by the node distances of the radial velocity field. Triangles represent half the wavelength, $\lambda=\Delta_{in}+\Delta_{out}$. Dashed line is half the wavelength with maximal growth rate resulting from a linear analysis [8] done for $\eta=0.5$. Crosses show half the wavelength at the marginal stability boundary [8] of CCF for $\eta=0.5$.

the wavelength λ_{max} of that infinitesimal perturbation of the basic CCF state
that grows fastest [8]. We finally should like to point out the interesting fact
that the outflow extension Δ_{out} in Fig. 12 lies for small ε_o almost on the
small-λ branch of the marginal stability curve of the basic CCF state.

6. CONCLUSION

We conclude by summarizing some aspects of structure formation via front propagation: In the Rayleigh-Bénard system of a horizontal fluid layer heated from
below the horizontally periodic structure of parallel convective rolls grows at
supercritical driving into the quiescent, horizontally homogeneous, unstable
conductive state. In the rotating Couette system of a fluid annulus between two
concentric cylinders the axially periodic structure of toroidal Taylor vortices
stacked around the inner, rotating one, grows at supercritical driving into the
axially homogeneous, unstable state of circular Couette flow.

Our numerically obtained solutions of the full hydrodynamic field equations
appropriate to these systems show that in both systems a uniquely selected,
stationary growth mode of the periodic state develops. Therein a uniquely
selected spatial front intensity profile which increases smoothly and monotonously from zero (homogeneous state) to the fully developed structured state
propagates with uniquely selected velocity into the homogeneous state. Also
the structural properties of the flow state unfolding under the propagating
front seem to be uniquely determined.

In the driving range $0.01\lesssim\varepsilon\lesssim0.2$ investigated here the propagation velocities
of convection (TVF) fronts are about 6% (15%) smaller than $c_A=2\sqrt{\varepsilon}\,\xi_0/\tau_0$, i.e.,
the stationary front propagation velocity resulting from the lowest-order
amplitude equation. The front intensity profiles are sharper by slightly larger
amounts than the ones obtained from the amplitude equation. But the typical front
extensions still scale reasonably well as $1/\sqrt{\varepsilon}$ in the driving range considered
here. In both systems the wavenumber of the periodic flow state growing along
the front is very close — in particular for small ε — to the wavenumber
$k_{max}(\varepsilon)$ of that infinitesimal perturbation of the homogeneous basic state that
grows fastest according to a linear analysis.

Note, however, that the TVF structure contains — in contradistinction to
the convective flow structure of parallel rolls — a substantial amount of
higher harmonics in its Fourier decomposition which reflect the asymmetry
between inflow and outflow extensions, Δ_{in} and Δ_{out}, and intensities, u_{in} and
u_{out}, respectively. Thus the lowest-order amplitude approximation involving
the first harmonic of the flow field only is less suited to describe the rather
complicated pattern evolution under a propagating front in the rotating Couette
system as compared to the Rayleigh-Bénard system where the growing structure is

more simple. In the former the asymmetry of the order-parameter field $u(r=r_1+d/2,z)$ between radial inflow and outflow intensities grows along the front. Simultaneously the node distances of u marking inflow domains, Δ_{in}, increase and the node distances marking outflow domains, Δ_{out}, decrease from their common value $\lambda/2$ in the extreme head of the front towards their bulk TVF value far behind the front.

Let us finally comment on the experimental observability of propagating fronts (experiments on propagating TVF fronts have already been done [8]). They can be generated in both systems in the same way as they were generated in our numerical simulation, namely by a sudden step-up of the driving, e.g., from a stationary subcritical state to a supercritical value. In the rotating Couette system the unavoidable Ekman vortices near the ends of the cylinders are also the ideal initial configuration for a subsequent growth inwards. In the Rayleigh-Bénard system we found a short lateral heat pulse applied through a sidewall a convenient and also experimentally realizable means to start convection near the wall.

To observe then a front propagating inwards, say, over a distance ℓ the driving ε and the experimental imperfections that break the translational invariance and thereby nucleate local growth of the flow structure have to be so small that the onset time, τ_{onset}, which this imperfection-triggered growth mode requires to develop sizeable amplitudes is larger than the propagation time, ℓ/c, of the front. In that case the quasihomogeneous flow in the bulk (which has adjusted its state diffusively to the new supercritical driving on the <u>fast</u> time scale of just one characteristic diffusion time, d^2/κ or d^2/ν, respectively) has not time enough to decay into the periodic state before the front arrives. Take as an example convection fronts travelling at $\varepsilon=0.05$ with velocity $c\sim2.15\ \kappa/d$ in a fluid of Prandtl number $\sigma=1$. To observe them over a distance of, say, 15d the onset time, τ_{onset}, for imperfection-triggered convection should exceed 7 vertical diffusion times.

This work was supported by the Stiftung Volkswagenwerk.

REFERENCES

1. <u>Cellular Structures in Instabilities</u>, ed. by J.E. Wesfreid and S. Zaleski, Lecture Notes in Physics, Vol. 210 (Springer, Berlin, Heidelberg 1984).

2. R.C. Di Prima and H.L. Swinney in <u>Hydrodynamic Instabilities and the Transition to Turbulence</u>, ed. by H.L. Swinney and J.P. Gollub, Topics in Applied Physics, Vol. 45 (Springer, Berlin, Heidelberg 1985).

3. J.K. Platten and J.C. Legros, <u>Convection in Liquids</u> (Springer, Berlin, Heidelberg 1984).

4. M. Lücke, M. Mihelcic, and K. Wingerath, Phys. Rev. Lett. 52, 625 (1984); Phys. Rev. A31, 396 (1985); and unpublished.

5. M. Lücke, M. Mihelcic, and B. Kowalski, Phys. Rev. A35 4001 (1987)

6. S. Chandrasekhar, Hydrodynamic and Hydromagnetic Stability (Dover, New York, 1981).

7. J.E. Welch, F.H. Harlow, J.P. Shannon, and B.J. Daly, Los Alamos Sci. Lab. Rep. LA-3425, 1966.

8. M.A. Dominguez-Lerma, G. Ahlers, and D.S. Cannell, Phys. Fluids 27, 856 (1984).

9. G. Ahlers and D.S. Cannell, Phys. Rev. Lett. 50, 1583 (1983).

10. M. Lücke and B. Kowalski, unpublished.

11. G. Pfister and I. Rehberg, Phys. Lett. 83A, 19 (1981).

12. D.G. Aronson and H.F. Weinberger, Advances in Mathematics (Academic, New York, 1978), Vol. 30, p. 33.

13. L. Kramer, E. Ben-Jacob, H. Brand, and M.C. Cross, Phys. Rev. Lett. 49, 1891 (1982); E. Ben-Jacob, H. Brand, G. Dee, L. Kramer, and J.S. Langer, Physica 14D, 348 (1985).

14. G. Dee and J.S. Langer, Phys. Rev. Lett. 50, 383 (1983); G. Dee, Physica 15D, 295 (1985).

15. L.A. Segel, J. Fluid Mech. 38, 203 (1969); A.C. Newell and J.A. Whitehead, J. Fluid Mech. 38, 279 (1969).

16. R. Graham and J.A. Domaradzki, Phys. Rev. A26, 1572 (1982).

17. M. Lücke in Nichtlineare Dynamik in kondensierter Materie, ed. by G. Eilenberger and H. Müller-Krumbhaar (Ferienkurs, Kernforschungsanlage Jülich, 1983).

18. G.P. Neitzel, J. Fluid Mech. 141, 51 (1984).

19. M.A. Dominguez-Lerma, G. Ahlers, and D.S. Cannell, Phys. Fluids 28, 1204 (1985).

20. W. Eckhaus, Studies in Nonlinear Stability Theory, (Springer, New York, 1965).

21. J. Niederländer, Diploma Thesis (Universität des Saarlandes, Saarbrücken, 1986), unpublished.

Oscillatory Doubly Diffusive Convection: Theory and Experiment

E. Knobloch[1], *A.E. Deane*[2], *and J. Toomre*[2]

[1]Department of Physics, University of California, Berkeley, CA 94720, USA
[2]Joint Institute for Laboratory Astrophysics, and
 Department of Astrophysical, Planetary and Atmospheric Sciences,
 University of Colorado, Boulder, CO 80309, USA

1. Introduction

Doubly diffusive convection in its various forms offers an excellent opportunity to study the dynamics near a Hopf bifurcation in a continuous system. Over the last year theoretical and numerical studies have been complemented by a number of new experiments on binary fluid mixtures revealing a richness of behavior that has in turn stimulated further theoretical effort. It is the purpose of this brief report to compare existing theory with the experiments, and indicate directions for further research.

The term doubly diffusive convection [1] is applied to convection in a fluid layer with a stabilizing solute gradient. Such a solute gradient may be maintained either through the imposition of appropriate boundary conditions or by means of the negative Soret effect [2–4]. The former is useful in modelling geophysical problems in which the Soret effect is generally negligible [3], while the latter is ideal for laboratory experiments. In these experiments a binary fluid mixture with a negative Soret coefficient (e.g. salt–water, ethanol–water or He^3–He^4) develops a stabilizing concentration gradient *in response* to an applied destabilizing temperature gradient. The mixtures are characterized by two internal parameters, the Prandtl number $\sigma \equiv \nu/\kappa_T$ and the Lewis number $\tau \equiv \kappa_S/\kappa_T$, where ν is the kinematic viscosity, and κ_S, κ_T, the solutal and thermal diffusivities. In addition, there are two external parameters, the thermal and solutal Rayleigh numbers R_T, R_S, which measure the strength of the destabilizing temperature and stabilizing concentration gradients, and which may be varied. In mixtures with a negative Soret coefficient the separation ratio $S(< 0)$ is used instead of R_S since $R_S = -S R_T$. Since $\tau < 1$, the conduction state can lose stability either at a Hopf bifurcation or a steady state bifurcation as R_T is increased [1,2], depending on the value of R_S (or S). The two possibilities are divided by a critical value R_{Sc} (or S_c) such that the linear stability problem has two zero eigenvalues when $R_T = R_{Tc}$ (see Fig. 1). The point $R_T = R_{Tc}$, $R_S = R_{Sc}$ (or $S = S_c$) is called the codimension–two point [5,6]. Its presence provides

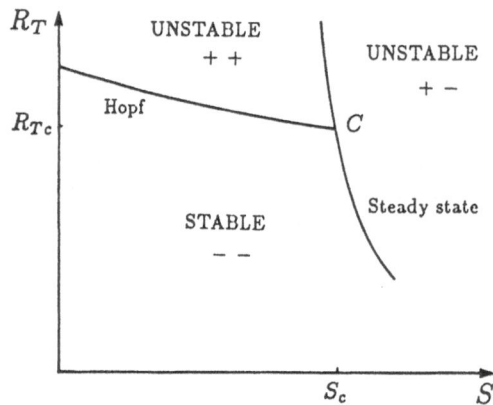

Fig. 1 Experimental linear stability diagram for a He^3–He^4 mixture in a porous medium showing the presence of the codimension–two point (C). The signs of the real parts of the two dominant eigenvalues are also indicated. After [15].

the key to the theoretical success described below. It is important, however, to note that in an infinite layer the wave numbers of the oscillatory and steady state modes that first become unstable are not necessarily the same so that this codimension–two point may not be observable.

2. Outline of the Theory

The theory is based on the presence of a translation symmetry in a horizontally infinite layer [7–9]. It is assumed that the solutions are two–dimensional and are spatially periodic with a period $2\pi/k$, and that in addition there is a reflection symmetry in any vertical plane. With these assumptions the system is equivariant under the group $O(2)$ of rotations and reflections of a circle.

2.1 The Hopf Bifurcation

As a consequence of the $O(2)$ symmetry the number of eigenvalues on the imaginary axis at a Hopf bifurcation is doubled to four [10], and two branches of solutions bifurcate simultaneously from the conduction solution: the standing waves (SW) and the travelling waves (TW). The bifurcation is described by the normal form [7–9]

$$\dot{A}_\ell = (\lambda + aA_r^2 + bA^2 + \cdots)A_\ell, \qquad (1a)$$

$$\dot{A}_r = (\lambda + aA_\ell^2 + bA^2 + \cdots)A_r, \qquad (1b)$$

$$\dot{\phi}_\ell = \Omega + \alpha A_r^2 + \beta A^2 + \cdots, \qquad (1c)$$

$$\dot{\phi}_r = \Omega + \alpha A_\ell^2 + \beta A^2 + \cdots, \qquad (1d)$$

where A_ℓ, A_r are the (real) amplitudes of left– and right–travelling waves in terms of which the streamfunction is given by

$$\Psi(x, z, t) = Re \left\{ A_\ell(t) e^{i[kx + \phi_\ell(t)]} + A_r(t) e^{i[kx - \phi_r(t)]} \right\} f(z) + \cdots . \quad (2)$$

Here $A^2 \equiv A_\ell^2 + A_r^2$, $\lambda \pm i\Omega$ are the eigenvalues of the linear stability problem each of which occurs twice, and $f(z)$ is an appropriate vertical eigenfunction. The Hopf bifurcation takes place at $\lambda = 0$ (i.e., $R_T = R_T^{(o)}$, say). There are four stationary solutions to (1a, b): The trivial state $(0,0)$, left–travelling waves $(A, 0)$, right–travelling waves $(0, A)$, and standing waves $(A/\sqrt{2}, A/\sqrt{2})$. All other solutions are transient. The corresponding bifurcation diagrams (A vs. λ) are shown in the (a, b) plane in Fig. 2 [7–9]. The (real) coefficients a, b must be computed from the partial differential equations. With stress–free boundary conditions and fixed temperature and concentration at top and bottom, b vanishes identically, and a higher order calculation is necessary [7–9,11]. However, with realistic boundary conditions $b \neq 0$, and we interpret existing experiments in the light of Fig. 2.

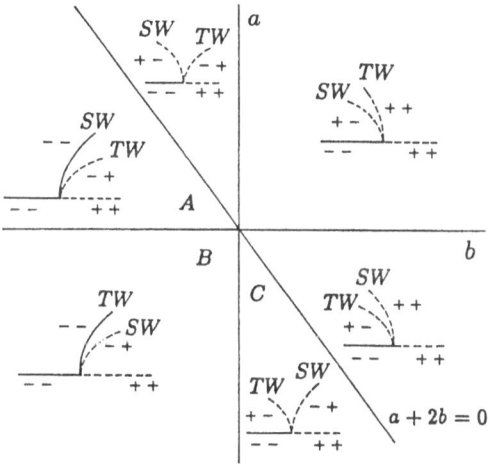

Fig. 2   Bifurcation diagrams (A vs. λ) in the (a, b) plane, indicating the regions A, B, C. Solid lines denote stable solutions, dashed lines unstable solutions. After [8].

2.2 The Codimension–Two Bifurcation

Owing to the $O(2)$ symmetry there are 4 zero eigenvalues at this bifurcation. Near R_{Tc}, R_{Sc} the streamfunction is given by

$$\Psi = Re \left\{ \epsilon v(\tilde{t}) e^{ikx} \right\} f(z) + O(\epsilon^3), \quad (3)$$

where v satisfies [9,12]

$$v'' = \mu v + \epsilon \nu v' + B|v|^2 v + \epsilon C(v\bar{v}' + \bar{v}v')v + \epsilon D|v|^2 v' + O(\epsilon^2), \qquad (4)$$

and the prime denotes differentiation with respect to $\tilde{t} = \epsilon t$. The parameters μ, ν are unfolding parameters proportional to linear combinations of $R_T - R_{Tc}$, $R_S - R_{Sc}$, and the coefficients B, C, D are real. There are 5 types of solutions, best described in terms of the real variables r, ϕ, where $v = re^{i\phi}$. These are pure conduction ($r = 0$), steady convection, SS ($r' = 0, \phi' = 0$), SW ($r' \neq 0$, $\phi' = 0$), TW ($r' = 0$, $\phi' \neq 0$) and modulated waves, MW ($r' \neq 0$, $\phi' \neq 0$). A complete classification of these solutions and their stability properties as a function of μ, ν is given in [12]. As before, with idealized boundary conditions $D = 0$, but $D \neq 0$ in the experiments.

3. Comparison of the Theory with Numerical Simulations and Experiments

In this section we compare the theoretical and numerical results obtained with periodic boundary conditions [7–12] with numerical simulations of thermosolutal convection in a finite geometry, and with three classes of recent laboratory experiments. We find that the idealized results are invaluable in interpreting these experiments.

3.1 Numerical Simulations of Thermosolutal Convection

Numerical simulations of two–dimensional thermosolutal convection have been conducted in order to explore more nonlinear behavior than can be analyzed by perturbation theory and to investigate the role played by sidewalls. We describe here results from two of these simulations. The calculations use finite–difference techniques [13] employing a computational grid of 16×256 points to describe flows within a very wide box in the horizontal, of aspect ratio $1 : 16$. The upper and lower surfaces are stress–free and at fixed temperature and salinity with the lower boundary hotter and more saline. The vertical end walls are impenetrable and stress–free. A fuller account of these results is given in [14].

Figure 3a shows the streamfunction at successive times for a case illustrating a TW. The motion is in the form of a pattern of rolls translating to the left. The rolls appear in the region on the right which is comparatively quiescent and travel to the region on the left where reflection from the walls creates a region of standing rolls. Figures 3b, c show the associated time trace of the thermal Nusselt number, N_T, and its power spectrum. The latter shows that the spectrum is broad–band, but with a peak at $\omega = 50$ (in units of radians per thermal diffusion time in the vertical). The TW is therefore essentially aperiodic, with the basic oscillation corresponding to

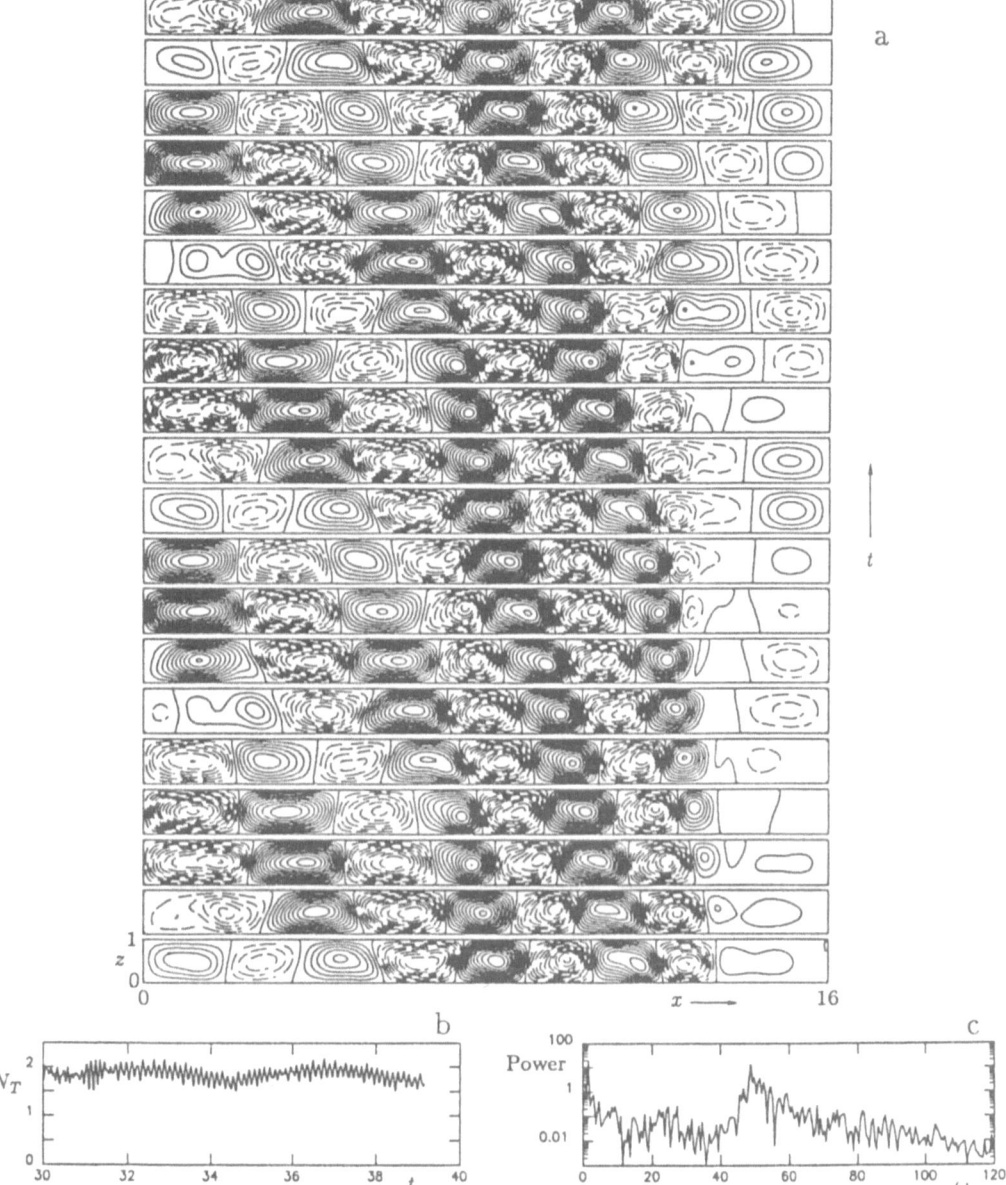

Fig. 3 Results from a numerical simulation of two–dimensional doubly diffusive convection within a wide horizontal box of aspect ratio $1 : 16$ for $R_T = 1.3 \times 10^4, R_S = 1.46 \times 10^4, \tau = 10^{-1/2}$ and $\sigma = 1$. A travelling wave (TW) solution is illustrated. (a) Contours of streamfunction are shown at successive times that increase upwards, spanning an interval $t = 38.5672(.0292)39.1220$ in thermal diffusion times. Solid contours indicate counterclockwise circulation of the fluid. (b) Time trace of the Nusselt number, N_T, as sampled near the top of the layer, and (c) its power spectrum.

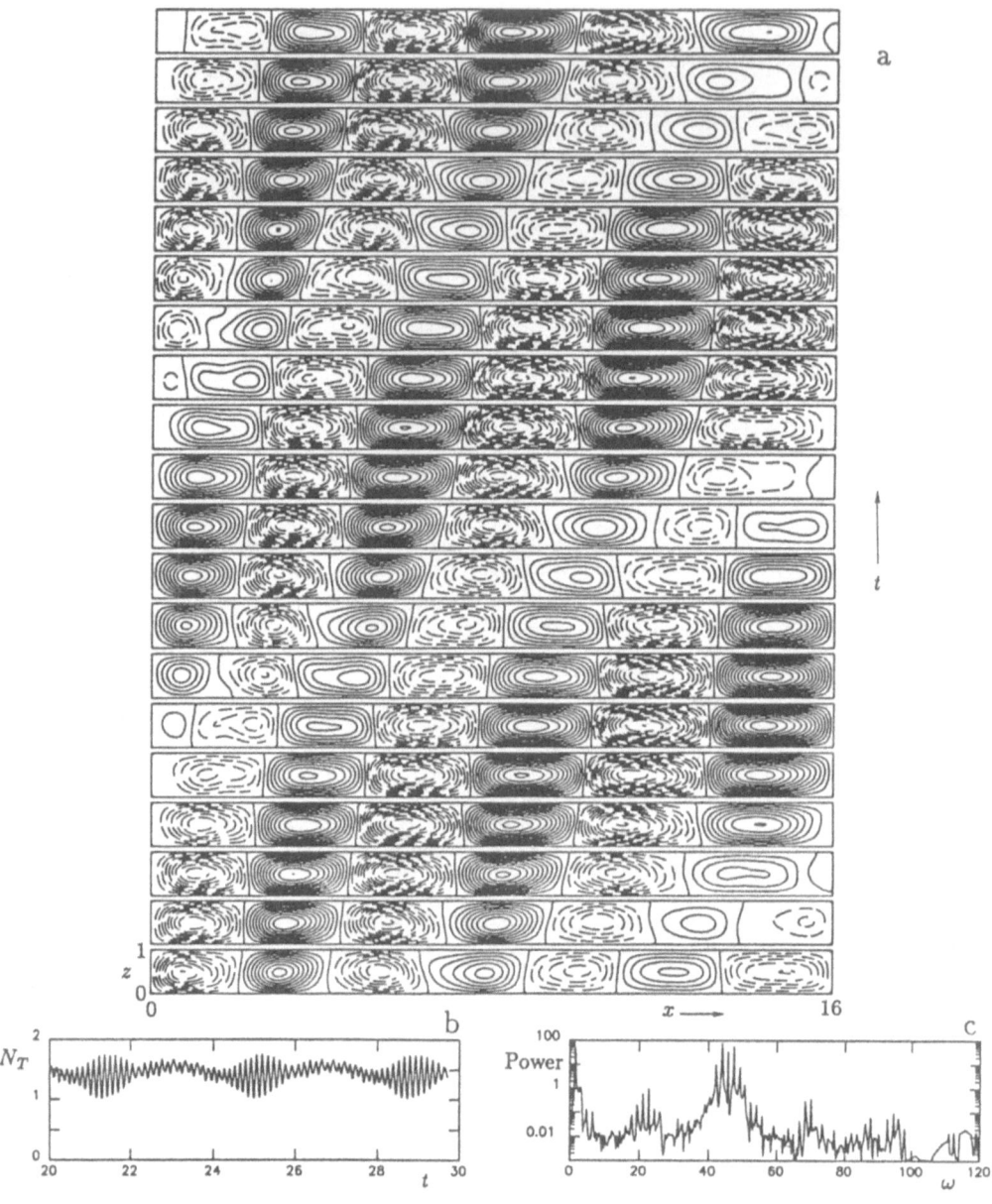

Fig. 4 As in Fig. 3, but illustrating a modulated wave (MW) solution for $R_T = 1.275 \times 10^4$, $R_S = 1.46 \times 10^4$, $\tau = 10^{-1/2}$, $\sigma = 1$ and aspect ratio $1 : 16$. The time interval in (a) spans $t = 29.0198(.0177)29.3561$.

the growth and decay of the standing rolls in the interaction region in the left portion of the layer.

Figure 4a shows a solution nearby in parameter space. Here the rolls travel to the right, but are modulated in amplitude. Again the interaction with reflections from the sidewalls produces a region of standing rolls. Figures 4b, c show the corresponding N_T and its power spectrum. This solution is a MW with three basic frequencies. The rolls translate with a frequency $\omega_1 = 22$, and are modulated at a frequency $\omega_2 = 44$. This produces a peak in the N_T power spectrum at $\omega = 44$. Such frequency locking is encouraged by the sidewalls [9]. There is in addition a third low frequency $\omega_3 = 1.9$ which yields the successive ballooning of N_T in Fig. 4b. Very similar behavior was found in the original experiments of Caldwell [4]. The other frequencies in the spectrum are sums and differences of these three frequencies.

These results show that there are clear counterparts in cells of finite size of TW, MW (as well as SW) predicted with periodic boundary conditions.

3.2 He^3–He^4 Mixture in a Porous Medium

In these experiments [15] on a normal He^3–He^4 mixture, flow visualization is not possible, and only N_T is measured. For SW the theory shows [16–18] that N_T oscillates with the frequency $\omega = 2\phi_r = 2\phi_\ell$. In contrast, for TW in a large aspect ratio container N_T is constant [8,9]. The experiments were carried out in a cell of aspect ratio $1:4:8$ for different values of $S < S_c < 0$, and reveal a supercritical bifurcation to an oscillatory N_T as R_T is increased. Comparison with convection in water in the same cell suggests that the resulting pattern consists of parallel rolls. Thus in this experiment we see a supercritical bifurcation to stable SW, as in the region labeled A in Fig. 2. For a fixed value of S close to S_c, the frequency ω was observed to approach zero as R_T increased further. This is in accord with the theory which predicts the formation of an infinite–period heteroclinic orbit in the dynamics [5,6]. As this orbit is approached ω vanishes as $-1/\ell n(R_T^{(SW)} - R_T)$ [5–9]. No attempt to verify this prediction has been made. For smaller values of S, N_T was seen to experience a period doubling bifurcation (cf. Fig. 3, [15]), which may be identified with a bifurcation to asymmetry in the flow as seen in earlier numerical studies of doubly diffusive convection [1,16–18]. With a further increase of R_T the frequency again vanished, possibly forming one of two homoclinic orbits [16–18]. For still smaller values of S, chaotic behavior was observed close to the initial Hopf bifurcation. As R_T increased, period halving was observed [19]. Similar behavior has been observed in numerical calculations [17,18], and related to the existence of a heteroclinic orbit joining two saddle–foci whose eigenvalues satisfy Shil'nikov's inequality. On a qualitative level there is therefore considerable agreement between the theory and experiment.

3.3 He³–He⁴ Bulk Mixture

In an experiment on a He³–He⁴ bulk mixture [20] in a significantly larger cell (aspect ratio 1 : 6.5 : 26), a convective state characterized by a time–independent N_T was found to bifurcate hysteretically from the conduction solution for $S < S_c$. This is consistent with the presence of a Hopf bifurcation only if this state is a TW and the bifurcation subcritical [11]. Since on theoretical grounds one expects the SW branch to bifurcate supercritically near S_c [6,9], the experiment corresponds to region C of Fig. 2. The unstable TW branch then acquires stability at a secondary saddle–node bifurcation. This interpretation is supported by the observation of constant N_T for TW in large aspect ratio cells with flow visualization [21]. The experiment also reveals, for sufficiently negative S, a crossover to SW, i.e., a bifurcation at $R_T^{(o)}$ to a stable state with an oscillatory N_T, and not a time–independent one. Within the framework of Fig. 2 this is only possible if as S is decreased there is a transition from region C for S near S_c where $b > 0$ to region A where $a + 2b < 0$, $a > 0$ so that the SW branch bifurcates supercritically, and is stable with respect to travelling wave disturbances. There are two possibilities by which this can be accomplished. The most likely is that as S decreases one traverses from region C to B via a tricritical

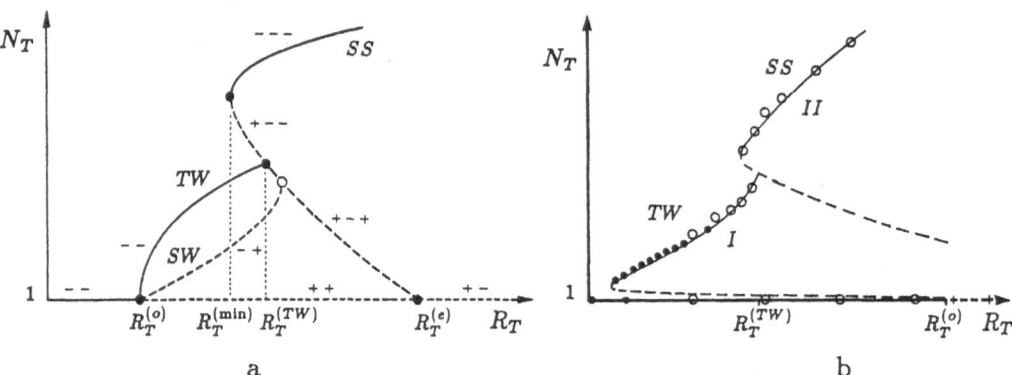

a   b

Fig. 5 (a) Theoretical bifurcation diagram showing a hysteretic transition between the TW and SS branches. Solid lines indicate stable solutions, dashed lines unstable solutions. After [11]. (b) Experimental bifurcation diagram for a He³–He⁴ bulk mixture. The circles indicate experimental points. After [20]. The solid and dashed lines drawn here are based on our theoretical interpretation using (a), and we speculate that I is a TW branch and II is a SS branch.

point ($b = 0$) followed by an interval of S values for which the bifurcation to TW is *supercritical* (region B) before crossing into region A. A second but less likely possibility is to take the "longer" counterclockwise path around the origin, in which case the tricritical point would not be observed. These conclusions are amenable to experimental investigation.

In the experiments a further hysteretic transition to another steady state may now be identified with the hysteretic transition from the TW branch to the branch of steady convection (SS), as shown in Figs. 5a, b. Such a transition is predicted by an analysis of a degenerate form of the codimension–two bifurcation [9,11,12], for which the coefficient $B = 0$ and higher order terms have to be added to (4). In the experiment the N_T on the second branch is not entirely time–independent, but exhibits low frequency modulation that could be due to three–dimensional effects.

3.4 Ethanol–Water Bulk Mixtures

Travelling waves were first *visualized* in an ethanol–water mixture heated from below [22]. Except for very weak modulation due to the presence of sidewalls, which became unobservable in larger aspect ratio cells [21], N_T was found to be constant as predicted by theory. Unlike the cryogenic experiments discussed above, these flows can be visualized using shadowgraph techniques, and thereby distinguished from steady convection. The experiments show that the TW bifurcate subcritically [21,22] from the conduction solution at $R_T^{(o)}$ showing that $b > 0$. In addition no stable SW are observed, even though the bifurcation to SW is expected to be supercritical near the codimension–two point, as in region C of Fig. 2. Consequently, for $R_T > R_T^{(o)}$ oscillatory transients are expected whose amplitude grows at a rate proportional to $R_T - R_T^{(o)}$. Such transients have been observed in the experiments [23].

In addition these experiments have revealed the possibility of a non-hysteretic transition between TW and SS as R_T is increased, with the drift speed of the rolls approaching zero linearly with the distance from the bifurcation [21], verifying a prediction of the analysis of the codimension–two bifurcation with $O(2)$ symmetry [9,12] The experimental bifurcation diagrams (Fig. 6b) are in complete qualitative agreement with one of the diagrams (Fig. 6a) obtained by analyzing the degenerate version of this bifurcation [11]. Similar diagrams have been obtained in numerical studies of thermosolutal convection [11]. In the experiments the slight hysteresis that is observed is attributable to three–dimensional effects which are particularly strong near the transition [21].

Calculations show [9] that under experimental conditions the coefficients B, C are $O(1)$, and $B > 0$, $C < 0$, while $D = 0$ with the boundary

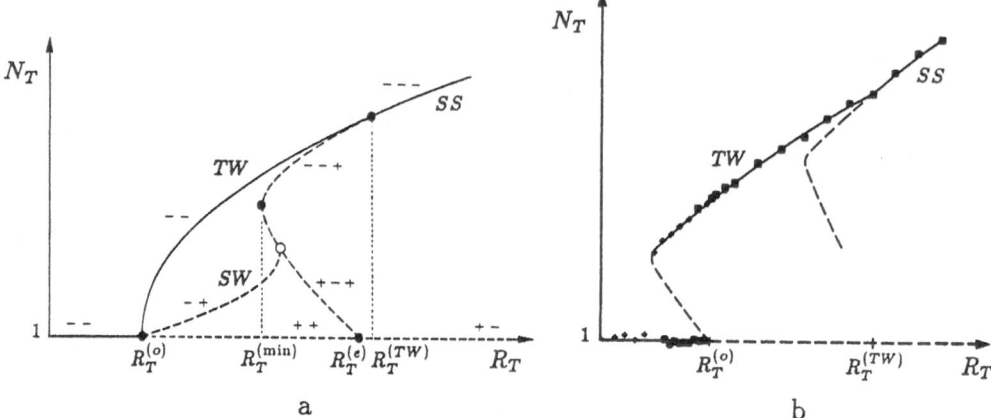

Fig. 6 (a) Theoretical bifurcation diagram showing a nonhysteretic transition between the TW and SS branches. After [11]. (b) Experimental bifurcation diagram for a water–ethanol mixture; the squares
indicate experimental points. After [21]. The solid and dashed
lines are based on the theoretical interpretation using (a).

conditions assumed. The experiments suggest that $D > 0$ but *small*. The
analysis of the appropriate degeneracy in the codimension–two bifurcation
then shows that the stable TW branch will lose stability at a secondary Hopf
bifurcation producing a stable MW [9,12], in which a roll pattern is modulated in time with an independent frequency as it translates. The theory
predicts [24] that beyond this bifurcation the streamfunction will be of the
form

$$\Psi = \epsilon r_o \{ \cos(\omega_o \tilde{t} + kx) + \frac{1}{2}\eta(1 - \frac{2}{\alpha}) \cos[(1 + \alpha)\omega_o \tilde{t} + kx + \theta]$$

$$+ \frac{1}{2}\eta(1 + \frac{2}{\alpha}) \cos[(1 - \alpha)\omega_o \tilde{t} + kx - \theta]\} f(z) + O(\epsilon^3, \epsilon \eta^2), \quad (5)$$

where r_o, ω_o are the amplitude and frequency on the TW branch at the
secondary bifurcation, while η, $\alpha \omega_o$ are the amplitude and frequency of the
oscillations about the TW appearing at this bifurcation. In addition θ is an
arbitrary phase. The signature of such a transition is therefore the appearance in the spectrum of Ψ, at a fixed reference point, of peaks at frequencies
$(1 \pm \alpha)\omega_o$ on either side of the principal peak at ω_o corresponding to the pure
TW. Such a transition, deduced from a spectrum of shadowgraph intensity,
has been reported in the latest of the experiments [25]. The experimental value $\alpha = 0.195$ is in qualitative agreement with the theoretical result
$\alpha = \sqrt{2D/C}$, since $|D| << |C|$. The bifurcation diagrams are compared in
Fig. 7. In addition, the theory predicts that the particle motions in such a
wave should be chaotic [24] and that the MW branch should terminate on

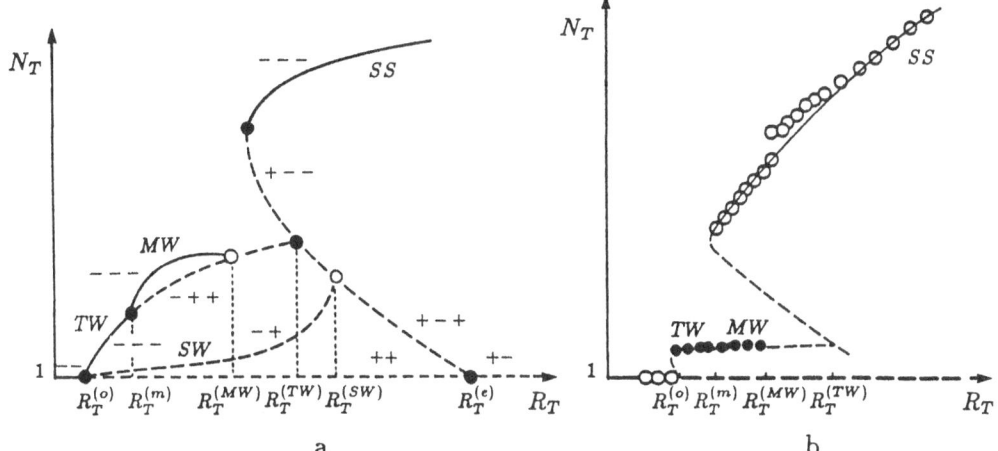

Fig. 7 (a) Theoretical bifurcation diagram showing a secondary bifurcation
from TW to MW. Open circles indicate global bifurcations. After
[9]. (b) Experimental bifurcation diagram for a water–ethanol bulk
mixture showing a secondary bifurcation from TW to MW at $R_T = R_T^{(m)}$. After [25].

the TW branch in a global bifurcation at which the new frequency vanishes
[9,12]. These predictions may be verifiable experimentally. The experiment
also shows that it is possible to find TW states that are spatially confined
to only a part of the cell with the remainder at rest [25,26]. Such states are
found after complicated transients in which both left– and right–travelling
waves coexist before one or the other decays. Similar states have been found
in the partial differential equations for thermosolutal convection [14].

4. Discussion

We have seen that the theory based on an approximate $O(2)$ symmetry of the system in a large aspect ratio cell has been remarkably successful
in accounting for a number of the experimental observations. The reason
for this success is due to the fact that many of the theoretical results (like
Fig. 2) depend only on the symmetry, and hence not on the details of the
assumed boundary conditions, provided these do not affect the symmetry. A
quantitative comparison with the experiments will become possible within
this theoretical framework on calculating the coefficients a, b etc. with the
correct boundary conditions. This work is currently in progress. In addition it is possible to study the effects of breaking the translation or reflection
symmetries [27] both of which are pertinent to the experiments.

The theory has, in its present form, neglected two interesting effects that will undoubtedly be studied in the near future: spatial modulation [28,29] in conjunction with the origin of defects in the travelling wave patterns, and three–dimensional phenomena observed in many of the experiments [21,22].

We are grateful to G. Ahlers and I. Rehberg for a discussion. The work at Berkeley was partly supported by the Applied and Computational Mathematics program of the Defence Advanced Research Projects Agency. The work at Colorado was partly supported by the National Aeronautics and Space Administration through grants NAGW–91 and NSG–7511 and contract NAS8–31958, and partly by the Naval Ocean Research Development Activity (NORDA). The Joint Institute for Laboratory Astrophysics is operated jointly by the University of Colorado and the National Bureau of Standards.

References:
1. H. E. Huppert and D. R. Moore: *J. Fluid Mech.* **78**, 821 (1976)
2. D. T. J. Hurle and E. Jakeman: *J. Fluid Mech.* **47**, 667 (1971)
3. D. R. Caldwell: *J. Fluid Mech.* **64**, 347 (1974)
4. D. R. Caldwell: *Deep–Sea Research* **20**, 1029 (1973)
5. E. Knobloch and M. R. E. Proctor: *J. Fluid Mech* **108**, 291 (1981)
6. H. R. Brand, P. C. Hohenberg and V. Steinberg: *Phys. Rev.* **A30**, 2548 (1984)
7. E. Knobloch: In *Proc. 1985 Joint ASCE–ASME Mechanics Conf.*, ed. by N. E. Bixler and E. A. Spiegel, FED Vol. **24**, 17 (1985)
8. E. Knobloch, A. E. Deane, J. Toomre, and D. R. Moore: In *Multiparameter Bifurcation Theory*, ed. by M. Golubitsky and J. Guckenheimer, *Contemp. Math.* **56**, 203 (Amer. Math. Soc., Providence, RI 1986)
9. E. Knobloch: *Phys. Rev.* **A34**, 1538 (1986)
10. M. Golubitsky and I. Stewart: *Arch. Rat. Mech. Anal.* **87**, 107 (1985)
11. A. E. Deane, E. Knobloch and J. Toomre: preprint
12. G. Dangelmayr and E. Knobloch: *Phil. Trans. Roy. Soc.*, London, in press.
13. D. R. Moore, R. S. Peckover and N. O. Weiss: *Comp. Phys. Comm.* **6**, 198 (1974)
14. A. E. Deane, E. Knobloch and J. Toomre: preprint
15. I. Rehberg and G. Ahlers: *Phys. Rev. Lett.* **55**, 500 (1985)
16. L. N. Da Costa, E. Knobloch and N. O. Weiss: *J. Fluid Mech.* **109**, 25 (1981)
17. D. R. Moore, J. Toomre, E. Knobloch and N. O. Weiss: *Nature* **303**, 663 (1983)
18. E. Knobloch, D. R. Moore, J. Toomre and N. O. Weiss: *J. Fluid Mech.* **166**, 409 (1986)
19. I. Rehberg and G. Ahlers: In *Multiparameter Bifurcation Theory*, ed. by M. Golubitsky and J. Guckenheimer, *Contemp. Math.* **56**, 277 (Amer. Math. Soc., Providence, RI, 1986)

20. G. Ahlers and I. Rehberg: *Phys. Rev. Lett.* **56**, 1373 (1986)
21. E. Moses and V. Steinberg: *Phys. Rev.* A**34**, 693 (1986)
22. R. W. Walden, P. Kolodner, A. Passner and C. M. Surko: *Phys. Rev. Lett.* **55**, 496 (1985)
23. P. Kolodner, A. Passner, C. M. Surko and R. W. Walden: *Phys. Rev. Lett.* **56**, 2621 (1986)
24. E. Knobloch and J. B. Weiss: preprint
25. R. Heinrichs, G. Ahlers and D. S. Cannell: preprint
26. E. Moses, J. Fineberg and V. Steinberg: preprint
27. G. Dangelmayr and E. Knobloch: these proceedings
28. M. C. Cross: preprint
29. H. R. Brand, P. S. Lomdahl and A. C. Newell: *Phys Lett.* A**118**, 67 (1986)

Pattern Selection in Anisotropic Systems

L. Kramer, E. Bodenschatz, W. Pesch, and W. Zimmermann

Physikalisches Institut der Universität Bayreuth,
D-8580 Bayreuth, Fed. Rep. of Germany

We examine the possible stationary patterns and their stability slightly above threshold in systems that are extended in two dimensions and possess an axial anisotropy. The universal amplitude equations used show interesting defect solutions which are discussed briefly. The connection to various experimental systems, such as electrohydrodynamic convection in liquid crystals, is outlined.

1. Introduction

The spontaneous formation of spatially periodic patterns in continuous systems is an intensely studied field. Much activity is devoted to the investigation of the structures and transitions in systems that are extended in two dimensions. Prominent among these is Rayleigh–Bénard convection of simple fluids in layers of large lateral dimensions /1–3/. Apart from boundary effects this quasi–two–dimensional system is isotropic in the plane of the layer and then the only periodic stationary structures that can appear at threshold are straight rolls, squares or hexagons with arbitrary orientation /1/. This should be contrasted with anisotropic quasi–two–dimensional systems having a preferred axis, which will be discussed here.

In order to display the universal properties in the simplest manner the treatment is kept mostly on a general phenomenological level. Thus in Sec.2 we first discuss the linear stability analysis that gives the possible stationary structures slightly above threshold (we mainly aim at continuous instabilities corresponding to forward bifurcations). Emphasis is put on the transition from "normal rolls", which are adapted to the symmetry, to oblique rolls, which are at a finite angle ($\neq \pi/2$) with respect to the preferred axis. In Sec.3 we discuss the behavior slightly above threshold of roll–solutions away from the normal–oblique transition by means of an amplitude equation. Thus the notion of two–dimensional linearly stable wavenumber bands and their stability limits is introduced. Transitions between neighboring states are mediated by nucleation and motion of point defects (dislocations). We briefly discuss the direction and velocity of motion of dislocations.

In Sec.4 the amplitude equation valid in the vicinity of the normal–oblique transition is presented. It describes the changeover between the two regimes and also has more complicated undulated solutions which are linearly stable. In the oblique–roll regime it describes line defects which we discuss briefly. Finally, in Sec.5 we discuss experimental as well as theoretical results for real systems like electrohydrodynamic convection in planar layers of nematic liquid crystals (EHC).

2. Linear analysis

The theoretical investigation of a pattern–forming instability usually starts with linear stability analysis of the basic unstructured state. Thus one has to linearize the underlying (hydro-)dynamic equations and look for mode-solutions. Since the linear equations are essentially translationally invariant in time (except possibly for a periodic external field, e.g. an ac voltage in EHC) and translationally invariant in their symmetry-plane (which we identify with the x-y-plane), the bounded solutions are of the form

$$u_j = e^{\sigma t}[U_j e^{i(qx+py)} + cc] \, v_j(z,\omega t) \qquad (1)$$

(cc = complex conjugate). Here ω is the frequency of the external field and the functions v_j are 2π-periodic in ωt. The U_j are phase factors which in most cases can be chosen equal to 1 or i. The u_j describe the various physical quantities involved. Solving the linear equations provides an implicit expression (determinantal condition) for the growth rate σ

$$f(\sigma;q^2,p^2;R,\omega,...) = 0. \qquad (2)$$

Here $R,\omega,...$ are the external control parameters. R is chosen as the "main" control parameter, which upon increase brings the system across the threshold of instability, where the eigenvalue σ with the largest real part becomes positive. In the following the second control parameter ω is not necessarily a frequency. Since we assume reflection symmetry in the x- and y-direction the wavenumbers q and p come in quadratically. We are here interested in steady bifurcations, so we look for real solutions σ.

A mode is neutrally stable when $\sigma = 0$. Equation (2) then leads to a relation

$$R = R_0(q^2,p^2;\omega,...), \qquad (3)$$

where the solution branch with the smallest R is to be taken. Equation (3) will be called the *neutral surface* in the 2-dimensional wavevector space. The threshold for instability R_c and the critical wavenumbers q_c, p_c are obtained by minimizing (3) with respect to q and p:

$$R_c(\omega,...) = \underset{q,p}{Min} \, [R_0(q^2,p^2;\omega,...)]. \qquad (4)$$

Clearly there are two general possibilities:

1. The minimum occurs at a symmetry point $q_c^2 \neq 0$, $p_c = 0$ or vice-versa. In this case we have normal rolls at threshold.

2. The minimum occurs at $q_c^2 \neq 0$ and $p_c^2 \neq 0$. In this case one has the two degenerate modes q_c,p_c and $q_c,-p_c$ (changing both signs does not change the solution). One now has two further possibilities, which are determined by the nonlinear interactions between the modes:

   a) Extinction of one mode is favored. Then one has oblique rolls near threshold.

   b) Coexistence of the modes is favored. Then one has a rectangular pattern with the sides parallel to x and y near threshold.

How does the transition from case 1 to case 2 occur as an external parameter, say ω, is changed while R is kept at R_c? For definiteness we choose $p_c = 0$ in case 1. The situation we are interested in is then described by a function

$R_0(q^2, p^2; \omega)$ that has a minimum with respect to q^2 at $q_{c1}(\omega)$ (>0). In addition at a particular value $\omega = \omega_z$ the term proportional to p^2 vanishes at this minimum $q_z = q_{c1}(\omega_z)$ (>0). Expanding around $q = q_z$ and $p = 0$ gives

$$R_0 = R_0(q_z^2, 0, \omega) + \bar{a}(q-q_z)^2 + 2\bar{b}(q-q_z)p^2 + \bar{c}p^4 \qquad (5)$$
$$+ \bar{d}(\omega-\omega_z)(q-q_z) + \bar{e}(\omega-\omega_z)p^2 + \dots$$

with $\bar{a} > 0$. The first term in Eq.(5) could also be expanded : $R_0(q_z^2, 0, \omega) = R_c(\omega_z) + r_1(\omega-\omega_z) + r_2(\omega-\omega_z)^2 + \dots$.

We consider the case of a continuous transition from case 1 to case 2 which requires $\bar{c} > 0$ and $\bar{b}^2 < \bar{a}\bar{b}$. For an isotropic system one would have $\bar{b} = 2q_z\bar{c}$, so we usually expect $\bar{b} > 0$. Minimization of Eq.(5) with respect to q and p leads to

$$q_{c1} = q_z - (\omega-\omega_z)\bar{d}/2\bar{a}, \quad p_{c1} = 0, \quad R_{c1} = R_0(q_z^2, 0, \omega) - (\omega-\omega_z)^2\bar{d}^2/4\bar{a}, \qquad (6)$$

or

$$q_{c2} = q_z - (\omega-\omega_z)(\bar{c}\bar{d}-\bar{b}\bar{e})/2(\bar{a}\bar{c}-\bar{b}^2),$$
$$p_{c2}^2 = -(\omega-\omega_z)(\bar{a}\bar{e}-\bar{b}\bar{d})/2(\bar{a}\bar{c}-\bar{b}^2),$$
$$R_{c2} = R_{c1} - (\omega-\omega_z)^2(\bar{a}\bar{e}+\bar{b}\bar{d})^2/4\bar{a}(\bar{a}\bar{c}-\bar{b}^2). \qquad (7)$$

The solution (6) represents normal rolls whereas the solution (7) gives oblique rolls which bifurcate from the normal rolls at $\omega = \omega_z$ in an ordinary pitch-fork bifurcation. The oblique rolls have a lower threshold on the side where they exist, i.e. where $(\omega-\omega_z)(\bar{a}\bar{e}-\bar{b}\bar{d}) < 0$. For reasons of continuity one always expects in a neighborhood of $\omega = \omega_z$ realization of case a), i.e. oblique rolls and not rectangles (which would be very elongated here). This can be shown in a more convincing way by the methods presented in the next sections.

Above threshold $(R > R_c)$ one has positive growth rates for q and p inside a two-dimensional wavenumber region (or regions). One here expects existence of periodic solutions. Some aspects of the post-threshold behavior are captured by the fastest-growing mode q_m, p_m, which corresponds to the growth rate

$$\sigma_m(R, \omega, \dots) = \underset{q, p}{\text{Max}} \left[\sigma(q, p; R, \omega \dots) \right]. \qquad (8)$$

Considerations analogous to those presented for R_c hold for σ_m.

3. Weakly nonlinear analysis for $\omega \neq \omega_z$

To study the possible structures and their stability slightly above threshold of a continuous (supercritical) instability it has proven useful to employ *envelope (or amplitude) equations* where the nonlinearities as well as (slow) spatial variations around the structure occurring at threshold are treated perturbatively. To obtain these equations we choose a suitable expansion parameter

$$\varepsilon(\omega,...) = [R - R_c(\omega,...)]/R_c(\omega,...). \tag{9}$$

For $\omega \neq \omega_z$ the neutral surface behaves parabolically in q and p in the vicinity of its minimum. To capture this behavior in the envelope equations one may introduce slow variables in the following manner

$$X = \varepsilon^{1/2}x, \quad Y = \varepsilon^{1/2}y, \quad T = \varepsilon t. \tag{10}$$

We expand the solutions of the full nonlinear equations as follows

$$u_j(X,Y,T;\vec{r},\omega t) = \varepsilon^{1/2}\,[u_j^{(0)} + \varepsilon^{1/2}u_j^{(1)} + \varepsilon u_j^{(2)} + ...], \tag{11}$$

where the $u_j^{(k)}$ are periodic in x and y with periods $2\pi/q_c$ and $2\pi/p_c$ and, if there is a time-periodic external field, 2π-periodic in ωt. At order $\varepsilon^{1/2}$ one obtains

$$u_j^{(0)} = [A(X,Y,T)\ U_j\ e^{i(q_c x + p_c y)} + cc]v_j(z,\omega t). \tag{12}$$

At order $\varepsilon^{3/2}$ one obtains a solvability condition of the form

$$T_0\partial_T A = [\xi_1^2\partial_X^2 + \xi_2^2\partial_Y^2 + 2\xi_1\xi_2 a\partial_X\partial_Y + 1 - |A|^2]A. \tag{13}$$

The coefficient of the nonlinear term in Eq.(13) has been chosen equal to 1, which is always possible by an appropriate choice of the normalization of the v_j in Eq.(12). In the normal-roll regime one has either $p_c = 0$ (no rapid y-variation) or $q_c = 0$ (no rapid x-variation). Then overall reflection symmetry requires either reflection symmetry with respect to Y or X, so that $a = 0$ in Eq.(13).

Equation (13) represents the envelope equation. Actually one does not have to go through the full analysis indicated above since all parameters of Eq.(13) are determined by the linear growth rate function (2) in the neighborhood of the threshold. To see this we note that Eq.(13) allows the particular solution

$$A = F\ \exp\,[i(QX/\xi_1 + PY/\xi_2)]\ f(T)/(1 + f^2(T))^{1/2}, \tag{14}$$

$$F = (1-Q^2-P^2-2aQP)^{1/2}, \quad f(T) = f_0\exp(F^2 T/T_0), \tag{15}$$

corresponding to time-dependent rolls with wavenumbers

$$q = q_c + \varepsilon^{1/2}Q/\xi_1 \quad, \quad p = p_c + \varepsilon^{1/2}P/\xi_2. \tag{16}$$

Clearly for $Q^2 + P^2 + 2aQP < 1$ the rolls grow and become stationary for $T \to \infty$. The initial growth rate for $f^2 \ll 1$ expressed in terms of the physical time t obtained from Eq.(15) must be identified with σ :

$$\sigma = F^2\varepsilon/T_0$$
$$= [\varepsilon - \xi_1^2(q - q_c)^2 - \xi_2^2(p - p_c)^2 - 2a\xi_1\xi_2(q - q_c)(p - p_c)]/T_0. \tag{17}$$

Thus we see that

$$T_0 = (d\sigma/d\varepsilon)^{-1}\Big|_{\varepsilon=0} = (R_c\, d\sigma/dR)^{-1}\Big|_{R=R_c}.$$ (18)

Similarly the parameters ξ_1, ξ_2 and a are determined by the curvatures of the neutral surface ($\sigma = 0$). The condition that $q = q_c$, $p = p_c$ is a minimum leads to $a^2 < 1$.

Suitable rescaling of time and a linear transformation of X,Y into new variables X',Y' brings Eq.(13) into the standard form

$$\partial_{T'}A = [\partial_{X'}^2 + \partial_{Y'}^2 + 1 - |A|^2]\, A.$$ (19)

We point out that the rhs of Eq.(19) is equivalent to the Ginzburg-Landau equation used to describe the stationary flow properties of superfluid ^{4}He /4/.

From Eqs.(13) or (19) a number of general properties can be deduced. It is fairly easy to test the linear stability of the stationary roll solutions (14) with $f \to \infty$ /5/. One finds that they are stable for

$$Q^2 + P^2 + 2aQP < 1/3.$$ (20)

This result shows that the stability limit is given by a direct generalization of the long-wavelength *Eckhaus instability* /6 - 8/ to two dimensions: There is a two-dimensional stable wavenumber band which is bounded by a paraboloid that is narrower than the neutral surface by a factor $1/\sqrt{3}$. Thus even in the normal-roll regime rolls with a (small) tilt-angle are, at least in ideal situations, linearly stable.

What happens when the system is quenched into the unstable wavenumber region? One can show that the modulation wavevector $\vec{K} = (K,L)$ of the most rapidly growing mode is proportional to $\vec{Q} = (Q,P)$ /5/. Usually one expects the system to end up in the state $\vec{Q} - \vec{K}$ /9/. This brings the system into a state where the initial Q and P are reduced by a common factor and thus moves the system towards the band-center $\vec{Q} = 0$ on a straight line. The factor is the same as in the ordinary one-dimensional Eckhaus instability (see e.g. /9,10/).

The states with $\vec{Q} \neq 0$ inside the range (20) are metastable. Transitions towards the band center $\vec{Q} = 0$ can occur by motion of dislocations across the system. The solutions of Eqs.(13) or (19) describing dislocations correspond exactly to the well-known vortex solutions in superfluid ^{4}He /4,11/. For $Q \neq 0$ and $P = 0$ the motion is in the Y-direction, i.e. it corresponds to a climb, as is also the case for isotropic systems /12/. More generally one sees from Eqs.(13) or (19) that the motion is in the direction perpendicular to the dimensionless wavevector displacement $\vec{Q}$. For $Q = 0$, $P \neq 0$ the motion then is in the X-direction and corresponds to a pure glide. Each such process brings the system on a straight line towards the band center. This behavior is completely different from that of an isotropic system where glide can occur only in nonpotential situations /13/.

The far-field of a static, isolated dislocation, which exists only at $\vec{Q} = 0$, is given by

$$A \sim [1 - (1-a^2)/|Z|^2]\, Z/|Z|,$$ (21)

$$Z = X/\xi_1 + (i\sqrt{1-a^2} - a)Y/\xi_2$$

In isotropic systems the velocity v of a dislocation scales as $Q^{3/2}$ for small Q /12,13/. This nonanalyticity of the dislocation motion is absent in the anisotropic case. Here v is proportional to $|Q|$. Calculations of the constant of proportionality are in progress.

Actually one expects some difference between the motion of dislocations parallel to the pattern (climb) and perpendicular to it (glide) from nonlinear effects which couple the slow and the rapid variations and which cannot be captured in an ε-expansion. Such "nonadiabatic" effects play a significant role in a variety of situations /8,14/.

In the stable wavenumber region the nucleation of dislocations can only occur by fluctuations or imperfections of the system. Clearly in the interior of the system they can only be produced in pairs with opposite polarity. The dislocation pair dissociates in the direction of motion discussed before. We plan to calculate the saddle-point solutions for the nucleation of dislocation pairs. The saddle-points merge with the periodic solutions at the Eckhaus-stability limit, similar to the situation in quasi-one dimensional systems /8/. Concerning the nucleation of dislocations by imperfections we point out that in a half-infinite system a boundary perpendicular to the rolls does not induce spontaneous production of dislocations in the stable range, although it reduces the barrier /10/. Other boundaries are known to have a pronounced influence on the wavevector band if they reduce the amplitude of the roll pattern /15/. Special boundary constructions ("selecting ramps") abolish the band completely (see I. Rehberg and H. Riecke, this volume).

At the (continuous) transition between normal and oblique rolls ξ_1 or ξ_2 vanish and Eq.(13) becomes inapplicable. In the situation where p_c bifurcates from 0 at $\omega = \omega_z$ one has in the vicinity of the transition the scaling law

$$\xi_2 = \xi_{\pm} \cdot |\omega - \omega_z|^{1/2}, \tag{22}$$

whereas ξ_1 and a remain finite at $\omega = \omega_z$ (on the normal-roll side a is of course trivially zero). As a result modulations in the Y-directions become rapid . The transition point may be called a Lifshitz point in analogy to a similar point in the context of equilibrium phase transitions /16/.

4. Weakly nonlinear analysis for $\omega \approx \omega_z$

In the vicinity of the normal-oblique transition the prefactors of the Y-derivatives in Eq.(13) become zero and one then needs an envelope equation containing higher derivatives. This is accomplished by using a different scaling for Y, namely the one employed in isotropic systems /7,17/ and also introducing a blown up version of the (small) parameter $\omega - \omega_z$

$$Y = \varepsilon^{1/4} y \quad , \quad \omega - \omega_z = \varepsilon^{1/2} \Omega. \tag{23}$$

The same type of procedure as before now leads at order $\varepsilon^{1/2}$ to Eq.(12) with $p_c = 0$ and at order $\varepsilon^{3/2}$ to

$$T_0 \partial_T A = [(\xi_1 \partial_X - i\xi_3^2 \partial_Y^2)^2 - b\xi_3^4 \partial_Y^4 + c\Omega\xi_3^2 \partial_Y^2 + 1 - |A|^2]A. \tag{24}$$

We emphasize that the ε used here is really defined with respect to the normal-roll threshold $R_c(\omega,...)$ as in Eq.(9) and not with respect to

$R_c(\omega_z,...)$. Equation (24) can also be used for negative ε. Then the term $+1$ in the square brackets has to be replaced by -1. Also the q_c in Eq.(12) is the critical normal roll wavenumber at the actual ω. Expanding q_c around ω_z as done for the neutral curve of Eq.(5) would lead to an additional term proportional to $i\Omega\partial_X A$ and a renormalization of c. For $b = c = 0$ Eq.(24) coincides with the simplest envelope equation for isotropic systems /7,17/.

The stationary roll-type solutions of Eq.(24) are now

$$A = F \exp[i(QX/\xi_1 + PY/\xi_3)], \quad F = [1 - (Q+P^2)^2 - c\Omega P^2 - bP^4]^{1/2}. \tag{25}$$

Their wavenumbers are

$$q = q_c + \varepsilon^{1/2}Q/\xi_1 \quad, \quad p = \varepsilon^{1/4}P/\xi_3 \tag{26}$$

so that the neutral surface $F = 0$ is given by

$$\varepsilon = \xi_1^2(q-q_c)^2 + [2\xi_1(q-q_c) + c(\omega-\omega_z)]\xi_3^2 p^2 + (1+b)\xi_3^4 p^4. \tag{27}$$

Comparison with Eq.(5) shows that

$$\bar{a} = R_c\xi_1^2 \quad, \quad \bar{b} = R_c\xi_1\xi_3^2 \quad, \quad \bar{c} = R_c(1 + b)\xi_3^4 \quad. \tag{28}$$

$$\bar{d} = -2R_c\xi_1^2(dq_c/d\omega) \quad, \quad \bar{e} = R_c\xi_3^2[c - 2\xi_1(dq_c/d\omega)].$$

Here all quantities are to be taken at $\omega = \omega_z$. From the overlap of the regions of validity of Eq.(24) and Eq.(13) for $\omega > \omega_z$ and $\omega < \omega_z$ one can find connections for the parameters of the equations in the different regimes. An envelope equation of the form of Eq.(22) with $c = 0$ was proposed before without noting that its validity is restricted to the very special situation $\omega = \omega_z$ /18,19/.

From Eq.(24) one can again obtain the regions of stability of the roll solution (25). The necessary condition for stability is

$$0 < \{F^2 - 2(Q+P^2)\}K^2$$

$$+ \{F^2[6(1+b)P^2 + 2Q + c\Omega] - 2P^2[2(1+b)P^2 + 2Q + c\Omega]^2\}L^2 \tag{29}$$

$$+ 4P\{F^2 - (Q+P^2)[2(1+b)P^2 + 2Q + c\Omega]\}KL$$

for arbitrary modulation wavenumbers K and L of the fluctuations /5/. Equation (29) can be shown to be identical to

$$(\partial_Q^2 V)K^2 + (\partial_P^2 V)L^2 + 2(\partial_Q\partial_P V)KL > 0 \tag{30}$$

where the potential for Eq.(24) is given by

$$V = \langle|(\xi_1\partial_X - i\xi_3^2\partial_Y^2)A|^2 + b|\xi_3^2\partial_Y^2 A|^2 + c\Omega|\xi_3\partial_Y A|^2 - |A|^2 + \tfrac{1}{2}|A|^4\rangle \tag{31}$$

with the solutions (25) inserted ($\langle...\rangle$ denotes a spatial average). A discussion of this criterion for the one-dimensional case was given before

/8/. The stability criterion (20) can be deduced in the same way. Equation (30) appears to be general for stability against long-wavelength modulations in potential systems, also far above threshold.

For normal rolls (P = 0) criterion (29) reduces to

$$Q^2 < 1/3 \quad \text{and} \quad Q > -c\Omega/2. \tag{32}$$

The first inequality is the Eckhaus criterion against longitudinal modulations. The second one insures stability against transverse modulations. For $\Omega = 0$ it coincides with the Zig-Zag-stability criterion well-known from isotropic systems /1,2/. For $c\Omega > 0$ the full stable band consists of one connected region centered around P = 0. For $c\Omega < -2/\sqrt{3}$ there are no stable states with P = 0 and one has two stable regions centered around the critical modes

$$Q_c = c\Omega/2b \quad , \quad P_c = \pm \ |c\Omega/2b|^{1/2}. \tag{33}$$

For an interval of values of $c\Omega$ slightly above $-2/\sqrt{3}$ one has three stable islands.

Besides the straight-roll solutions (25), Eq.(24) admits stationary solutions which describe more complicated periodic structures. By an approximate variational method and a numerical Galerkin procedure rectangular structures as well as undulated oblique and normal rolls were found /5/. The latter bifurcate from the normal rolls at $\Omega = 0$. Only the undulated normal rolls turn out to be linearly stable in some range $c\Omega_-(\varepsilon) < c\Omega < c\Omega_+(\varepsilon) < 0$.

They are always less stable than the oblique rolls, i.e. the potential is higher for the undulated rolls than for the oblique rolls. In the stable range the undulation has a quite large amplitude (see Fig.1). We also find more complicated linearly stable undulated structures with meandering regions (Fig.1).

We have also studied models which correspond to Eq.(24) with additional anisotropic nonlinear terms that become important for larger ε. One can then find situations where the undulations remain linearly stable down to quite small amplitudes.

For $c\Omega < 0$, Eq.(24) admits stationary domain-wall solutions which separate oblique rolls with $P = P_+ > 0$ from those with $P = P_- < 0$. In anology with the

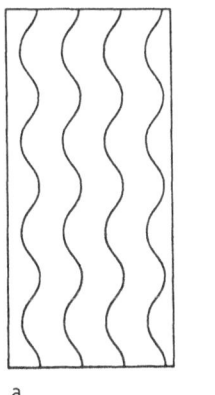 

a   b

Fig.1:
Roll patterns for $u^{(0)}$ (see Eq.12)) calculated from Eq.(24) with (26), b = 0.45, $c\Omega = -0.65$ and two different periodicities:
a) $q = 0.98/\xi_1$, $p = 0.3/\xi_3$
b) $q = 0.96/\xi_1$, $p = 0.1/\xi_3$.

The patterns correspond to the separated stability regimes as discussed in Fig.3 of Ref.5.

stationary point defects (dislocations) discussed before, which exist only at $Q = Q_c$, $P = P_c$, these line defects exist only when a special relation between Q_+, P_+ and Q_-, P_- is fullfilled. To deduce this relation it is useful to investigate the conservation law following from the gauge-invariance of Eq.(24)

$$T_0 (A^* \partial_T A - A \partial_T A^*) = \vec{\nabla}. \vec{j}, \tag{34}$$

$$j_x = 2i\xi_1 \; \text{Im} \; \{A^* (\xi_1 \partial_X - i\xi_3^2 \partial_Y^2) A\}, \tag{35}$$

$$j_y = 2i\xi_3 \; \text{Im} \; \{c\Omega\xi_3 A^* \partial_Y A + i\xi_3 [(\partial_Y A^*) - A^* \partial_Y]$$
$$\cdot [\xi_1 \partial_X - i(1 + b)\xi_3^2 \partial_Y^2] A\}. \tag{36}$$

Consider, for example, a domain-wall in the X-direction. Then we expect $Q_+ = Q_-$, since otherwise the phase is not coherent at the wall. Integrating Eq.(34) over a rectangle and using Gauss' theorem leads to $j_y^+ = j_y^-$, or, evaluating j_y far away from the wall,

$$\partial_P (F^4) \Big|_+ = \partial_P (F^4) \Big|_-, \tag{37}$$

where F is given in Eq.(25). This leads to a relation between P_+ and P_-. There is one solution with $P_+ = -P_- = P_m$, belonging to the maximum of F^4. In the other solutions P_+ and P_- are changed in the same direction, which corresponds approximately to a rigid rotation of the structure. For $Q_+ \neq Q_-$ we expect domain walls which are not parallel to the X-direction.

5.Discussion

Several pattern-forming systems with the axial anisotropy considered in this paper have been or are being studied experimentally and theoretically. A classic example is Rayleigh-Bénard convection in a conducting fluid in the presence of a magnetic field. Here convection sets in with the rolls along the magnetic field so that one has normal rolls /20,21/. This system appears well suited to test the predictions about the stability of (slightly) oblique rolls above threshold presented in Sec.3. One merely has to rotate the convection cell or the magnetic field by the desired angle after normal rolls have formed. Oblique rolls can be obtained spontaneously by adding a steady rotation of the layer /22/. This, however, destroys reflection symmetry. Similarly, Taylor vortex flow represents in the small-gap limit an anisotropic quasi-two-dimensional system (preferred axis = cylinder axis) without reflection symmetry. For counter-rotating cylinders one obtains spiral vortex structures, the analog of oblique rolls /23,24/, which now move in the rest frame.

Alternatively one has the pattern-forming instabilities in fluids with an intrinsic anisotropy like ferrofluids or liquid crystals. We will only discuss nematic liquid crystals. The intrinsic anisotropy, here represented by the director $\vec{n}$, has to couple to an external axial field acting in the plane of the layer.

A well-studied example is the hydrodynamic instability of homeotropically oriented nematics subjected to a time-periodic elliptic shear /25,26/. Here the force that drives the instability also provides the external anisotropy. At threshold, one apparently has always oblique rolls whose angle connot be varied continuously. The structure and motion of dislocations were studied fairly extensively /27/ and some pattern-selection properties were analysed theoretically /28/.

The most interesting systems are the convective instabilities in nematic liquid crystals with planar orientation of the director. Here the external field is provided by the alignment of the director at the upper and lower plates. For the electrically driven case (EHC), the normal rolls, which are aligned perpendicularly to the director, are known for a long time /29,30/. The possibility of oblique rolls at threshold and the Lifshitz-point-like transition from normal to oblique rolls were demonstrated clearly only recently /31/ and explained subsequently /32,33/. In this system the frequency of the applied voltage or a magnetic field serve as a secondary control parameter. The calculations have been performed for simplifying stress-free boundary conditions at the top and bottom plates /32/ and for realistic rigid boundary conditions /33/. Above threshold also undulated rolls, meandering roll patterns and, for higher voltages, more complicated structures and finally transition to turbulence are observed /31/. As yet there is no quantitative agreement between theory and experiment on the behavior near the Lifshitz point. In the oblique roll regime one almost always observes domain walls that separate regions with the two degenerate structures ("Zig-Zag") /31/. The structure and motion of dislocations were investigated /31,34/.

In thermally driven convection of planar nematics one also appears to find transitions from normal to oblique rolls /35/. Here a magnetic field applied in the direction of the unperturbed director serves as a secondary control parameter.

Study of the wavevector selection may also be possible on the electrically or magnetically driven static periodic structures in planar nematics with wavevector at threshold parallel to the undistorted director. They replace the ordinary homogeneous Freedericksz transition when either the twist elastic constant is sufficiently small /36,33/ or when the flexoelectric · effect is sufficiently large /37/. The flexoelectric effect presumably also plays an important role for the electrohydrodynamic instability, at least for very low and very high frequencies (dielectric regime) /38/.

This work was supported by the Deutsche Forschungsgemeinschaft (Sonderforschungsbereich 213, Bayreuth). One of us (L.K.) wishes to thank R. Ribotta from the Laboratoire de Physique des Solides in Orsay (France) for his hospitality and financial support from Université de Paris-Sud.

References

1. F. H. Busse: Rep. Prog. Phys. <u>41</u>, 28 (1978); and in <u>Hydrodynamic Instabilities and the Transition to Turbulence</u>, 2nd, ed., H. L. Swinney and J. P. Gollub, eds. (Springer, Berlin 1986).

2. M. C. Cross and A. C. Newell: Physica <u>10D</u>, 299 (1984).

3. M. S. Heutmaker, P. N. Fraenkel and J. P. Gollub: Phys. Rev. Lett. <u>54</u>, 1369 (1985); G. Ahlers, D. S. Cannell, and V. Steinberg: Phys. Rev. Lett. <u>54</u>, 1373 (1985); P. Le Gal, A. Pocheau and V. Croquette: Phys. Rev. Lett. <u>54</u>, 2501 (1985) and <u>55</u>, 1094 (1985).

4. V. L. Ginzburg and L. Pitaevskii: Zh. Exp. Teor. Fiz. 34, 1240 (1958) [English transl.: Sov. Phys. JETP 7, 858 (1958)].

5. W. Pesch and L. Kramer: Z. Phys. B63, 121 (1986).

6. W. Eckhaus: Studies in Nonlinear Stability Theory (Springer, Berlin ,1965) p.63ff.

7. A. C. Newell and J. A. Whitehead: J. Fluid Mech. 38, 279 (1969).

8. L. Kramer and W. Zimmermann: Physica 16D, 221 (1985).

9. The full Eckhaus-evolution for the one-dimensional case was studied recently by H. Schober, W. Zimmermann and L. Kramer: to be published; experiments were performed by M. Lowe and J. P. Gollub: Phys. Rev. Lett. 55, 2575 (1985).

10. E. Bodenschatz and L. Kramer: Physica D, in press.

11. W. F. Vinen: in Quantum Fluids, D. F. Brewer, ed. (North-Holland, Amsterdam, 1966) p.74.

12. E. D. Siggia and A. Zippelius: Phys. Rev. A24. 1036 (1981).

13. Y. Pomeau, S. Zaleski and P. Manneville: Phys. Rev. A27, 2710 (1983).

14. Y. Pomeau: in Cellular Structures in Instabilities, J. E. Wesfreid and S. Zaleski, eds. (Springer, Berlin , 1984) p.207; H. Riecke: Europhys. Lett. 2, 1 (1986).

15. M. C. Cross, P. G. Daniels, P. C. Hohenberg and E. D. Siggia: J. Fluid. Mech. 127, 155 (1983); Y. Pomeau and S. Zaleski: J. Physique 42, 515 (1981); L. Kramer and P. C. Hohenberg: Physica 13D, 357 (1984); P. C. Hohenberg, L. Kramer and H. Riecke: Physica 15D, 402 (1985).

16. R. M. Hornreich, M. Luban and S. Shtrikman: Phys. Rev. Lett 35, 1678 (1975); Physica 86A, 465 (1977).

17. L. A. Segel: J. Fluid Mech. 38, 203 (1969).

18. E. D. Belotskii and P. M. Tomchuk: Zh. Eksp. Teor. Fiz. 88, 1634 (1985) [English transl.: Sov. Phys.-JETP 61, 974 (1985)].

19. S. Chang-Qing and Lin Lei: 11th Int. Liquid Crystal Conference, Berkeley, 1986 (to be published in Mol. Cryst. and Liq. Cryst.)

20. S. Chandrasekhar: Hydrodynamic and Hydromagnetic Stability (Clarendon Press, Oxford, 1961) p.186.

21. B. Lehnert and N. C. Little: Tellus 9, 97 (1957). For recent experiments on the influence of a horizontal magnetic field on Rayleigh-Bénard convection see S. Fauve, C. Laroche and A. Libchaber: J. Physique Lett. 42, L-455 (1981); 45, L-101 (1984).

22. I. A. Eltayeb: Proc. R. Soc. Lond. A326, 229 (1972).

23. E. R. Krüger, A. Gross and R. C. Di Prima: J. Fluid Mech. 24, 521 (1966); H. A. Snyder: Int. J. Non-Linear Mechanics 5, 659 (1970).

24. C. D. Andereck, S. S. Liu and H. L. Swinney: J. Fluid Mech. <u>164</u>, 155 (1986).

25. P. Pieranski and E. Guyon: Phys. Rev. Lett. <u>39</u>, 1281 (1977); E. Guazzelli and E. Guyon: J. Physique <u>43</u>, 985 (1982); E. Guazzelli and A.-J.Koch: J. Physique <u>46</u>, 673 (1985).

26. E. Dubois-Violette and F. Rothen: J. Physique <u>10</u>, 1039 (1978); J. Sadik, F. Rothen, W. Bestgen and E. Dubois-Violette: J. Physique <u>42</u>, 915 (1981); A.-J. Koch, F. Rothen, J. Sadik and O. Schöri: J. Physique, <u>46</u>, 699 (1985).

27. E. Guazzelli, E. Guyon and J. E. Wesfreid: Phil. Mag. <u>A48</u>, 709 (1983); E. Dubois-Violette, E. Guazzelli and J. Prost: Phil. Mag. <u>A48</u>, 727 (1983).

28. E. Guazzelli, G. Dewel, P. Borckmann and D. Walgraef: to be published.

29. R. Williams: J. Chem.Phys, <u>39</u>, 384 (1963); A. P. Kapustin and L. S. Larinova: Kristallografia <u>9</u>, 297 (1964); E. F. Carr: Mol. Cryst. Liq. Cryst. <u>7</u>, 253 (1969).

30. W. Helfrich: J. Chem. Phys. <u>51</u>, 4092 (1969); E. Dubois-Violette, P. G. de Gennes and O. Parodi: J. Physique <u>32</u>, 305 (1971); S. A. Pikin: Zh. Eksp. Teor. Fiz. <u>60</u>, 1185 (1971) [English transl.: Sov. Phys.-JETP <u>33</u>, 641 (1971)]; P. A. Penz and G. W. Ford: Phys. Rev. <u>A6</u>, 414 (1972).

31. A. Joets and R. Ribotta: in Cellular Structures in Instabilities, J. E. Wesfreid and S. Zaleski, eds. (Springer, Berlin 1984) p.294; J. Physique <u>47</u>, 595 (1986); R. Ribotta, A. Joets and Lin Lei: Phys. Rev. Lett. <u>56</u>, 1595 (1986).

32. W. Zimmermann and L. Kramer: Phys. Rev. Lett. <u>55</u>, 402 (1985).

33. E. Bodenschatz, W. Zimmermann and L. Kramer: to be published.

34. R. Ribotta and A. Joets: in Cellular Structures in Instabilities, J. E. Wesfreid and S. Zaleski, eds. (Springer, Berlin, 1984) p.249.

35. A. Joets, L. Kramer, R. Ribotta, J. Salan and W. Zimmermann: to be published.

36. F. Lonberg and R. B. Meyer: Phys. Rev. Lett. <u>55</u>, 718 (1985); E. Miraldi, C. Oldano and A. Strigazzi: Phys, Rev. <u>A34</u>, 4348 (1986); W. Zimmermann and L. Kramer: Phys. Rev. Lett. <u>56</u>, 2655 (1986); U. D. Kini: J. Physique <u>47</u>, 693 (1986); <u>47,</u> 1829 (1986).

37. Yu. P. Bobylev and S. A. Pikin: Zh. Eksp. Teor. Fiz. <u>72</u>, 369 (1977) [English transl.: Sov. Phys.-JETP <u>45</u>, 195 (1977)]; Y. P. Bobylev, V. G. Chigrinov and S. A. Pikin: J. Physique Coll. <u>40</u>, C3-331 (1979);M. I. Barnik, L. M. Blinov, A. N. Trufanov and B. A. Umanskii: Zh. Eksp. Teor. Fiz <u>73</u>, 1936 (1977) [English transl.: Sov. Phys.-JETP <u>46</u>, 1016 (1977)]; J. Physique <u>39</u>, 417 (1978).

38. N. V. Madhusudana, V. A. Raghunathan and K. R. Sumathy: to be published.

RAYLEIGH NUMBER RAMPS CAUSE MOVING CONVECTION PATTERNS

I. Rehberg and H. Riecke

Physikalisches Institut, Universität Bayreuth,
Postfach 101251, D-8580 Bayreuth, Fed. Rep. of Germany

Summary

In this paper the induction of moving (drifting) patterns by wavelength selection is investigated. In the first part the theory for selection by ramps (i.e. control parameters which vary slowly in space) and the resulting motion is reviewed. Particular emphasis is put on qualitative features which should appear in most pattern forming systems if their dynamics is not derived from a minimizing potential. For experimental relevance of this theory the influence of imperfections (pinning centers) has to be taken into account. In the second part the experimental investigation of moving convection patterns in a convection cell with ramps is presented. In agreement with the theoretical predictions it is found that
— Drifting patterns are observable,
— Due to the pinning centers drifting starts at a value R_m of the control parameter well above the onset of convection R_c,
— The drift velocity approaches zero at R_m.

Introduction

In pattern–forming systems like Rayleigh–Benard convection and Taylor–Couette flow the wavelength of the patterns is not unique, the patterns are stable within a certain band. The related question of a "preferred" wavelength cannot be answered in the usual experimental set–ups as the distance between the two ends fixes the wavelength. In order to get rid of these ends one is led to the idea of a subcritical ramp /1/: The driving force is decreased slowly in space from supercritical to subcritical values, as indicated by the figure in the title. As was shown theoretically such ramps lead to a perfect selection of a single wavelength /1/. Experiments in Taylor–Couette flow /2,3/ with conical cylinders clearly confirm this vanishing of the wavelength band: A single wavelength is selected in the limit of an infinitely long ramp. It has been pointed out theoretically, however, that this wavelength is not unique /1,4/. In Rayleigh–Benard convection it depends on the combination of the geometrical (varying layer thickness) and thermal (varying temperature difference) part of the ramp. If so, different combinations at the left and right end of a convection cell could lead to different wavelengths. Such a situation is not stable, the patterns move from the small wavelength end to the long wavelength end driven by a phase diffusion process. This paper presents the experimental investigation of this phenomenon.

1. THEORY

1.1 Smooth Ramps in Rayleigh-Bénard Convection

Two-dimensional convection can be described by

$$R_0 \, \partial_x T - \nabla^4 \psi = 0, \qquad \left[\partial_t + (\partial_z \psi \, \partial_x - \partial_x \psi \, \partial_z) \right] T - \nabla^2 T = 0. \tag{1}$$

Here the Overbeck-Boussinesq-approximation and an infinite Prandtl number have been assumed and all quantities have been made dimensionless by using typical scales. The fluid is assumed to lie in the x-y-plane and the velocity $\underline{v} = (v_x, v_y)$ is given by the streamfunction ψ via $v_x = \partial_z \psi$ and $v_z = -\partial_x \psi$. The temperature is denoted by T and the Rayleigh number by R_0.

As indicated in the title we wish to analyze the convection pattern that arises for boundary conditions which depend smoothly and slowly on the x-coordinate. Therefore a slow spatial scale $X = \alpha x$ with $\alpha \ll 1$ is introduced. For simplicity the fluid is now assumed to be contained between an upper flat plate (z=0) and a lower plate at z=-G(X). Similarly, the temperature is taken to be constant (T=0) at z=0 and X-dependent at z=-G(X) (T=H(X)G(X)). Thus H(X) is the local temperature gradient in the non-convective state. To transform the region contained between the plates to a rectangle, new coordinates are introduced which are orthogonal to order $O(\alpha)$,

$$\zeta = z/G, \qquad \xi = x + \alpha \, (\zeta^2 - 1) \, G \, G'/2. \tag{2}$$

Here $G' = \partial_X G$. The boundary conditions now read

$$T = 0 \qquad \text{at } \zeta = 0, \qquad T = H(X) \, G(X) \quad \text{at } \zeta = -1,$$

$$\psi = \partial_\zeta^2 \psi = 0 \quad \text{at } \zeta = -1, \, 0. \tag{3}$$

For simplicity stress-free boundary conditions for ψ have been assumed. Rigid boundary conditions are expected to give qualitatively the same results. Note that with the coordinates (2) the usual free boundary conditions are in fact given by (3), which was not the case for the coordinates used in /4/.

How do we now obtain moving patterns? As mentioned above they result if in the *same* region two *different* wave numbers (or *disjoint* wave-number bands) are selected by suitable selection mechanisms. Therefore we briefly review wave-number selection by smooth ramps /1,4/.

Introducing θ via $T = HG(\theta - \zeta)$, θ and ψ are expanded in α,

$$\psi(\eta, X, \zeta, \tau) = \psi_0 + \alpha \, \psi_1 + ..., \qquad \theta(\eta, X, \zeta, \tau) = \theta_0 + \alpha \, \theta_1 + ... \tag{4}$$

Both, ψ and θ are 2π-periodic in the phase η. In addition to the slow time τ a slow phase ϕ is introduced

$$\tau = \alpha^2 t, \qquad \eta = \alpha^{-1} \phi(X, \tau)/G. \tag{5}$$

Using $\partial_\xi = (q/G)\partial_\eta + \alpha \partial_X$ (1) is regained in $O(\alpha^0)$ with the local Rayleigh number

143

$R(X)=R_0HG^4$ replacing R_0. Thus, to this order any local wave number $q(X)$ is possible as long as it stays within the band of stable solutions. Due to the singularity of the linearized operator $\mathcal{L}$ of (1) a solvability condition arises at order $O(\alpha)$, which is obtained by projecting the equations in $O(\alpha)$ onto the adjoint translational null mode (V_1^+,V_2^+) of $\mathcal{L}$ /4/,

$$\tau_0 \, \partial_\tau \phi \;=\; D \, \partial_X q \;+\; (B + E) \, \partial_X R/R \;+\; (C - 4B) \, \partial_X G/G. \tag{6}$$

Because the coefficients depend on $q=G\partial_\zeta \eta=G\partial_X(\phi/G)$, (6) is a nonlinear diffusion equation for the phase ϕ, describing the effect of the ramped boundary conditions on the statics and slow dynamics of the periodic pattern. The coefficients are given by /5/

$$\tau_0 = \langle V_2^+ \partial_\eta \theta_0 \rangle,$$

$$D = - \langle V_1^+ \left[R\partial_q \theta_0 - 4q(\partial_\zeta^2 + q^2\partial_\eta^2)\partial_\eta \partial_q \psi_0 - 2(\partial_\zeta^2 + 3q^2\partial_\eta^2)\partial_\eta \psi_0 \right]$$
$$V_2^+ \left[\partial_\zeta \psi_0 \partial_q \theta_0 - (\partial_\zeta \theta_0 - 1)\partial_q \psi_0 - (1 + 2q\partial_q)\partial_\eta \theta_0 \right] \rangle,$$

$$B = - \langle V_1^+ R(\theta_0 - \zeta) + V_2^+ \left[(\theta_0 - \zeta)\partial_\zeta \psi_0 - 2q\partial_\eta \theta_0 \right] \rangle,$$

$$C = - \langle V_1^+ \left[R(1 - \zeta\partial_\zeta)\theta_0 + 4q(\partial_\zeta^2 + q^2\partial_\eta^2)\partial_\eta \psi_0 \right] + V_2^+ \left[(\theta_0 - \zeta)\partial_\zeta \psi_0 - 2q\partial_\eta \theta_0 \right] \rangle$$

$$E = - \langle V_1^+ \left[R \partial_R^2\theta_0 - 4q(\partial_\zeta^2 + q^2\partial_\eta^2)R\partial_R\partial_\eta \psi_0 \right] +$$
$$V_2^+ \left[(1 - \partial_\zeta \theta_0)R\partial_R\psi_0 + \partial_\zeta \psi_0 R\partial_R\theta_0 - 2qR\partial_R\partial_\eta \theta_0 \right] \rangle, \tag{7}$$

where $\langle ... \rangle$ denotes the average $\int_0^{2\pi}\int_{-1}^0 ...d\zeta\,d\eta$. The expressions for B and C differ from those given in /4/ due to the different coordinate system. In addition, the fact has been used that the basic flow need not be calculated explicitly to order $O(\alpha)$.

The phase diffusion equation (6) shows that for ramps which become subcritical in some region the wave-number band collapses to a single wave number /1,4/. Consider now a system with a homogeneous part of length L which is connected to two different subcritical ramps at both ends. If these ramps select different wave numbers q_1 and q_2, say, then a wave-number gradient is established in the homogeneous region, which leads to phase diffusion (see (6) with R and G constant). This, however, never comes to an end, as the wave numbers at both ends are fixed by the ramps, and a moving pattern with velocity

$$v = D \, \partial_X q/q\tau_0 \tag{8}$$

is obtained. The only requirement for a system to support this dynamics is the existence of different ramps which select *different* wave numbers. For

Rayleigh–Benard convection this amounts to the condition that neither of the coefficients B+E and C–4B in (6) vanish identically. Close to threshold this requirement is in fact fulfilled as a calculation using the amplitude approximation shows /4,5/ (far from threshold and for rigid boundary conditions the coefficients have to be computed numerically, which has been done for pure thermal ramps only /6/). This can be seen in detail for the family of ramps given by /4/

$$\delta\ \partial_X G/G + (1 - \delta)\ \partial_X H/H = 0, \quad \delta = const. \tag{9}$$

The static phase diffusion equation can then be written as

$$(\varepsilon - 3p^2)\ \partial_\varepsilon p = 2\beta\ (\varepsilon - p^2) - p + 7p^2/\sqrt{48},$$

$$2\beta = (4b - c)\ \frac{1 - \delta}{4 - 5\delta} - b - 7/\sqrt{48}. \tag{10}$$

with

$$\varepsilon = (R - R_c)/R_c, \ R_c = 27\pi^4/4, \ p = \sqrt{8/3\pi^2}\ (q - q_c), \ q_c = \pi/\sqrt{2}.$$

$$B = - b\ (\varepsilon - p^2), \ b = 0.60, \quad C = - c\ (\varepsilon - p^2), \ c = 1.54, \tag{11}$$

The solutions to this equation with $p(\varepsilon=0)=0$ are given in fig. 1 for various values of β. It clearly shows that the selected wave number is *non–universal*, i.e. it depends on the ramp chosen. In fact, the slope of the selected wave number is given by $\partial_\varepsilon p=1/\beta$, which can be varied from $-\infty$ to ∞ by a suitable choice of δ.

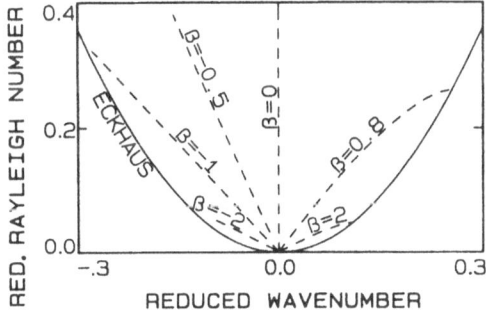

Fig. 1
Selected wave numbers
for different ramps
as given by (10)

Thus, according to this analysis moving patterns are possible in RB–convection and the threshold R_m for the motion coincides with the threshold R_c for the appearance of the pattern itself. The wave–number gradient $\partial_X q \approx (q_2 - q_1)/\alpha L$ and therefore also the velocity v grows linearly when increasing the Rayleigh number.

1.2 Imperfect Wave–number Selection by Pinning Ramps

How much of the analysis in sec.1 carries over to a real experiment where the basic requirement of this analysis, the perfectly smooth variation of the control parameters, is not really met? To clarify this it is useful to look at the influence of the ramp on the basic state. This state is purely conductive for constant boundary conditions. For $\partial_X H \neq 0$, however, the ramp induces a large-scale convection with

$$v_X^B(\zeta,X) = - \alpha R \partial_X H (5\zeta^4 - 10\zeta^2 + 7/3)/120H + O(\alpha^2) \tag{12}$$

$$v_z^B(\zeta,X) = \alpha^2 \partial_X (R G \partial_X H/H) (\zeta^5 - 10\zeta^3/3 + 7\zeta/3)/120. \tag{13}$$

In particular, if H has a localized region with large curvature $\partial_X^2 H$, the vertical velocity v_z contains a peak-like term, which can lead to a pinning of the periodic convective pattern. In this section the influence of such *pinning centers* on wave-number selection is reviewed. It has been shown that in the usual set-ups they are the predominant cause for deviations from the perfect selection displayed in fig. 2 and lead to finite band widths Δq /7,5/. Because drifting of the pattern sets in only if the wave-number bands of different ramps do not overlap, its onset is expected to lie above the convection threshold $(R_m > R_c)$.

For simplicity the analysis is demonstrated using an extension of the Swift-Hohenberg model /8/,

$$\partial_t \psi = \left[R - (\partial_X^2 + 1)^2 \right] \psi - \psi^3 + G (1 - R + G^2), \tag{14}$$

where the additional term $G(1-R-G^2)$ has been introduced to obtain a non-trivial basic state

$$\psi^B = G + \alpha^2 \frac{2q^2 \partial_X^2 G}{R - 1 - 3G^2} \tag{15}$$

similar to that in RB-convection (13). Again, the control parameters R and G are assumed to vary slowly and smoothly in space in two regions I and III. However, in region II, which separates I and III, they are allowed to vary more rapidly. As non-adiabatic effects are responsible for the wave-number bands a separate slow scale X *must not* be introduced. Instead, for R we require

$$\partial_X^n R = O(\alpha^n) \text{ in I and III}, \quad \partial_X^n R = O(\alpha) \text{ in II}, \quad \alpha \ll 1, \tag{16}$$

and analogously for G. This is, for instance, the case if two smooth ramps are joined in II with different slopes. In /7,5/ it is shown that the rapid variation in II leads to a matching condition for the wave numbers in the two phase diffusion regions I and III. Similarly to (6) it is obtained by projecting (14) in order $O(\alpha)$ with $\psi = \psi_0 + \alpha\psi_1 + ..$ onto the null mode. However, as ψ_1 is not 2π-periodic in the phase η in II, boundary terms appear, which are removed by a second integration over all of region II $(\eta_1 < \eta < \eta_2)$. The resulting matching condition reads

$$D\left[q_2 - q_1 + \pi(\partial_\eta q_2 - \partial_\eta q_1) \right] = \frac{1}{q} \int_{\eta_1}^{\eta_2} d\eta \int_{\eta}^{\eta+2\pi} d\eta' \left[\rho(\eta'+\eta_0)\partial_\eta R + \gamma(\eta'+\eta_0)\partial_\eta G \right] \tag{17}$$

where $q_1 = q(\eta_1)$ etc. and D, ρ and γ are functionals of ψ_0 similar to A,B,C and E in (6) /7/. The crucial point of (17) is the dependence of the r.h.s on the phase η_0 of the solution ψ_0 relative to the ramp. This is illustrated in fig. 2 for a ramp of the form

$$R = R_1 - \Theta(-x)\, \Delta R' x, \qquad G = G_1 - \Theta(-x)\, \Delta G' x, \qquad (18)$$

with $\Theta(x)$ being the step function ($R_1 = 0.5$, $G_1 = 0.1$, $\Delta R' = -0.005$, $\Delta G' = 0.0005$). This ramp becomes smoothly subcritical for $x \to -\infty$ and therefore yields perfect selection in region I ($q \approx q_c = 1$). The wave number q in III is given by the matching condition (17) and therefore depends on η_0, the phase of ψ_0 at $x=0$. The function $q(\eta_0)$ is 2π-periodic. Due to the harmonic content of ψ_0 also higher harmonics are apparent (compare also with the experimental results for Taylor vortex flow, fig. 10 in /3/). Thus, although the wave number in I is perfectly selected, a wave-number band Δq arises in III through the η_0-dependent influence of the pinning center.

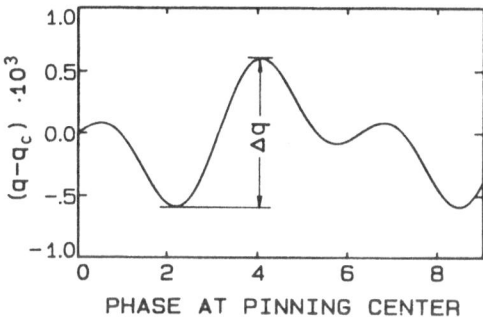

Fig. 2
Wave number q as a function of the phase η_0 of ψ_0 at $x=0$.

q is 2π-periodic.
Parameters:
$R_1 = 0.5$ $G_1 = 0.1$
$\Delta R' = -0.005$ $\Delta G' = 0.0005$
(see (18))

A characteristic feature of wave-number bands produced by a pinning basic state is their dependence on the control parameter: close to threshold they *grow* with *decreasing* control parameter. This is shown in fig. 3, where the matching condition is compared with a numerical simulation of (14) ($\Delta R' = -0.001, -0.005$, $\Delta G' = 0.0005$, $G_1 = 0.1$). This ("anomalous") behavior of the band width is also observed in Taylor vortex flow /2,3/. In these experiments a characteristic cross-over from the anomalous to the normal band is observed, which is also reproduced by the matching condition (17) when applied to a further extension of the model (14) /9/. Figure 3 shows very good agreement for larger $R - R_c$. For too small values of $R - R_c$ the validity of (17) breaks down due to the divergence of the correlation length at threshold.

To obtain moving patterns the wave-number bands of different ramps may not overlap. Therefore the pinning basic state will lead to a shift in the onset of motion to larger values of R as illustrated in fig. 5 below. As before, however, the velocity should still grow linearly from 0 when increasing R beyond onset. Experimentally, one is interested in minimizing the pinning in order to maximize the velocity. This is done by smoothing out the transition from the ramp to the homogeneous region. The matching condition shows that this very efficiently reduces the band width. Figure 4 gives the result for rounded ramps of the form

$$R = \tfrac{1}{2}R'\left[x - d\,\ell n(2\,\cosh\tfrac{x}{d})\right] + R_1, \qquad \text{i.e.} \quad \partial_x R = \tfrac{1}{2}R'(1 - \tanh\tfrac{x}{d}), \tag{19}$$

with $G=0$, $R_1=1$, $R'=0.1$. These ramps lead in amplitude approximation to

$$\Delta q = \left|\tfrac{1}{4}R'\right|\,\frac{\pi d\,\sinh(\pi d)}{\cosh(2\pi d)-1}\,. \tag{20}$$

Thus, already a rounding over a quarter of a wavelength ($d=1.5$) reduces the band width by a factor of 6. A similar result was obtained in a numerical simulation of Taylor-Couette flow /10/.

Figure 5 summarizes the results of the theoretical analysis qualitatively. In RB-convection close to threshold any wave number in the stable band can be selected by a suitable ramp. For realistic ramps with a finite length and with pinning centers each selection curve is expanded to an anomalous band, which

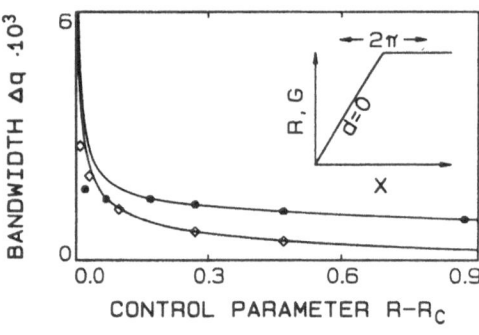

Fig. 3
Wave-number band for ramp with discontinuous slope.

The lines represent (17). The symbols are obtained by a numerical simulation of the model (14).
Parameters:
$\Delta G'=0.0005$, $G_1=0.1$

solid circles: $\Delta R'=-0.005$
open diamonds: $\Delta R'=-0.001$

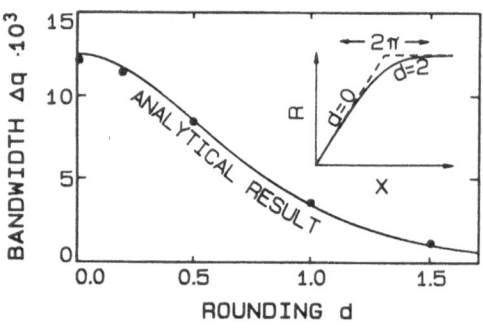

Fig. 4
Wave-number band for ramp with continuous slope

The solid circles are obtained by numerical integration of model (14). The line represents (20).
Parameters:
$G=0$, $R_1=1$, $R'=0.1$

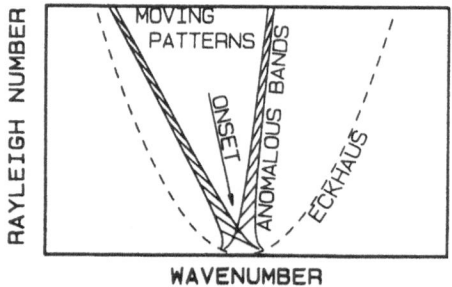

Fig. 5
Schematic explanation for the existence of moving patterns. Dashed regions indicate the bands selected by the two different ramps

widens close to the convection threshold R_c. Therefore the bands will overlap there. The patterns will start to move above a threshold $R_m > R_c$ where this overlap ceases to exist. The exact position of this critical point thus depends sensitively on the smoothness of the ramps. The velocity v increases linearly with R close to R_m.

2. EXPERIMENTAL RESULTS

Governed by the theoretical arguments summarized above, a convection cell was designed in order to observe the predicted drifting convection patterns. The shape of the convection box is cut out of a copper plate of 1.5mm thickness. The height in the middle of the cell is 3mm over a length of 18mm, thus allowing at least 3 pairs of convection rolls within this bulk part. The subcritical ramps to the left and right of this part are obtained by cutting a parabolic curve out of the copper as indicated in fig. 6. The height of the cell in the ramps is decreased from 3mm to 1mm over a distance of 26mm. The cell is heated from below by means of constant electrical power. The top of the cell is attached to an aluminium block regulated to a constant temperature (20°C) by means of a water circuit. Cyclohexan was used as the working fluid. Note that this design leads to a combination of a geometrical and a thermal ramp because the copper at both sides forms a thermal short-circuit. This thermal ramp can be influenced on the right hand side by an additional heater.

The method used for the flow visualisation is the shadowgraph technique: Density gradients in the fluid caused by temperature gradients lead to light intensity modulations. This light intensity is measured with a photodiode which can be moved parallel to the convection cell under computer control by a stepping motor.

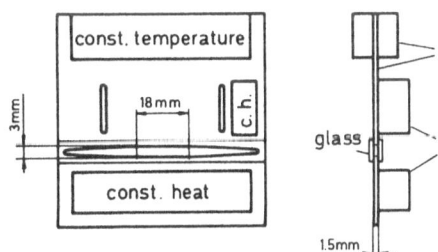

Fig. 6
Convection cell
(not to scale)

The height of the convection cell is 3mm, the length of its bulk part 18mm. Note the additional heater on the right hand side (c.h.).

Figure 7 shows scans of the light intensity at various temperature differences measured in the bulk part of the cell (position 0 corresponds to the middle of the cell) The scan at 2.5K temperature difference shows no modulation, this difference is clearly below the convection threshold (modulations of the light intensity caused by spatial inhomogenities of the light source are taken away by image division). The measurement at 9.9K is clearly above the convection threshold — intensity peaks are visible. These are caused by the colder (sinking) parts of the fluid. Their position is not constant, but a function of the temperature difference. The peaks move to the right when increasing this difference, but they tend to come back to their original position when decreasing this difference, as indicated by the dashed lines. The difference between the solid and dashed curves is expected to go to zero when increasing the waiting time between successive temperature steps.

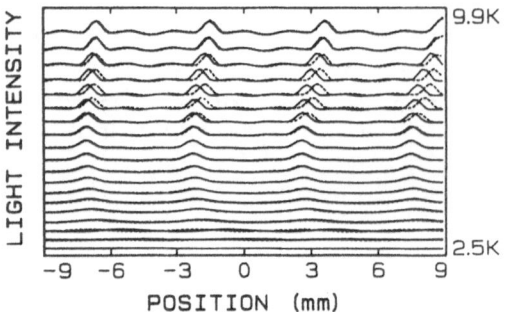

Fig. 7
Intensity profiles for
various temperature
differences in the range
between 2.5K and 9.9K.

Intensity peaks (caustics)
correspond to the location of
cold sinking fluid.

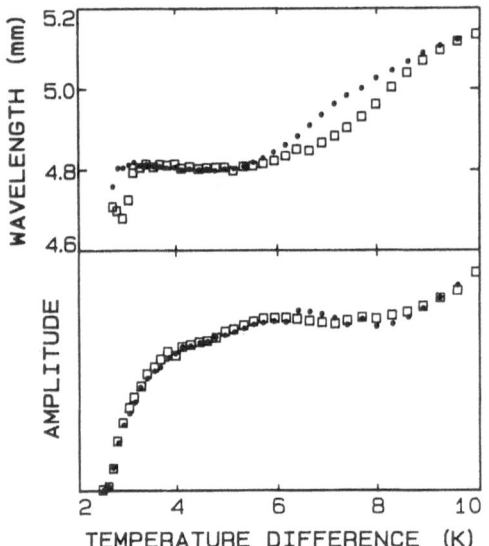

Fig. 8
Amplitude of the light
intensity modulation and
wavelength of the
convection patterns as a
function of the
temperature difference

By spectral analysis the amplitude of the light intensity modulation and the wavelength of the convection pattern can be extracted from the profiles shown in the above figure. The amplitude is expected to be proportional to the temperature modulations in the fluid in the limit of small modulations, i.e. close to the convection threshold. The lower part of fig. 8 shows the result. Because the temperature modulation is known to grow according to a square-root law, the same would be expected for the intensity modulation in the neighborhood of the threshold. Measurements like this allow a determination of the critical temperature difference for the onset of convection (2.63K). The upper part of fig. 8 shows the wavelength of the convection patterns. Here a difference between data points obtained when increasing the temperature (open squares) and decreasing the temperature (solid circles) is observed. This indicates that the waiting time between consecutive data points (30 minutes) is long enough for the amplitude of the convection to reach its equilibrium value, but too short for the phase diffusion process to adjust the length and position of the convection pattern. Like in the previous figure, the difference between the two curves is expected to go to zero for a longer waiting time. The important feature of figure 8 is the smooth variation of the wavelength. This is made possible by the ramps and unlike the discontinuous jumps of the wavelength observed in convection boxes with sharp corners.

The measurements shown in figs. 7,8 are done without the side wall heating. No moving patterns are observed in that case, except for the transients when reaching the equilibrium position. If the symmetry of the two ramps is broken by means of the additional heater, drifting convection patterns become possible. This is demonstrated in fig. 9. The heater is located at the left end (negative positions). Thus the temperature difference is already subcritical at a position of −15mm, while on the other end the critical position seems to be close to +30mm. The profiles are taken within time intervals of 15 minutes and plotted on top of each other. The movement of the patterns from the right to the left is clearly visible.

As indicated by fig. 9 the velocity of the moving patterns is not constant but rather a periodic function of time. One way to measure the mean velocity is to locate a photodiode at a fixed position and to measure the frequency of the time-periodic intensity modulations. These frequencies are plotted in fig. 10 as a function of the reduced temperature difference ε. The onset of motion is located at $\varepsilon=0.4$, i.e. well above the onset of convection. The frequency goes down to small values at this point. More measurements are needed to evaluate the detailed structure of the frequency curve at this point. The data presented are consistent with the statement that the frequency would go to zero monotonically when approaching R_m. Measurements close to this point obviously become time-consuming − the lowest measured frequency corresponds to a period of 20 hours.

Increasing the temperature difference leads to an increase of the travelling speed of the convection patterns. If ε is too large, however, this movement has to come to an end because the whole cell is supercritical then. A discussion of this second transition was not attempted here. It would presumably require an inclusion of the lateral boundary conditions.

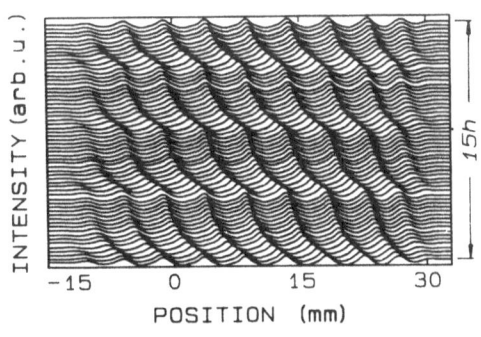

Fig. 9
Light intensity scans
at different times.
The time interval
between two scans is
15 minutes

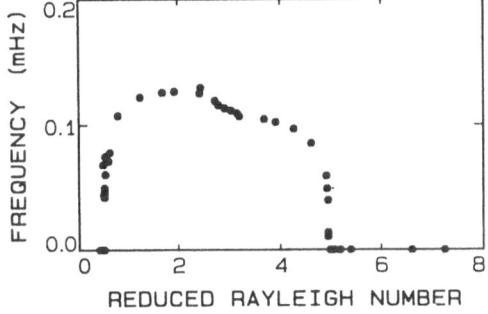

Fig. 10
Frequency of the light
intensity modulations
measured by a stationary
photodiode

The periodic signal is caused
by the drifting convection
patterns. Its frequency is
proportional to the velocity.

In this paper drifting patterns have been discussed using phase diffusion theory for the wavelength selection by ramps. This theory explains the experimental observations, provided that the effect of pinning centers is taken into account.

Acknowledgement

The theoretical work by H.R., which is part of his Ph.D. thesis, has been done under the supervision of L. Kramer. H.R. also thanks H.-G. Paap for his contributions to the choice of coordinates. The experimental work of I.R. was started following suggestions of F.H. Busse and has been done with the help of E. Bodenschatz, B. Winkler and S. Rasenat.

References

1. L. Kramer, E. Ben-Jacob, H. Brand, M.C. Cross:
   Phys. Rev. Lett. 49, 1891 (1982)
2. D.S. Cannell, M.A. Dominguez-Lerma, G. Ahlers:
   Phys. Rev. Lett. 50, 1365 (1983)
3. M.A. Dominguez-Lerma, D.S. Cannell, G. Ahlers:
   Phys. Rev. A34, 4956 (1986)
4. L. Kramer, H. Riecke: Z. Phys. B59, 245 (1985)
5. H. Riecke: Ph.D. thesis, Universität Bayreuth (1986)
6. J.C. Buell, I. Catton: preprint (1985)
7. H. Riecke: Europhys. Lett. 2, 1 (1986)
8. J. Swift, P.C. Hohenberg: Phys. Rev. A15, 319 (1977)
9. H. Riecke: to be published
10. M. Lücke: unpublished

Interactions of Stationary Modes in Systems with Two and Three Spatial Degrees of Freedom

M. Neveling, D. Lang, P. Haug, W. Güttinger, and G. Dangelmayr

Institute for Information Sciences, University of Tübingen,
D-7400 Tübingen, Fed. Rep. of Germany

Abstract. This paper deals with steady state mode-interactions in three-dimensional directional solidification and in two-dimensional thermohaline convection with variable aspect ratios. We show that a solidifying material can exhibit three coupled stationary cellular modes which occur at codimension-three bifurcation points and discuss the associated bifurcation diagrams. The thermohaline convection problem gives rise to a double tricritical point identifiable with a codimension-four bifurcation point. To obtain structurally stable configurations, non-Boussinesq effects have to be taken into account which give rise to temporally oscillating states. Mixed mode patterns and hysteretic behaviour between different unstable modes are modeled and simulated using cellular automata concepts.

1. Introduction

Bifurcation occurs in a nonlinear evolution equation when a variation of a parameter causes qualitative changes in the behaviour of the solutions. The fundamental or primary bifurcations are steady state and Hopf bifurcations where, respectively, stationary and temporally oscillating modes branch off a basic state when a distinguished bifurcation parameter passes through a critical value. The type of bifurcation depends on the eigenvalues of the linearized equation: A zero eigenvalue leads to a stationary solution whereas a pair of imaginary eigenvalues induces temporally oscillating behaviour. Variation of an extra parameter may change the eigenvalue structure and thus may lead to multiple bifurcations which induce mode interactions. Typically, the distinguished bifurcation parameter describes some external force, e.g. the applied stress in buckling problems, the imposed temperature gradient in convection systems or the pulling velocity in directional solidification. In order that interactions between distinguishable modes can take place, the system must be discrete. This is always the case in systems with finite spatial extension or in infinite systems with periodic boundary conditions. In the finitely extended case the extra parameter which induces mode interaction is often given by a characteristic length of the system, e.g., the length of a buckling column [3], the width of the sample in directional solidification [7,11] or aspect ratios in the buckling of plates [16] and in convection systems [14].

In this paper we are concerned with interactions of steady state modes in systems of finite spatial size. A paradigm of such interactions is provided by two-dimensional convection in a rectangular box. Here a countable set of critical values $l_c(k)$ ($k=1,2,\ldots$) for the aspect ratio l exists such that, if $l_c(k-1)<l<l_c(k)$, the first instability caused by increasing the Rayleigh number leads to a stationary convection pattern with k rolls [14]. At a critical value $l_c(k)$ a multiple, codimension-two bifurcation from the conduction solution occurs: The basic state loses stability to two different stationary modes simultaneously and the system encounters steady state mode interactions. One of the main results of [14] is the appearance of stable mixed mode patterns, i.e., a superposition of k and $k+1$ rolls, for sufficiently small Prandtl numbers. The normal form for the bifurcations describing steady state mode interactions is in principle capable of producing also transitions to oscillatory solutions; however, these do not appear in ordinary convection systems. Another example for stationary mode interactions is provided by two-dimensional directional solidification [11,15,17]. Here a planar solidifying interface loses stability in favor of a cellular interface whose wavelength depends on the width of the sample.

153

When the width crosses certain critical values, the wavelength of the unstable mode changes, and near a critical width the system encounters steady state mode interactions similarly to the convective roll transitions induced by a varying aspect ratio in Rayleigh Bénard convection. There is, however, one important difference between the solidifying and the convective system. Namely, the normal form describing interactions between small cellular wavenumbers admits tertiary bifurcations to temporary oscillating interfaces [11] whereas such oscillations are not possible in the conventional Rayleigh Bénard convection system [14]. These oscillating interfaces were found in numerical simulations [19].

Steady state mode interactions are meanwhile well understood in systems with one or two spatial dimensions [1,4,8,16], however, except for systems with spherical geometries [5], little is known about the corresponding three-dimensional problem. In Section 2 a first attempt is made towards an understanding of the variety of possible interactions of steady states in three-dimensional rectangular boxes. We choose the equations of directional solidification as model equations, but the formalism also applies to other physical systems, e.g., to Rayleigh Bénard convection. While in two-dimensional systems at most two modes may typically become unstable simultaneously, the three-dimensional problem admits an arbitrary finite number of unstable modes. This endows the problem with a very high complexity and therefore offers a rich variety of possible bifurcation behaviour. Among the various phenomena we consider here interactions between three different stationary modes and describe their behaviour in terms of bifurcation diagrams.

In section 4 the Boussinesq equations for ordinary Rayleigh Bénard convection are discretized and simulated as a cellular automaton. Starting from arbitrarily chosen initial configurations, the automaton evolves for Rayleigh numbers slightly above criticality towards a k-roll pattern where k depends on the aspect ratio.

2. Three-dimensional cellular patterns in directional solidification

We consider a 3-dimensional directional solidification system consisting of a long sample of binary alloy with rectangular cross section which is pulled, at constant velocity v, through a temperature gradient established between hot and cold contacts (Figure 1). The liquid and solid phases of the sample are separated by an interface. Using a one-sided model [15], neglecting convection in the liquid part and choosing suitable non-dimensionalized variables, the equations governing the temporal evolution of the system are (in a frame of reference moving in the z-direction at the interface velocity) given by [11]

$$\nabla^2 W + vW_z - W_t = 0, \quad z \geq s(x,y,t) \tag{1a}$$

$$W = -\overset{.}{s} - \frac{1}{u} H(s) \tag{1b}$$

$$s_t + v + W_z - s_x W_x - s_y W_y = 0 \quad \text{at } z=s(x,y,t) \tag{1c}$$

$$W_x = 0 \text{ at } x = 0,b_1; \quad W_y = 0 \text{ at } y=0,b_2 \tag{1d}$$

$$\lim_{z \to \infty} W(x,y,z,t) = -1, \tag{1e}$$

where the subscripts denote partial differentiation. Here $W(x,y,z,t)$ ($0 \leq x \leq b_1$, $0 \leq y \leq b_2$) is proportional to the solute concentration, u is a constant depending on the various physical parameters involved and the interface is located at $z=s(x,y,t)$ with mean curvature

$$H(s) = \frac{2s_x s_y s_{xy} - (1+s_x^2)s_{xx} - (1+s_y^2)s_{yy}}{(1+s_x^2 + s_y^2)^{3/2}} . \tag{2}$$

Equation (1a) describes the solute diffusion, Eq. (1b) corresponds to the Gibbs-Thomson boundary condition at the interface, the Rankine-Hugeniot condition (1c) relates the speed of diffusion to that of the interface and (1d,e) represent the

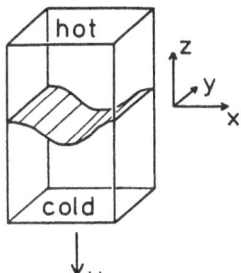

Fig. 1. The directional solidification system

boundary conditions. The system (1) is highly nonlinear in virtue of the curvature term H(s) in (1b). We assume that the interface moves slowly compared with the speed of solute diffusion. This corresponds to a quasi-stationary approximation [15] in which the W_t-term in (1a) is neglected. This allows us to represent W as a functional of s [11,12].

Equations (1) admit a basic planar interface solution given by

$$W_0(z) = e^{-vz} -1, \quad s = 0. \tag{3}$$

In order to find other solutions, close to the basic state, we represent s by a double cosine series,

$$s(x,y,t) = \sum_{m,n=0}^{\infty} \varepsilon_{mn}(t)\cos(mk_1\sqrt{u}x)\cos(nk_2\sqrt{u}y), \tag{4}$$

where $k_j = \pi/b_j\sqrt{u}$ (j=1,2). In the quasi-stationary approximation, W becomes a function of the amplitudes ε_{mn}, so that (1) reduces to an infinite system of equations for the ε_{mn}, viz.,

$$\dot{\varepsilon}_{mn} = a_{mn}\varepsilon_{mn} + \sum_{k=2}^{\infty} M_{mn}^k(\{\varepsilon_{ij}\}), \tag{5}$$

where M_{mn}^k is a homogeneous polynomial of degree k. The coefficients of the M_{mn}^k can in principle be calculated to any desired order. However, only a few of them are needed. The calculations are similar to those of the corresponding two-dimensional problem presented in [7,11] and will be omitted. The linear coefficients are given by

$$a_{mn} = (v-1-m^2k_1^2-n^2k_2^2) \left\{\frac{v}{2} +\left[\frac{v^2}{4}+ u(m^2k_1^2 +n^2k_2^2)\right]^{1/2}\right\} - v^2. \tag{6}$$

The planar interface loses stability if one or more eigenvalues a_{mn} of the linearization of (5) pass through zero from negative values. To illustrate the appearance of unstable modes, we set $(K_1,K_2)=(mk_1,nk_2)$ and consider K_1,K_2 first as continuous variables for the moment. Then the equation $a_{mn}\equiv a(K^2,v) = 0$ $(K^2=K_1^2+K_2^2)$ defines, for fixed u, a surface $S(u)$ in (K_1,K_2,v)-space which is invariant under rotation of (K_1, K_2). A section of this surface is shown in Figure 2. When v exceeds a critical value v_c, then the surface $S(u)$ intersects the plane v=const in two circles with radii $R_i(v) < R_0(v)$ such that $a(K^2,v)$ is positive inside the sector between the circles and negative outside. For an infinitely extended system (K_1,K_2) are continuous variables; however, the presence of the sidewalls here forces (K_1,K_2) to lie on a rectangular lattice generated by $(k_1,0)$ and $(0,k_2)$. Hence a bifurcation occurs if one or more lattice points are located on the circles and all other lattice points are outside the sector. The condition that a particular point (K_1,K_2) is on a circle leads to the equation $K_1^2 + K_2^2 = R^2$ where $R = R_i$ or R_0 (Figure 3). If (K_1,K_2,R) are rationally dependent, this equation reduces to the determination of Pythagorean triples. Since for any N there exists an integer C such that the Pythagorean equation $A^2+B^2=C^2$ is satisfied for at least N different pairs of integers (A,B), we can, by

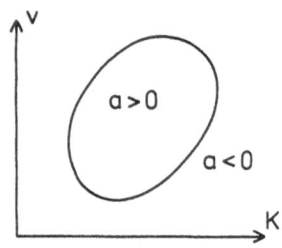

Fig. 2. Section of the surface
$a(K_3^2v) = 0$

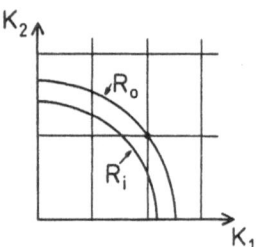

Fig. 3. The lattice in (K_1,K_2)-plane
and the circles R_i,R_o

suitable choices of K_1,K_2, in principle achieve that any prescribed number of modes becomes unstable simultaneously. To see this in an elementary way consider two different Pythagorean triples $a^2+b^2=c^2$ and $r^2+s^2=t^2$ where a,b,c,r,s,t are relatively prime. Then we have $(at)^2+(bt)^2=(ct)^2=(rc)^2+(sc)^2$, so that (at,bt,ct) and (rc, sc,ct) are different Pythagorean triples with the same hypotenuse ct. Obviously this construction can be repeated ad infinitum, i.e., we can generate any desired number of different Pythagorean triples. In a forthcoming paper we will derive and analyze normal form representations for such multi-mode instabilities by using group theory, here we confine ourselves to a description of three-mode interactions emerging as codimension-three bifurcations.

Assuming that three modes (m_j,n_j) $(j=1,2,3)$ become unstable simultaneously, one can perform a Liapunov-Schmidt reduction [8] which reduces the set of equations $a_{mn}\epsilon_{mn} + M_{mn}^K = 0$ for the steady states of (5) to three coupled nonlinear equations for the amplitudes $(X,Y,Z)\equiv(\epsilon_{m_1n_1},\epsilon_{m_2n_2},\epsilon_{m_3n_3})$. It can be shown that, if the three-mode instability has codimension three (i.e., three parameters are fixed), then for most choices of the m_j,n_j these equations take the form

$$X(X^2+kY^2+pZ^2-\lambda-\alpha) = 0$$

$$Y(eX^2+Y^2+gZ^2-\lambda-\beta) = 0 \qquad\qquad (7)$$

$$Z(fX^2+lY^2+Z^2-r\lambda) = 0.$$

In (7) terms of higher order have been neglected which can be justified rigorously by using normal form transformation techniques of singularity theory. The coefficients e,f,k,l,p,q,r depend on the Taylor coefficients of the M_{mn}^K and are considered as "modal parameters" which have to satisfy certain non-degeneracy conditions. The parameters λ and (α,β) are, respectively, proportional to $v-v_0$ and to a linear combination of (k_1-k_{10},k_2-k_{20}), (v,k_1,k_2) being the values of (v,k_1,k_2) at the three-mode instability point. We regard λ as a distinguished bifurcation parameter in the sense of imperfect bifurcation theory and (α,β) as unfolding or imperfection parameters [8]. Varying (α,β) away from $(0,0)$ destroys the three-mode instability at $\lambda=0$. The (α,β)-plane is divided into a number of regions such that in each region a structurally stable bifurcation diagram occurs, i.e. a variation of (α,β) inside a particular region does not produce a qualitative change in the bifurcation diagram.

The equations (7) possess the following types of solutions:

   (a) Single-mode solutions, determined by two zero and one nonzero amplitude. We denote them by S_X,S_Y,S_Z if, respectively, $X\neq0$, $Y\neq0$, $Z\neq0$.
   (b) Two-mode solutions, determined by one zero and two nonzero amplitudes. They are denoted by S_{XY}, S_{YZ}, S_{XZ} where the subscripts refer to the nonzero amplitudes.
   (c) The three-mode solution S_{XYZ} along which all three amplitudes are nonzero.

In order to describe the main features of the bifurcation diagrams corresponding to Eq. (7), we resort to a special case, namely, the simultaneous instability of the modes $(m,n) = (7,0),(0,5),(4,4)$, whose amplitudes we denote by X,Y,Z. The conditions $a_{mn}=0$ yield three equations that must be satisfied by the parameters u,v,k_1,k_2. If u is fixed, we obtain numerical values for v,k_{10},k_{20} and, consequently, also for the modal parameters $e,f,...$. Two particular structurally stable diagrams corresponding to $u=27.5$ and $u=50$ are shown schematically in Figures 4(a) and (b), respectively. In this Figure the signs associated to a particular subbranch refer to its stability assignment, i.e., (---) means a stable subbranch whereas all other assignments correspond to unstable subbranches. For $u=27.5$ we observe mode jumping and hysteresis between the S_{YZ} and S_{XZ} branches which both occur stably in some range of the bifurcation parameter. For $u=50$ the S_Z-branch loses stability when λ increases, but it is not clear in which of the stable branches S_X, S_Y the system settles down. In either case we have hysteretic behaviour. We wish to emphasize that a number of other different bifurcation diagrams is organized by (7) for both values of u. The form of the interface for a S_{XZ}-solution is shown in Figure 5.

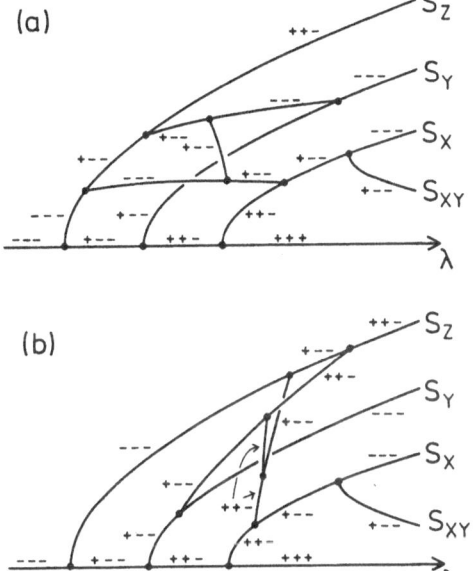

(a)

(b)

Fig. 5. Contour-plot of a S_{XZ}-interface

Fig. 4. Two structurally stable bifurcation diagrams corresponding to (7) for (a) $u=27.5$ and (b) $u=50$

3. Mode interactions induced by varying the aspect ratio in thermohaline convection

In this section we discuss steady state mode interactions for the two-dimensional Boussinesq equations describing convective transitions in binary mixtures driven by a destabilizing temperature gradient and a stabilizing concentration gradient. In nondimensionalized variables the basic equations for thermohaline convection in a box $\{(x,z):0 \leq x \leq 1, 0 \leq z \leq 1\}$ are [13]

$$\frac{1}{\sigma} [\nabla^2 \psi_t + J(\psi, \nabla^2 \psi)] = + R\theta_x - R_s s_x + \nabla^4 \psi \tag{8a}$$

$$\theta_t + J(\psi, \theta) = \psi_x + \nabla^2 \theta \tag{8b}$$

$$s_t + J(\psi, s) = \psi_x + \tau \nabla^2 s - \tau \nabla^2 \theta \tag{8c}$$

$$\psi = \psi_{zz} = \theta = s = 0 \quad \text{for } z = 0,1 \tag{8d}$$

$$\psi = \psi_{xx} = \theta_x = s_x = 0 \quad \text{for } x = 0,1 \tag{8e}$$

where $J(f,g)=f_xg_z - f_zg_x$. Here, ψ is the nondimensionalized stream function, θ and s describe deviations of the temperature and concentration fields from the basic conduction state and R,σ,R_s,τ are, respectively, the thermal Rayleigh number, the Prandtl number, the solutal Rayleigh number and the Lewis number [13]. The equations (8d,e) represent the simplest boundary conditions, with temperature and concentration fixed at the top and bottom, no sideways heat and salt fluxes, and no tangential viscous stresses.

The basic conduction state is given by $\psi=\theta=s=0$. Setting $U=(\psi,\theta,s)^T$, this state loses stability to exponentially growing solutions of the form $e^{pt}U_k(x,z)$ (Re$p>0$) when R exceeds the critical value

$$R_c^k(1) = (R_s/\tau)(1+\tau) + (\pi^4/k^2 1^4)(k^2+1^2)^3. \tag{9}$$

The unstable mode $U_k(x,z)$ is given by

$$U_k(x,z) = \begin{pmatrix} (k^2+1^2)\pi^2 \sin(k\pi x/1) \\ 1^2 \cos(k\pi x/1) \\ [1^2(1+\tau)/\tau] \cos(k\pi x/1) \end{pmatrix} \sin\pi z \ . \tag{10}$$

In most finitely extended systems with one or two spatial dimensions it is possible to adjust the width so that two consecutive modes become unstable simultaneously. For ordinary convection this situation was analyzed in [14]. Here we discuss the corresponding problem for thermohaline convection. Setting $R_c^k(1)=R_c^{k+1}(1)$ yields a simple equation for 1, with solution

$$1 = 1_k \equiv [k(1+k)]^{1/3} \{k^{2/3}+(1+k)^{2/3}\}^{1/2}. \tag{11}$$

This is a countable set of critical aspect ratios 1_k, $k=1,2,\ldots$ such that, if $1_{k-1} <1<1_k$, the first instability when R is increased occurs at $R_c^k(1)$ and leads to a k-roll pattern corresponding to the unstable mode $U_k(x,z)$. If $\bar{1}=1_k$, then the first instability sets in at $R=R_k\equiv R_c^k(1_k)$ and both modes U_k,U_{k+1} become unstable simultaneously. Thus close to 1_k we can expect steady-state mode interactions.

In order to analyze (8) in a vicinity of a critical aspect ratio, we set

$$U(t,x,z) = X(t)U_k(x,z) + Y(t)U_{k+1}(x,z) + W(X,Y;x,z), \tag{12}$$

and perform a center manifold reduction [10] of (8) for (R,1) close to $(R_k,1_k)$. Here, W represents a superposition of all linearly damped modes whose amplitudes depend on (X,Y) but are at least of second order in (X,Y). The center manifold reduction yields a two-dimensional system for the amplitudes (X,Y) which, owing to the $Z(2)\times Z(2)$-symmetry of (8), has the form

$$\dot{X} = AX^3 + BXY^2 + \alpha X$$
$$\dot{Y} = CY^3 + DX^2Y + \beta Y, \tag{13}$$

where
$$\alpha = (k^2 1_k^2\sigma/\pi^2 a_k^4)(R-R_k) + (2\sigma\pi^2(2a_k^2-31_k^2)/1_k^3)(1-1_k)$$
$$\beta = ((k+1)^2 1_k^2\sigma/\pi^2 a_{k+1}^4)(R-R_k) + (2\sigma\pi^2(2a_{k+1}^2 - 31_k^2)/1_k^3)(1-1_k) \tag{14}$$

Here, $a_k^2=k^2+1_k^2$ and $a_{k+1}^2 = (k+1)^2 +1_k^2$. In Eq. (13) terms of fifth and higher order have been neglected. The main task is to compute the coefficients A,B,C,D. We state here the results for A,C, which are given by

$$A = -\ (k^4 1_k^2\sigma)/(8a_k^6 \pi^2)\cdot M_k, \quad C = -\ ((k+1)^4 1_k^2\sigma)/(8a_{k+1}^6\pi^2)\cdot M_k \tag{15}$$

where
$$M_k = R_k - (R_s/\tau^3)(1+\tau)(1+\tau^2). \tag{16}$$

158

The coefficients B,D are given by huge expressions which have to be evaluated numerically and thus will not be written down explicitly. The normal form (13) has been analyzed by several authors in different contexts (e.g. [8,10]) and is well understood. A study of its dynamics involves locating the fixed points, analyzing their stability and using the divergence test to locate possible limit cycles. The behaviour of (13) depends crucially on the signs of A,B,C,D and on the determinant AC-BD so that a number of different cases has to be distinguished. The system (13) possesses three types of steady states: the trivial solution T corresponding to the conduction state, single-mode solutions S_x, S_y with $(X \neq 0, Y=0)$, $(X=0, Y \neq 0)$ and mixed mode solutions S_{xy} where both X and Y are nonzero. If A and C have equal signs, oscillations do not occur, if they are opposite a limit cycle may be created at a Hopf bifurcation along S_{xy}. From (15) we infer that A and C cannot have opposite signs, but vanish simultaneously if $M_k=0$ which is readily solved for R_S. In contradistinction A and C are always negative in ordinary convection [14] (cf. (16) with $R_S=0$).

The simultaneous vanishing of two cubic normal form coefficients is ungeneric and must be considered as an artefact of the Boussinesq approximation or the stress-free boundary conditions. Our objective is therefore to determine physically meaningful perturbations of (8) that split the zeros of A and C. This involves symmetry considerations: The $Z(2) \times Z(2)$-symmetry of (13) originates from the reflection symmetries $(x,\psi) \rightarrow (1-x,-\psi)$ and $(z,\psi,\theta,s) \rightarrow (1-z,-\psi,-\theta,-s)$ of the original system (8). We denote the first of these symmetries by (rl) and the second by (tb). Incorporating x-independent non-Boussinesq effects to the system (e.g., a temperature-dependent thermal conductivity) generally breaks (tb), but the leading terms of the system governing the flow on the center manifold still have the form (13) for $k \geq 2$ because (rl) is still present and has its origin in a "hidden O(2)-symmetry" [1,2,4,9]. Inspection shows, however, that x-independent perturbations shift, but do not split the zeros of A and C. Thus we have to include x-dependent perturbations which, although breaking the (rl)-symmetry, have to preserve the form of (13). This requires a very careful selection among the possible perturbations because a generic breaking of the (rl)-symmetry will destroy the form of (13) completely. An appropriate perturbation is provided by inhomogeneous heating, chosen so that the nondimensionalized temperature distribution at the bottom of the container is

$$T|_{z=0} = T_0 + \varepsilon b [\cos 2(k+1)\pi x/l_k],$$ (17a)

where ε is a small parameter, T_0 is the reference temperature at the bottom and b is a constant of order O(1). We include also a weak dependence of the thermal conductivity χ on the temperature by setting

$$\chi = \chi_0 [1+\varepsilon c (T-T_0)].$$ (17b)

The inhomogeneous perturbation (17a) and the non-Boussinesq-effect (17b) produce additional terms of order ε and ε^2 on the right-hand sides of (8a,b,c) and consequently modify the system (13). We omit the details and just give the results. Our choice of perturbations implies that the structure of (13) is preserved but the coefficients A,B,C,D become functions of ε. In particular, the "critical coefficients" A,C are now given by

$$A(\varepsilon) = - (k^4 l_k^2 \sigma)/(8a_k^6 \pi^2)(M_k + P\varepsilon + Q\varepsilon^2)$$ (18)

$$C(\varepsilon) = - ((k+1)^4 l_k^2 \sigma)/(8a_{k+1}^6 \pi^2)(M_k + P\varepsilon + R\varepsilon^2) .$$ (19)

The expressions for P,Q,R are summarized in the appendix. In (18) and (19) we have expanded A and C up to order ε^2 because only at that order their zeros are split. We define $\bar{R}_S(\varepsilon) = R_{S,0} + R_{S,1}\varepsilon$ by the condition that $M_k + P\varepsilon = 0(\varepsilon^2)$ for $R_S = R_S(\varepsilon)$ and set $R_S = \bar{R}_S(\varepsilon) + d\varepsilon^2$, i.e., we explore a very small neighborhood of $\bar{R}_S(\varepsilon)$ when $\varepsilon \ll 1$ is fixed and d is varied. Then the coefficients A and C are $0(\varepsilon^2)$ and it is obvious that, for suitable choices of b,c,d, A and C can have either signs. However, because A and C are very small, fifth order terms must be included and our normal form (13) is to be replaced by

$$\dot{X} = FX^5 + BXY^2 + \gamma X^3 + \alpha X$$

$$\dot{Y} = EY^5 + DX^2Y + \sigma Y^3 + \beta Y.$$

(20)

In (20), $A \equiv \gamma$ and $C \equiv \delta$ are now considered as unfolding or imperfection parameters which can be varied freely in a neighborhood of the origin. The system (20) is a normal form of a *codimension four bifurcation*. The new coefficients E,F are just the butterfly coefficients computed in [2], p specified to $1=1_k$. We will not write them down explicitly (see [2]) because only their signs are relevant: both F and E are negative. The behaviour of the steady states is then completely determined by the terms given in (20), but it is not yet clear whether further quintic or even seventh order terms are necessary to describe the appearance of limit cycles properly. We will not discuss the extended normal form (20) here, but note that it admits the bifurcation diagram of Figure 6. The point labelled t in this diagram marks a tertiary Hopf bifurcation where a limit cycle is created from the mixed mode branch. We cannot say yet whether the branch of oscillatory solutions bifurcates sub- or supercritically, hence this branch is not shown in the diagram. These results show clearly that a combination of two non-Boussinesq effects can lead to oscillatory convection if $(R,1,R_S)$ is close to $(R_k,1_k,\bar{R}_s(\varepsilon))$. We observe that, when the inhomogeneous term in (17a) is replaced by a linear combination of $\cos(k\pi x/1_k)$ and $\cos((k+1)\pi x/1_k)$, then a number of further terms have to be added to (20) which destroy the $Z(2) \times Z(2)$-symmetry completely and make the problem much more complicated. Our choice of perturbations yields the simplest structure for the resulting normal form.

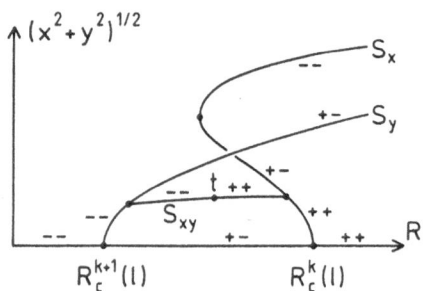

Fig. 6. A bifurcation diagram organized by the normal form (20)

4. Simulations of the Boussinesq equations

When simulating a system of field equations on a digital computer, time and space become discrete and an approximation scheme for the operators involved has to be constructed. If the field interactions are local this leads naturally to the concepts of cellular automata [18] which provide a useful tool for testing analytical predictions. They allow us to follow the evolution of the system from an arbitrary initial configuration towards an attractor, which may be a stable steady state, a temporally oscillating state or a strange attractor. In the presence of several competing attracting states the simulations can provide some insight into their basins by testing a sample of initial configurations. There are many ways to construct the local rules of a cellular automaton that simulates a given system of partial differential equations. Rules for the Navier Stokes equations have been constructed in [6] on the basis of lattice gas automata.

In this section we discuss the results of simulation experiments of the Boussinesq equations describing convection in a finite layer of fluid heated from below. Our objective is to visualize the different convection patterns predicted in [14] in dependence of the aspect ratio and the Rayleigh and Prandtl numbers. When the Boussinesq equations are approximated by a finite difference scheme and then discretized we obtain rules for a cellular automaton. At each site of this cellular automaton there exists a large, though finite number of states because we are approximating a continuous system. Thus the cellular automaton should be called "real valued" in contrast to "elementary cellular automata" where at each site only a small number of states is available [18].

The Boussinesq equations for ordinary convection in a rectangular box are given by (8) with s=0 and (8c) omitted. It is convenient to introduce an auxiliary field ϕ, defined by

$$\phi = \nabla^2 \psi, \tag{21a}$$

and to replace (8a) by

$$\frac{1}{\sigma} [\phi_t + J(\psi,\phi)] = R\theta_x + \nabla^2\phi. \tag{21b}$$

Using first order approximations we utilize a forward difference scheme in time and a central difference scheme in space for both first and second derivatives. Then equations (8b) and (21b) take the form

$$\phi_{t+1} = \phi + (\delta t/\delta s^2)\{\sigma[\frac{1}{2}R\delta s(\theta_{i+1}-\theta_{i-1}) + \phi_{i-1} + \phi_{i+1} + \phi_{j-1} + \phi_{j+1}-4\phi]$$
$$- \frac{1}{4}(\psi_{i+1}-\psi_{i-1})(\phi_{j+1}-\phi_{j-1}) + \frac{1}{4}(\psi_{j+1}-\psi_{j-1})(\phi_{i+1}-\phi_{i-1})\} \tag{22a}$$

$$\theta_{t+1} = \theta + (\delta t/\delta s^2)\{\frac{1}{2}\delta s(\psi_{i+1}-\psi_{i-1}) + \theta_{i-1} + \theta_{i+1} + \theta_{j-1} + \theta_{j+1}-4\theta$$
$$- \frac{1}{4}(\psi_{i+1}-\psi_{i-1})(\theta_{j+1}-\theta_{j-1}) + \frac{1}{4}(\psi_{j+1}-\psi_{j-1})(\theta_{i+1}-\theta_{i-1})\} \tag{22b}$$

Here, the discretized fields are labeled by integers (t,i,j) which refer to the co-ordinates $t_0+t\delta t$, $x_0+i\delta s$, $z_0+j\delta s$, where (t_0,x_0,z_0) are some reference values and $\delta t, \delta s$ denote the discrete step lengths in time and space. We interpret $(\psi_{tij}, \phi_{tij}, \theta_{tij})$ as the state of a cellular automaton at time t and site (i,j). The equations (22) are considered as local rules determining the state of site (i,j) at time t+1 from those of the neighboring sites at time t. Only the subscripts differing from (t,i,j) have been explicitly written down in (22). The boundary conditions (8d,e) are realized by setting the desired values on the border of the lattice after each time step. Unfortunately (22a) determines only the states of ϕ and θ at time t+1 and not that of ψ. To obtain ψ, (21a) must be inverted in the finite difference scheme. This yields a relation of the form

$$\psi_{tij} = \sum_{k,l} A_{ij,kl} \phi_{tkl}, \tag{23}$$

where the matrix $A \equiv \{A_{ij,kl}\}$ represents the discretized version of the inverse Laplacean. Since A involves sites (k,l) far from (i,j), (23) introduces a non-local dependence which, however, does not cause serious problems. Equations (22), (23) provide a complete description of the temporal evolution of our system from a given initial configuration. The states at a given time t are displayed graphically by assigning to each site (i,j) the velocity vector.

Recall from the preceding section that the basic conduction state loses stability simultaneously to a k-roll and to a (k+1)-roll pattern if $l=l_k$ (Eq. (11)) and if R exceeds the critical value R_k. The behaviour of the steady states depends essentially on the Prandtl number σ (see [14]). To each k there exists a critical value σ_k so that for $\sigma<\sigma_k$ the mixed mode pattern is stable and only one stable steady state is present for each R (Figure 7(a)). For $\sigma>\sigma_k$ the mixed mode pattern is unstable. However, there is a range of R-values where both single-mode patterns coexist stably which gives rise to a hysteresis (Figure 7(b)). For example, when l is between $l_4 = 6.33$ and $l_5 = 7.75$, $\sigma<\sigma_4 = 0.17$ and R is slightly above $R_4 = 660.51$ one expects the 4-roll pattern of Figure 8(a). Keeping l and σ fixed but increasing the Rayleigh number induces a transition from the 4-roll pattern to a family of mixed-mode patterns which terminates in the 5-mode pattern of Figure 8(c). A mixed-mode pattern with amplitude ratio 1:1 is shown in Figure 8(b).

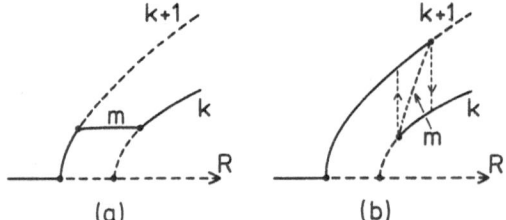

Fig. 7. Schematic bifurcation diagrams near (R_k,l_k) for (a) $\sigma<\sigma k$ and (b) $\sigma>\sigma k$. k+1, k and m refer to the (k+1)-roll and mixed-mode patterns, respectively. Solid branches are stable and broken branches are unstable.

(a)   (b)

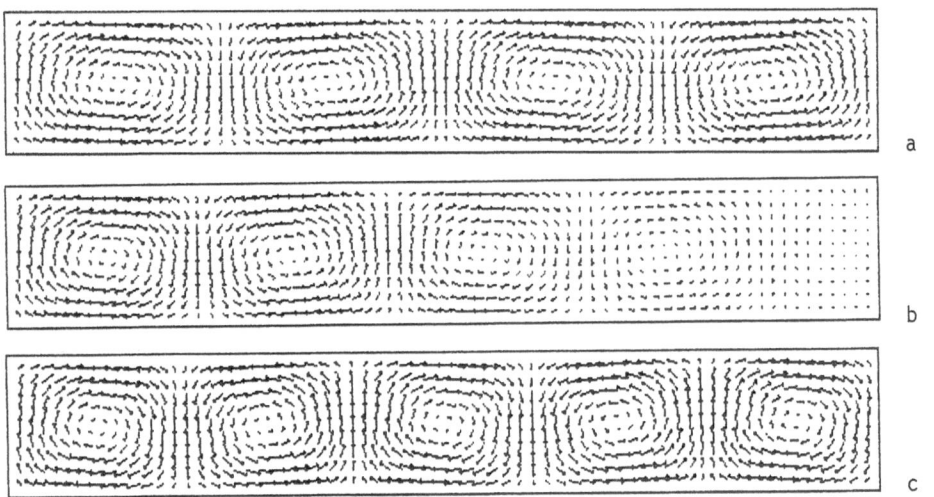

a

b

c

Fig. 8. Stable patterns expected near $l=l_4$ and $\sigma<\sigma_4$. (a) and (c) are pure-mode patterns and (b) is a mixed-mode pattern with amplitude ratio 1:1.

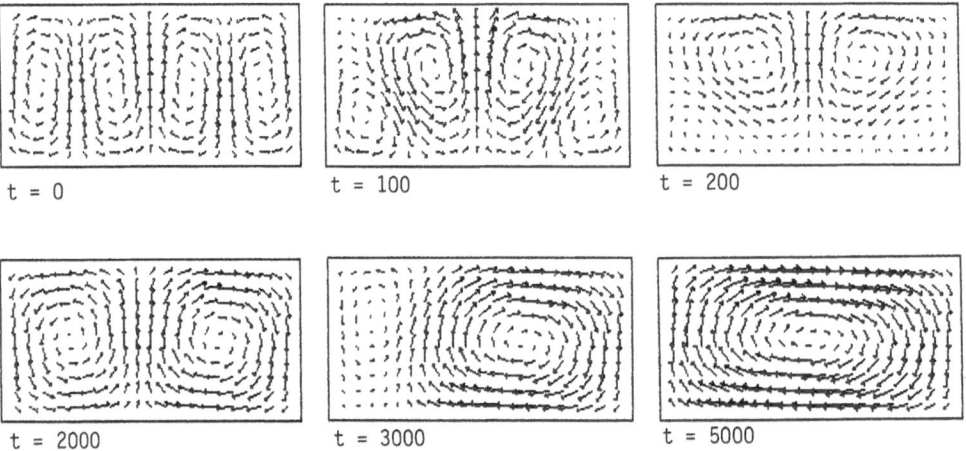

t = 0   t = 100   t = 200

t = 2000   t = 3000   t = 5000

Fig. 9. Evolution towards a stable 1-roll pattern from a 4-roll initial configuration ($R=820$, $l=1.82$, $\sigma=0.1$)

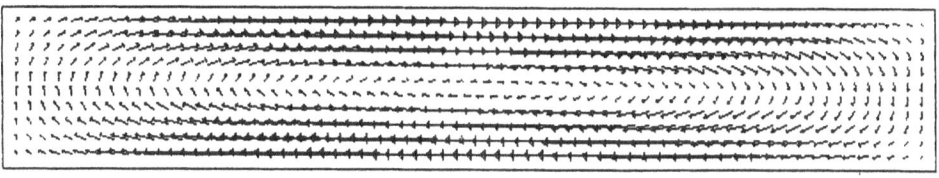

t = 0

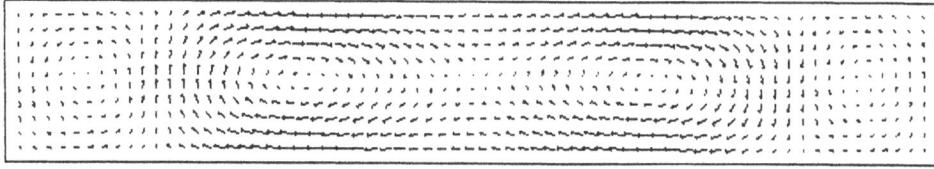

t = 2000

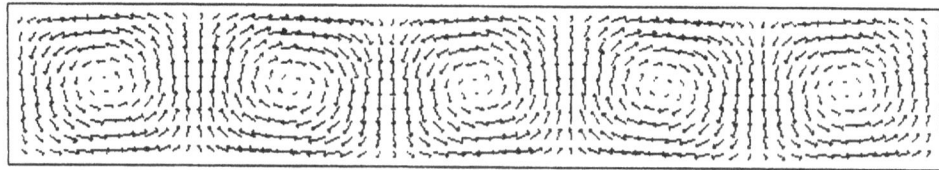

t = 3500

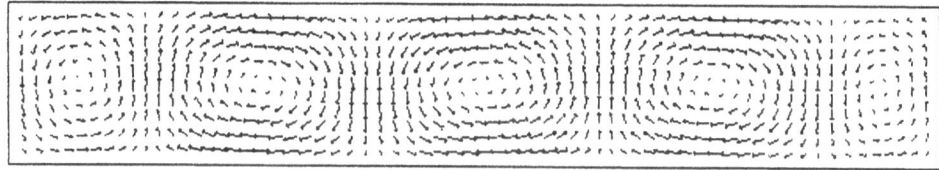

t = 6000

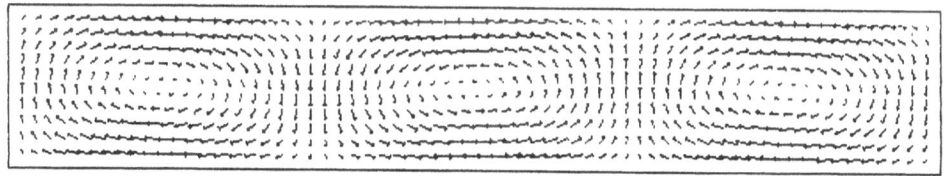

t = 10000

Fig. 10. Evolution towards a stable 3-roll pattern from a 1-roll initial configuration (R = 820, l = 6.00, σ = 0.1).

We performed a number of simulations in the parameter region where the analysis of [14] predicts stable mixed-mode patterns, however, we did not find them. The reason seems to be that the critical values σ_k are affected by the discretization and that the range of R where mixed-mode patterns stably exist is too small. On the other hand the division of the l-axis into intervals corresponding to different numbers of stable rolls is observed and in good agreement with the analytical pre-dictions. Figure 9 shows the temporal evolution towards a stable 1-roll pattern from an initial configuration with 4 rolls. First the two outer rolls are compressed (t = 100, 200) and then disappear leaving a transient 2-roll pattern (t = 2000). When time increases further the left roll of this 2-roll pattern is compressed

while the right roll expands (t = 3000) until the system settles down in the stable
1-roll pattern (t = 5000).

The evolution from 1 roll to 3 rolls is shown in Figure 10. The most interesting
feature here is the transient 5-roll pattern (t = 3500) which evolves towards the
stable 3-roll pattern (t = 10 000) by compressing the two outer rolls.

5. Conclusions

We have presented new results concerning interactions of steady state modes near
multiple bifurcations in finitely extended systems. In Section 2 we considered
pattern formation in systems with three spatial dimensions. Although the analysis
was carried out specifically for directional solidification, the formalism is also
applicable to other physical systems, e.g. to convection in boxes. We demonstrated
the appearance of three mode interactions near a codimension three bifurcation and
exemplified the resulting bifurcation behaviour in terms of bifurcation diagrams.
This situation was analyzed for "generic modes" leading to bifurcation equations
of the form of Eq. (7). Other combinations of unstable modes may lead to rather
different bifurcation equations similar to, though more complicated than the case
of two-mode interactions in systems on a line and with non-flux boundary conditions
at the ends [1,4]. There is one important feature in systems defined in three-
dimensional boxes which is not present in the lower-dimensional cases, namely, the
possibility of multi-mode instabilities. As in spherically symmetric systems the
number of unstable modes can be as high as we wish. A bifurcation analysis of
multi-mode instabilities requires subtle symmetry considerations involving the
concept of hidden symmetries [9] which deserves further study.

In section 3 we investigated thermohaline Boussinesq convection in a finite two-
dimensional box and with stress-free boundary conditions. It turned out that two
cubic coefficients of the normal form describing the simultaneous instability of two
consecutive modes vanish simultaneously at a critical value of the solutal Rayleigh
number. The point in parameter space corresponding to this degeneracy may be interpreted as a "double tricritical point". The simultaneous vanishing of two coefficients when only one parameter is varied is ungeneric and must be viewed as an
artefact of the Boussinesq approximation and the boundary conditions. To achieve
genericity we included non-Boussinesq effects, chosen such that the $Z(2) \times Z(2)$-
symmetry of the normal form is preserved to leading order. This allowed us to
describe the double tricritical point in terms of the codimension-four normal form
(20) in which also quintic terms are taken into account. The system (20) provides
a starting point for further investigations. In particular, tertiary Hopf bifurcations on the mixed mode branch and the disappearance of limit cycles in heteroclinic and homoclinic connections have to be studied. If genuine non-Boussinesq
effects are included, Eq.(20) will be subject to perturbations that break the
$Z(2) \times Z(2)$-symmetry completely. The task is then to clarify generic symmetry breaking
of (20) in terms of a further symmetry-breaking parameter, e.g., the amplitude of a
critical mode which occurs in the Fourier decomposition of a general inhomogeneous
heating term.

In the simulations of Section 4 we visualized the roll patterns predicted in
[14] for two-dimensional Boussinesq convection in a finite box. The division of
the l-axis into intervals corresponding to different numbers of stable rolls was
observed, however, mixed-mode patterns were not found. Analogeous simulations for
the thermohaline problem, specifically near a double tricritical point will be discussed elsewhere.

Acknowledgments

It is a pleasure to acknowledge helpful discussions with D. Armbruster, W. Güttinger,
E. Knobloch and U. Kolb, and the support of the Stiftung Volkswagenwerk.

References

1. D. Armbruster & G. Dangelmayr, Math. Proc. Cambr. Phil. Soc. 101 (1987), 167
2. D. Armbruster & M. Neveling, J. Nonequ. Thermodyn. (1987, in press)
3. G. Dangelmayr, in "Structural Stability in Physics", W. Güttinger & H. Eikemeier (eds.), Springer 1979, pp. 84-103
4. G. Dangelmayr & D. Armbruster, Contemporary Mathematics 56 (1986), 53
5. R. Friedrich & H. Haken, this volume
6. U. Frisch, B. Hasslacher & Y. Pomeau, Phys. Rev. Lett. 56 (1986), 1505
7. C. Geiger, W. Güttinger & P. Haug, in "Complex Systems - Operational Approaches", H. Haken (ed.), Springer 1985, pp. 279-299
8. M. Golubitsky & D. Schaeffer, "Singularities and Groups in Bifurcation Theory, vol. 1", Springer 1984
9. M. Golubitsky, J. Marsden & D. Schaeffer, in "Partial Differential Equations and Dynamical Systems", W. Fitzgibbon (ed.), Pitman Press 1984, pp. 181-210
10. J. Guckenheimer & P. Holmes, "Nonlinear Oscillations, Dynamical Systems and Bifurcation of Vector Fields", Springer 1983
11. P. Haug, Phys. Rev. A (1987, in press)
12. M. Kerzsberg, Phys. Rev. B 27 (1983), 6796
13. J.K. Platten & G. Chavepeyer, Adv. Chem. Phys. 32 (1977), 581
14. E. Knobloch & J. Guckenheimer, Phys. Rev. A 27 (1983), 408
15. J.S. Langer, Rev. Mod. Phys. 52 (1980), 1
16. D. Schaeffer & M. Golubitsky, Comm. Math. Phys. 69 (1979), 209
17. L.H. Ungar & R.A. Brown, Phys. Rev. B 29 (1984), 1367
18. S. Wolfram, "Theory and Applications of Cellular Automata", World Scientific, Singapore, 1986
19. M.J. Bennett, R.A. Brown & L.H. Ungar, this volume

Appendix. The coefficients P,Q,R occurring in (18) and (19) are given by

$$P = c(R_k - R_S/2\tau^2)(1+2\tau+2\tau^2)$$

$$Q = K + Q_1, \quad R = K + R_1 + R_2,$$

where

$$K = c^2\{R_S(\tau-1)/4\tau - (R_S/4\tau^2 R_k)(1+2\tau+2\tau^2) + \frac{1}{4}R_k\}$$

$$Q_1 = bcL\{(R_S/\tau^3)(1+\tau) - 2a_k^6\pi^4 l_k^6\}$$

with

$$L = \frac{k^2\tau l_k^4(R_S - R_k)(5a_{k+1}^2 - 4l_k^2)}{4(2k+1)[k^2 l_k^4(R_S(1+\tau) - R_k\tau) - 2a_k^6\pi^4\tau]},$$

$$R_2 = bc(\frac{5}{4} - l_k^2/a_{k+1}^2)\{(R_S/\tau^2)(1+\tau+\tau^2) - R_k\},$$

and R_1 is given by the expression for Q_1 with (k^2, a_k^2) and $((k+1)^2, a_{k+1}^2)$ interchanged.

The Basic $(n,2n)$-Fold of Steady Axisymmetric Taylor Vortex Flows

R. Meyer-Spasche[1] *and M. Wagner*[2]

[1]Max-Planck-Institut für Plasmaphysik,
  Boltzmannstraße, D–8046 Garching, Fed. Rep. of Germany
[2]Institut für Mathematik III, Freie Universität,
  Arnimallee 2–6, D-1000 Berlin 33, Germany

We study steady axisymmetric flows in a wide gap between concentric cylinders. End effects are neglected (periodic boundary conditions). With both cylinders rotating, there are four parameters in the problem: The Reynolds number Re, the axial period λ, the radius ratio η, and the rotation rate μ. We solve the Navier-Stokes equations for such flows numerically, using the very reliable methods described earlier: Discretization by Fourier decomposition in the axial direction and centered finite differences in the radial direction; systematic variation of Re and λ by using the method of continuation with Gauss-Newton iterations. η and μ are mostly kept fixed at $\mu = 0$ and $\eta = 0.727$, the wide gap value of the Burkhalter/Koschmieder experiments and of numerical investigations.

The curve of neutral stability of Couette flow in the (Re, λ)-plane can be found by solving a 'linear' eigenvalue problem (1st eigenvalue, Taylor 1923). We found that flows bifurcating from Couette flow at the 2nd eigenvalue in the radial direction play a role in the investigation of the common basic Taylor vortex flows bifurcating at the 1st eigenvalue. Such flows consist of two adjacent vortices in the radial direction, which gives rise to free stagnation points. Here we report on a fold and on a curve of bifurcation points which seem to be generated by a nonlinear interaction of the two eigenvalue curves and their periodic repetitions.

1. Introduction

We consider steady axisymmetric Taylor vortex flows between long, concentric cylinders. The inner cylinder with radius $\hat{r} = R_1$ is rotating at the speed $\omega_1 R_1$, and the outer cylinder with radius $\hat{r} = R_2$ at the speed $\omega_2 R_2$. End effects are neglected (periodic boundary conditions). With these assumptions the Navier-Stokes equations used to compute these flows read in dimensionless form

$$u_r + \frac{1}{r}u + w_z = 0, \tag{1a}$$

$$\Delta u - \frac{1}{r^2}u - p_r = Re(uu_r + wu_z - \frac{v^2}{r}), \tag{1b}$$

$$\Delta v - \frac{1}{r^2}v = Re(uv_r + wv_z + \frac{uv}{r}), \tag{1c}$$

$$\Delta w - p_z = Re(uw_r + ww_z). \tag{1d}$$

Here, lengths have been scaled by R_1, velocities by $\omega_1 R_1$, the Reynolds number is given by

$$Re := \omega_1 R_1^2 / \nu \,,$$

ν being the kinematic viscosity of the fluid, and Δ is the Laplacian,

$$\Delta := \frac{\partial^2}{\partial r^2} + \frac{1}{r} \frac{\partial}{\partial r} + \frac{\partial^2}{\partial z^2} \,.$$

There are four independent parameters in the problem: The radius ratio $\eta := R_1/R_2$, the axial period λ, the rotation rate $\mu := \omega_2/\omega_1$, and Re. The dimensionless gap width $\delta := (R_2 - R_1)/R_1$ is related to η by $\delta = \eta^{-1} - 1$, the wave number k is related to the period λ of the flow by $\lambda = 2\pi/k\delta$, and the Taylor number T is related to Re by $T = Re^2 \cdot$ const. With these definitions we can formulate the boundary conditions as

$$u(1,z) = u(1+\delta,z) = 0\,,$$
$$w(1,z) = w(1+\delta,z) = 0\,, \qquad -\frac{\pi}{k} \leq z \leq \frac{\pi}{k}\,, \tag{2a}$$
$$v(1,z) = 1, v(1+\delta,z) = \mu/\eta\,,$$

$$u\left(r,-\frac{\pi}{k}\right) = u\left(r,\frac{\pi}{k}\right)\,,$$
$$v\left(r,-\frac{\pi}{k}\right) = v\left(r,\frac{\pi}{k}\right)\,, \qquad 1 \leq r \leq 1+\delta\,, \tag{2b}$$
$$w\left(r,-\frac{\pi}{k}\right) = w\left(r,\frac{\pi}{k}\right) = 0\,.$$

The boundary conditions (2) enforce $\mathbf{v} \cdot \mathbf{n} = 0$, i.e. no fluid can leave the domain of computation. Equations (1), (2) are well suited to model flows in the middle portion of long cylinders; see, for instance, [22, 14, 19]. We solve the problem posed by (1), (2) numerically, using the TAYPERIO code as described earlier. Here we give a short account of the numerical methods used. The details can be found in earlier publications, mostly in [14, 15]. To discretize eqs.(1), (2), we use a Fourier decomposition in the axial direction and equidistant centered finite differences of second order in the radial direction. This gives us a nonlinear algebraic system

$$\mathbf{F}(\mathbf{x}, k, Re) = 0. \tag{3}$$

Here the vector $\mathbf{x}$ denotes the unknowns, i.e. the values of the Fourier components in the radial grid points, while k and Re are the parameters introduced earlier. η and μ are not mentioned explicitly in (3) since we keep them fixed during our computations: $\eta = 0.727$ and $\mu = 0$. These values have been used before [16, 7, 15, 9]. They allow steady axisymmetric flows up to $Re \geq 9Re_{cr}$ for $\lambda = 2$, as has been shown both experimentally [3] and theoretically [10]. Equations (1), (2) are antimetric with respect to the plane $z \equiv 0$. Our Fourier decomposition is such that all computed solutions exhibit this sort of symmetry. This is a restriction [4]. But the number of solutions left is still so large that this restriction might even be considered to be advantageous.

We investigate the solution set of (3) with the method of continuation as developed by Keller [12], both with respect to Re and with respect to the wave number k. In this method, the pseudo arc length s is used to parametrize the curve

$$\mathbf{y}(s)^t = (\mathbf{x}(s)^t, Re(s)) \text{ or } \mathbf{y}(s)^t = (\mathbf{x}(s)^t, k(s))\,.$$

The enlarged system

$$G(\mathbf{y}(s), s) = 0 \qquad (4)$$

is then solved by the (simplified) Gauss-Newton method. This allows us to compute a whole connected component of the solution set over the (Re, λ)-space whenever one solution in that component is found. For the algorithm it does not make any difference if the computed flows are stable or not. During these investigations we actually completely ignore the important question of stability. We focus on mathematical properties of the equations; see [16].

When (4) has been solved and $\mathbf{y}$ is known, the reduced pressure p, the velocity components u, v, w and the stream function ψ are computed in a numerically stable way. ψ is defined by

$$u = \frac{1}{r}\frac{\partial \psi}{\partial z} \ , \ w = -\frac{1}{r}\frac{\partial \psi}{\partial r} \ . \qquad (5)$$

Most of the computations were performed with M2 = 33 radial grid points (i.e. M = 31 inner radial grid points) and N Fourier components, N = 12 or N = 8, on the Cray 1 of the ZIB, Freie Universität, Berlin.

2. The $(n, 2n)$ double bifurcation point and the basic $(n, 2n)$-fold

The curve of neutral stability of Couette flow in the (Re, λ)-plane with minimum $(Re_{cr}, \lambda_{cr} \approx 2)$ intersects with its periodic repetition [15] with minimum $(Re_{cr}, 2\lambda_{cr} \approx 4)$ in a point (Re_{24}, λ_{24}). Both neutral curves consist of bifurcation points, and (Re_{24}, λ_{24}) is a double bifurcation point. For $\eta = 0.727$ its values are $\lambda_{24} \approx 2.9, Re_{24} \approx 238 \approx 1.09 Re_{cr}(0.727)$, and for $\eta = 0.5$ the values are $\lambda_{24} \approx 2.86$ and $Re_{24} = 74.2 \approx 1.09 Re_{cr}(0.5)$. When normalized with $Re_{cr}(\eta)$, the neutral curves for different η, $0.5 \leq \eta \leq 1.$, do not differ considerably [13; 19, Fig.7; 7].

In [15] one of the authors and H.B. Keller studied the structure of families of solutions or flows with respect to continuous variation of both parameters λ and Re for $\eta = 0.727, \mu = 0$. Varying λ with fixed $Re = \sqrt{1.5}Re_{cr}$ (or $T = 1.5T_{cr}$), they found that the 2-vortex solution bifurcating from Couette flow on the low-λ side of the neutral curve continuously changes with increasing λ into a 4-vortex solution which passes a fold point near $\lambda = 2.68$ and then undergoes a period-halving bifurcation to its periodic repetition near $\lambda = 2.59$. When they started with maximum λ at the other side of the neutral curve, they found essentially the same thing, but without fold. The period-halving bifurcation occurs at $\lambda \approx 2.8$. In Fig.1a we repeat the bifurcation diagram given in [15, Fig.1].

These things are partially explained by the analysis of the solution set of eqs (1), (2) in a neighborhood of the point (Re_{24}, λ_{24}) defined earlier. Using perturbation methods, Andreichikov [1] discussed qualitatively what can happen for arbitrary η and μ. In the case $\eta = 0.5, \mu = 0$ he gave a detailed analysis by numerical evaluation of the formulas derived for the general case. In Fig.1b we repeat Fig.4 of [1], with adapted nomenclature. Besides the neutral curves for one pair of vortices and for two pairs of vortices (or for n and $2n$ pairs in the n-th repetition), it gives the loci of the two subharmonic bifurcations

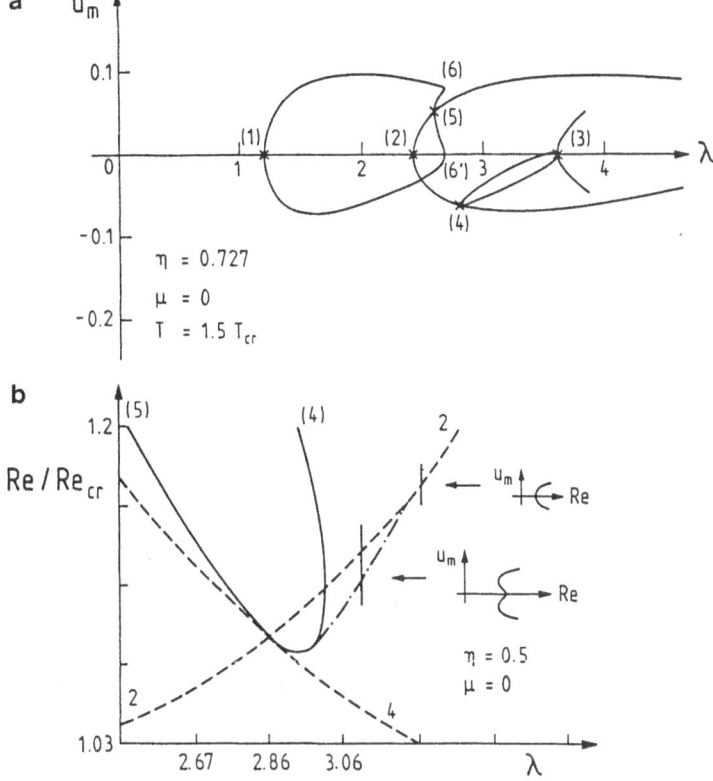

Fig.1: a) Variation of the radial velocity at the midpoint, $u_m := u(1 + \frac{\delta}{2}, 0)$, with the period λ [15, Fig.1]. Points (1), (2), (3): Bifurcation of 2-vortex flow and two periodic repetitions from Couette flow ($u_m \equiv 0$). Point (3) is the double bifurcation point (Re_{26}, λ_{26}), a 2-vortex flow bifurcates there as well. Points (4), (5): wavelength-halving bifurcation points. (6) is a point on the basic (2,4)-fold, while (6') is the corresponding point on the branch of solutions shifted by $\lambda/2$. The computations were performed with M = 31 inner radial grid points and N = 6 Fourier components. b) Andreichikov's analysis [1, Fig.4] of the neighborhood of (Re_{24}, λ_{24}). It shows that (Re_{24}, λ_{24}) is a double bifurcation point according to linear analysis, but a triple bifurcation point when fully analyzed. - - - : Neutral curves for 2 and 4 vortices. —— : Curve of the subharmonic bifurcation points marked (4) and (5) in part a) of this figure. - · - : fold generated by subcritical bifurcation. Either non-existent or very short for $\eta = 0.727$.

in the (Re, λ)-plane. It also shows that there is a λ-interval adjacent to λ_{24} where λ-periodic Taylor vortex flows bifurcate subcritically from Couette flow. This generates a fold where the solution surface bends back to the supercritical domain. The size of this λ-interval is strongly η-dependent. While it is easy to detect this 'subcritical fold' numerically for $\eta = 0.5$ [8], we did not find it for $\eta = .727$. In the narrow gap limit $\eta \to 1$, $\mu \to 1$ it has even been proved [17] that bifurcation from Couette flow is always supercritical (see also [8, p.714f]).

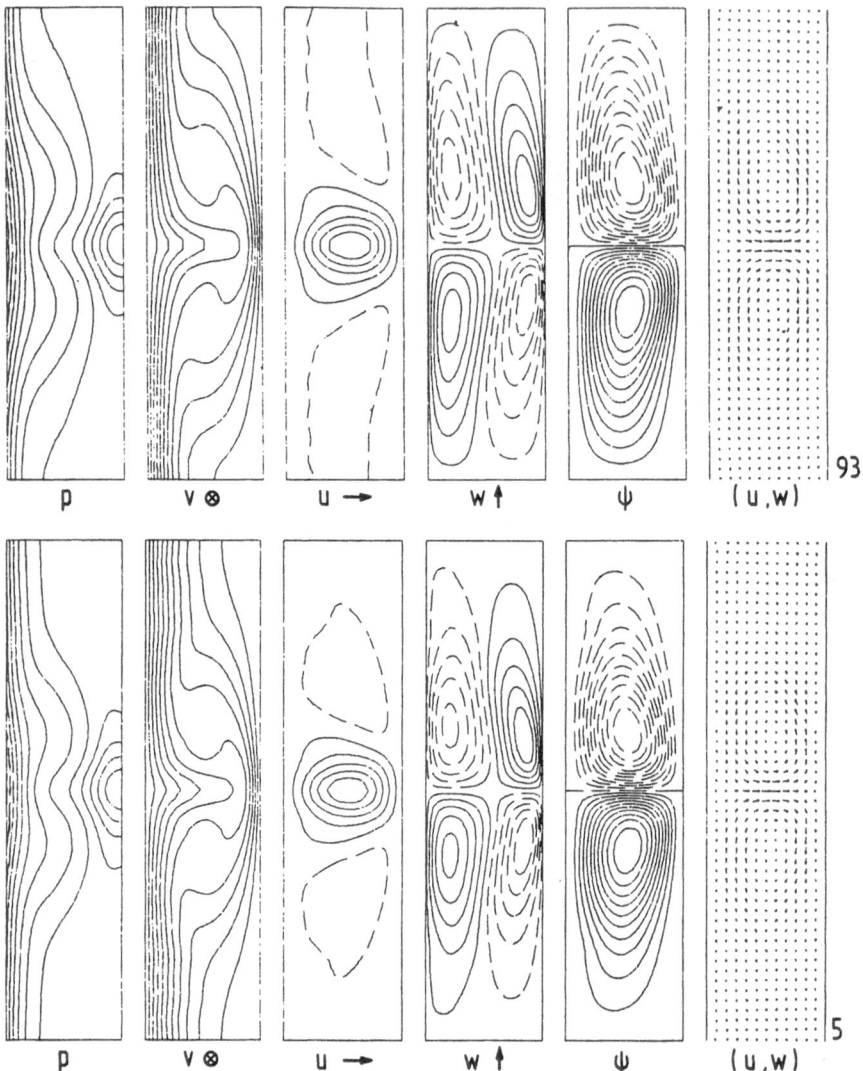

Fig.2: Level lines of reduced pressure p, velocity components v, u, w, stream function ψ, and (u, w) velocity field in the (r, z)-plane for the two solutions marked '93' ($\lambda = 3.9008$) and '5' ($\lambda = 4.1539$) in Fig.3 and in [16]. $Re = 800 \approx 3.65 Re_{cr}, \eta = 0.727, \mu = 0$; M = 31, N = 12.

The fold marked (6), (6′) in Fig.1a, however, seems to exist for all η commonly investigated, and also for a μ-interval. It has been detected for $\eta = 0.5$ [8, Fig.6; 19, Table I], $\eta = 0.75$ [19], $\eta = 0.95$ [18, Fig.1a] in the case $\mu = 0$, and also for $\eta = 0.5, \mu = -0.2$ [11, Fig.2a] and $\eta = 1, \mu = 1$ (?) [4, Fig.6]. Because of its existence for such a wide parameter range and because of its connection to the basic Taylor vortex flow we call it the *basic* $(n, 2n)$-fold, following [7] in part. Why is this (non-generic) fold not contained in Andreichikov's analysis? The reason could be that it does not exist

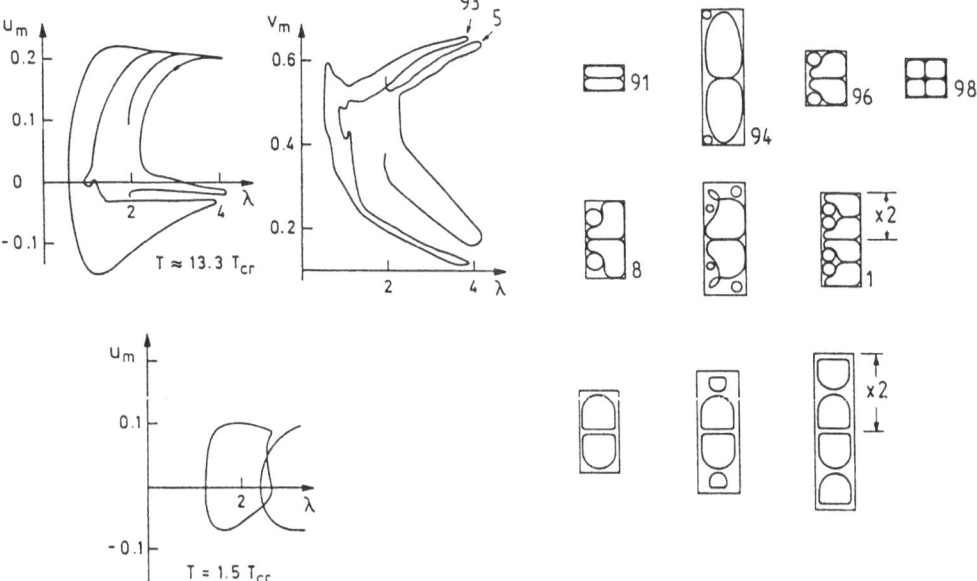

Fig.3: Radial velocity u_m versus λ and angular velocity $v_m := v(1 + \frac{\delta}{2}, 0)$ versus λ for parts of two $T = 13.3T_{cr}$ branches and for the $T = 1.5T_{cr}$ branch of Fig.1; $\eta = 0.727, \mu = 0$. Note that both u_m versus λ plots are drawn in the same scale. The change of flow pattern on these branches is only sketched here. Detailed pictures of the solutions on the $T = 13.3T_{cr}$ branches are given in [16]. M = 31, N = 6 for $T = 1.5T_{cr}$ and M = 31, N = 12 for $T = 13.3T_{cr}$.

in a small neighborhood of (Re_{24}, λ_{24}), but emanates from the bifurcation curve (5) for a slightly larger Reynolds number. This has been observed by M. Paffrath in the case $\eta = 0.95$ for $Re \approx 1.03Re_{24}$ [18, Fig.1a], and is also suggested by unpublished results of M. Henderson in connection with the investigations of [9] for $\eta = 0.727$.

In [7] the basic $(n, 2n)$-fold has been computed for $\eta = 0.727, \mu = 0$ by a fold-following algorithm up to $Re^\delta \approx 290$ or $Re \approx 3.53Re_{cr}(T \approx 12.3T_{cr})$, and then numerical difficulties were encountered. These could be due to the fact that there is another fold nearby. The solutions for $Re = 3.65Re_{cr}$ labeled '5' and '93' in [16] and in Fig.3 are very similar to each other. But they seem to belong not only to different branches, but even to different connected components of the solution set (there might, however, be a point in parameter space where these components meet in a bifurcation point). In Fig.2 we display the structure of both solutions, and in Fig.3 we show portions of both branches in two different projections.

Our experience is that the change of structure along a branch usually does not occur in the fold point, but somewhere else. The exact locus in parameter space is usually affected by discretization errors. This includes: If the change of structure happens close to the fold point, then it might depend on the discretization used on which side of the fold it occurs. In the case of the basic $(n, 2n)$-fold under consideration here, we also

171

found the locus of the fold to be strongly affected by discretization errors of the Fourier approximation. Figure 5 giving this fold has thus to be read with care; see also Fig.6. On the other hand, all studies of the discretization error of the code TAYPERIO [14, 8, 15, 21] have confirmed that the discretization error affects only details, but not the main structures observed, as long as there are enough Fourier components used so that solutions do not completely disappear [15]. The main effect of the discretization error is a shift of whole structures in parameter space, and possibly also a deformation of these structures. This will be discussed in more detail elsewhere.

Since there is a close relation between Taylor vortex flows and convection layers between two infinitely extended parallel plates [6, 5], the analysis of the bifurcation pattern in a neighborhood of the corresponding double bifurcation point is of interest here (see [4, 5, 2] for treatment with periodic boundary conditions). Briefly, the main result is: There are a pure-mode 2-roll solution and its periodic repetition, a 4-roll solution, and then two [20, p.29] antimetric mixed-mode solutions called 'transition solutions', and also asymmetric mixed-mode solutions called 'mean component solutions'. The two branches of transition solutions each bifurcate from a 2-roll solution and from a 4-roll solution and thus connect them [4]. On one of them the little rolls come in where the flow goes inward. We call these solutions the 'inward transition solutions'. (In [11]they were called 'alternating cell solutions'). On the other branch the little rolls come in where the flow goes outward. We call these solutions the 'outward transition solutions'.

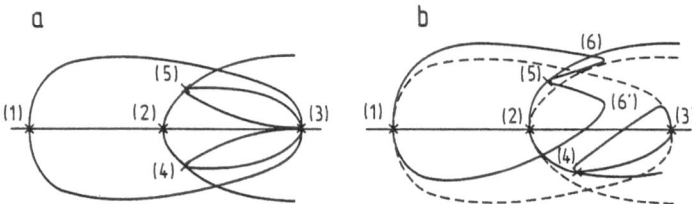

Fig.4: Sketch tentatively summarizing how the breaking of the midplane symmetry changes the bifurcation diagram according to [4, 5, 2, 20]. a) Symmetric case (Bénard problem). Loop (1)-(3)-(1): pure-mode 2-roll solutions, bifurcating from the trivial solution in points (1) and (3) on the neutral curve. The upper and lower branches contain solutions shifted by $\frac{\lambda}{2}$. In the text, these 'twins' are treated as one solution, but they are in fact two [15]. The periodic repetition of this loop bifurcates in point (2). Loop (5)-(3)-(5): inward transition solutions. The branch bifurcates in point (5) from the 4-roll branch and meets in point (3) both the 2-roll branch and the trivial solutions. The solutions are mixed-mode solutions with 2 large and 2 small rolls. The small rolls disappear/grow where the flow is directed inward. Loop (4)-(3)-(4): outward transition solutions. This time the rolls disappear/grow where the flow is directed outward. b) Unsymmetric case (Taylor problem). The triple bifurcation point (3) is broken up into the bifurcation point (3) and the two fold points (6), (6'). The bifurcations (4) and (5) now occur at different values of the period λ.

Their structures are similar to the ones given in a sketch of [15, Fig.2]. Convection rolls are symmetric with respect to the midplane between the two plates. This symmetry is broken by the curvature and by the different speeds of the two cylinders in the Taylor problem. In Fig.4 we tentatively sketch how to transform the conclusions drawn in [4, 5, 2, 20] for the Taylor problem into the nomenclature of Fig.1a.

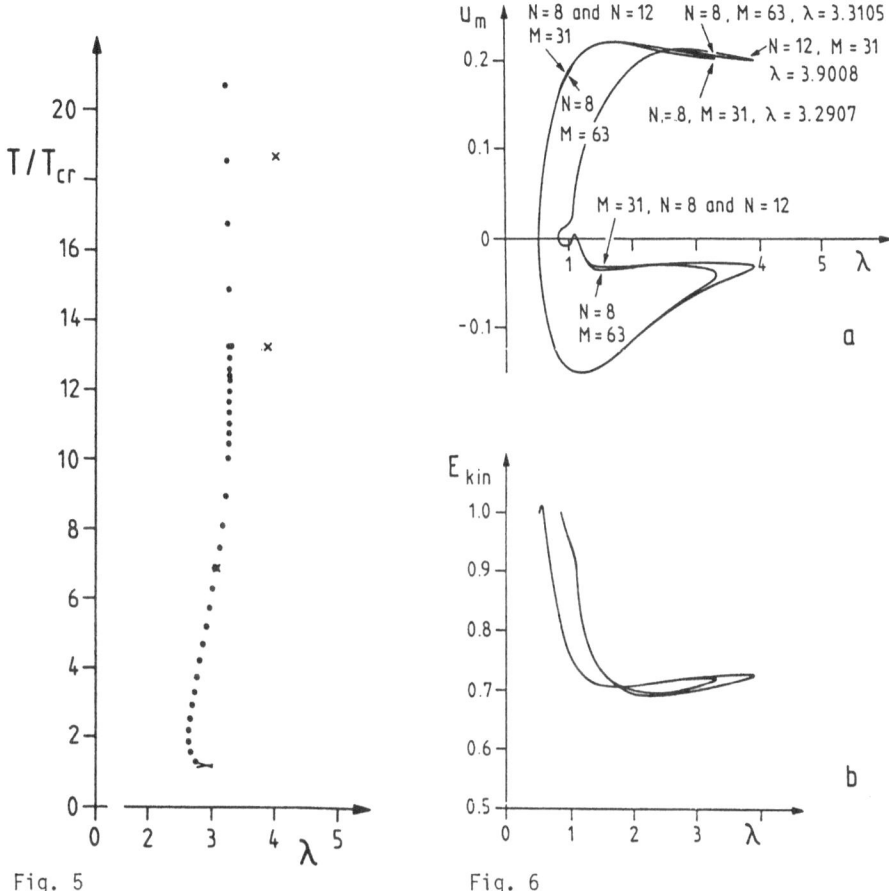

Fig. 5   Fig. 6

Fig.5: Basic $(n, 2n)$-fold for $n = 2$. The fold points were computed by λ-continuation for each fixed T. $\bullet$: $M = 31$ radial grid points, $N = 8$ Fourier components. $\circ$: $M = 63$, $N = 8$. $\times$: $M = 31$, $N = 12$. $N = 12$ is not sufficient for larger T: For $T = 13.3T_{cr}$, all 12 Fourier components are of the same order of magnitude, 10^{-2}.

Fig.6: Study of discretization error for the basic 2-vortex branch for $Re = 800$. The neighborhood of the fold turns out to be very sensitive to cut-off errors in the Fourier expansion. a) u_m versus λ plot. b) Kinetic energy E_{kin} versus λ. E_{kin} is normalized so that $E_{kin} = 1$ for the Couette flow with the same Re.

3. The basic 2-vortex surface

As has been pointed out in Section 2, we still cannot make a final statement about the basic $(n, 2n)$-fold for higher Reynolds numbers. The form of the fold, its locus in parameter space, and maybe even its existence for higher Re are subject to discretization errors. The existence of the two surfaces meeting in the fold, however, and also the structure of the solutions on these surfaces, do not seem to be affected by discretization. In Fig.6 we show a cut through the surface for $Re = 800 \approx 3.65 Re_{cr}$ in two different projections as computed with three different discretization errors. For all three discretizations we found the same patterns: bifurcation of a 2-vortex flow from Couette flow on the neutral curve; growth of both vortices with a growing difference between the maximum outward and the maximum inward speed; generation of two additional

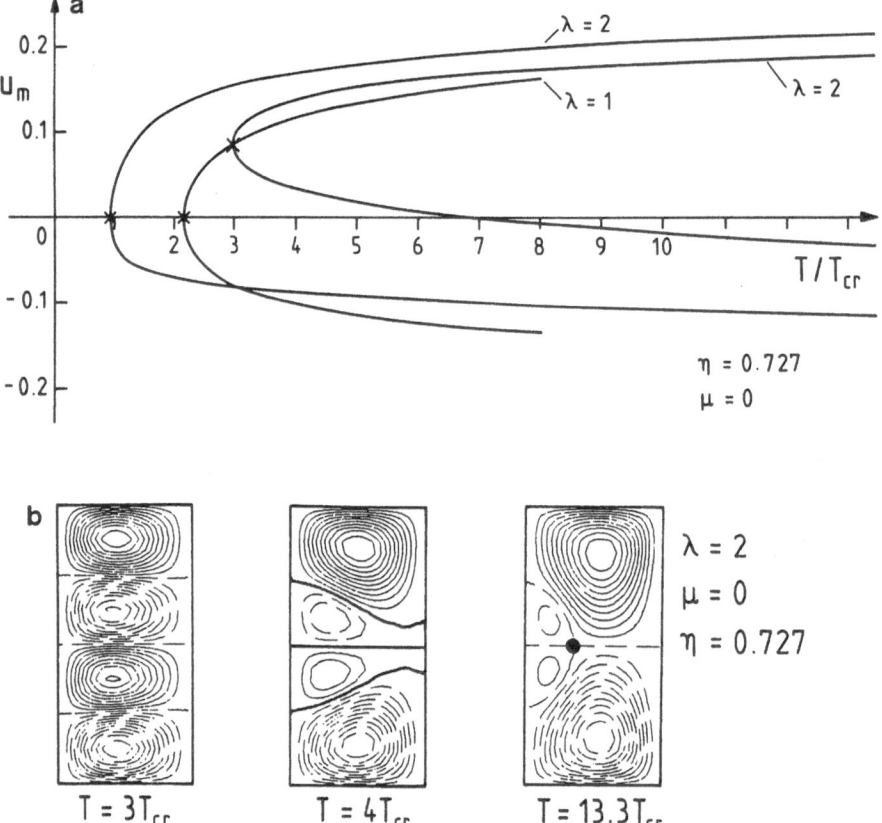

Fig.7: a) u_m versus T/T_{cr} plot for the basic 2-vortex branches with $\lambda = 2$ and $\lambda = 1$, and a transition solution branch bifurcating from the periodic repetition of the $\lambda = 1$ branch. b) Level lines of the stream function ψ for three mixed-mode solutions on the branch shown in part a. • : free stagnation point. Computations with M = 31, N = 8.

vortices in the inward region and thus formation of an inward transition solution; formation of a double-vortex flow; bifurcation to Couette flow on the eigencurve generated by the second eigenvalue of the linear differential operator related to Couette flow [16].

The formation of the double-vortex flow did not occur for $Re = \sqrt{1.5}Re_{cr}$; see Fig.3. We thus now investigate how the surfaces of these two cuts for fixed Re are connected for fixed λ and varying Re. Figure 7a shows a more complete version of a

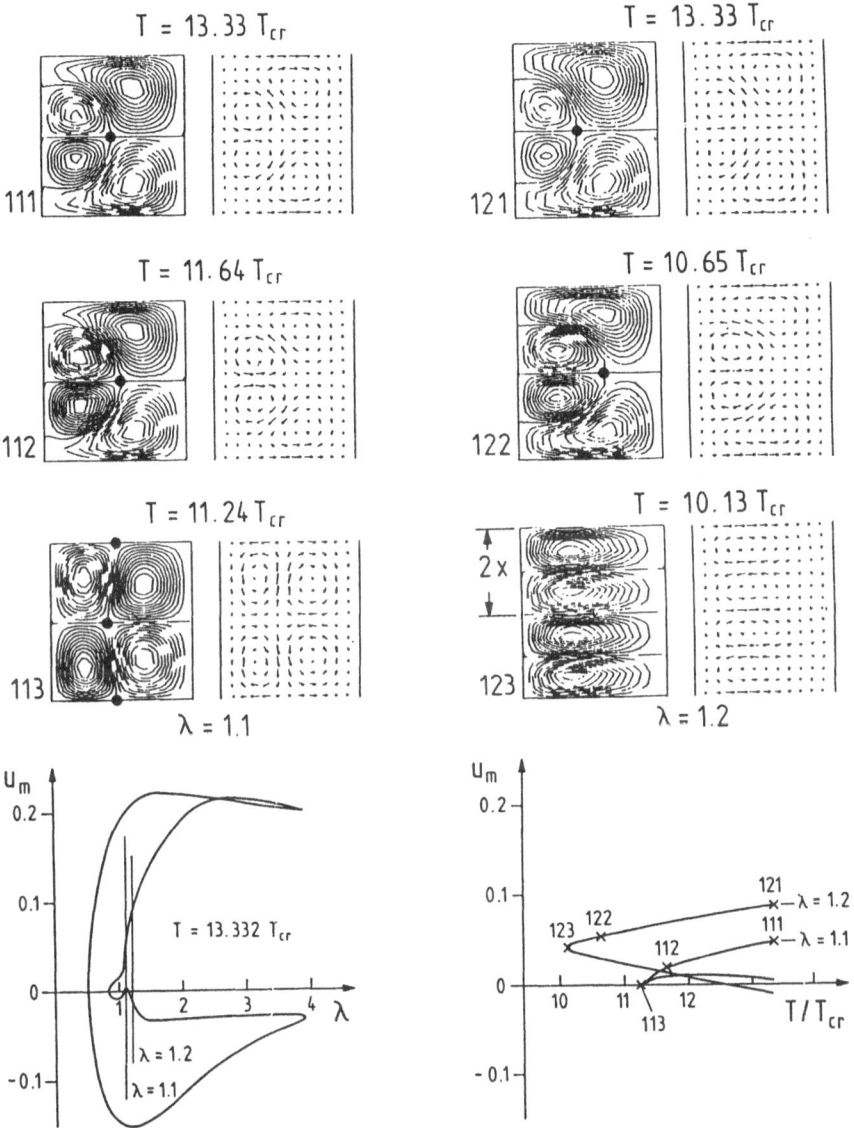

Fig.8: Further investigation of the surface. While there is still subharmonic bifurcation for $\lambda = 1.2$, there is a continuous change into a double-vortex flow and bifurcation to Couette flow for $\lambda = 1.1$. • : free stagnation point. M = 31, N = 12.

figure already given in [15, Fig.6]. In Fig.7b we show three solutions on the $\lambda = 2$ branch connecting the inward transition solution on the $Re = 800$ branch of Fig.6 with the 2-vortex solution for $\lambda = 1$ in a subharmonic bifurcation. Figure 8 gives the same type of investigation for λ-values close to the 'nose' in Fig.6. For $\lambda = 1.2$ there is still subharmonic bifurcation. For $\lambda = 1.1$, however, there is bifurcation of the double-vortex flow to Couette flow. Figure 9 gives the bifurcation curve of the double-vortex flow from Couette flow together with the common neutral curve, and periodic repetitions of both curves. Note that both curves intersect with the first periodic repetition of the neutral curve. These two intersection points (Re_{24}, λ_{24}) and (Re_{d4}, λ_{d4}) seem to be connected by a curve of subharmonic bifurcations: wavelength-halving bifurcations between the inward transition solution with period λ and the basic 2-vortex flow with period $\lambda/2$. This result should be checked with a bifurcation-following algorithm. Until now, we know only four such bifurcation points: For $T = 1.5T_{cr}$ (point (5) of Fig.1a), for $T = 4T_{cr}$ [21], for $\lambda = 2$ (Fig.7), and for $\lambda = 1.2$ (Fig.8). It would be very interesting to see an analysis of the bifurcation pattern in a neighborhood of the double bifurcation point (Re_{d4}, λ_{d4}). Does the curve of the other subharmonic bifurcations ((4) in Fig.1) also go through this point?

If we now complete Fig.9 in our mind by all the periodic repetitions of the curves given there and if we imagine that there will also be curves of higher-order subharmonic

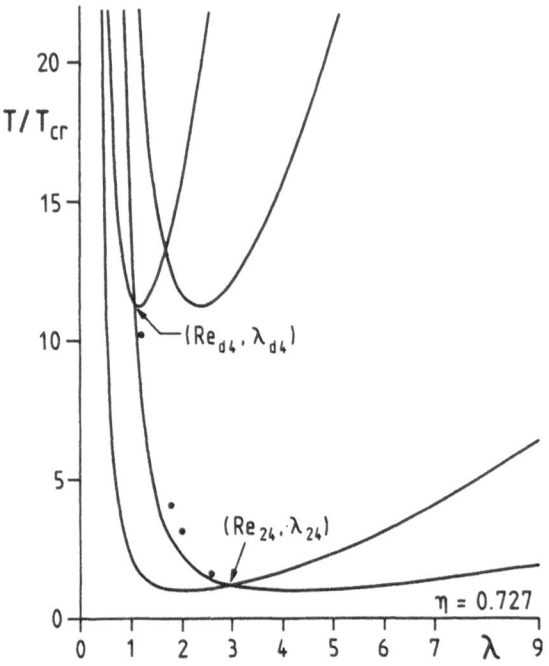

Fig.9: Neutral curve for Couette flow (first eigenvalue) and bifurcation curve of the double-vortex flow (second eigenvalue) in the (T, λ)-plane, together with their first periodic repetitions. • : subharmonic bifurcation as described in the text.

bifurcations and folds, and they all intersect and generate new curves of bifurcation points and folds where they intersect ..... . Then it's not surprising anymore that there are so many different solutions, and branches as complicated as the one displayed in [16, Fig.3].

4. Acknowledgement

Most of this work was done while one of the authors (R. M.-S.) was visiting Freie Universität Berlin. Both authors would like to express their thanks to R. Gorenflo for valuable discussions and for providing financial support from FU-FPS "Modellierung und Diskretisierung". We thank H. Specht, D. Armbruster, F. H. Busse, M.E. Henderson, C.A. Jones, H.B. Keller, D. Lortz, H. Riecke, Ph. Saffman, I. Stewart and L. Ungar for stimulating discussions. A number of these discussions took place during this meeting - special thanks are due to W. Güttinger and G. Dangelmayr for providing this opportunity.

5. References

1. I.P. Andreichikov (1975) Branching of secondary modes in the flow between rotating cylinders. Fluid dynamics (1977), translated from Izv. Akad. Nauk SSSR, Mekh.Zhidk. Gaza, No. 1, 47 – 53

2. D. Armbruster, O(2)-symmetric bifurcation theory for convection rolls. To be published

3. J.E. Burkhalter and E.L. Koschmieder (1974) Steady supercritical Taylor vortices after sudden starts. Phys. Fluids **17**, 1929 – 1935

4. F.H. Busse and A.C. Or (1986) Subharmonic and asymmetric convection rolls. ZAMP **37**, 608–623

5. F.H. Busse, Transition to asymmetric convection rolls. Proc. "Bifurcation", Dortmund 1986. (ed. Küpper, Seydel, Troger), Birkhäuser, in press

6. S. Chandrasekhar (1961) **Hydrodynamic and Hydromagnetic Stability**. Oxford University Press

7. N. Dinar and H.B. Keller, Computations of Taylor vortex flows using multigrid continuation methods. To be published

8. G. Frank and R. Meyer-Spasche (1981) Computation of transitions in Taylor vortex flows. ZAMP **32**, 710 – 720

9. M.E. Henderson and H.B. Keller, Complex bifurcation. To be published

10. C.A. Jones (1981) Nonlinear Taylor vortices and their stability. JFM **102**, 249 – 261

11. C.A. Jones (1982) On flow between counter-rotating cylinders. JFM **120**, 433 – 450

12. H.B. Keller (1977) Numerical solution of bifurcation and nonlinear eigenvalue problems. In: **Applications of Bifurcation Theory** (ed. P. Rabinowitz), 359 – 384

13. K. Kirchgässner (1961) Die Instabilität der Strömung zwischen rotierenden Zylindern gegenüber Taylorwirbeln für beliebige Spaltbreiten. ZAMP **12**, 14 – 30

14. R. Meyer-Spasche and H.B. Keller (1980) Computations of the axisymmetric flow between rotating cylinders. J. Comp. Phys. **35**, 100 – 109

15. R. Meyer-Spasche and H.B. Keller (1985) Some bifurcation diagrams for Taylor vortex flows. Phys. Fluids **28**, 1248 – 1252

16. R. Meyer-Spasche and M. Wagner, Steady axisymmetric Taylor vortex flows with free stagnation points of the poloidal flow. Proc. "Bifurcation", Dortmund 1986. (ed. Küpper, Seydel, Troger), Birkhäuser, in press

17. S.N. Ovchinnikova and V.I. Yudovich (1974) Stability and bifurcation of Couette flow in the case of a narrow gap between rotating cylinders. Prikl. Mat. Mekh. **38**, No.6

18. M. Paffrath (1986) Das Taylorproblem und die numerische Behandlung von Verzweigungen. Ph. D. dissertation in mathematics, U Bonn

19. H. Riecke and H.-G. Paap (1986) Stability and wave vector restriction of axisymmetric Taylor vortex flows. Phys. Rev. A **33**, 547

20. H. Riecke (1986) Wellenlängeneinschränkung in quasi-eindimensionalen strukturbildenden Systemen. Ph. D. dissertation in physics, U Bayreuth

21. H. Specht and M. Wagner, Diplomarbeit in Mathematik, FU Berlin, in preparation

22. G.I. Taylor (1923) Stability of a viscous liquid contained between two rotating cylinders. Phil. Trans. Roy. Soc. A, vol CCXXIII, 289 – 343

Part III

Interfacial Patterns

Nonlinear Interactions of Interface Structures of Differing Wavelength in Directional Solidification

M.J. Bennett[1], *R.A. Brown*[1], *and L.H. Ungar*[2]

[1]Department of Chemical Engineering,
 Massachusetts Institute of Technology, Cambridge, MA 02139, USA
[2]Department of Chemical Engineering,
 University of Pennsylvania, Philadelphia, PA 19104, USA

Abstract

The solution structure of the cellular patterns formed during the directional solidification of a material containing a dilute impurity is studied by numerical analysis of a simple model. Solution families corresponding to cells of different spatial wavelength arise as the growth velocity is increased above the value for the onset of the instability. The resonant interaction between these steady-state shapes leads to both stable and unstable time-periodic oscillations. Only some of the oscillatory interface structures computed here have been predicted by normal mode analyses of codimension-two bifurcations. The interactions between cells with different spatial structures limit the range of validity of asymptotic expansions to a velocity range too small to be reliably measured experimentally.

1. Introduction

When a metal or semiconductor containing an impurity is solidified, impurities rejected during solidification build-up ahead of the solidification front, and can destabilize it [1]. In directional solidification, where the average rate of solidification is set by pulling the solidifying material through an imposed temperature gradient at a constant velocity, these instabilities of the planar solidification front take the form of bifurcations to sinusoidally varying interfaces of different wavelength [2]. Linear analysis gives the critical velocity for the onset of instability as a function of the wavelength of the disturbance [3].

For solidification, as for many physical systems exhibiting pattern formation, the wavelength of the disturbance to which stability is first lost is easily calculated, but at solidification rates above the critical rate the planar interface is unstable to disturbances with a range of wavelengths. Long wavelength instabilities restrict the range of wavelengths for which deformed interfaces are stable, but do not select a unique wavelength. In spite of much excellent work, in systems where natural and forced convection drive pattern formation, a general description of the interactions between solutions of different wavelength has not been found and wavelength selection is not well understood [4-6].

Numerical calculations of finite amplitude interface shapes require selection of a specific computational domain. Since only melt/solid interfaces with wavelengths that are integer fractions of the width of the computational sample can be represented, this restriction has the effect of selecting a discrete set of velocities at which stability is lost to perturbations of different wavelength. At each critical velocity, a family of solutions with a sinusoidal interface of a different wavelength bifurcates from the family of planar solutions. The families with nonplanar interfaces, although of different symmetry (i.e. different dominant wavelength or "mode"), do not remain independent but intersect or connect via mixed-mode families at finite amplitude.

One method for understanding the structure of the interactions of the families in a discrete computational sample is to study the secondary bifurcations arising from the double points which occur when solidification conditions are picked so that two families with interfaces of different wavelength evolve from the planar state at the same growth rate. Both steady and oscillatory interface families have been predicted to arise from codimension-two interactions in directional solidification [7]. Such secondary steady and time-periodic solutions arising from double points between solutions of different spatial symmetry are predicted for systems governed by a wide class of equations [8]. In fact, multiple bifurcations between solution families of different spatial wavelength occur in almost all systems governed by partial differential equations, and have been studied in - among other systems - Taylor-Couette flow [9] and natural convection [10].

Solidification is a particulary rich example of interactions between multiple wavelengths because the marginally stability curve is typically very flat; at growth rates only slightly above the critical value associated with the most dangerous wavelength, there are cellular shapes with many different wavelengths. The system is thus extremely near the parameter values for a codimension-two bifurcation and any increase in the growth rate above the critical value results in the nonlinear interaction of different spatial modes. This paper presents nonlinear calculations which demonstrate the interaction of different nonlinear modes near the onset of instability. The interactions of many modes close to the onset of instability has dramatic implications for efforts to compare asymptotic analysis with experiment: We predict many transitions between melt/solid interfaces of different wavelength to occur within even a few percent change in the solidification rate above the value for onset. Because solidification rates have not been controlled with greater than ten percent accuracy, this casts doubt on whether the predicted onset wavelength can be accurately measured experimentally.

2. Model and Solution Method

We view unidirectional solidification of a binary melt from a reference frame attached to a planar melt/solid interface moving at the imposed average solidification rate, V. The temperature in the melt and solid is described in the Cartesian coordinate system shown in Fig. 1 and the melt/solid interface is located at $y = h(x,t)$.

We assume that the latent heat of solidification is small and that the thermal conductivities of the melt and solid are similar enough that the melt/solid interface does not significantly disturb the imposed temperature gradient. These assumptions lead to the solutal model of directional solidification, which is described in more detail elsewhere [11,12]. The shape of the melt/solid interface

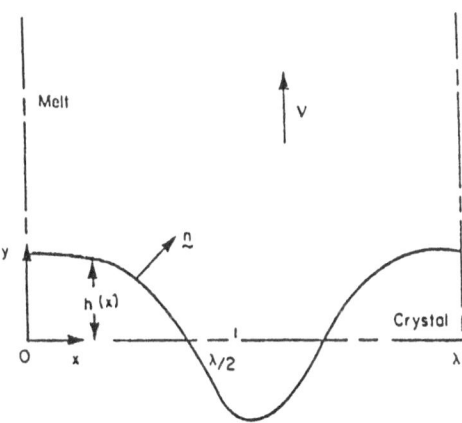

Fig. 1. Schematic of the melt/solid interface and coordinate system used.

is then set solely by the deformation of the solute composition field and its interaction with the imposed temperature field

$$T(y) = T_m^o + y G \quad , \qquad (2.1)$$

where T_m^o is the melting temperature of the planar interface located at $y = 0$, and G is the imposed temperature gradient. All terms here and below are non-dimensionalized by scaling length, temperature and concentration with the critical wavelength (λ_c), the melting temperature of a flat interface in the absence of solute (T_m) and the bulk solute concentration (c_∞), respectively. The solute conservation equations are:

in the melt

$$\nabla^2 c + P \, \partial c/\partial y = \partial c/\partial t \qquad (2.2a)$$

and in the solid

$$R_m \nabla^2 c + P \, \partial c/\partial y = \partial c/\partial t \qquad (2.2b)$$

where $R_m = D_S/D$ is the ratio of diffusivities in the solid and melt and $P = \lambda_c V/D$ is the solutal Peclet number. Solute is conserved across the melt/solid interface:

$$\underline{n}.\nabla c - R_m \, \underline{n}.\nabla c = \underline{n}.\underline{e}_y P(1+\partial h/\partial t)c(k-1) \quad . \qquad (2.3)$$

For two-dimensional interfaces represented as $h=h(x,t)$ in Cartesian coordinates (x,y),

$$\nabla^2 = \partial^2/\partial x^2 + \partial^2/\partial y^2, \quad \nabla = \underline{e}_x \partial/\partial x + \underline{e}_y \partial/\partial y, \qquad (2.4)$$

and the interface $h = h(x,t)$ has normal

$$\underline{n} = (\underline{e}_y - \underline{e}_x h_x)/(1 + h_x^2)^{1/2} \quad . \qquad (2.5)$$

The temperature at the interface is given by

$$T_I = 1 + mc + \Gamma \, 2 \, H \quad , \qquad (2.6)$$

where m is the liquidus slope of the phase diagram, Γ is the surface free energy, and the curvature of the interface is given by

$$2 \, H = h_{xx}/(1+h_x^2)^{3/2} \quad . \qquad (2.7)$$

As throughout the paper, the critical wavelength (λ_c) has been used to nondimensionalize all lengths.

Equating the interface temperature given by Eqn. (2.6) with the temperature given by Eqn. (2.1) gives a relation between interface position, curvature, and concentration. The surface free energy, Γ, is typically two or three orders of magnitude smaller than G or m. Because the surface free energy sets the wavelength of the interface by preventing the formation of cellular patterns with short length scales, the low surface free energy allows interface shapes with a range of wavelengths, making the marginal stability curve very flat. As is shown below, the exceedingly flat marginal stability curve gives solidification a behavior somewhat different from that of many other continuum transport and fluid dynamics systems.

The equations are completed by using reflective symmetry boundary conditions on the sides of the computational domain, x=0 and x=λ. The equations are mapped to a fixed domain, so that the interface position occurs as an ordinary variable, are discretized using a Galerkin/finite element method and are solved using Newton's method. Stable and unstable steady solution families are tracked in solidification rate using a numerical perturbation analysis of the above equations with all time dependence removed. Stability is tested and time-varying states are calculated using a fully implicit integration method applied to the transient equations. See Ungar and Brown [13] for details.

3. Results

Linear analysis of the stability of a planar solidification front gives the solidification rate at which the interface is unstable to perturbations of given wavelength. Because our numerical calculations are done in a fixed computational domain, only interface shapes with wavelengths that are integer fractions of the computational domain are computed. Calculating all of the different interconnecting interface families and determining their stabilities would provide a complete map of the system's behavior. However, this is intractable for large computational domains because there are too many interface families. One way of avoiding this work is simply to do transient (time-dependent) calculations. However, transient calculations only find stable interfaces, making it very hard to understand the solution structure and hence the interactions between families of different wavelength. An alternative approach, local perturbation around the point of loss of stability, and particularly around points where two modes simultaneously become unstable, provides much information about local behavior. We will, however, show that the range of validity of these analyses is strongly limited.

We adopt a third strategy, and conduct numerical investigations of systems carefully selected to permit the desired level of complexity. The range of solidification velocities and of wavelengths to which a planar interface is unstable depends on the properties of the system being solidified. In particular, the range of instability is sharply limited when the solute concentration is low or the segregation coefficient is near unity (there is little solute rejection). The width of the computational domain can also be expanded or contracted to allow or exclude interfaces of different wavelength. We thus select solidification conditions to obtain the desired number of interface families.

Calculations are presented below for parameters typical of succinonitrile containing impurities of differing segregation coefficients. To better show the full range of solution behaviors obtained, we present results both for no diffusivity in the solid (System 1) and for equal diffusivity in the melt and solid (System 2). A complete list of parameters is given below in Table 1.

TABLE 1. Dimensionless parameter values used in calculations.

	System 1	System 2
k	0.865	0.1
R_m	0.0	1.0
G	1.7×10^{-4}	2.1×10^{-3}
m	-6.7×10^{-4}	-6.7×10^{-4}
Γ	2.2×10^{-6}	1.6×10^{-7}
nondimensionalized with length:		
λ_c	8.41 microns	117 microns

Marginal stability curves for segregation coefficients of $k = 0.865$, 0.8, and 0.1 are shown in Fig. 2. For high values of k, the planar interface is stable to all infinitesimal perturbations. For slightly lower segregation coefficients, such as 0.865 for this succinonitrile system, the range of wavelengths which cells can have is severely limited.

At a segregation coefficient of 0.865, as Fig. 2 shows, the planar interface is unstable to disturbances with wavelength between roughly $0.4\lambda_c$ and $1.2\lambda_c$, where λ_c denotes the critical wavelength. The lower portion of the marginal stability curve is shown in Fig. 3; the critical wavelengths and solidification rates at which the planar interface becomes unstable to disturbances which fit in a computational domain of twice the critical wavelength are marked . A computational domain of half of the critical wavelength only allows cells of wavelength λ_c and $\lambda_c/2$. These solutions are denoted by circles on Fig. 3.

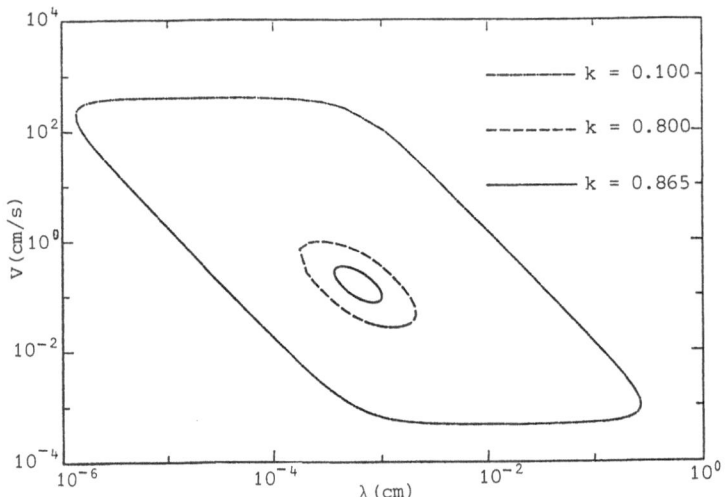

Fig. 2. Marginal stability curves for segregation coefficients of k = 0.8 and 0.865 for R_m=0.0 and k=0.1 for R_m=1.0.

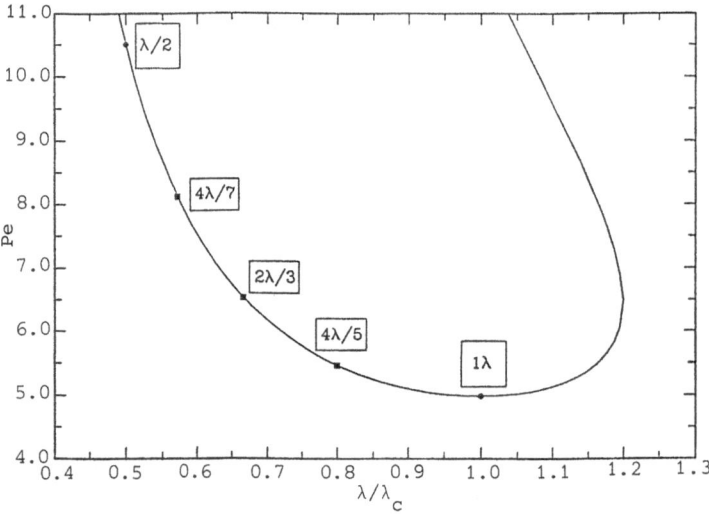

Fig. 3. Marginal stability curve for System 1 showing where the system loses stability to different wavelengths.

The amplitude of the interfaces which fit in a computational domain of half the critical wavelength are plotted as a function of growth rate in Fig. 4. The λ_c family of interfaces bifurcates supercritically (to higher values of P) from the planar family and continues until it ends on the λ_c family. Although the shapes in the λ_c family are initially stable, they lose stability at a Hopf bifurcation before reaching the λ_c family. Time-dependent calculations show the family of time-periodic solutions evolves subcritically and is unstable. The time scale of the oscillatory solution is of order one in terms of diffusion units, as is expected for a Hopf bifurcation.

The Hopf bifurcation described above arises from the interaction of the λ_c and $\lambda_c/2$ modes, and is of the form predicted by Haug [7] in a normal mode analysis. Haug used a normal mode analysis for two modes which have almost the same critical

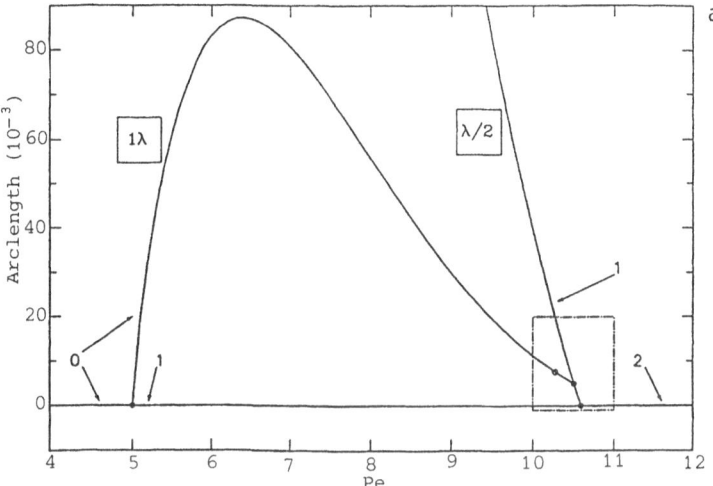

Fig. 4a. Bifurcation diagram of amplitude (measured by interface arclength) vs. dimensionless growth velocity (Pe) for System 1. Solutions are constrained to lie in a computational domain of half the critical wavelength. The number of positive eigenvalues is given on each curve.

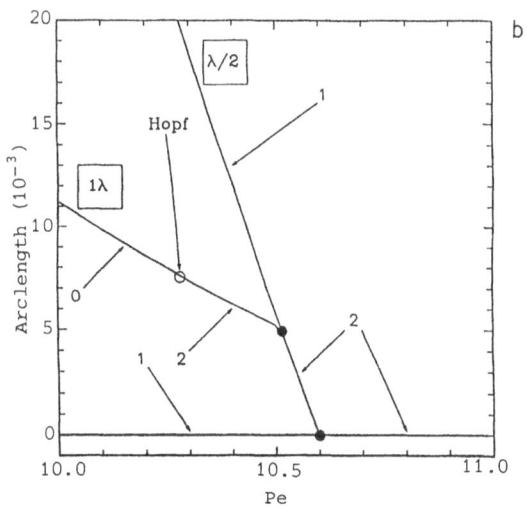

Fig. 4b. Enlarged picture of the boxed region of Fig. 4a.

185

growth velocity to show that Hopf bifurcations can arise in a certain class of solidification problems when the first bifurcation is supercritical and it connects without limit points to a subcritically bifurcating family. This is the case for the calculations in Fig. 4a.

A loose argument predicting the existence of the Hopf bifurcation can be simply made by looking at the signs of the eigenvalues corresponding to the two dominant modes, and assuming that a single eigenvalue changes sign at each simple bifurcation point. The number of unstable eigenvalues along each branch is indicated on Fig. 4b. There are no positive eigenvalues on the left side of the critical wavelength family, but there must be two positive eigenvalues on the right side of the curve. Therefore, two eigenvalues must pass into the left half of the complex plane somewhere along the critical wavelength family. Since no other steady families are present, this must be a Hopf bifurcation.

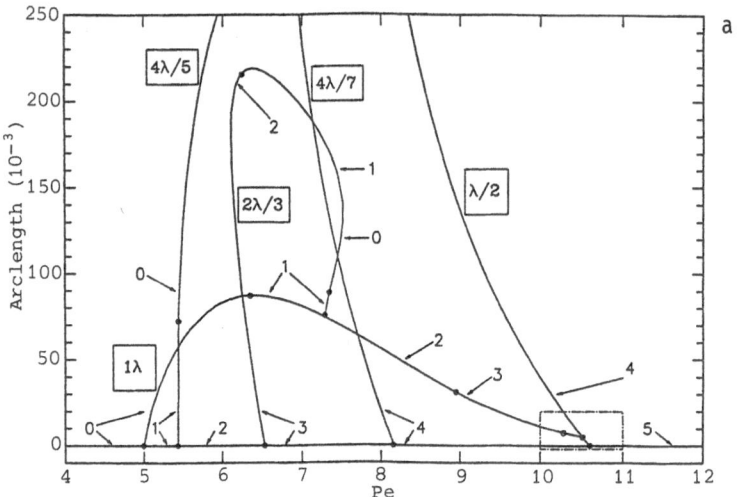

Fig. 5a. Bifurcation diagram for same systems as Fig. 4, but calculated on a domain of twice the critical wavelength. The number of positive eigenvalues is given on each curve. Bifurcation points are indicated by solid dots.

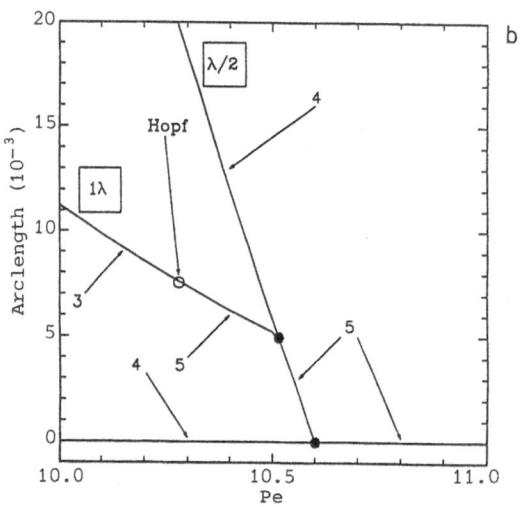

Fig. 5b. Enlarged picture of the boxed region in Fig. 5a.

The calculations represented by Fig. 4 were made using a computational domain of half the critical wavelength in order to limit the number of modes present. A computational domain of twice the critical wavelength allows additional cells of 2/3, 4/5, and 4/7 the critical wavelength, as shown by the marginal stability curve as in Fig. 3. The corresponding bifurcation diagram computed for this larger sample is shown in Fig. 5. There are now many interface families, all connected by "mixed mode" families that arise from secondary bifurcations. Although the Hopf bifurcation still exists in this computational domain, it now occurs off of a solution family that is already unstable.

Haug examined the interactions between two interface shapes with spatial wavelengths related by the ratio of n to (n+1), where n is an integer, and showed that secondary bifurcations arise. The interactions in Fig. 5 between the interface family with the critical wavelength and families of 1/2, 2/3 and 4/5 the critical wavelength correspond to interactions between modes 1 and 2, 2 and 3, and 4 and 5, respectively in the analysis of Haug.

Haug's analysis has several limitations: it only predicts what solution structures may arise; it does not predict what solution structures will be seen for specific parameter values, and it does not predict whether or not the time-periodic solutions will be stable. Also, it is only strictly valid when the dominant nonlinear interactions involve only two modes. When interface shapes of different wavelength are present, the analysis can, and as we show below does, fail.

4. Interface Shapes Arising from Multiple Interacting Modes

At higher values of the segregation coefficient, k, the marginal stability curve becomes flatter and more interface shapes of different wavelength exist close to the onset of instability. A marginal stability curve for System 2 is shown in Fig. 6. The curve is very flat; within one percent of the critical velocity at which stability is lost, families of wavelength between roughly $0.4\lambda_c$ and $3\lambda_c$ lose stability.

As would be expected from the marginal stability curve, families of different wavelength abound. Figure 7 shows a partial diagram of the different families. Of particular interest is the Hopf bifurcation off of the λ_c family. The bifurcating family is stable in this case. Unlike the Hopf bifurcation presented above, it arises from the interaction of multiple modes. It cannot occur in a computational domain of half the critical wavelength. In fact, it occurs under conditions for which Haug's analysis predicts no Hopf bifurcation.

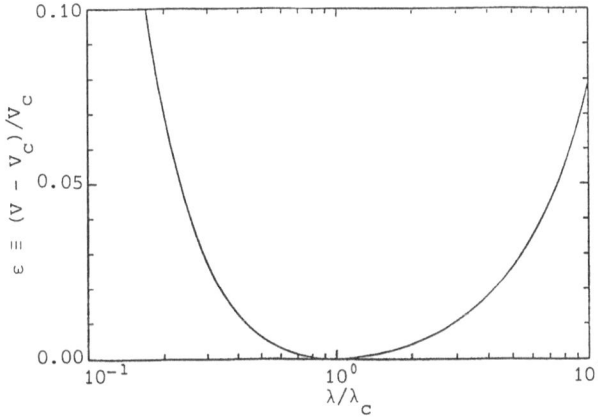

Fig. 6. Marginal stability curve for System 2.

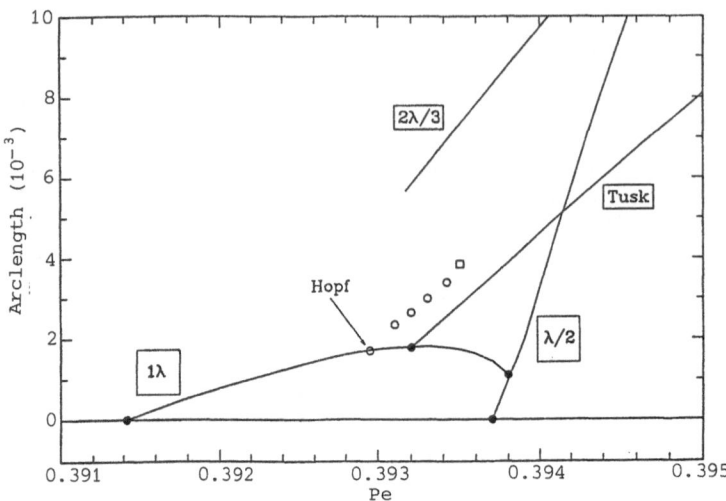

Fig. 7. Bifurcation diagram for system 2 showing solutions in a computational domain of twice the critical wavelength.

The solution growing out of the Hopf bifurcation oscillates between solutions with two and with three cells in the computational domain. (See Fig. 8.) These interfaces oscillate on a time scale which is very long compared to the diffusion time. It is possible that they are caused by an additional nonlinear interaction that causes zero-speed traveling waves, rather than a conventional Hopf bifurcation.

5. Discussion

The flat marginal stability curve typical of solidification problems leads to the presence of many solution families of different wavelength close to the onset of instability. Since any pair of these families can be brought together at a codimension-two point on the planar family by appropriate alteration of the computational domain, it is no surprise that secondary bifurcations occur. The secondary bifurcations can directly connect the bifurcating families (as is the case between any family and a family of half its wavelength) or indirectly connect the bifurcating families (as is the case between the other solution families shown). We have also found that the modes of different wavelength can interact to give, among other phenomena, time-periodically oscillating interfaces. Similar interactions are present in other systems such as Taylor-Couette flow, but occur much farther from the onset; for typical parameter values, the marginal stability curve for solidification problems is two orders of magnitude flatter than it is for Taylor-Couette flow [9].

The interaction of modes of different wavelength limits the range of validity of local asymptotical analyses. One cannot assume that all other modes are small compared to the mode to which stability is first lost. Worse, one cannot even assume that the first two wavelengths dominate; many wavelengths are important, even close to onset. Thus asymptotic analyses will typically fail less than one percent above criticality. Multiple length scale analyses which look at the effect of long wavelength perturbations on phase behavior [3] have an equally sharply limited range of validity, again because of the strong interactions of the short wavelength modes. Only a fully nonlinear analysis [5] can determine how the band of wavelengths of possible stable cells is limited by the Eckhaus instability.

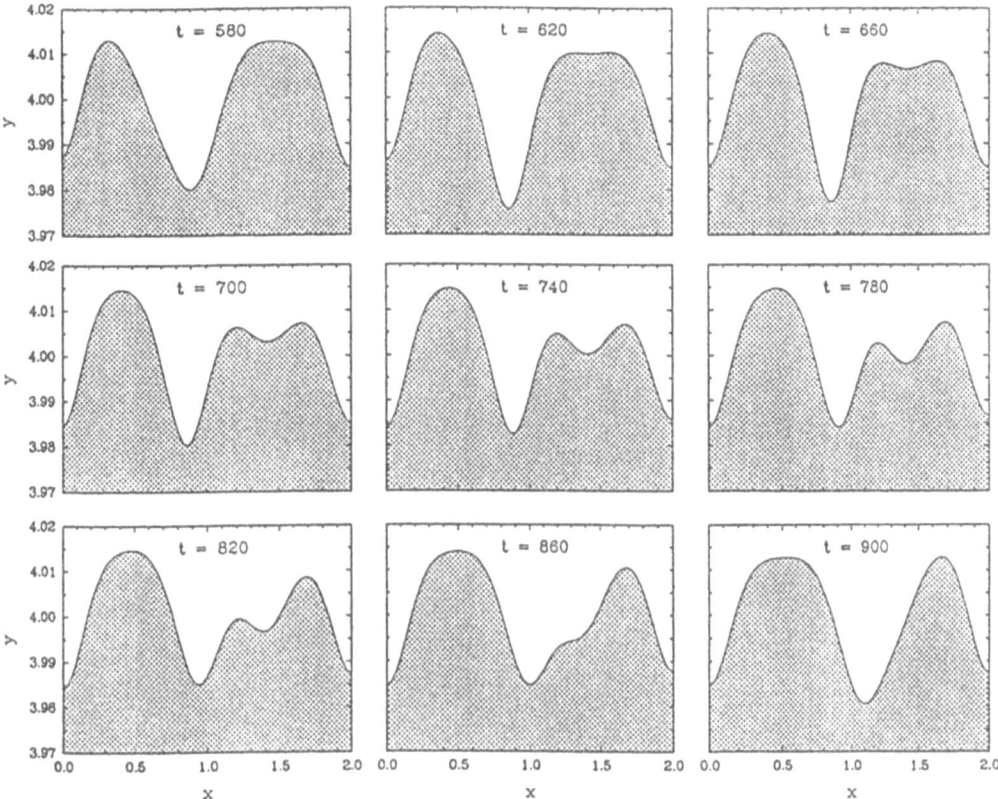

Fig. 8. Sequence of interface shapes from the oscillatory solution at Pe = 0.3932 on Fig. 7 showing cell creation and destruction.

Actual solidification systems are much larger than the critical wavelength for the onset of instability, and typically have very shallow marginal stability curves. There will therefore be a tangled profusion of steady and unsteady solutions of different wavelength, mostly unstable. Most experiments can only control the growth velocity to within twenty to thirty percent. This should make it very difficult to get quantitative agreement between predictions of asymptotic theories and experimental measurements.

6. References

1. Rutter, J.W. and B. Chalmers, Can. J. Phys. 31, 15-39 (1953).

2. Wollkind, D.J. and L.A. Segal, Trans. R. Soc. Lond. 268, 351-380 (1970).

3. Mullins, W.W. and R.F. Sekerka, J. Appl. Phys. 35, 444-451 (1964).

4. Newell, A.C. and J.A. Whitehead, J. Fluid Mech. 38, 279-303 (1969).

5. Rieke, H. and H.-G. Paap, Phys. Rev. A33, 547-553 (1986).

6. Cross, M.C. et al., Phys. Rev. Let 45(11), 898-901 (1980).

7. Haug, P., Cellular solidification as a bifurcation problem, preprint (1986).

8. Bauer, L. H.B. Keller, and E.L. Reiss, **SIAM Rev.** 17, 101-122 (1975).

9. Meyer-Spasche, R.and H.B. Keller, **Phys. Fluids** 28(5), 1248-52 (1985).

10. Yamaguchi, Y., C.J. Chang and R.A. Brown, **Philos. Trans. R. Soc. Lond. A** 312, 519-52 (1984).

11. Ungar, L.H. and R.A. Brown, Physical Review B 29, 1367-1380 (1984).

12. Ungar, L.H. and R.A. Brown, Physical Review B 31, 5931-5940 (1985).

13. Ungar, L.H. and R.A. Brown, Finite Element Methods for Unsteady Solidification Problems, J. Computat. Physics, submitted (1986).

14. Langer, J.S., **Rev. Mod. Phys.** 52(1), 1-28 (1980).

Acknowledgements

This work was funded in part by the NASA Microgravity Science Program and by a Camille and Henry Dreyfus Teacher-Scholar Award to RAB.

Velocity Selection in Dendritic Growth

B. Caroli[1], *C. Caroli, and B. Roulet*

Groupe de Physique des Solides de l'E.N.S., associé au C.N.R.S.,
Université Paris 7; 2, place Jussieu, F-75251 Paris Cedex 05, France
[1]Also: Departement de Physique, UER des Sciences Fondamentales
et Appliquées, F-80039 Amiens, France

Growth of a pure solid from its undercooled melt gives rise to dendritic front
patterns, characterized by the propagation of a needle-shaped tip advancing at
constant velocity and the persistent emission of side branches. The systematic experimental studies of Glicksman et al. /1/ on succinonitrile have shown that, in
materials with microscopically rough solid-liquid interfaces, dendrite tips are
smooth and quasi-paraboloidal. The tip velocity V and radius of curvature ρ are
both well-defined reproducible functions of the undercooling T_M-T_∞ (where T_M is
the melting temperature and T_∞ the - constant - temperature in the liquid far ahead
of the front). Note that, in the frame moving at velocity V, while the tip is stationary, the side-branches, which are emitted quasi-periodically, globally move
backwards. So, a dendrite is an intrinsically unstationary pattern. A full understanding of dendrite shapes (e.g. branch spacing, degree of self-similarity, branch
coarsening,...) is still a completely open problem. Up to now, most theoretical
investigations have only aimed at predicting from first principles the tip velocity
and radius.

Solidification of a pure material is controlled by the dynamical balance between two processes : (i) latent heat is produced on the moving front, thus warming
the material close to it; (ii) in order for solidification to proceed, this heat
must be evacuated by diffusive transport away from the interface.

The relevant dynamics is described by the following set of equations /2/ :
(i) in the volume of each phase (i ≡ S,L), the temperature fields obey the heat
diffusion equation :

$$\frac{\partial T_i}{\partial T} = D_i \ \nabla^2 T_i \tag{1}$$

where D_i is the thermal diffusion coefficient of phase i.
(ii) at the interface (Z ≡ ζ(x,y,t))
   - Continuity of temperature

$$T_S = T_L \tag{2}$$

   - Heat balance equation

$$(K_S\vec{\nabla T}_S - K_L\vec{\nabla T}_L)\cdot\vec{n} = L \ \vec{v}\cdot\vec{n} \tag{3}$$

where L is the specific latent heat, K_i the thermal conductivity of phase i, $\vec{v}$ the
local front velocity and $\vec{n}$ the unit vector normal to the front.

   - Conditions (2), (3) must be supplemented with an equation describing the
kinetics of atomic attachment to the solid. For materials with microscopically
rough interfaces, this kinetics is fast enough for an assumption of local thermodynamic equilibrium to be valid. For a curved interface, this equilibrium is described by the Gibbs-Thomson equation :

$$T_i = T_M - \frac{\gamma T_M}{L} \mathcal{H} \tag{4}$$

where γ is the interfacial tension and $\mathcal{H}$ is the front curvature.

The shape and motion of the interface are then to be obtained by solving eqs (1) to (4) supplemented with the appropriate boundary conditions far from the front. Due to the non-linearities present in eqs (3)(4) (e.g., for a 2-D system, $\mathcal{H}\{\zeta(x,t)\} = -\zeta''/(1+\zeta'^2)^{3/2}$; $n_z = (1 + \zeta'^2)^{-1/2}$) this is in general an extremely difficult problem.

The first important breakthrough in the problem of dendritic growth has been Ivantsov's solution of the "needle-crystal" problem in the limit of zero surface tension /3,4/. A needle-crystal is defined as a needle-like solid of constant shape growing at constant velocity V along its axis of symmetry. In 3 dimensions, these solutions are paraboloids. In 2-D systems, they reduce to parabolas described, in the tip frame, by

$$\zeta_{IV}(x) = -\frac{x^2}{2\rho} \tag{5}$$

where the tip radius is related to the velocity by the Ivantsov relation :

$$\Delta = 2\sqrt{p}\ e^p \int_{\sqrt{p}}^{\infty} dy\ e^{-y^2}. \tag{6}$$

Δ is the dimensionless undercooling, defined by

$$\Delta = \frac{T_M - T_\infty}{L/C_p} \tag{7}$$

and $p = (\rho V/2D)$ is the dimensionless Péclet number.

So, it is seen that this zero surface tension analysis does not, in contradiction with experiments, predict unique V and ρ at a given undercooling, but a continuous family of solutions characterized by a given value of the product ρV.

This shortcoming of the Ivantsov solution can be understood from a very simple dimensional argument : in the absence of capillarity, there are only two lengths in the problem, namely the tip radius ρ and the diffusion length $l = 2D/V$. So, the only quantity which can be determined by a given Δ is the "dimensionless length" $\rho/l = p$.

Retaining the Gibbs-Thomson term in equation (4) introduces in the problem a new length, $d_0 = (\gamma T_M C_p/L^2)$. This suggests that capillarity is an essential physical ingredient in the mechanism which operates to select separately unique values of V and ρ.

The capillary length d_0 is typically of the order of Ångströms, i.e. very small compared with ρ and l. It was therefore natural to treat the Gibbs-Thomson term as a perturbation of the Ivantsov zeroth-order solution - a point of view supported by the fact that experimental results for ρV are in good agreement with the Ivantsov relation. Calculations of needle-crystal shapes treating capillarity as a standard perturbation, as well as the study of their linear stability /5/ failed to produce any selection mechanism. However, they resulted in the formulation by Langer and Müller-Krumbhaar of a "marginal stability" ansatz, according to which the point of operation of the dendrite should be given by the relation

$$\sigma \equiv \frac{d_0}{p\rho} \equiv \frac{d_0 V}{2Dp^2} = \sigma^* \tag{8}$$

where σ^* should be a universal Δ-independent number ($\cong 0.02$).

Although the marginal stability hypothesis proved to provide a very fruitful and satisfactory criterion for analyzing experimental data (see figure 15 of ref /5/), it has not been possible to justify it on any clear theoretical basis.

So, the problem of the nature of the selection mechanism remained open until about two years ago, when the introduction and analysis of phenomenological local models /6,7/ led to the important new idea that capillarity acts, in this problem, as a singular perturbation. Numerical and, later, analytical studies /8,9/ have shown that, in these models, properly taking into account this singular character leads to a solvability condition to be satisfied by needle-crystals, which has at most a discrete set of solutions. That is, in these local models, capillarity breaks up the continuous zero surface tension families.

Whether this result carries over to realistic solidification models is a non-trivial problem. Indeed, the full solidification problem is non-local and retarded : the motion of a point of the front at time t is determined by the whole history of heat release on all points of the front at all previous times. Thus, stationary front shapes do not obey (as assumed by local models) a differential equation, but an integro-differential one.

The simplest formulation of the problem is obtained in the symmetric model /9/ which assumes that the solid and liquid phases have identical thermal properties. From now on, we will only consider two-dimensional systems. In the symmetric model, equations (1) to (4) then result /10/ in the following closed equation for the stationary front :

$$\Delta - \frac{d(\theta)}{\rho}\mathcal{H}\{\zeta(x)\} = \frac{p}{2\pi}\int_0^\infty \frac{d\tau}{\tau}\int_{-\infty}^\infty dy \exp\left\{-\frac{p}{2\tau}\left[(y-x)^2 + \overline{(\zeta(y)-\zeta(x)-\tau)^2}\right]\right\} . (9)$$

$\mathcal{H}$ is the 2-D curvature, and

$$d(\theta) = d_0(1 - \alpha\cos(4\theta)). \tag{10}$$

That is, expression (10) allows for a possible fourfold anisotropy of the surface tension. θ is the angle between the local normal to the front and the growth axis Oz.

In equation (9) the length unit has been chosen to be the tip radius ρ.

Three different analytical methods for solving equation (9) have been proposed.

The first attempt /11/ was directly inspired by the boundary-layer approach and, for this reason, deals with the large undercooling $(1 - \Delta \ll 1)$ i.e., equivalently, large Péclet number limit ($p \propto (1 - \Delta)^{-1} \gg 1$). It takes advantage of the smallness of p^{-1} to transform equation (9) into a formal (linear) differential equation of infinite order, linearized in the departure of the local curvature from that of the Ivantsov parabola. It then appears that, although the large-p limit could a priori be thought to be quasi-local, this equation has WKB-like solutions for which it cannot be truncated. Imposing that the resulting shape be well-behaved $(\lim_{x\to\pm\infty} (\zeta(x) - \zeta_{IV}(x)) = 0)$ then produces a solvability condition which has no solution for an isotropic system.

The two other methods were first set up to treat the opposite, completely non-local limit $p \to 0$ - which is much closer to actual experimental conditions.

The first one, due to Ben Amar and Pomeau /12/ makes use of a mathematical method developed by Kruskal and Segur /9/ to treat the local "geometrical model". They linearize the non-local term in the $p = 0$ limit of equation (9) /13/, but retain the non-linearity in the curvature term in the vicinity of the singular points $x = \pm i$ of the Ivantsov curvature $\mathcal{K}_{IV} = (1 + x^2)^{-3/2}$. In this "inner region" (which plays in this problem a role analogous to that of a turning point in the WKB method) the front equation is solved by a scaling method. This solution, which can be matched asymptotically to the Ivantsov parabola for $x \to \pm\infty$, contains a term small beyond all orders in the small parameter $\sigma = d_0/p\rho$, from which the slope $\zeta'(x=0)$ can be calculated. Imposing that the front profile be smooth ($\zeta'(o) = 0$) then yields a solvability condition. Again, this has no solution, for isotropic systems. Ben Amar and Pomeau showed that, in the presence of an anisotropic surface tension (equation (10)), it has a discrete set of solutions the fastest (and, most likely, physically selected /8,14/) of which is characterized, at small anisotropies, by

$$\sigma = \sigma^*(\alpha) \propto \alpha^{7/4} \ . \tag{11}$$

Barbieri, Hong and Langer /15/ recently devised, in the same p=0 limit, a method which, although in principle not so well controlled as the Ben Amar-Pomeau one, is, for all practical purposes, equivalent and leads to identical results. Since it is in this treatment that the existence of a solvability condition shows up most easily, we will now briefly outline its principle.

Taking the p=0 limit /13/ of equation (9) and performing a complete linearization in $\zeta_1(x) = \zeta(x) - \zeta_{IV}(x)$ leads to

$$\mathcal{L}\zeta_1 \equiv \sigma \frac{d^2\zeta_1}{dx^2} - \frac{3\sigma x}{1 + x^2} \frac{d\zeta_1}{dx}$$
$$- \frac{(1+x^2)^{3/2}}{2\pi A(x)} \int_{-\infty}^{\infty} dy \ \frac{(x+y)(\zeta_1(x) - \zeta_1(y))}{(x-y) \left[1 + \frac{(x+y)^2}{4}\right]} = \sigma \tag{12}$$

where $A(x) = 1 - \alpha + \frac{8\alpha x^2}{(1+x^2)^2}$ (13)

is the anisotropy factor resulting from equation (10).

Equation (12) is a linear inhomogeneous equation. As is well known, such an equation has a solution only if the corresponding Fredholm condition is satisfied, i.e. if the r.h.s. inhomogeneous term is orthogonal to the null-space of the adjoint operator $\mathcal{L}^+$ associated with $\mathcal{L}$. One thus gets the following solvability condition :

$$\Lambda(\sigma,\alpha) = \int_{-\infty}^{\infty} dx \ \sigma\tilde{\zeta}_1(x) = 0 \tag{14}$$

where $\tilde{\zeta}_1$ is a null-eigenvector of $\mathcal{L}^+$:

$$\mathcal{L}^+ \tilde{\zeta}_1(x) = 0 \ . \tag{15}$$

Equation (15) is still non-local. However, Barbieri et al., following Shraiman's work /16/ on viscous fingering, were able to solve it approximately in the following way : they assume (and check a posteriori) that it has WKB-like solutions of the form $\tilde{\zeta}_1(x) = \exp\{S_\pm(x)/\sqrt{\sigma}\}$, where $S_\pm(x)$ has a stationary phase point $x = \pm i$ with Re $\{S_\pm(x_\pm)\}$ finite and negative. One may then take advantage of this structure to fold the integration contour in the non-local term of $\mathcal{L}^+\tilde{\zeta}_1$ into the complex plane, and show that, up to transcendantally small corrections, the only important contribution comes from integrating around the pole $y = x$ of the integration kernel. So, for such singular solutions /17/, equation (15) reduces to a second order linear differential equation /18/, which is then solved by a WKB method.

194

One can then calculate $\Lambda(\sigma,\alpha)$ explicitly, and finds :
- for $\alpha = 0$ (isotropic system)

$$\Lambda(\sigma,0) \approx \sigma^{-1/28} \exp\left(-\frac{a}{\sqrt{\sigma}}\right) \neq 0 \qquad (16)$$

- for $\alpha \neq 0$, equation (14) has a denumerably infinite set of solutions, the fastest of which obeys, at small α, equation (11).

It has recently been possible to apply the method of Barbieri et al. to the general case of finite Péclet numbers /19/. Surprisingly enough, it is found that the resulting solvability condition is p-independent - a result which can be checked to be coherent with the predictions of the p >> 1 method of ref. /11/. This entails that, in the small-σ (i.e., equivalently, small anisotropy) limit, the velocity of the needle-crystal should scale as p^2.

So, all the existing analytical approaches agree to predict that, in two dimensions at least, the presence of a finite anisotropy is necessary for needle-crystals to exist. This has been confirmed by the results of numerical studies of the full needle-crystal equation /20, 21, 22, 23/.

The theoretical analysis described above are concerned with needle-crystals, i.e. stationary growth shapes. A question then naturally arises : is there anything in common between the tip shapes thus predicted and those occurring in the side-branched dendritic growth mode ? A numerical answer to this question has been given most recently by Saito, Goldbeck-Wood and Müller-Krumbhaar /24/. Starting from a 2-D Ivantsov parabola, they study its evolution in the one-sided model (no diffusion in the solid) by direct forward time integration of eqs. (1-4). They do find that anisotropy is necessary to obtain dendritic growth, and the tip behavior then obtained is in excellent agreement with the above predictions ($\sigma^*(\alpha) \propto \alpha^{7/4}$; $V \propto p^2$ for constant α) for the needle-crystal.

So, theory and numerics agree to produce a coherent set of predictions about tip behavior, which should now be checked against actual physical experiments.

The main qualitative idea which has emerged from the study of the 2-D needle-crystal problem is clearly the (previously unsuspected) seemingly essential role of anisotropy.

Experimental results, as already mentioned, agree with the non-trivial prediction that the parameter σ is indeed Δ-independent. However, it seems that the observed σ's depend only weakly - if at all - on the strength of anisotropy.

Of course, experiments deal with three-dimensional dendrites /25/, while the theoretical results described above are two-dimensional. However, it has been possible to extend them to 3-D systems, either isotropic /19, 27, 28/ or with an anisotropy restricted to be symmetric around the growth axis /15/. It is found that, for such systems, the condition for existence of a needle-crystal is strictly identical to that for 2-D systems, a result which does not bring us closer to a fit with experimental data. It is of course possible that the introduction of axial anisotropy, which has not been feasible yet, could modify considerably the structure of the results.

Finally, another very important question remains open, namely that of the nature of the physical mechanism responsible for side-branching. Until recently, it has been thought that this could be described as corresponding to the existence of weakly unstable oscillatory modes of deformation of the needle-crystal. This calls for studying the local stability of stationary solutions. Recently, a first attempt by Kessler and Levine /14/ has led them to predict linear stability for the 2-D needle-crystal. This result, if confirmed, leaves open several possibilities, among which :

- the branching mechanism might be different in two and three dimensions ;
- dendrite branches may be generated by noise amplification /29/ ;
- the coupling between growth and hydrodynamic motion in the liquid might be of some importance.

The above discussion eloquently shows that, although considerable progress has been made in the past few years, in the present state of the art we are still far from a true understanding of dendritic growth.

References

1. M.E. Glicksman, R.J. Shaefer and J.D. Ayers : Metall. Trans. A7, 1747 (1976)
   S.C. Huang and M.E. Glicksman : Acta Metall. $\underline{29}$, 701 and 717 (1981)
2. This purely thermal model neglects hydrodynamic motion in the liquid as well as elastic and plastic deformations in the solid. These effects, though they may be non-negligible quantitatively, are very unlikely to have a qualitative influence on the selection of dendritic shapes.
3. G.P. Ivantsov : Dokl. Akad. Nauk SSSR 58, 567 (1947)
4. G. Horway and J.W. Cahn : Acta Metall. 9, 695 (1961)
5. For a review, see J.S. Langer in Rev. Mod. Phys. $\underline{52}$, 1 (1980) and references therein
6. R. Brower, D. Kessler, J. Koplik and H. Levine : Phys. Rev. A29, 1335 (1984)
7. E. Ben-Jacob, N. Goldenfeld, J.S. Langer and G. Shön : Phys. Rev. A29, 330 (1984)
8. D. Kessler, J. Koplik and H. Levine : Phys. Rev. A30, 3161 (1984)
   E. Ben-Jacob, N.D. Goldenfeld, B.G. Kotliar and J.S. Langer : Phys. Rev. Lett. 53, 240 (1984)
9. J.S. Langer : Phys. Rev. A33, 435 (1986)
   J.S. Langer and D.C. Hong : Phys. Rev. A34, 1462 (1986)
   M. Kruskal and H. Segur : Preprint A.R.A.P. Tech. Memo. 85-25
   R.F. Dashen, D.A. Kessler, H. Levine and R. Savit : Physica D 21, 371 (1986)
10. J.S. Langer : Acta Metall. $\underline{25}$, 1121 (1977)
11. B. Caroli, C. Caroli, B. Roulet and J.S. Langer : Phys. Rev. A33, 442 (1986)
12. M. Ben-Amar and Y. Pomeau : Europhysics Lett. 2, 307 (1986)
13. P. Pelcé and Y. Pomeau : Dendrite in the small undercooling limit, to appear in Studies in Applied Mathematics
14. D.A. Kessler and H. Levine : Stability of dendritic crystals, preprint (1986)
15. A. Barbieri, D.C. Hong and J.S. Langer : preprint NSF-ITP 86-65
16. B.I. Shraiman : Phys. Rev. Lett. $\underline{56}$, 2028 (1986)
17. One cannot exclude the possibility for equation (15) to have other solutions with different structures. Note, however, that the existence of such extra solutions for ζ_1 could only result in further restriction of the initial family.
18. Note that the method described here cannot be used to solve directly the equation for the front shape : $\mathcal{L}\zeta_1 = \sigma$; indeed, if one looks for solutions of the form $\zeta_1 \sim \exp \{\Sigma_\pm(x)/\sqrt{\sigma}\}$ and performs the above-mentioned reduction of the non-local term, one finds that Re $\Sigma_\pm(x_\pm) > 0$, which means that the approximations involved in the local reduction are, in this case, not legitimate.
19. B. Caroli, C. Caroli, C. Misbah and B. Roulet : to be published
20. M. Ben-Amar and B. Moussallam : Numerical results on two-dimensional dendritic solidification (preprint)
21. D.I. Meiron : Phys. Rev. A33, 2704 (1986)
22. D.A. Kessler, J. Koplik and H. Levine : Phys. Rev. A33, 3352 (1986)
23. Computational limitations have not permitted to explore the region of very small α's. This prevents detailed quantitative comparison with the analytic predictions for $\sigma^*(\alpha)$ which - due to the linear approximation involved - are valid only in the small σ (i.e. small α) limit. However, the finite p results of ref. /22/ indicate that the p-dependence of σ^* does decrease with the magnitude of the anisotropy parameter.
24. Y. Saito, G. Goldbeck-Wood and H. Müller-Krumbhaar : Dendritic crystallization: numerical study of the one-sided model (preprint, 1986)

25. Let us mention that, even when dendrites are grown in thin layers /26/, capillary constraints at the solid-liquid-confining plates contacts impose a curvature of the front in the small dimension of the sample, which prevents the dendrite from being truly two-dimensional.

26. H. Honjo, S. Ohta and Y. Sawada : Phys. Rev. Lett. 55, 841 (1985)
27. M. Ben-Amar and Y. Pomeau : Theory of needle crystals (preprint, 1986)
28. B. Caroli, C. Caroli, C. Misbah and B. Roulet : J. de Physique 47, 1623 (1986)
29. R. Pieters and J.S. Langer : Phys. Rev. Lett. 56, 1948 (1986).

Interfacial Pattern Formation: Dynamical Effects

D.A. Kessler[1] *and H. Levine*[2]

[1]Department of Physics, University of Michigan,
 Ann Arbor, MI 48109, USA
[2]Schlumberger-Doll Research, Old Quarry Road,
 Ridgefield, CT 06877, USA

In this talk, we focus on the dynamical aspects of pattern forming systems. These include the formation of dendritic sidebranches as well as the tip-splitting instability. We study these phenomena through first computing the steady-state pattern (via the "solvability" mechanism) and then solving the associated linear stability problem. The results to date suggest that both of the aforementioned effects must be treated by including the effects of finite amplitude perturbations on the deterministic growth equations.

In the past several years, a considerable amount of progress has been made towards elucidating the mechanisms whereby interfacial patterns are formed [1]. Of particular interest is the resolution of the long standing problems of width selection for the Saffman-Taylor finger (formed during multiphase fluid displacement) and of velocity selection for the free space dendritic crystal. This progress occurred by the recognition that surface tension, acting in a non-perturbative manner, creates a solvability condition which selects a discrete set of allowed solutions from the continuous family of steady-state patterns found by solving the macroscopic equations of motion.

This talk will review recent attempts to extend our understanding of pattern formation past the determination of the steady-state structure. We would like to address the stability of these structures to perturbations. If these perturbations are taken to be infinitesimal in magnitude, this leads to the problem of diagonalizing a linear stability operator. If these perturbations have finite amplitude, a non-linear analysis is required. The need to study this issue is clear from the experiments; steady-state structures do break down either partially via emitting sidebranches or completely via undergoing tip splitting.

Let us briefly review the evolution equations for two-dimensional dendritic solidification and multiphase fluid displacement. In dendritic solidification of a pure material, the rate of interface motion is determined by the rate at which latent heat can be diffused away from the phase boundary [2]. Experimentally, the control parameter is the undercooling at large distances from the growing crystal, measured in units of the latent heat over specific heat:

$$\tilde{\Delta} = \frac{T - T_m}{L/c_p}.$$

(1)

The dimensionless $u \equiv c_p T/L$ obeys the diffusion equation with the moving interface acting as a heat source:

$$\dot{u} = D\nabla^2 u + \int v_n(s)\delta(\vec{x} - \vec{x}(s))ds$$

(2)

where v_n is the normal velocity at arclength s. Finally the interface is taken to be in local thermodynamic equilibrium. The latter leads to the well-known Gibbs-Thomson formula,

$$u(\vec{x}(s)) = -\gamma(1 - \epsilon \cos 4\theta(s))\kappa$$

(3)

where θ is the angle made by the interface normal with the crystal axes (assuming four-fold symmetry). κ is the curvature and γ is the surface tension $\bar{\gamma}$ made dimensionless by use of a reference velocity v_0, $\gamma = v_0 T_m c_p \bar{\gamma}/2DL^2$. The strength of the crystal anisotropy is described by the parameter ϵ.

An analogous set of equations describes multiphase displacement by a fluid of negligible viscosity in a Hele-Shaw cell [3]. This cell is simply a pair of parallel glass plates restricting the flow to occur in a narrow gap between the plates. Under these conditions, Stoke's law can be approximated by the simple relation

$$\vec{v} = -\frac{b^2}{12\mu}\vec{\nabla}p$$

(4)

with gap thickness b, viscosity μ and pressure p. Incompressibility leads to Laplace's equation for the pressure,

$$\nabla^2 p = 0.$$

The control parameter here is the asymptotic flow velocity v_∞. The boundary conditions at the interface are conservation of fluid which requires $v_n(s) = -\vec{\nabla}p \cdot \hat{n}$, and the capillary pressure drop equation which yields $p(\vec{x}(s)) = -\bar{\gamma}\kappa$. The only dimensionless parameter governing the process is the dimensionless surface tension

$$\gamma = \frac{\bar{\gamma}}{12\mu v_\infty}(\frac{b}{a})^2$$

(5)

with channel width $2a$. γ is sometimes treated as the inverse of a dimensionless parameter called the capillary number.

Why do these systems generate patterns? The answer lies in the fact that simple shapes such as planes or circles will be unstable due to the Mullins-Sekerka instability [4]. Simply put, sharp protruding bumps are more efficient at radiating heat (or have enhanced pressure gradients) and therefore have large growth velocities. To see this, we consider a small perturbation around a planar interface in the Hele-Shaw cell (see Fig. 1 for the geometry):

$$x(t) = t + e^{iky}\delta_k(t),$$
$$p(x,y) = -x + p_k e^{-kx} e^{iky}. \tag{6}$$

The capillary equation requires $p_k = \delta_k(1 - \gamma k^2)$ and the velocity condition then leads to the stability formula

$$\omega_k \equiv \dot{\delta}/\delta = k\delta_k(1 - \gamma k^2). \tag{7}$$

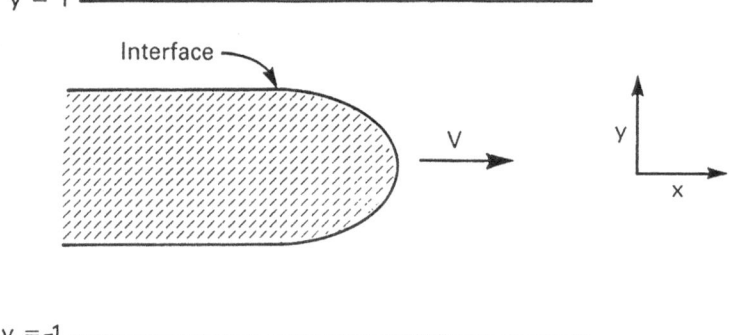

Fig. 1. Geometry of the Saffman-Taylor problem.

For small wavevectors, ω_k is positive and perturbations grow in time. This would lead to an arbitrarily disordered interface were it not for the restabilization at large k due to the surface tension. This competition between macroscopic destabilizing tendencies and microscopic stabilization at smaller length scales underlies all interfacial pattern formation.

The simplest structure that can be formed in these systems is a steady-state pattern which translates at a uniform velocity v_0 while retaining its shape. Fig. 2 shows a Saffman-Taylor finger [5] which is a good example of this phenomenon. Most of the recent work has focused on such steady-state structures and on the selection problem. Specifically, SAFFMAN and

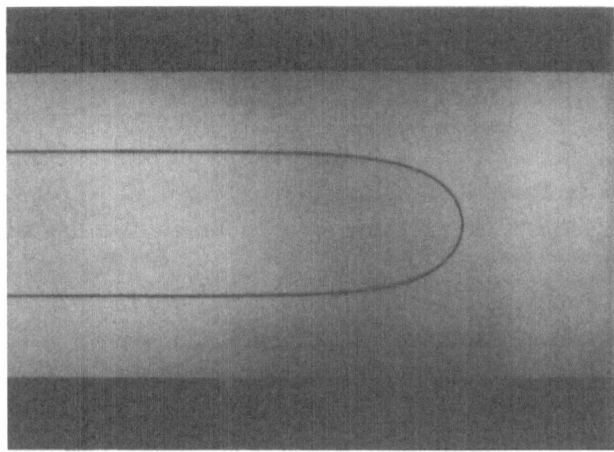

Fig. 2. The Saffman-Taylor finger; the plates are 5 cm wide and the gap thickness is $800\mu m$ (Courtesy A. Libchaber).

TAYLOR [5] showed that if $\gamma = 0$, solutions could be found which moved at arbitrary velocity. From fluid conservation, v_0 equals $1/\lambda$ where λ is the asymptotic width and so the choice of the proper solution is usually referred to as the width selection problem. It was shown later [6] that a perturbative expansion in γ did not break this continuous degeneracy. Experimentally and in numerical simulations [7], however, a single width is chosen for fixed γ. A similar conundrum existed in solidification for the IVANTSOV [8] family of parabolic needle crystals [9].

The resolution of these paradoxes was made possible by the discovery of "microscopic solvability", initially within the context of simple phenomenological models of pattern formation [10]. The idea is that surface tension leads to an additional boundary condition near the tip of the pattern that is not satisfied for arbitrary velocity. Most crucially, the magnitude of this effect is essentially singular in γ; that is, the mismatch near the tip is proportional to $e^{-c/\sqrt{\gamma}}$ and is hence incalculable by doing perturbation theory. This was first shown numerically [11,12] and later formed the basis of powerful analytic techniques for solving the selection problem [13,14]. The results of all these analyses are that only a discrete set of possible widths satisfy the solvability constraint and that for small γ, the shape obeys the scaling

$$\lambda - 1/2 \sim \gamma^{2/3}. \tag{8}$$

Similarly for needle crystals, $\gamma \sim \epsilon^{7/4}$ and there are no solutions at all for isotropic surface tension.

At this point, several questions remain unanswered. The solvability condition allows for a discrete set of steady-state solutions; this is a vast improvement over the continuum but is still not unique. Also, we have yet to account for sidebranching, the process which turns needle crystals into full-fledged dendrites (see Fig. 3). In fact, it seems at first glance unlikely that the selection of needle crystal shapes should have anything whatsoever to do with dendrites replete with non-steady branching. Numerical simulations show, however, that the steady-state solution does accurately determine the velocity and tip shape of full dendritic patterns. A last issue concerns instabilities such as tip-splitting which occur at small γ and eventually lead to disordered, possibly fractal,patterns.

To address these issues, we recast the evolution dynamics as an integro-differential equation for the moving interface. For solidification, this can easily be accomplished by making use of the free space Green's function for the diffusion equation:

$$(\nabla^2 - \frac{\partial}{\partial t})G = -\delta(\vec{x} - \vec{x}')\delta(t - t')$$

$$G = \frac{exp[-(\vec{x} - \vec{x}')^2/4(t - t')]}{4\pi(t - t')}$$

to write

$$\tilde{\Delta} - \gamma(1 - \epsilon\cos(4\theta))\kappa = \int_{-\infty}^{t} dt' \int ds' G(\vec{x}(s,t); \vec{x}(s',t'))v_n(s',t'). \tag{9}$$

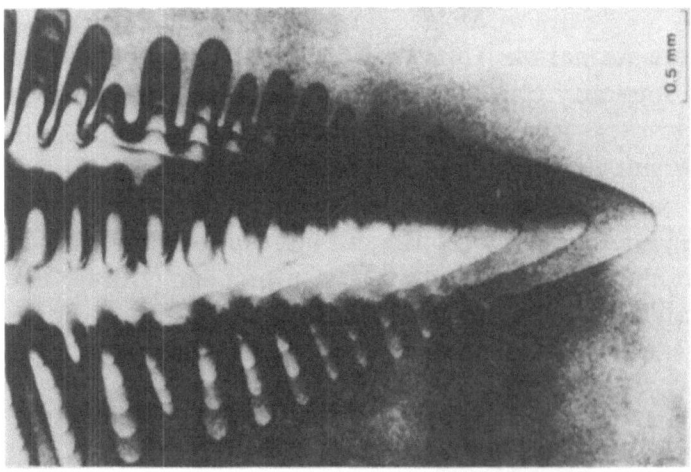

Fig. 3. Time-exposure sequence of the solidification of succinonitrile
(Courtesy M. E. Glicksman, Rensselaer Polytechnic Institute).

Next, we make use of the known steady-state solution to linearize the above equation [15]

$$\vec{x}(s) = \vec{x}_0(s_0) + v_0 \hat{x} t + \hat{n}_0 \delta(s_0, t) \qquad (10)$$

where δ is small and the subscript "0" refers to the steady-state shape. Substituting (10) into (9), we can derive an equation for δ. Finally, we can take the limit of small $\tilde{\Delta}$ (experimentally $\tilde{\Delta} \sim 0.01 - 0.1$) to approximate the full diffusive Green's function by the quasistatic one. This leads to an eigenvalue equation of the schematic form

$$L[\delta(s)] = \omega \delta(s) \qquad (11)$$

where L is a singular integro-differential operator. We are mostly concerned with modes for which Re $\omega \geq 0$ which correspond to instabilities and modes which are stable but have decay rates close to zero and may therefore be excited by finite amplitude perturbations. A similar formulation exists for the multiphase flow system [16].

Let us first focus on continuum modes of the stability problem; these modes are defined by requiring the eigenvector to asymptotically approach a plane wave $\delta \sim e^{ks}$ at large s. That such modes can be found is a simple consequence of the fact that the steady-state interface approaches a plane far from the tip. Substitution of this form into the eigenvalue equation (11) can be shown to yield

$$\omega_k = -k + \gamma k^2 \sqrt{2k - k^2} \qquad \text{(dendrites)}$$
$$\omega_k = -k/\lambda - \gamma k^3 \tan(k\lambda) \qquad \text{(Saffmann} - \text{Taylor)} \,.$$

Notice that in both of these expressions the term corresponding to the original Mullins-Sekerka instability in (7) is absent. This is due to the fact that the normal velocity of the interface goes to zero as we move away from the tip. The above expressions mean that perturbing the system with some infinitesimal noise at spatial wavevector $k = iq$ will lead to an amplitude with decay rate Re $\omega \sim -\gamma q^3$ and the noise will die away. As far as the continuum is concerned, the pattern is stable [17].

The discrete modes must be treated numerically. We have done this using several different methods for both of these systems. A typical plot of the discrete spectrum is presented in Fig. 4 where we have varied λ at fixed γ for the Saffman-Taylor finger. As we pass through the allowed widths λ_i, one mode crosses $\omega = 0$; this mode is just the translation zero mode. At the smallest λ, all other modes have Re ω negative. At the next allowed width, the spectral flow guarantees that one mode will now

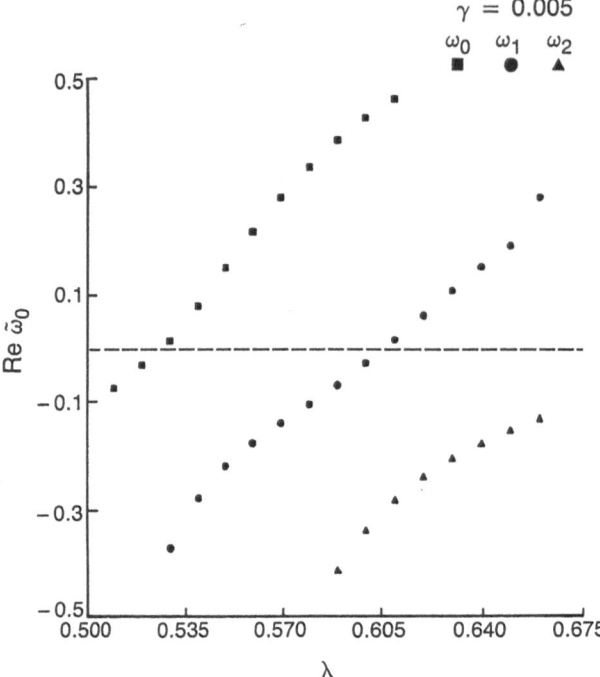

Fig. 4. Spectral flow for three lowest discrete eigenvalues, Saffman-Taylor finger at $\gamma = 0.005$.

occur at positive growth rate and hence the pattern will be unstable. As we reach each successive allowed width, one additional instability will appear. The upshot of this result is that there is a *unique* steady-state structure stable with respect to infinitesimal noise. This allows the prediction of the width (or the dendritic velocity) as a function of γ; the results agree with the simulations and with actual experiments whenever the two-dimensional models we are using are adequate descriptions of the physical processes.

So, if the structures are stable what is sidebranching? The answer can be seen by using the linearized equation (11) and imposing a fixed frequency finite amplitude noise source at the tip. The response for a typical set of parameters is presented in Fig. 5a. Note that the initial perturbation is amplified by an extremely large factor as it moves away from the tip. Fig. 5b demonstrates that the amplification is highly selective; broadband noise near the tip will give rise to sidebranches with a very narrow distribution of wavevectors. Our picture requires that non-linear effects stabilize sidebranches formed by this amplification process. The predictions that come out of this approach are [18] that a) the sidebranch wavelength should scale as the tip radius as a function of $\tilde{\Delta}$ (with a calculable prefactor which is

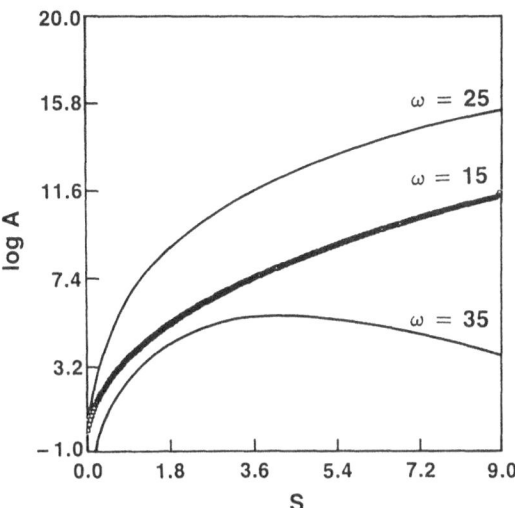

Fig. 5a. Temporary amplification of monochromatic noise at a dendritic
tip, as a function of arclength down the finger.

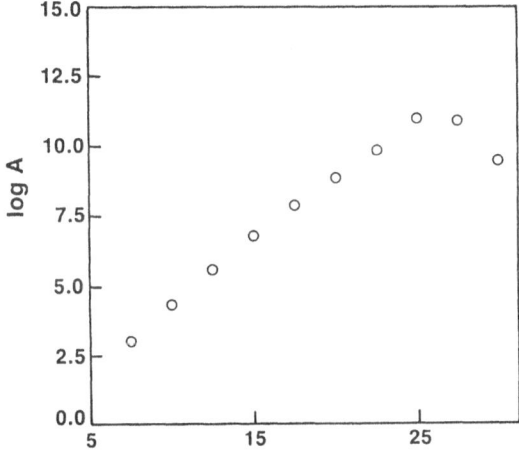

Fig. 5b. Amplification vs. frequency at a fixed distance ($x = 3$) from the
tip.

approximately equal to 3.1 at $\epsilon = 0.1$) and that b) sidebranches are fixed
in the laboratory frame of reference. Both of these predictions are qualitatively valid and there is even some evidence of the noisy character of
the branches that such an approach invariably suggests. Also, this idea
neatly solves the paradox of why the steady-state calculation agrees with
the growth of full dendrites. An approach based on a subcritical bifurcation
to a limit cycle might have trouble explaining this agreement but cannot be

ruled out and would in fact give rise to the same predictions for sidebranch wavelength.

In Figs. 6a and 6b we show a similar computation for the Saffman-Taylor finger. The amplification is much smaller and the peak amplitude occurs near the tip. This leads to the conclusion that this system will not sidebranch but might for large enough noise exhibit a tip-region instability.

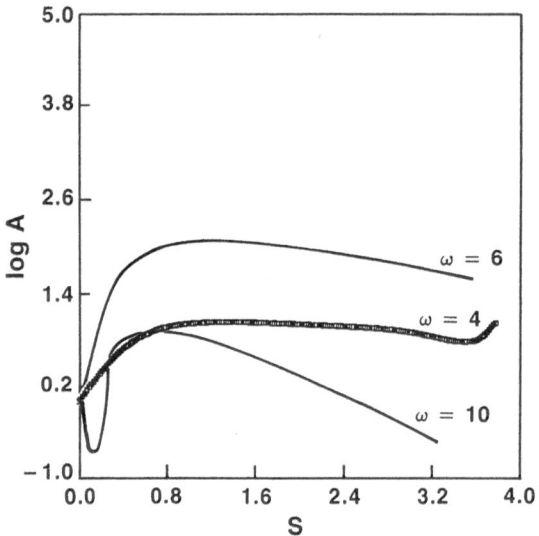

Fig. 6a. Temporary amplification of monochromatic noise at a Saffman-Taylor finger, ($\lambda = 0.523$, $\gamma = 5.724 \times 10^{-3}$) as a function of arclength down the finger.

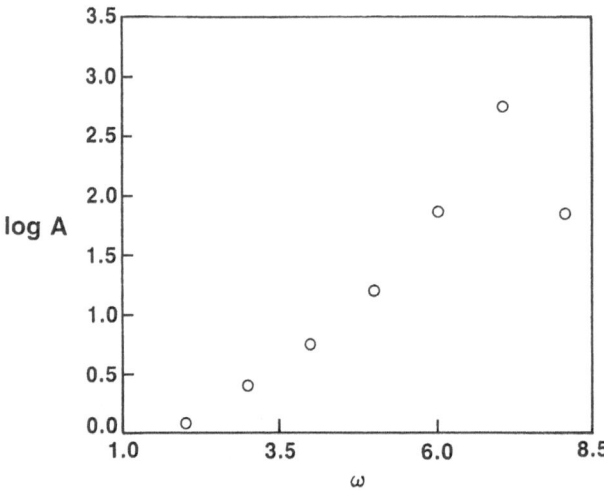

Fig. 6b. Amplification vs. frequency at a fixed distance from the tip.

This instability can be studied by going back to the linear stability problem and adding noise to the system. The added noise causes the discrete modes to move to positive Re ω and one can predict the sequence of instabilities as $\gamma \to 0$ [19]. This calculation also shows that the noise threshold goes to zero as $e^{-c/\sqrt{\gamma}}$. This result can be interpreted as a breakdown of steady-state motion whenever the noise amplitude is bigger than the "solvability" signal as discussed above.

The idea that tip-splitting and sidebranching are both non-linear effects occurring in different parts of the parameter space has led to a variety of experimental demonstrations of the transition from dendritic to disordered growth. This has been done by etching grooves in the Hele-Shaw plates to mimic crystal anisotropy [20], by using anisotropic fluids [21] and by growing crystals on a rough surface [22]. One may hope for an eventual understanding of fractal patterns of the type seen in DLA [23] to emerge from the study of a sequence of instabilities but there are no good ideas at present as to how to push in that direction. It is worthwhile to note that the ideas presented here are consistent with a recent discussion of the importance of stochasticity in the DLA algorithm [24].

As in all non-linear systems, it is easier to understand ordered solutions and their breakdown than it is to understand the nature of the resulting disorder. This remains an unsolved problem possibly akin to trying to understand turbulence by focusing on its onset. As far as ordered structures are concerned, we need to apply our methodology to other systems such as directional solidification, electrochemical deposition and rising bubbles. We do believe, however, that we have developed enough of an understanding of the basic mechanisms at work to be able to make detailed predictions and proceed to handle the real-world complications in actual experiments.

After completion of the work, we received a preprint from Saito, Goldbeck-Wood and Müller-Krumbhaar who present numerical simulations of the two-dimensional diffusion equation. Their results are in complete quantitative agreement with the ideas presented here both regarding the velocity selection problem and the determination of sidebranch wavevectors.

References

1. For a recent review, see D. Kessler, J. Koplik, and H. Levine, *Pattern Selection in Fingered Growth Phenomena*, to appear in *Advances in Physics*

2. D.P. Woodruff, *The Solid-Liquid Interface*, Cambridge University Press, 1973

3. H.J.S. Hele-Shaw, *Nature*, **58**, 334 (1898)

4. W.W. Mullins and R.F. Sekerka, *J.Appl. Phys.*, **34**, 323 (1963)
5. P.G. Saffman and G.I. Taylor, *Proc. Roy. Soc.*, **A 245**, 312 (1958)
6. J.W. McLean and P.G. Saffman, *J. Fluid Mech.*, **102**, 455 (1981)
7. G. Trygvasson and H. Aref, *J. Fluid Mech.*, **136**, 1 (1983); L.M. Schwartz and A.J. DeGregorio, *J. Fluid Mech.*, **164**, 383 (1986)
8. G.P. Ivantsov, *Dokl. Akad. Nauk SSSR*, **58**, 567 (1947)
9. For a review of early attempts at resolving this paradox, see J.S. Langer, *Rev. Mod. Phys.*, **52**, 1 (1980)
10. For a discussion of local models of pattern formation, see D.A. Kessler, J. Koplik, and H. Levine, *Physiochem. Hydro.*, **6**, 507 (1985)
11. D.A. Kessler and H. Levine, *Phys. Rev.*, **A 33**, 2634 (1986)
12. D.A. Kessler and H. Levine, *Phys. Rev.*, **B 33**, 7687 (1986); D. Meiron, *Phys. Rev.*, **A 33**, 2704 (1986); D.A. Kessler, J. Koplik, and H. Levine, *Phys. Rev.*, **A 33**, 3352 (1986); M. Ben-Amar and B. Moussalam, to appear
13. B.I. Shraiman, *Phys. Rev. Lett.*, **56**, 2028 (1986); D.C. Hong and J.S. Langer, *Phys. Rev. Lett.*, **56**, 2032 (1986); R. Combescot, T. Dombre, V. Hakim, Y. Pomeau, and A. Pumir, *Phys. Rev. Lett.*, **56**, 2036 (1986)
14. D.A. Kessler, J. Koplik, and H. Levine, in *Proc. NATO A.R.W. Patterns, Defects and Microstructure*, (1986, to appear); A. Barbieri, D.C. Hong, and J.S. Langer, to appear; M. Ben-Amar and Y. Pomeau, *Europhys. Lett.*, **2**, 307 (1986)
15. For more details, see D. Kessler and H. Levine, *Phys. Rev. Lett.*, **57**, 3069 (1986)
16. D. Kessler and H. Levine, *Discrete Set Selection of Saffman-Taylor Fingers*, submitted to Phys. Fluids
17. For a simple WKB approach to this result, see D. Bensimon, L. Kadanoff, S. Liang, B.I. Shraiman, and C. Tang, *Rev. Mod. Phys.*, **58**, 977 (1986); P. Pelce, *Dynamique des Fronts Courbés*, Thesis Université de Provence (1986)
18. D.A. Kessler and H. Levine, *Determining the Wavelength of Dendritic Sidebranches*, in preparation
19. D. Bensimon, *Phys. Rev.*, **A 33**, 1302 (1986)
20. E. Ben-Jacob, R. Godbey, N.D. Goldenfeld, J. Koplik, H. Levine, T. Mueller, and L.M. Sander, *Phys. Rev. Lett.*, **55**, 1315 (1985)
21. A. Buka, J. Kertesz, and T. Vicsek, *Nature*, **323**, 424 (1986)
22. H.Honjo, S. Ohta and M. Matsushita, *Irregular Fractal-like Growth of Ammonium Chloride*, submitted to J. Phys. Soc. Japan (1986)
23. T.A. Witten and L.M. Sander, *Phys. Rev.*, **B 27**, 5696 (1983)
24. C. Tang, *Phys. Rev.*, **A 31**, 1977 (1985)

Part IV

Diffusion Limited Aggregation
and Fractals

Role of Fluctuations in Fluid Mechanics and Dendritic Solidification

H.E. Stanley

Center for Polymer Studies and Department of Physics,
Boston University, Boston, MA 02215, USA

Abstract: *Our purpose is to review certain recent advances in understanding the role of fluctuations in fluid mechanics and dendritic solidification; many of these represent joint work of the author and J. Nittmann. If one understands completely the simple Ising model, then one understands virtually all systems near their critical points— although the detailed descriptions of many such systems require a suitably-chosen variant of the Ising model (such as the XY or Heisenberg model). By analogy, we shall argue here that if one understands completely the simple diffusion-limited aggregation (DLA) model or the closely related dielectric breakdown model (DBM), then one understands the role of fluctuations in a range of fluid mechanical systems, as well as in dendritic solidification. The detailed descriptions of some such systems require suitably chosen variants, such as DBM with anisotropy and noise reduction.*

The overall theme I'll develop is that recent work on relatively simple *non-deterministic* models has some utility for describing experimentally observed phenomena in fluid mechanics and dendritic growth. I'll first make the case that we can approach these experimental subjects of classic difficulty with the same spirit that has been used in recent years to approach problems associated with phase transitions and critical phenomena. This approach is to carefully choose a microscopic model system that captures the essential physics underlying the phenomena at hand, and then study this model until we understand "how the model works." Then we reconsider the phenomena at hand, to see if an understanding of the model leads to an understanding of the phenomena. Sometimes the original model is not enough, and a variant is needed, and we shall see that this is the case here also. Fortunately, however, we shall see that the same underlying physics is common to the model and its variants.

We begin, then, with the classic Ising model.

(a) The Ising Model and Its Variants

The first time I heard a lecture on the Ising model, the speaker apologized for having what was termed "the Ising disease" (an appellation attributed to Montroll). The Ising model was proposed 67 years ago (Lenz 1920) and its solution for a one-dimensional lattice occurred 62 years ago (Ising 1925). However, at that time no one knew that the Ising model describes a wide range of materials near their critical points. Over 1000 papers have been published on this model, but only since 1977 have we known that if one understands the Ising model thoroughly, one understands the essential physics of virtually all 3-dimensional materials systems near thermal critical points. This is because other systems are simply variants of the Ising model. For example, most systems are related to special cases of the n-vector model, which in turn is a simple Ising model in which the spin variable s has not one component but rather n separate components s_j: $s \equiv (s_1, s_2, \ldots, s_n)$.

The Ising model solves the puzzle of how it is that nearest-neighbor interactions of **microscopic** length scale 1Å "propagate" their effect cooperatively to give rise to a correlation length ξ_T of **macroscopic** length scale near the critical point (Fig. 1a). In fact, ξ_T increases without limit as the coupling $K \equiv J/kT$ increases to a critical value $K_c \equiv J/kT_c$,

$$\xi_T \sim A \left(\frac{K - K_c}{K_c} \right)^{-\nu_T}. \tag{1a}$$

The "amplitude" A has a numerical value on the order of the lattice constant a_o. A snapshot of an Ising system shows that there are fluctuations on all length scales from a_o ($\cong$ 1Å) to ξ_T (which can be from $10^2 - 10^4$Å in a typical experiment).

Attempts to simplify the essential problem of propagation of order from one spin to its neighbors by making mean-field type of truncations (such as the Weiss approximation, the Bethe approximation, and the Kasteleyn-van Kranendonk constant coupling approximation) fail to describe 3-dimensional systems near their critical points.

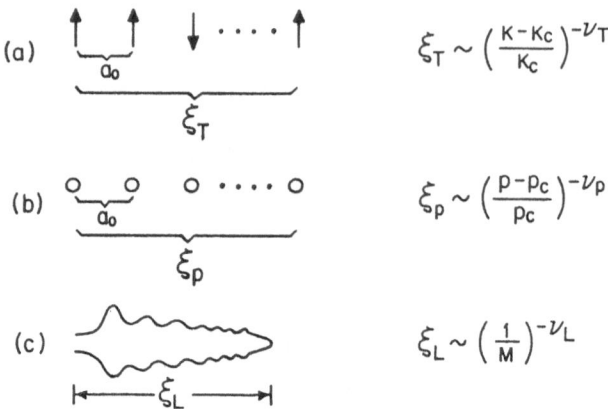

$$\xi_T \sim \left(\frac{K - K_c}{K_c}\right)^{-\nu_T}$$

$$\xi_p \sim \left(\frac{p - p_c}{p_c}\right)^{-\nu_p}$$

$$\xi_L \sim \left(\frac{1}{M}\right)^{-\nu_L}$$

Fig. 1: Schematic illustration of the analogy between (a) the Ising model, which has fluctuations in spin orientation *on all length scales* from the microscopic scale of the lattice constant a_o up to the macroscopic scale of the thermal correlation length ξ_T, (b) percolation, which has fluctuations in characteristic size of clusters *on all length scales* from a_o up to the diameter of the largest cluster—the pair connectedness length ξ_p, and (c) the DLA/DBM problem, whose clusters have fluctuations *on all length scales* from the microscopic length $d_o = \gamma/L$ (γ is the surface tension and L the latent heat) up to the diameter of the cluster ξ_L. Also shown, on the right side, is the analogy between the scaling behavior of the three length scales ξ_T, ξ_p, and ξ_L.

To describe the specific heat near the λ point of ^{4}He, one finds that the Ising model is not appropriate. This is because the order parameter in ^{4}He is not a one-dimensional variable with only two values (up or down), but rather a two-dimensional object with an amplitude and a phase. Accordingly, the Ising model has to be replaced by a "variant" for which the one-dimensional Ising spins are replaced by two-dimensional XY spins.

(b) Random Site Percolation on a Lattice, and Its Variants

In its simplest form, one randomly occupies a fraction p of the sites of a d-dimensional lattice (the case $d = 1$ is shown schematically in Fig. 1b). Again, phenomena occurring on the local 1Å scale of a lattice constant are "amplified" near the percolation threshold $p = p_c$ to a macroscopic length ξ_p.

Here p plays the role of the coupling constant K of the Ising model. When p is small, the characteristic length scale is comparable to 1Å. However when p approaches p_c, there occur phenomena on all scales ranging from a_o to ξ_p, where ξ_p increases without limit as $p \to p_c$

$$\xi_p \sim A \left(\frac{p - p_c}{p_c} \right)^{-\nu_p}. \tag{1b}$$

Again, the amplitude A is roughly 1Å.

It is by now a well-known piece of "magic" that each phenomenon of thermal critical phenomena has a corresponding analog in percolation, so that the percolation problem is sometimes called a geometric or "connectivity" critical phenomenon. Any connectivity problem can be understood by starting with pure random percolation and then adding interactions, or whatever. Thus, e.g., we understand why the critical exponents describing the divergence to infinity of various geometrical quantities (such as ξ_p) are the same regardless of whether the elements interact or are non-interacting (Kertész et al 1985; Geiger and Stanley 1982b). This has been predicted on theoretical grounds and confirmed by detailed numerical simulations. Similarly, the same connectivity exponents are found regardless of whether the elements are constrained to the sites of a lattice or are free to be anywhere in a continuum (see, e.g., Gawlinski and Stanley 1981 for $d = 2$, and Geiger and Stanley 1982a,b for $d = 3$).

(c) The Laplace Equation and Its Variants

Is there some lesson to be learned for fluid mechanics from our experience with thermal and geometric critical phenomena? We don't know the answer to this question, but J. Nittmann and I have been exploring this possibility in recent months. Just as *variants* of the Ising and percolation problems were found to be useful in describing a rich range of thermal and geometric critical phenomena, so we have found that *variants* of the original Laplace equation are useful in describing patterns in fluids mechanics and dendritic growth (Table 1).

Table 1

A "Rosetta stone" connecting the physics underlying (a) an electrical problem (dielectric breakdown), (b) a fluid mechanics problem (viscous fingering), and (c) a diffusion problem (dendritic solidification).

(a) electrical	(b) fluid mechanics	(c) solidification
electrostatic potential: $\phi(r,t)$	pressure: $P(r,t)$	concentration: $c(r,t)$
electric field: $E \propto -\nabla\phi(r,t)$	velocity: $v \propto -\nabla P(r,t)$	growth rate: $v \propto -\nabla c(r,t)$
conservation: $\nabla \cdot E = 0$	$\nabla \cdot v = 0$	$\nabla \cdot v = 0$
Laplace equation: $\nabla^2\phi = 0$	$\nabla^2 v = 0$	$\nabla^2 v = 0$

In the Ising model, we place a spin on each pixel (site) of a lattice. In percolation we allow each pixel to be occupied or empty. In fluid mechanics, we assign a number—call it ϕ—to each pixel. Generally we shall understand ϕ to be the pressure at this region of space.

The spins in an Ising model interact with their neighbors. Hence the state of one Ising pixel depends on the state of all the other pixels in the system—up to a length scale given by the thermal correlation length ξ_T. The "global" correlation between distant pixels in an Ising simulation arises from the fact that neighboring pixels at i and j have a "local" exchange interaction J_{ij}. Similarly, the correlation in connectivity between distant pixels in the percolation problem arises from the "propagation" of local connectivity between neighboring pixels. In fluid mechanics, the pressure on each pixel is correlated with the pressure at every other pixel because the pressure obeys the Laplace equation.

One can calculate an equilibrium Ising configuration by "passing through the system with a computer" and flipping each spin with a probability related to the Boltzmann factor. Similarly, one can calculate the pressure at each pixel by "passing through the system" and re-adjusting the pressure on each pixel in accord with the Laplace equation.* If we were to arbitrarily flip the configuration of a single pixel in the Ising problem (from $+1$ to -1), we would significantly influence the equilibrium configuration of the system out to a length scale on the order of ξ_T. Similarly, if we were to arbitrarily impose a given pressure on a single point of a system obeying the Laplace equation, we would drastically change the resulting pattern out to a length scale that we shall call ξ_L.

Does ξ_L obey a "scaling form" analogous to Eqs. (1a) and (1b) obeyed by the functions ξ_T and ξ_p for the Ising model and percolation? We believe that the answer to this question is "yes," although our ideas on this subject remain somewhat tentative and subject to revision.

The best way to see the fluctuations inherent in structures grown according to the Laplace equation is to first introduce some specific models. There are two models that were once thought to be fully equivalent, although it is now recognized that the actual patterns produced by each have a different "susceptibility to lattice anisotropy" (Ball 1986). The first of these models is diffusion limited aggregation (DLA). Here one releases a random walker from a large circle surrounding a seed particle placed at the origin. When the random walker touches a perimeter site of the seed, it "sticks" (i.e., the perimeter site becomes a cluster site), and we have a cluster of mass $= 2$. A second random walker is then released. This process continues until a large cluster is formed. Initially the "mass" M of clusters was typically 10^3 to 10^4. However it has become possible to make very

---

* There is an intimate connection between the diffusion equation and the random walk problem (see, e.g., Chandrasekhar 1943).

fast algorithms, and the largest cluster to date has a mass of 4×10^6 (Meakin 1986a).

The dielectric breakdown model (DBM) differs from DLA in that nothing happens until the random walker touches a cluster site, at which time the perimeter site it was just on at the previous step is transformed into a cluster site. Not surprisingly, this tiny local change in boundary conditions does not affect the "critical exponents" of this problem—DLA and DBM have the same value of the fractal dimension d_f describing how the cluster mass depends on cluster diameter L: $M \sim L^{d_f}$.† In both thermal critical phenomena (or percolation) the length L introduced when we have a finite system size scales the same as the correlation lengths ξ_T (or ξ_p). Hence for DLA we expect that there will be fluctuations on length scales up to ξ_L, where ξ_L itself increases with the cluster mass according to

$$\xi_L \sim A\left(\frac{1}{M}\right)^{-\nu_L} \qquad [\nu_L = 1/d_f]. \qquad (1c)$$

Here the amplitude A is again on the order of 1Å. Note that (1c) is analogous to (1a) and (1b) if we think of $M \to \infty$ as being analogous to $K \to K_c$. This reasoning is common in polymer physics, where we relate the radius of gyration R_g of a polymer to the mass through an equation of the form of (1c), $R_g \sim (1/M)^{-1/d_f}$. Note that $\nu_L = 1/d_f$

---

† The difference in boundary conditions *does* affect the rate at which the asymptotic behavior shows up (Ball 1986). For example, for DLA the screening will be more severe: as soon as a random walker steps on a perimeter site, the walker is stopped and the perimeter site becomes a cluster site. However for the DBM a random walker is free to walk on perimeter sites with impunity: only when the walker steps on a **cluster** site does the walker stop walking. Hence in the DBM the walkers can better penetrate the fjords of the system, so in overall appearance DBM clusters appear to have thicker branches and to be more "compact." The critical exponent $\nu_L = 1/d_f$ is not changed since it depends not on the density but on the *rate* at which the density decreases as the mass increases.

plays the role of the critical exponents ν_T and ν_p of (1a) and (1b). Suppose we test this idea, qualitatively, by examining the largest DLA clusters in detail. We find that indeed there are fluctuations in mass on length scales less than, say, the width W of the side branches. If one makes a log-log plot of W against mass M, one finds the same slope $1/d_f$ that one finds when one plots the diameter against M.

Evidence for Similarity of Viscous Fingering Patterns and Laplace Equation (DLA/DBM) Patterns

In the remainder of this talk, we'll describe in some detail the sorts of results we obtain from variants of the Laplace equation. First, it is necessary to describe the simplest system that produces patterns resembling interesting objects found in nature. Consider, e.g., the classic Saffman-Taylor viscous fingering problem. Here one injects a low-viscosity fluid into a medium filled with high viscosity fluid. In the limit that the viscosity ratio between the high and low viscosity fluids can be taken to be zero, we can assume that the pressure everywhere inside the low viscosity fluid is a constant: $P(i) = 1$ for $i \in$ [cluster of pixels occupied by low-viscosity fluid]. The pressure everywhere else in the system will have a value given by the solution of the Laplace equation. This problem is modelled by the dielectric breakdown model or DBM (Niemeyer et al. 1984) or diffusion-limited aggregation model or DLA (Witten and Sander 1981). These two models have in common that both are solutions to the Laplace equation for the case in which the pressure is zero at infinity and $P = 1$ on an object called the cluster.

Daccord has made accurate measurements on the fractal dimension of viscous fingers in both lateral (Nittmann et al 1985) and radial (Daccord et al 1986) geometries (Fig. 2). He reduced the length scale normally imposed by surface tension by using liquids with zero interfacial tension—the two fluids were water and a viscous aqueous solution of polysaccharide (Fig. 3). He found that the resulting patterns are indeed fractal, with a fractal dimension identical to that of DLA/DBM (Fig. 4). Måløy et al (1985) found analogous behavior where the cell itself introduced the randomness: he accomplished this

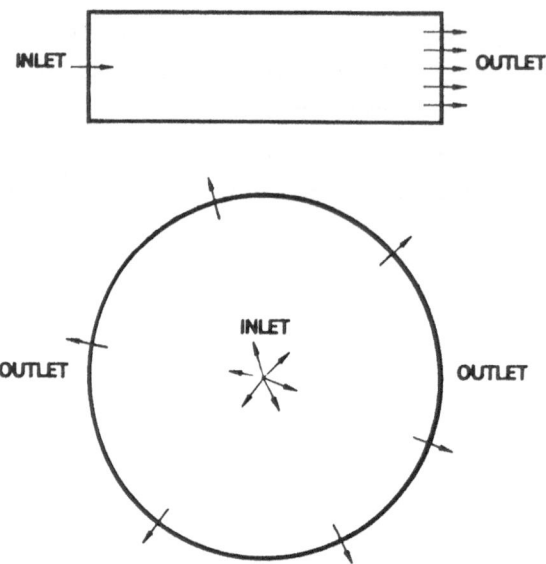

Fig. 2: Schematic illustration of the lateral and radial Hele-Shaw cells. Shown are top views. The spacing between the plates is typically 1 mm or less. From Daccord et al. (1986).

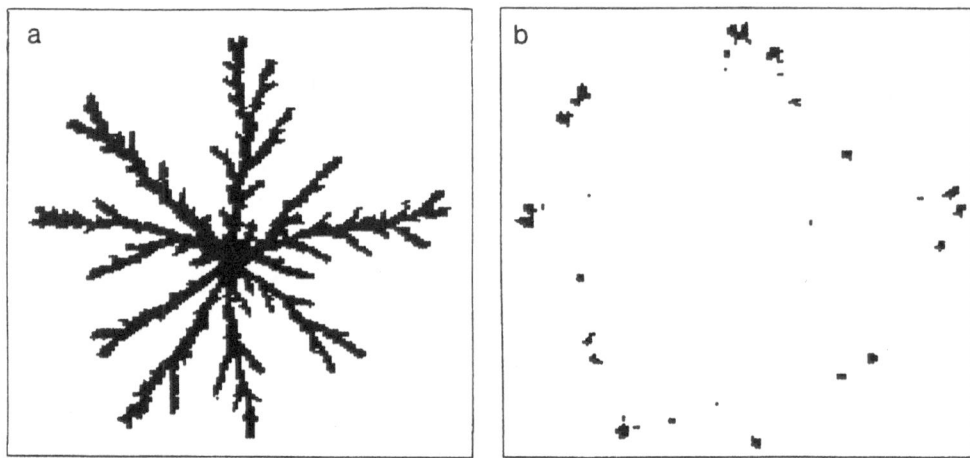

Fig. 3: The growth region of a radial viscous finger, a typical experimental pattern for which DLA is the appropriate model. The finger at time $t = t_o$ is shown in (a), while (b) displays the difference between the pattern at $t = t + \Delta t$ and $t = t$, obtained experimentally by simply subtracting the images of the same finger photographed at slightly different times. After Daccord et al. (1986).

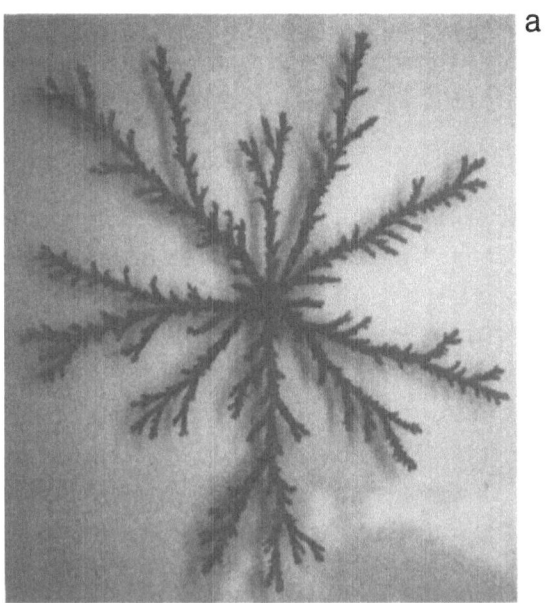

a

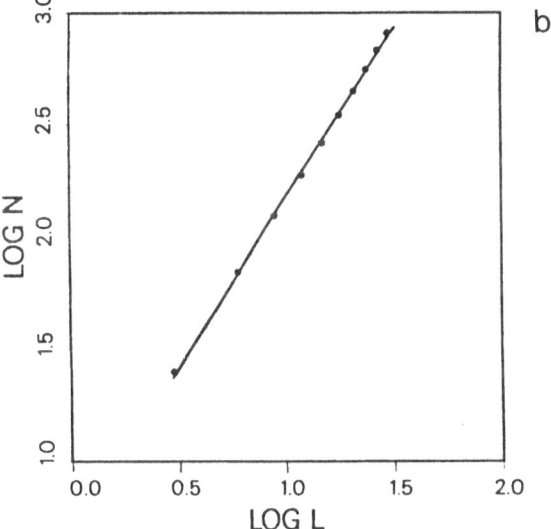

b

Fig. 4: Analysis of the fractal dimension typical of a radial viscous finger by the sandbox method (N is the number of occupied pixels in a $L \times L$ sandbox whose center is on an occupied pixel). The slope of the straight line shown is $d_f = 1.70 \pm 0.05$, while for DLA d_f is believed to be about 1.71 (from Daccord et al. 1986).

by placing glass beads inside the cell at random. Chen and Wilkinson (1985) imposed the randomness by studying viscous fingering inside a network of glass tubes whose diameter L was randomly chosen from a probability distribution $\pi(L)$.

Not only is the fractal dimension the same for the fluid mechanics problem and for the Laplace patterns, but **so also are the multifractal properties the same**. Multifractals arise when one defines some quantity on all the pixel sites. Perhaps the simplest example is that of a charged needle: if we assign to every pixel a number equal to the electric field, then the set $\{E_i\}$ of field values for the perimeter sites of the needle form a multifractal set. The distribution $n(E)$ giving the number of perimeter pixels with electric field E is characterized, like all distribution functions, by its moments

$$Z(q) = \sum_E n(E)E^q. \tag{2}$$

As might be anticipated for a self-similar system, these moments scale with the mass M (or with the diameter L)

$$Z(q) \sim M^{\sigma(q)} \sim L^{-\tau(q)}. \tag{3a}$$

Since $M \sim L^{d_f}$, the exponents σ and τ are related by the fractal dimension d_f,

$$\sigma = \frac{\tau}{d_f}. \tag{3b}$$

For thermal and geometric critical phenomena, exponents analogous to the $\sigma(q)$ and $\tau(q)$ can be defined by considering a large L × L system at the critical point $[K = K_c]$ (or $p = p_c$). One finds that the ratio of two successive exponents is a constant "gap," so that there is no new information obtained by studying higher moments of the distribution. Connected with this simplicity is the fact that there is only one independent exponent in finite size scaling at the critical point (a second exponent arises if we wish to relate quantities that describe the approach to the critical point).

In percolation, these exponents have geometric interpretations:

(i) $y_h = d_f$, the fractal dimension of the incipient infinite cluster (the largest cluster found in a box of edge L at $p = p_c$), and

(ii) $y_T = d_{\text{red}}$, the fractal dimension of the red bonds that occur inside the largest spanning cluster (red bonds are singly connected bonds: when cut, the cluster falls into two pieces).

Relation (i) was noted by Stanley (1977) while (ii) was proved by Coniglio (1982).

In the case of the moments Z_q, there is an infinite hierarchy of exponents in the sense that the ratio $\tau(q+1)/\tau(q)$ depends on q:

$$\frac{\tau(q+1)}{\tau(q)} = D(q). \tag{4}$$

For the case of a long thin needle, the exponent $D(q)$ sticks at the value $3/2$ for small q, but for q above a critical value $q = q_c$, $D(q)$ becomes "unstuck" and varies continuously with q.

The same considerations apply to the fluid mechanics problem. Here the analog of the electric field $E \propto \nabla V$ is the growth probability $p_i \propto \nabla P$, where the index i runs over all perimeter sites i. Thus p_i is the probability that site i is the next to be added to the cluster. If we think of random walkers (Fig. 5), then p_i is the hit probability (the probability that site i is the next to be hit by a random walker). Clearly the set p_i play a vital role in determining the dynamics of growth, since if we know all the p_i for every perimeter site i at a given time t, then we can predict (in a statistical sense) the state of the system at time $t+1$.

Recently, considerable attention has focussed on the question of how a DLA aggregate grows. Such growth phenomena are **completely** characterized by assigning to each perimeter site i the number p_i, the probability that site i is the next to grow. Theoretical evidence has been advanced recently to suggest that the numbers p_i form a multifractal set: this set cannot be characterized by a single exponent (as in the case of the DLA aggregate itself) but rather an infinite hierarchy of exponents is required. The physical basis for this fact is that the hottest tips of a DLA aggregate grow much faster than the deep fjords (which hardly grow at all); hence the **rate of change** of the p_i differs greatly when i is a tip perimeter site than when i is a fjord perimeter site.

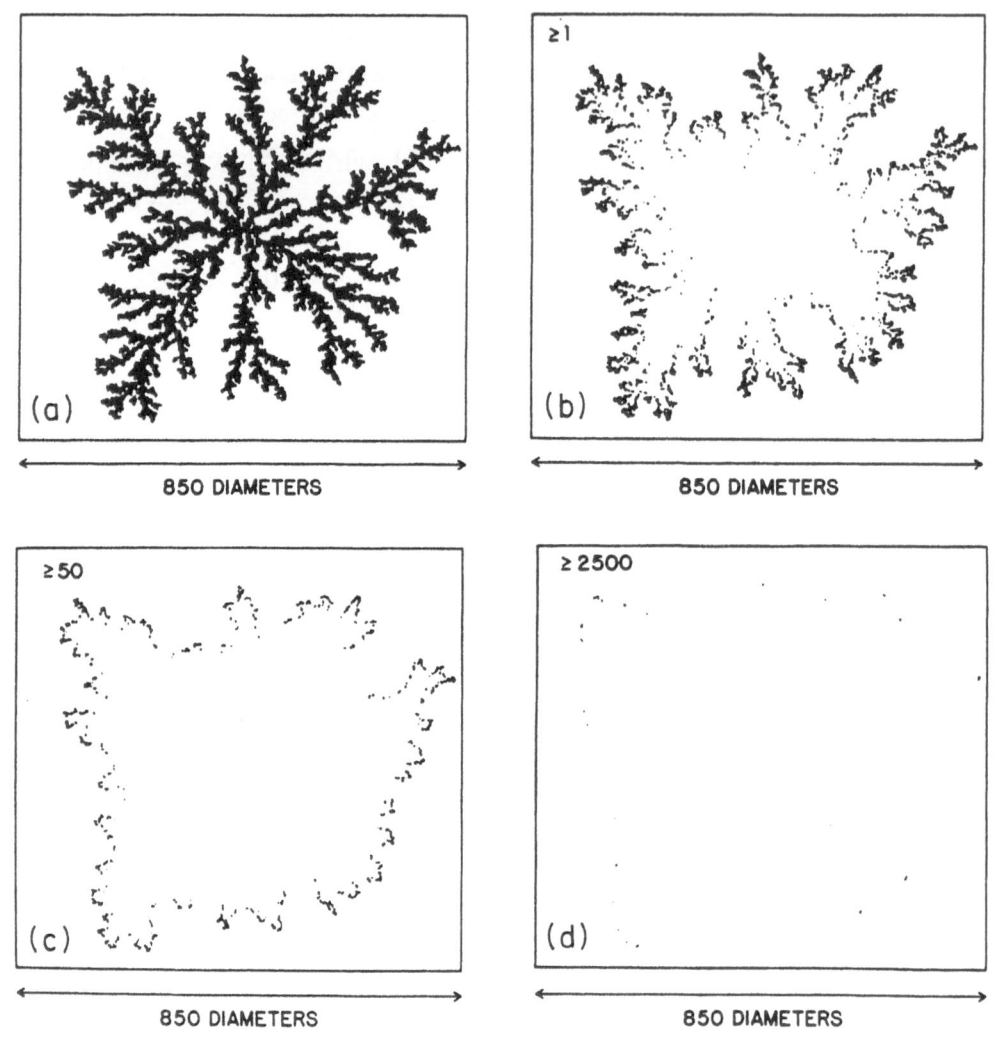

Fig. 5: This figure illustrates the harmonic measure for a 50,000 particle off-lattice 2d DLA aggregate. Figure 5a shows the cluster. Figure 5b shows all 6803 perimeter sites which have been contacted by at least one of 10^6 random walkers (following off-lattice trajectories). Figure 5c shows all of those perimeter sites which have been contacted 50 or more times and Fig. 5d shows those sites which have been contacted 2500 or more times. The maximum number of contacts for any perimeter site was 8197 so that $p_{max} = 8.2 \times 10^{-3}$. After Meakin et al (1986).

Although there have been theoretical calculations of the multi-fractality of DLA (Meakin et al. 1985,1986; Halsey et al. 1986), there had been no experimental tests of these predictions. We have recently carried out the first such tests, and found experimental confirmation of the broad outlines of the theory of multifractals (Nittmann et al. 1987).

There are many experimental realizations of DLA, and for the present work we will focus upon two-dimensional fractal viscous fingers since it is possible to study the real-time growth using a movie camera and to digitize precisely the observed time development of the DLA fractal. By subtracting two successive "snapshots" we can obtain an accurate estimate of the appropriate normalized growth probability p_i for each perimeter site of the finger (Fig. 3).

We first calculated the distribution function $n(p)$, where $n(p)dp$ is the number of perimeter sites with p_i in the range $[p_i, p_i + dp_i]$. This curve has a long tail extending to the extremely small values of p_i for perimeter sites deep inside fjords. We found good agreement between the experimental $n(p)$ for viscous fingers (Fig. 6a) and the corresponding theoretical $n(p)$ calculated for DLA (Fig. 6b).

We next formed the moments $Z_q = \Sigma(p_i)^q$ which are characterized by the hierarchy of exponents τ_q defined through $Z_q = L^{-\tau_q}$, where L is a characteristic linear dimension. The experimental results (Fig. 7a) show that when q is large, τ_q is linear in q but for q small there is downward curvature in τ_q, showing that the fjords are characterized by different growth rates than the tips. It is conventional to also calculate the Legendre transform with respect to q of τ_q: $-f(\alpha) = \tau(q) - q\alpha$ where $\alpha = d\tau/dq$. Downward curvature in $\tau(q)$ corresponds to upward curvature in $-f(\alpha)$ [Fig. 8a]. The experimental data of Figs. 7a and 8b compare favorably with the theoretical DLA model calculations shown in Figs. 7b and 8b.

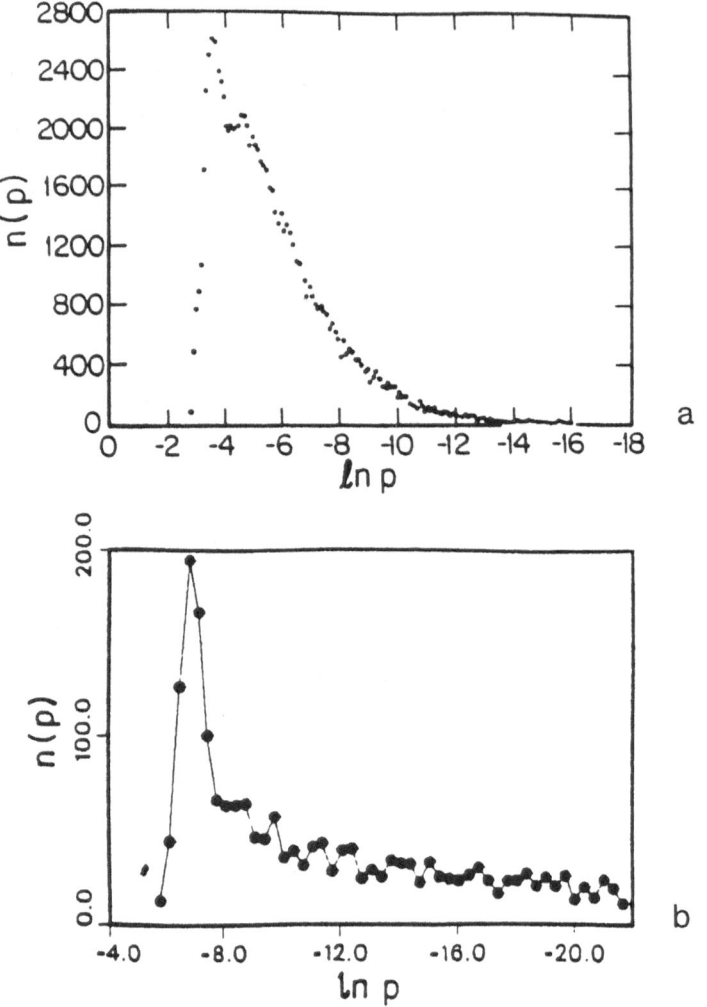

Fig. 6: Comparison between the distribution functions $n(p)$ for simulated (a) and "experimental" (b) viscous fingering patterns. Here $n(p)\delta p$ is the number of perimeter sites with growth probabilities in the range $[p, p + \delta p]$. The simulated patterns and their growth probabilities were obtained using the dielectric breakdown model. The growth probabilities for the experimental patterns were obtained by numerically solving the Laplace equation in the vicinity of a digitized representation of the pattern with absorbing boundary conditions on the sites occupied by the pattern. Similar results were obtained for large α (corresponding to the "tips") by directly subtracting two successive experimental patterns. After Amitrano et al (1986) and Nittmann et al (1987).

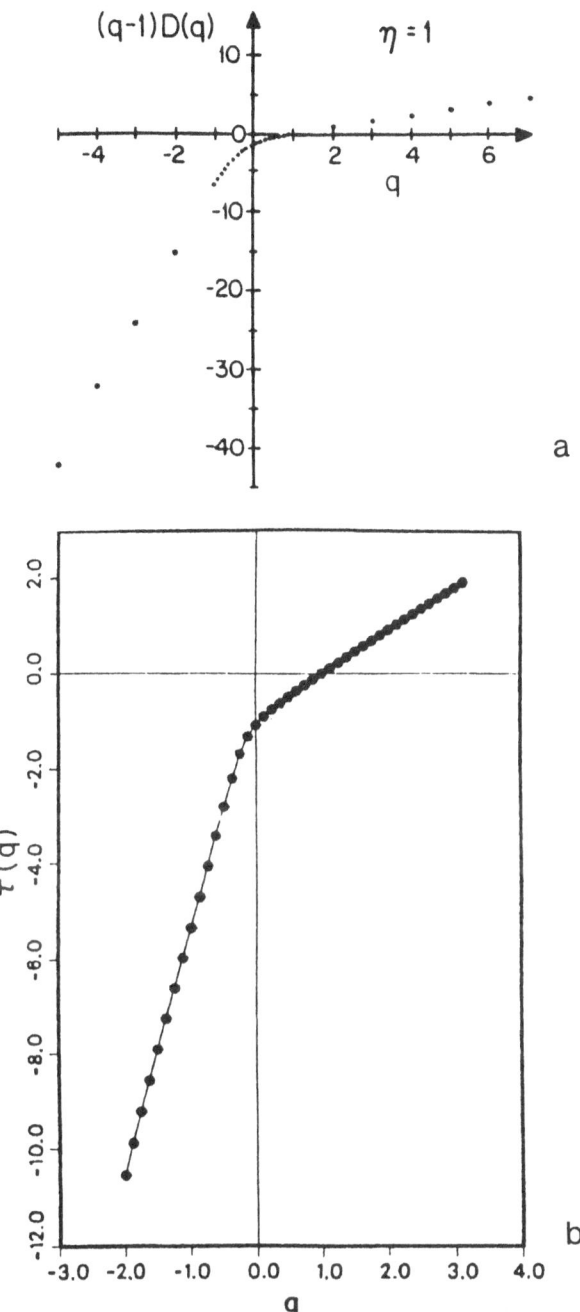

Fig. 7: Comparison of the critical exponents $\tau(q) = (q-1)D(q)$ for the (a) theoretical and (b) "experimental" viscous fingering patterns. In both cases, $\tau(q)$ was obtained numerically (see caption to Fig. 6). After Amitrano et al (1986) and Nittmann et al (1987).

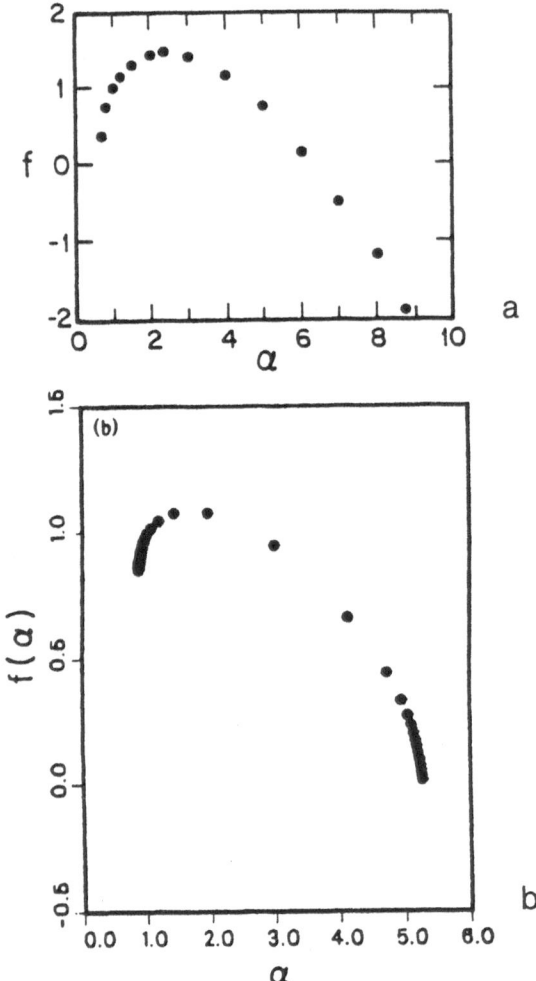

Fig. 8: Comparison between (a) theoretical and (b) "experimental" plots of the function $f(\alpha)$. After Amitrano et al (1986) and Nittmann et al (1987).

"Dendritic Solidification": Variants of the Fluid Mechanical Models

By analogy with the Ising model and its variants, we can modify DLA/DBM to describe other fluid mechanical phenomena. One of the most intriguing of these concerns a variation of the viscous fingering phenomenon in which there is present anisotropy. Ben-Jacob et al (1985) imposed this anisotropy by scratching a lattice of lines on their Hele-Shaw cell. They found patterns that strongly resemble snow crystals! If viscous fingers are described by DLA, then can the Ben-Jacob patterns be described by DLA with imposed anisotropy?

Nittmann and Stanley (1986) attempted to answer this question; specifically, they attempted to reproduce the Ben-Jacob patterns with suitably modified DLA. A scratch in a Hele-Shaw cell means that the plate spacing b is increased along certain directions, and the permeability coefficient k relating growth velocity to ∇P is proportional to b^2 ($k \propto b^2$). Hence Nittmann and Stanley calculated DLA patterns for the case in which there was imposed a periodic variation in the k. It is significant that their simulations reproduce snow crystal type patterns, just like the experiments. These simulations relied for their efficacy on the presence of noise reduction.

Noise Reduction

The original DLA and DBM models are prototypes of completely chaotic systems. No discernable pattern emerges. If there is a weak anisotropy, we expect that the resulting pattern reflects this anisotropy. For example, if the simulations are carried out on a lattice, then the presence of the lattice imposes a weak anisotropy (e.g., on a square lattice, it is more likely that particles attach to the westernmost tip if they approach from the west than from the north or south). This weak anisotropy is not visually apparent unless large clusters are grown. However, the largest DLA clusters made (Meakin 1986a), with mass about 4 million sites, clearly display the anisotropy (Fig. 9). Unfortunately, no one can afford the computer resources to make such "mega-DLA" clusters each time we wish to model a new phenomenon. Noise reduction is a computational trick that seems to have the property that it speeds up the attainment of this asymptotic limit. In the absence of noise reduction, a perimeter site becomes a cluster site whenever it is chosen (e.g., whenever a random walker lands on that site).

"Noise reduction" means that we associate a counter with each perimeter site; each time that site is chosen, the counter increments by one. The perimeter site becomes a cluster site only after the counter reaches a pre-determined threshold value termed s (Tang

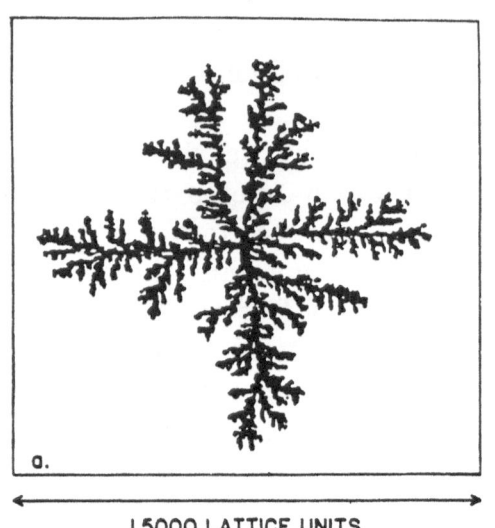 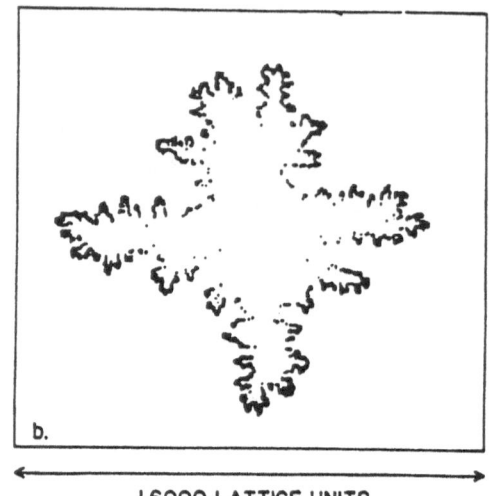

a.	b.

← I 5000 LATTICE UNITS → ← I 6000 LATTICE UNITS →

Fig. 9: A huge DLA cluster with a mass of 4 million sites grown on a square lattice. Shown is only the last 5% of the growth. In reality, there is structure on all scales less than the width W of the 4 arms. Moreover, W scales with cluster mass as $W \sim (1/M)^{-1/d_f}$, just in the same way as the quantity ξ_L defined in Eq. (1c). The spontaneous appearance of side branches is reminiscent of experimental dendritic growth patterns such as those shown in Fig. 13. After Meakin (1986a).

1985, Nittmann and Stanley 1986, Kertész and Vicsek 1986). When $s = 1$, we recover the original noisy DLA. Growth is dominated by the stochastic randomness in the arrival of random walkers. If s is very large, then growth is determined by the actual probability distribution.

For example, suppose we start with a large disc as a seed particle (instead of a single site). The growth probability at all points on the disc surface will be equal, assuming a continuum. By the D'Arcy growth law this disc should evolve in time into a larger disc. On the other hand, for ordinary DLA ($s = 1$), as soon as a random walker touches a single perimeter site on the disc, this site will become part of the cluster and the disc will lose its circular symmetry. The growth probabilities will all be re-calculated, and the perimeter sites close by the one that just grew will have higher growth probabilities. Thus the disc with a single site added to it will be more likely to grow in the direction of that single site. At a later time we will almost certainly not find a cluster with circular symmetry.

Clearly if s is very large, then the initial growth will preserve an almost circular structure. This is because before the first site is added to the circular seed, all the perimeter sites will acquire large numbers in their counters ($s - 1$, $s - 2$, etc.). After the first site is added, these additional perimeter sites will be very close to the threshold for growing while the new perimeter sites that were born when the first cluster site is added will all have counters initialized at zero. A typical cluster grown in this fashion is shown in Fig. 10; actually this cluster is grown on a square lattice with first and second neighbor interactions, not on a continuum. However Meakin et al. (unpublished) have found an almost identical pattern for the continuum case.

At first sight, there is little economy in computational speed, since one needs "s times as many" random walkers to reach a given cluster size. Thus to grow a cluster with merely 4000 sites with $s = 1000$ requires almost as much time as to generate a mega-DLA with 4,000,000 sites and $s = 1$. Fortunately, there is a way around this problem. Instead of using random walkers to solve the Laplace equation (to sample the growth probabilities p_i on each perimeter site), we can directly solve the Laplace equation numerically. This is the approach used when the dielectric breakdown model was first proposed (Fig. 11). Whether one calculates the growth probabilities by sending in random walkers or by solving the Laplace equation is immaterial: the difference between DLA and DBM is the boundary conditions, not the method of calculation.

The advantage of the Laplace equation approach when s is large is obvious: one need re-solve the Laplace equation only after a site is actually added to the cluster. In between adding sites, one simply chooses random numbers weighted by the growth probabilities of each perimeter site. This is a relatively rapid procedure for the computer, compared with its counterpart of sending random walkers.

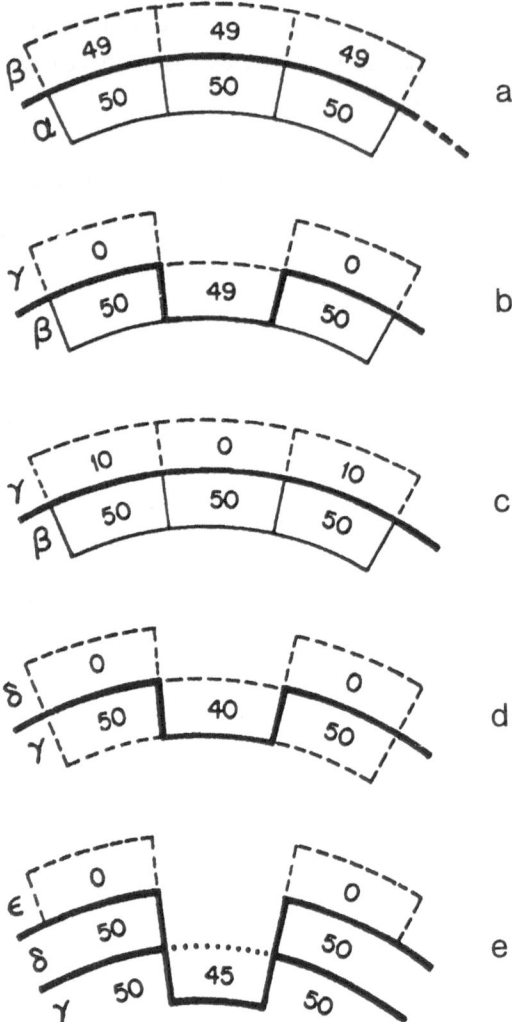

Fig. 10: Schematic illustration of the difference between an outward ('positive') and an inward ('negative') interface fluctuation. A positive fluctuation tends to be damped out rather quickly, as mass quickly attaches to the side of the extra site that is added. On the other hand, a negative fluctuation grows, in the sense that mass accumulates on both sides of the tiny notch. The notch itself has a lower and lower probability of being filled in, as it becomes the end of a longer and longer fjord. This is the underlying mechanism for the tip-splitting phenomenon when no interfacial tension is present. a shows the advancing front (row α) of a cluster with $s = 50$. The heavy line

Continued on p. 231.

separates the cluster sites (all of which were chosen 50 times) from the perimeter sites (all of which have counters registering less than 50). In a, no fluctuations in the counters of these three sites have occurred yet, and all three perimeter counters register 49. b shows a negative fluctuation, in which the central perimeter site is chosen slightly less frequently than the two on either side; the latter now register 50, and so they become cluster sites in row β. The perimeter site left in the notch between these two new cluster sites grows much less quickly because it is shielded by the two new cluster sites. For the sake of concreteness, let us assume it is chosen 10 times less frequently. Hence by the time the notch site is chosen one more time, the two perimeter sites at the tips have been chosen 10 times (c). The interface is once again smooth (row γ), as it was before, except that the counters on the three perimeter sites differ. After 40 new counts per counter, the situation in d arises. Now we have a notch whose counter lags behind by 10, instead of by 1 as in b. Thus the original fluctuation has been amplified, due to the tremendous shielding of a single notch. Note that no new fluctuations were assumed: the original fluctuation of 1 in the counter number is amplified to 10 solely by electrostatic screening. This amplification of a negative 'notch fluctuation' has the effect that the tiny notch soon becomes the end of a long fjord. To see this, note that e shows the same situation after 50 more counts have been added to each of the two tip counters, and hence (by the 10 : 1 rule) 5 new counts to the notch counter. The tip counters therefore become part of the cluster, but the notch counter has not yet reached 50 and remains a perimeter site. The notch has become an incipient fjord of length 2, and the potential at the end of this fjord is now exceedingly low. Indeed it si quite possible that the counter will never pass from 45 to 50 in the lifetime of the cluster. In our simulations we can see tiny notch fluctuations become the ends of long fjords, and all of the above remarks on the time-dependent dynamics of tip splitting are confirmed quantitatively. After Nittmann and Stanley (1986).

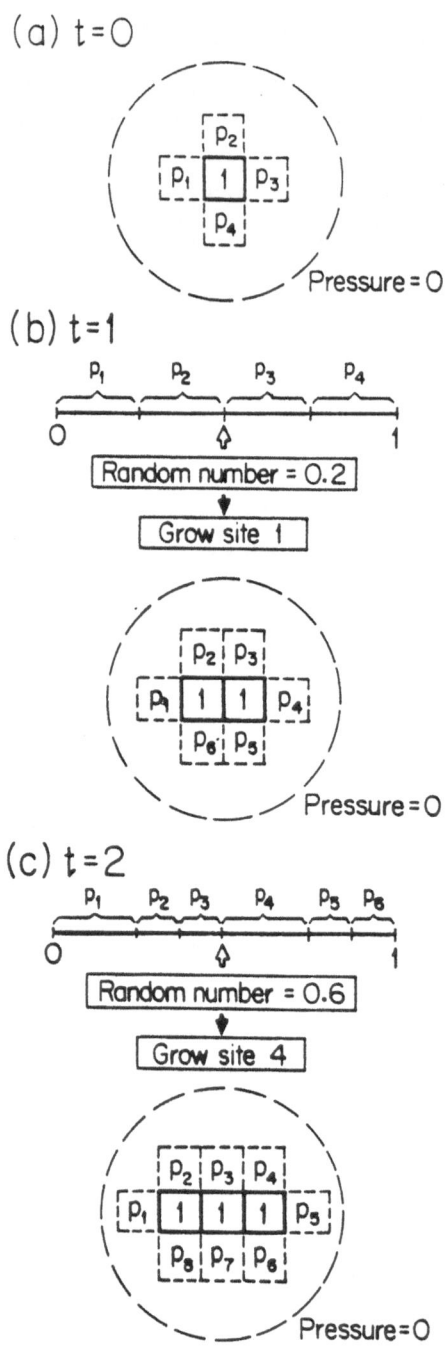

(a) t=0

P₂
P₁ 1 P₃
P₄

Pressure = 0

(b) t=1

P_1 P_2 P_3 P_4

0   1

Random number = 0.2

Grow site 1

P_2 P_3
P_1 1 1 P_4
P_6 P_5

Pressure = 0

(c) t=2

P_1 P_2 P_3 P_4 P_5 P_6

0   1

Random number = 0.6

Grow site 4

P_2 P_3 P_4
P_1 1 1 1 P_5
P_8 P_7 P_6

Pressure=0

Fig. 11: Schematic illustration of the first steps in the generation of a DLA cluster by solving directly the Laplace equation on a square lattice. After Nittmann and Stanley (1986).

"Snow Crystals"

Of course, real dendritic growth patterns (such as snow crystals) do not occur in an environment with periodic fluctuations in $k(x,y)$. Rather, the *global* asymmetry of the pattern arises from the *local* asymmetry of the constituent water molecules. Can this local asymmetry give rise to global asymmetry? Buka et al.(1986) replaced the Ben-Jacob experiment (isotropic fluid, anisotropic cell) by the reverse: isotropic cell but anisotropic fluid! To accomplish this, they used a nematic liquid crystal for the high viscosity fluid. Thus the analog of the water molecules in a snow crystal are the rod-shaped anisotropic molecules of a nematic. This experiment shows that the underlying anisotropy can as well be in the fluid as in the environment.

Snow crystal formation is thought to involve mainly the aggregation of tiny ice particles and droplets of supercooled water. To the extent that snow crystals grow by accreting water molecules previously in the vapor or liquid phase, the growth rate is thought to be limited by the diffusion away from the growing snow crystal of the latent heat released by these phase changes. Under conditions of small Peclet number, the diffusion equation describing the space and time dependence of the temperature field $T(\mathbf{r},t)$ reduces to the Laplace equation. Thus a reasonable starting point is DLA, independent of whether we wish to focus on particle aggregation, heat diffusion, or both.

DLA reflects well the randomness inherent in a wide range of growth processes, including colloidal aggregation, it fails to describe dendritic solidification. While the deterministic models of snow crystals produce patterns that are much too "symmetric," the DLA approach suffers from the opposite problem: DLA patterns are too "noisy." That DLA is too noisy has long been recognized as a defect of this otherwise physically appealing model. Recently, an approach has been proposed (Nittmann and Stanley 1987a) that retains the "good" features of DLA and at the same time produces patterns that resemble real (random) snow crystals.

Firstly, we introduce (Nittmann and Stanley 1987a) controlled amounts of noise reduction of the same sort used previously for both

DLA and for DBM. It is believed that noise-reduced DLA is in the same universality class as ordinary DLA—i.e., it has the same fractal dimension d_f, the only difference being an increase in the character-istic local length scale W. One advantage of setting $s > 1$ is that the asymptotic ("mass" $= \infty$) behavior shows up much sooner than if $s = 1$. We do not explicitly introduce anisotropy—the only anisotropy present is the six-fold anisotropy arising from the underlying triangu-lar lattice.

The patterns obtained (Nittmann and Stanley 1987a) have the same general features for all values of s greater than about $s = 100$—the effect of increasing s seems mainly to be that of increasing the width W of the fingers and sidebranches. The fjords between the 6 main branches contain much empty space. Some snow crystals have such wide "bays" but some do not. A better model would seem to require some tunable parameter that enables the complete range of snow crystal morphologies to be generated. We have found one such parameter, η, that has the desired effect of reducing the difference in the ratio of the growth probabilities between the tips and fjords. Specifically, we relate by the rule $p_i \propto (\nabla \phi)^\eta$ the growth probability p_i (the probability that perimeter site i is the next to grow) to the potential ϕ (e.g., ϕ may be the temperature $T(\mathbf{r})$ at point $\mathbf{r}$, or the probability that a tiny ice particle is at point $\mathbf{r}$). Our model is thus the analog for DLA of the "η model."

We used η to tune the balance between tip growth and fjord growth and found growth patterns that resemble better the wide range of experimentally observed snow crystal morphologies (Nittmann and Stanley 1987a). To what does the case $\eta \neq 1$ correspond? For $\eta = k$ (k = positive integer), we have a model (Meakin 1986b) in which a site grows only if it is chosen k times in succession ($k = 1$ is pure DLA). It is possible that we have a situation not altogether different from the classic n-vector model of isotropically-interacting n-dimensional classical spins: this model makes physical sense only if n is a positive integer, yet its study for other values of n has led to rich insights—particularly the cases $n = 0$ (the dilute polymer chain limit), $n = \infty$ (the spherical model) and $n = -2$ (the mean field limit). Similarly, the Q-state Potts model makes physical sense only if Q is an integer above 1, yet the cases $Q = 0$ (random resistor network), $Q = 1$ (percolation) and $Q = 3/2$ (a spin glass model) are of great interest.

The fractal dimension d_f is believed independent of the value of the noise reduction parameter s (s renormalizes the cluster mass). We confirmed this belief. However, we found d_f does depend on η. The most reliable estimates were obtained by first calculating estimates of d_f for a sequence of increasing cluster masses, and then extrapolating this sequence to infinite cluster mass. Our values for d_f agreed remarkably well with values we obtained by digitizing photographs of experimentally observed snow crystals. Of course this preliminary study (Nittmann and Stanley 1987a) does not completely "solve" the snow crystal problem:

(i) The initial seed of a snow crystal is almost certainly hexagonal (i.e., quasi-2-dimensional), since this is the local geometry that water molecules take when they form hexagonal ice I_h. Are DBM-type considerations (small growth probability near the center of a plate-like structure) sufficient to explain why a snow crystal remains quasi-2-dimensional as it continues growing? Why does its thickness remain less than its width? It is perhaps appropriate to mention that no adequate explanation has yet been advanced for why a snow crystal remains quasi-2-dimensional throughout its growth, despite the fact that the "assembly plant" is certainly 3-dimensional. Intuition on this subject stems from experience not only from critical phenomena but also from recent theoretical and experimental work on pattern formation, where it was found that even minute amounts of anisotropy are sufficient to stabilize structures of lower effective dimension.

(ii) What are the microscopic mechanisms that give rise to the feature that real snow crystals contain branches (and sidebranches) which are much more than one molecule thick? Is noise reduction relevant, or is noise reduction merely a "computational trick" that allows one to see the asymptotic form of a DLA cluster using reasonable masses? (E.g., on a square lattice, the same cross-like pattern for a mass of 5,000 sites seen in noise-reduced DLA with a noise-reduction parameter of $s = 500$ is also seen in ordinary "noisy" DLA ($s = 1$)

provided the mass is allowed to increase to roughly 5,000,000 sites! We know that DLA is obtained even if the incoming random walkers have a sticking probability that is less than one. Hence we anticipate that DLA might possibly describe a modest range of phenomena with structural re-arrangement. What is the actual sticking probability for newly arriving water molecules in real snow crystals? Is a value of the sticking probability less than unity sufficient to account for the fact that the arms and sidebranches of real snow crystals have macroscopic thickness?

(iii) Are those real snow crystals which possess relatively compact cores with ramified dressing on their surfaces products of different environments of assembly, or did melting and structural re-arrangement take place after formation? Can one mimic the effect of the changing environments in which a given snow crystal is actually assembled? Do these correspond to varying parameters such as η or γ *in the course of the growth process*? To study this effect, we generated patterns with values of η and γ that change during the growth process–e.g., we might choose $\eta \ll 1$ for an initial fraction f of the growth (thereby creating a hexagonal core), and $\eta = 1$ thereafter (thereby creating a ramified exterior portion).

(iv) Does the presence in the clouds of a wind whose direction and speed varies randomly (both in time and in space, with characteristic time scales and length scales that are microscopic) imply that the actual trajectories of water molecules and water droplets might more resemble those of some extremely "pathological" path than those of a conventional DLA type random walk? We know that the random walk trajectories of DLA correspond exactly to the present electrostatic growth model, the DBM with DLA boundary conditions. What are the trajectories in "real space" corresponding to a choice of the η parameter below unity? One can speculate that a Lévy flight with tunable fractal dimension may be related to the path of a real ice particle buffeted around in a cloud.

(v) How significant, in practice, is the role played by diffusion of latent heat away from the growing aggregate in determining the actual structure of a snow crystal? We know that this phenomenon

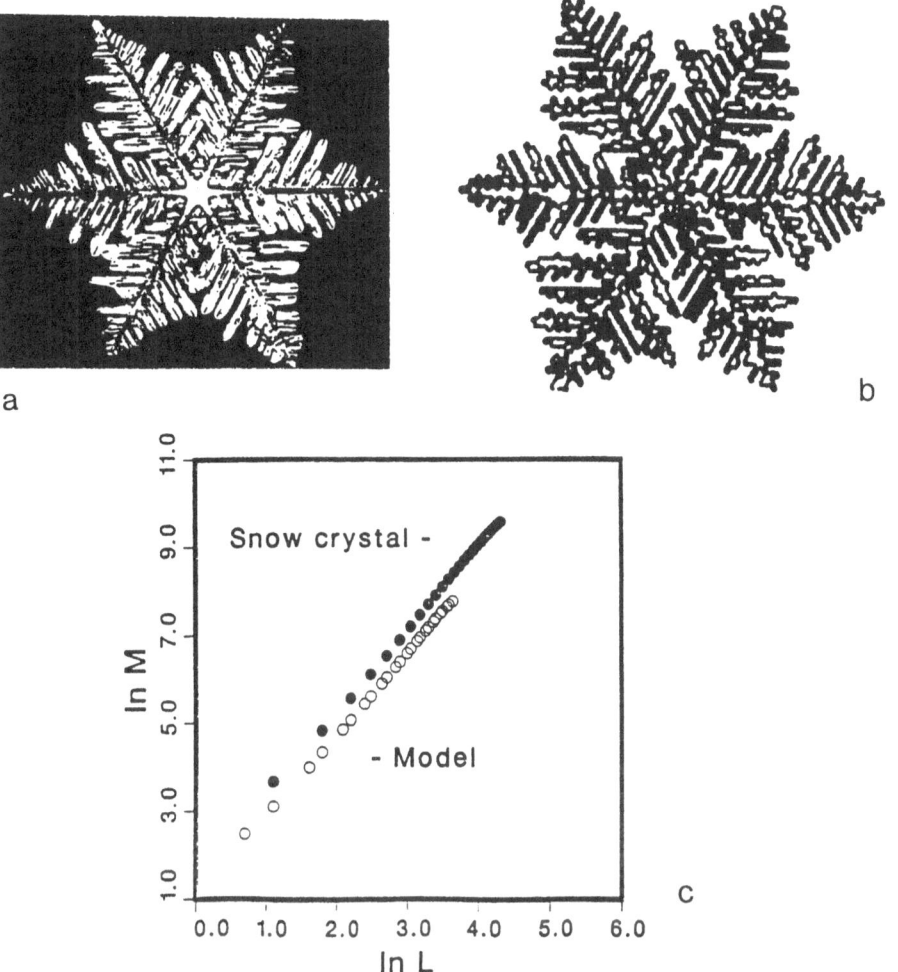

a

b

c

Fig. 12: (a) A typical snow crystal from the collection of 2453 photographs assembled in Bentley and Humphreys (1962). Other experimental examples may be found in Nakaya (1954) and LaChapelle (1969). (b) A DLA simulation with noise reduction parameter of $s = 200$ and non-linearity parameter $\eta = 0.5$. (c) Comparison between the fractal dimensions of (a) and (b) obtained by plotting the number of pixels inside an $L \times L$ sandbox logarithmically against L. The same slope, $d_f = 1.85 \pm 0.06$, is found for both. The experimental data extend to larger values of L, since the digitizer used to analyze the experimental photograph has 20,000 pixels while the cluster has only 4000 sites. After Nittmann and Stanley (1987a).

is of paramount importance in dendritic growth of crystals from a liquid phase. How significant is the role played by the capillary length $d_o = \gamma/L$ in vapor phase deposition of water molecules onto a growing snow crystal? (Here L is the latent heat.) An ideal model might encompass *both* the diffusion of heat away from the snow crystal *and* the aggregation of particles toward the snow crystal.

(vi) Are real snow crystals sometimes fractal objects? This intriguing question has been the object of considerable discussion in recent years. Our growth patterns are fractal, for all positive values of η. We found (Nittmann and Stanley 1987a) that the fractal dimension d_f is independent of the value of the noise reduction parameter s (s seems to mainly renormalize the cluster mass), but d_f does depend on η. We also found that these values for d_f agreed well with values we obtained by digitizing the corresponding photographs of experimentally observed snow crystals (Fig. 12).

Dendritic Growth of NH$_4$Br

Dendritic crystal growth has been a field of immense recent progress, both experimentally and theoretically. In particular, Dougherty et al (1987) have recently made a detailed analysis of stroboscopic photographs, taken at 20 second intervals, of dendritic crystals of NH$_4$Br (Fig. 13a). They have found three surprising results: (i) the sidebranches are non-periodic at any distance from the tip, with random variations in both phase and amplitude, (ii) sidebranches on opposite sides of the dendrite are essentially uncorrelated, and (iii) the rms sidebranch amplitude is an exponential function of distance from the tip, with no apparent onset threshold distance. Some of these results are apparently at variance with predictions from recent theories (Saito et al 1987; Ben-Jacob et al 1984; Kessler et al 1984).

How can we understand these new experimental facts? Many existing models reflect the essential physical laws underlying the growth phenomena, but fail to find a tractable mechanism to incorporate the effects of noise on the growth. Growth of a dendrite from solution is controlled by the diffusion of solute towards the growing dendrite. In the limit of small Peclet number, the diffusion equation reduces

238

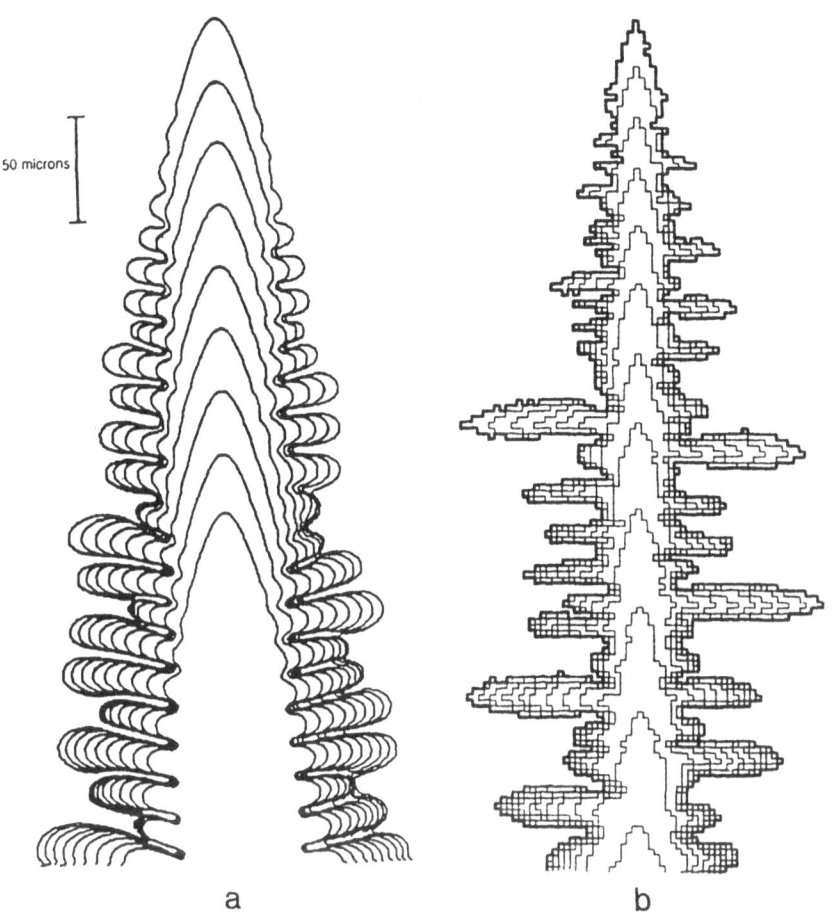

50 microns

a   b

Fig. 13: (a) Experimental pattern of dendritic growth, measured for
NH$_4$Br by Dougherty et al. (1987). (b) DLA simulation with noise
reduction parameter $s = 200$ (after Nittmann and Stanley 1987b).

to the Laplace equation (as mentioned above). The Laplace equation for a moving interface (the growing dendrite) brings to mind
the diffusion limited aggregation model (DLA). Growth patterns produced by the various DLA simulation algorithms do *not* resemble
dendritic growth patterns: DLA patterns are much too chaotic in
appearance. We shall discuss here a related model (Nittmann and
Stanley 1987b) whose asymptotic structure does resemble the patterns found experimentally—both in broad qualitative features and
in quantitative detail. The picture that emerges is one of Laplacian
growth, where noise arises from the fact that there are concentra-

tion fluctuations in the vicinity of the growing dendrite (these are estimated to be roughly $\pm 10^5$ NH$_4$Br molecules per cubic micron).

Our starting point is the observation that minute amounts of anisotropy become magnified as the mass of a cluster increases. In fact, even the weak anisotropy of the underlying lattice structure can become so amplified that clusters of 4,000,000 particles take on a cross-like appearance (cf. Fig. 1 of Meakin 1986b). A real dendrite has a mass of roughly 10^{16} particles; it is impossible to generate clusters of this size on a computer, since even clusters of size 10^6 require hundreds of hours on the fastest available computers. Fortunately, there is a computational trick—termed *noise reduction*—that speeds the convergence of the pattern toward its asymptotic "infinite mass" limit. The patterns we obtained with noise-reduced DLA resemble Fig. 1 of Dougherty et al (1987), reproduced in Fig. 13a.

A typical result (Nittmann and Stanley 1987b) for a mass of 4000 particles is shown in Fig. 13b. After each 333 particles are added, a contour is drawn:

(i) It is apparent from the "stroboscopic" representation of Fig. 13b that the distance between successive tip positions is a decreasing function of the mass; in fact, we find that $log\ x_{tip}$ is linear in $log\ M$ with slope 2/3. This result is consistent with the belief that $d_f = 1.5$ for DLA with anisotropy.

(ii) The tip is remarkably parabolic: specifically, when we form $(y_c - y_o)^2$ (where y_c is the contour, and y_o is the centerline of the dendrite) and plot this on linear graph paper as a function of $x - x_{\text{tip}}$, we obtain a straight line with an R value of 0.997.

(iii) The sidebranches are non-periodic at any distance from the tip, with random variations in both phase and amplitude. To demonstrate this, we have analyzed our simulations in exactly the same mathematical fashion as Dougherty et al analyzed the experimental dendrite patterns.

An open theoretical question concerns the microscopic origin of the sidebranching phenomenon. One current hypothesis predicts that the sidebranch amplitude would be periodic and the two sides of the dendrite should have correlated sidebranching. Dougherty et al (1987) noted that their experimental data are not consistent with this hypothesis, and we can make similar remarks for the present model. A second hypothesis views sidebranching as a result of the noise arising from concentration fluctuations. To test this hypothesis, Dougherty et al (1987) plot the sidebranch amplitude as a function of $x - x_{\text{tip}}$, the distance from the tip. They found that the sidebranch amplitude decreases as the distance variable $x_{\text{tip}} - x$ decreases, and shows no sign of a threshold distance below which the amplitude is zero. Moreover, they found that close to the tip the sidebranch amplitude is roughly linear on semi-log paper. If we plot y_c, the amplitude, which should scale roughly as the square root of the area under the peak if the peak maintains its shape as a function of $x - x_{\text{tip}}$; we find exactly the same exponential growth of sidebranch amplitude with distance from the tip.

In summary, we have developed a model in which noise reduction is used to tune the effect of noise, and cubic anisotropy is introduced through the use of an underlying square lattice. The resulting patterns obtained strongly resemble the experimental patterns of Dougherty et al (1987), both in their *qualitative* appearance and in the same degree of *quantitative* detail studied experimentally. Sidebranching arises from the fact that an approximately flat interface in the DLA problem grows trees (which resemble "bumps" in the presence of noise reduction); these compete for the incoming flux of random walkers. If one tree gets ahead, it has a further advantage for the next random walker and so gets ahead still more. Thus some sidebranches grow while others do not. The characteristic spacing λ between sidebranches scales with the dendrite mass with the same exponent $2/3$ that characterizes the growth of dendrite length x_{tip}. Moreover, the patterns we obtain are reasonably independent of details of the simulation in that similar patterns are obtained when we vary the surface tension parameter σ over a modest range; we can also alter the boundary conditions of the model with some latitude and even allow for non-linearity in the growth process $(\eta \neq 1)$.

The significance of the present findings is that the essential physics embodied in the DLA model—previously used to describe fluid-fluid displacement phenomena ("viscous fingering")— seems sufficient to describe the highly uncorrelated (almost random) dendritic growth patterns recently discovered from the experiments and quantitative analysis of Dougherty et al (1987).

Summary

We have argued that it is worth exploring all the consequences of a straightforward physical model. Our optimism is based on the success of the Ising model and percolation in the past. We must be mindful that substantial variants of the original model may be called for. In our case, e.g., anisotropy must be introduced or else the pattern bears absolutely no resemblance to dendritic growth. Also, noise reduction must be introduced or else the computer time becomes prohibitive.

This modest work perhaps raises more questions than it answers, but it nonetheless might stimulate further investigation of the basic physics of random systems that must be better understood in order to explain experimentally observed non-symmetric dendritic growth patterns and fluid mechanics patterns. The reader interested in more details than provided here may consult recent books on the subject (Family and Landau 1984; Boccara and Daoud 1985; Pynn and Skjeltorp 1986; Stanley and Ostrowsky 1986; Pietronero and Tosatti 1986; Stanley 1987).

Acknowledgements

I'd like to thank the Organizing Committee for inviting me. The results I report here are largely due to interactions with A. Coniglio, G. Daccord, P. Meakin, E. Touboul, T. A. Witten and, most especially, J. Nittmann.

AMITRANO, C., CONIGLIO, A., and DI LIBERTO, F., 1986, Phys. Rev. Lett., **57**, 1016.
BALL, R., 1986, *Physica*, **140A**, 62
BEN-JACOB, E., GODBEY, R., GOLDENFELD, N. D., KOPLIK, J., LEVINE, H., MUELLER, T., and SANDER, L. M., 1985, *Phys. Rev. Lett.*, **55**, 1315.
BEN-JACOB, E., GOLDENFELD, N., KOTLIAR, B. G., LANGER, J. S., 1984, *Phys. Rev. Lett.*, **53**, 2110.

BENTLEY, W. A., and HUMPHREYS, W. J., 1962, *Snow Crystals* (Dover, NY).

BOCCARA, N. and DAOUD, M. (eds), 1985, *Physics of Finely Divided Matter* [Proceedings of the Winter School, Les Houches, 1985] (Springer Verlag, Heidelberg).

BUKA, A., KERTÉSZ, J., and VICSEK, T., 1986, *Nature*, **323**, 424.

CHANDRASEKHAR, S., 1943, *Rev. Mod. Phys.*, **15**, 1.

CHEN, J. D. and WILKINSON, D., 1985, *Phys. Rev. Lett.*, **55**, 1985.

CONIGLIO, A., 1982, *J. Phys. A*, **15**, 3829.

DACCORD, G., NITTMANN, J., and STANLEY, H. E., 1986, *Phys. Rev. Lett.*, **56**, 336.

DOUGHERTY, A., KAPLAN, P. D., and GOLLUB, J. P., 1987, *Phys. Rev. Lett.* **58**, 1652.

FAMILY, F. and LANDAU, D. P. (eds), 1984, *Kinetics of Aggregation and Gelation* (Elsevier, Amsterdam).

GAWLINSKI, E.T., and STANLEY, H. E., 1981, *J. Phys. A.*, **14**, L291-L299.

GEIGER, A., and STANLEY, H. E., 1982a, *Phys. Rev. Lett.*, **49**, 1749.

GEIGER, A., and STANLEY, H. E., 1982b, *Phys. Rev. Lett.*, **49**, 1895.

HALSEY, T. C., MEAKIN, P., and PROCACCIA, I., 1986, *Phys. Rev. Lett.*, **56**, 854.

ISING, E., 1925, *Ann. Physik*, **31**, 253-258.

KERTÉSZ, J., and VICSEK, T., 1986, *J. Phys. A*, **19**, L257.

KERTÉSZ, J., STAUFFER, D., and CONIGLIO, A., 1985, *Ann. Israel Phys. Soc.* (Adler, Deutscher and Zallen, eds), p. 121-148.

KESSLER, D., KOPLIK, J., and LEVINE, H., 1984, *Phys. Rev. A*, **30**, 3161.

LaCHAPELLE, E. R., 1969, *Field Guide to Snow Crystals* (U. Washington Press, Seattle).

LANGER, J. S., 1980, *Rev. Mod. Phys.*, **52**, 1.

LENZ, W., 1920, *Phys. Z.*, **21**, 613-5.

MÅLØY, K. J., FEDER, J., and JØSSANG, T., 1985, *Phys. Rev. Lett.*, **55**, 2688.

MEAKIN, P., STANLEY, H. E., CONIGLIO, A., and WITTEN, T. A., 1985, *Phys. Rev. A*, **32**, 2364.

MEAKIN, P., CONIGLIO, A., STANLEY, H. E., and WITTEN, T. A., 1986, *Phys. Rev. A*, **34**, 3325-3340.

MEAKIN, P., 1986a, *J. Theor. Biol.*, **118**, 101.

MEAKIN, P., 1986b, *Proc. Israel Conference on Fracture*.

NAKAYA, U., 1954, *Snow Crystals* (Harvard Univ Press, Cambridge).

NIEMEYER, L., PIETRONERO, L., and WEISMANN, H. J., 1984, *Phys. Rev. Lett.*, **52**, 1033.

NITTMANN, J., DACCORD, G., and STANLEY, H. E., *Nature* **314**, 141 (1985).

NITTMANN, J., and STANLEY, H. E., 1986, *Nature* **321**, 663-668.

NITTMANN, J., and STANLEY, H. E., 1987a, *J. Phys. A* **20**, xxx

NITTMANN, J., and STANLEY, H. E., 1987b *J. Phys. A* , **20**, xxx

NITTMANN, J., STANLEY, H. E., TOUBOUL, E., and DACCORD, G., 1987, *Phys. Rev. Lett.*, **58**, 619.

PIETRONERO, L. and TOSATTI, E. (eds), 1986, *Fractals in Physics* (North Holland, Amsterdam).

PYNN, R. and SKJELTORP, A. (eds.), 1986, *Scaling Phenomena in Disordered Systems* (Plenum, N.Y.).

SAITO, Y., GOLDBECK-WOOD, G., and MÜLLER-KRUMBHAAR, H., 1987, *Phys. Rev. Lett.*, **58**, 1541.

STANLEY, H. E., 1977, *J. Phys. A*, **10**, L211-L220.

STANLEY, H. E., 1987 *Introduction to Fractal and Multifractal Phenomena* (to be published).

STANLEY, H. E., and OSTROWSKY, N. (eds), 1986, *On Growth and Form: Fractal and Non-Fractal Patterns in Physics* (Martinus Nijhoff, The Hague).

TANG, C., 1985, *Phys. Rev. A.*, **31**, 1977.

WITTEN, T. A., and SANDER, L. M., 1981, *Phys. Rev. Lett.*, **47**, 1400.

A Stochastic Model for Arborescent and Dendritic Growth

J. Nittmann

Dowell Schlumberger, F-42003 St. Etienne Cedex 1, France

A simple non-deterministic gradient-governed growth model is described and used to study the growth of interfaces as observed in a wide range of physical systems such as in fluid displacement in porous media, particle aggregation, dielectric breakdown and solidification. The Dielectric Breakdown Model is extended to include noise reduction, a finite viscosity ratio, spatial anisotropy and surface tension. This set of physical parameters plays a dominant role in the pattern-forming process.

1. Introduction

The evolution of fluid-fluid [1] and liquid-solid [2] interfaces frequently leads to non-trivial structures ranging from an ordered to a disordered appearance. Much work is currently in progress to understand the physical mechanisms involved in the establishment of certain patterns. Some typical examples of astonishing pattern formation in nature are found in snow crystals, lightning, eroded river beds, viscous fingers, caves and clusters of dust particles.

Recently, a simple computer model was proposed which brought new insight into the growth of random structures : diffusion limited aggregation [3] (DLA). The arborescent structures obtained by adding randomly diffusing particles to a seed has rapidly led to experimental evidence of this growth mechanism in very different physical systems [4].

The purpose of this work is to demonstrate that by extending the concept of diffusion limited aggregation a whole range of physical systems can be studied.

2. Modelling Fractal Growth of Viscous Fingers

We develop the model from the view of a fluid dynamicist but shall give cross references to solidification and aggregation where appropriate.

If a less viscous fluid displaces a more viscous but miscible fluid, e.g. in a Hele-Shaw cell, the interface between the two fluids will develop non-uniformly as small fluctuations grow rapidly to macroscopic size [5]. In Fig.1 two typical viscous finger patterns are shown which appear when water displaces a non-Newtonian (Fig.1a) and a Newtonian fluid (Fig.1b) in a Hele-Shaw cell. Both fluids are miscible with water. As one can see in Fig.1b, where three intervals of growth are indicated by different exposure times, the leading finger tips grow by *splitting* giving rise to characteristic *arborescent* structures.

The randomness of the growth and the unrepeatability of a particular pattern initially discouraged people from investigating this tip splitting instability. However, the application of the concepts of *fractal scaling* and *self-similarity* [6] has made new progress possible. Viscous fingers were shown to obey fractal scaling [7]. Hence this important instability phenomenon was quantified for the first time, by its fractal dimension d_f. As the fingers grow the volume V of water pumped scales with a characteristic length (radius) of the structure as

$$V \propto L^{d_f}, \tag{1}$$

where $d_f = 1.7$ independent of flow rate and for a wide range of miscible fluids.

What is the physical mechanism giving rise to *tip splitting* and how can we understand the growth? We shall start from the governing equations: In a radial Hele-Shaw cell with radius R and plate separation b, the pressure p and the velocity $\mathbf{u}$ of a Newtonian fluid obey the following equations:

$$\mathbf{u} = -\frac{b^2}{12\mu}\nabla p, \tag{2}$$

$$\nabla.\mathbf{u} = 0. \tag{3}$$

a   b

Fig.1: The viscous fingering phenomenon in a Hele-Shaw cell.
(a) Water displacing a polymer solution. (b) Water displacing fluidose

Equations (2) and (3) can be combined to give

$$\nabla \frac{1}{\mu} \nabla p = 0. \tag{4}$$

If two fluids (μ_1, μ_2) are present in the system the pressure at the interface has, in addition, to satisfy

$$\frac{1}{\mu_1} \nabla p_1 = \frac{1}{\mu_2} \nabla p_2. \tag{5}$$

Let us assume for the moment that the less viscous fluid has negligible viscosity compared with the more viscous fluid $(\mu_2 = 0)$. The pressure is therefore constant in the displacing fluid and satisfies $\nabla^2 p = 0$ within the displaced fluid. This set of equations, with the boundary conditions (R_I is the radius of the interface)

$$p = p_1 \text{ at } r \le R_I \text{ and } p = p_0 \text{ at } r = R, \tag{6}$$

has been studied [8] for an ideal system free of fluctuations. Calculations show that a perturbed interface develops a single finger structure without showing tip splitting.

In order to find tip splitting it is essential that random fluctuations are present in the system. We have therefore replaced equation (2) by a stochastic growth law [9]:

$$Pr\{u_i\} = \frac{\nabla p_i}{\sum_l \nabla p_l} = \nabla p_i^*. \tag{7}$$

We present the interface by a discrete number of sites u_i belonging to the cluster representing the displacing fluid. The number of sites is l. For each potential growth site we calculate the local pressure gradient ∇p_i normalized by the sum over all gradients. Each gradient ∇p_i^* represents part of the interval between 0 and 1. We now choose a number randomly between 0 and 1 and the site whose gradient corresponds to this random number is advanced. Naturally, a site with a large gradient is more like to be chosen than one with a small gradient. Statistically we set the probability of growth $Pr\{u_i\}$ of a site i equal to its pressure gradient ∇p_i^*.

Equation (7) differs fundamentally from Darcy's law (equation 2). Although for the case of stable flow, equations (2) and (7) are equivalent and give identical flow evolutions, they differ significantly in the unstable case. Equation (7) introduces fluctuations in the growth which evolve to macroscopic size leading to the formation of structure. The patterns in Fig.1 are a manifestation of the noise inherent in the experiment. Although the same physical mechanism of growth is visible, the structures are not identical, indicating that rather different fluids were used.

The Laplace equation for the pressure (potential) field can be simply written in a finite difference representation as

$$p_i = \frac{\sum_n p_n}{n}. \tag{8}$$

The index n corresponds to the number of neighbour sites to which growth can advance and is 4, 6 or 8 depending on the lattice used. So far this model is identical to the Dielectric Breakdown Model [10]. A typical DBM structure is shown in Fig.2a, which has a fractal dimension equal within experimental errors to viscous fingering. We notice, however, that Fig.2a looks "noisier" than Fig.1. Evidently equation (7) has introduced too much noise. It is interesting to notice, however, that Fig.2a strikingly resembles fractal fingering in a porous medium which differs from a Hele-Shaw cell [11]. A priori, a fluid element in a Hele-Shaw cell can flow in any direction with roughly equal probability. This, however, is not the case in many porous media. Owing to a random pore and throat structure each pore is only connected to a certain number of throats, some of which can actually be dead ends. Therefore the flow of invading fluid *fluctuates* strongly in space as it chooses to penetrate the throats of less resistance. In general the flow patterns of the displacing fluid reflect the irregularities of the porous media in form of flow patterns indistinguishable from Fig.2a.

How can we tune the noise in our model? First, let us look at DBM. At a time step δt, one site, out of all possible growth sites, is allowed to grow. After this site has been advanced the potential field is recalculated in the whole system. Therefore, the new growth has changed, although modestly, the local gradients of all other sites and, in particular, has slightly increased its own

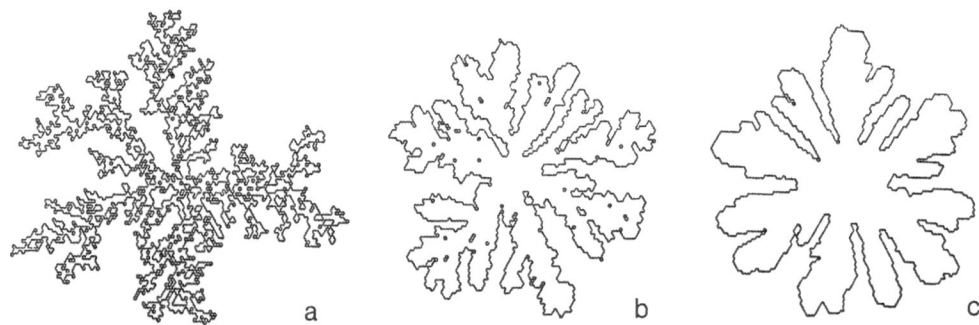

Fig.2: Tunable noise model, (a) s = 1, (b) s = 10, (c) s = 100.

probability of growing again. The fact that the growth is related to a probability and not strictly to the gradient as proposed by equation (2) introduces large fluctuations in the growth, as it is likely e.g. never to choose a site of very small gradient, which in a continous approach would grow, although slowly. To reduce the noise we follow an idea proposed by Tang [12]:

We calculate the local pressure gradients as before. Again we throw a die and choose a site a. However this site is not advanced immediately. We store the information that the site a was chosen in the form of a counter $c(a)$ and continue to throw a die in order to find new sites $b, c, d, ...$ to advance. If we choose a again we put the counter $c(a)$ up by one. If finally $c(a)$ or $c(b)$, etc. exceeds a chosen parameter value s the site is finally chosen to grow. Only now do we recalculate the potential everywhere. It is important to notice that for the next series of throws of the die we keep the counter for all sites at their last values except for the one we have just grown, which has the counter reset to zero.

The change of the structures as a function of s can be seen in Fig.2. Figs. 2b-2c show viscous fingering with the noise parameter $s = 10$ and $s = 100$.

3. Spatial Anisotropy and Snowflake Pattern Formation

Equation (8) determines the potential distribution in an isotropic medium. However, many systems show spatial anisotropy in practice e.g. porous media in the form of natural fractures, crystals due to lattice anisotropy or a Hele-Shaw cell due to artificially engraved linear grooves [13]. This additional spatial anisotropy requires modified equations (2) and (4). The constant permeability coefficient $k = b^2/12\mu$ is replaced by a permeability function $k(x, y)$:

$$\mathbf{u} = -\frac{k(x, y)}{\mu} \nabla p, \tag{9}$$

$$\nabla k(x, y) \nabla p = 0. \tag{10}$$

In the finite difference representation equations (9) and (10) read

$$Pr\{u_i\} = \frac{k_{ij} \nabla p_i}{\sum_l k_{lm} \nabla p_l}, \tag{11}$$

$$p_i = \frac{\sum_n k_{in} p_n}{\sum_n k_{in}}. \tag{12}$$

The sum in equation (11) is taken over all interface sites whereas the sum in equation (12) is taken over the nearest neighbour sites. The function $k(x,y)$ is chosen in the following way [9]: We superpose on our initial lattice with lattice spacing $l_1 = 1$ a second lattice with a larger spacing l_2, e.g. for the presented results we chose $l_2 = 3l_1$. Any site on the second lattice has a permeability k which is a free parameter. In Figs.3a - 3c we assumed k to be 1.316 (Fig.3a), $k = 2.0$ (Fig.3b) and $k = 10.0$ (Fig.3c). We see clearly how the anisotropy manifests itself with increasing $k(x,y)$. We notice an important change of growth mechanism. Although viscous fingering in an isotropic medium is clearly characterized by tip splitting, anisotropy is now stabilizing the advancing tips giving rise to *dendritic* structures characterized by stable tips with irregularly spaced side branches.

In order to obtain tip-stable growth, using DBM, it was essential to introduce spatial anisotropy. Kertész and Vicsek [14] found, however, that introducing noise reduction into DLA leads to a tip-stable growth and on a triangular lattice to snowflake pattern formation. DLA differs from DBM only in that the potential of the neighbour sites of the interface is the same as on the cluster [15]. In DBM the neighbour sites are by definition outside the cluster and their potential satisfies the Laplace equation. This means that in DLA the gradients of growth are actually calculated one unit length ahead of the interface. This small difference leads, when noise reduction is introduced, to a remarkable difference between DLA and DBM as we demonstrate in Fig.4. Calculating the gradients ahead of the interface has enhanced the influence of the anisotropy of the underlying lattice. Both Figs.4a and 4b are grown with four nearest neighbours only, $s = 200$. DLA shows a remarkable dendritic appearance with stable tips and random side branching. In contrast, DBM shows strong evidence of tip splitting as observed in viscous finger experiments (Fig.1). Some anisotropic

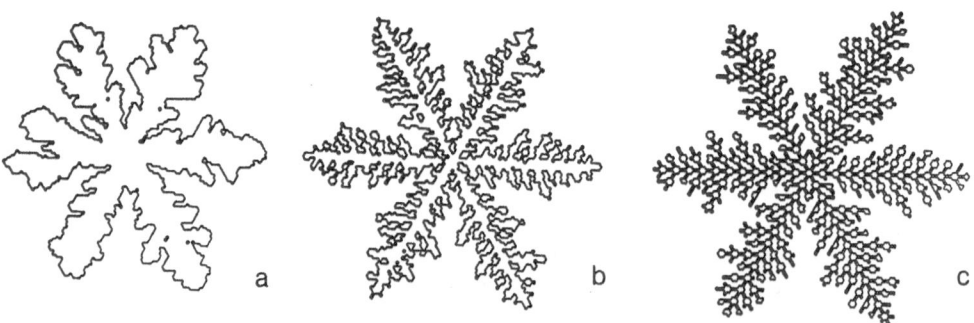

Fig.3: Tunable noise model ($s = 200$) with spatial anisotropy, (a) $k = 1.316$, (b) $k = 2.0$ and (c) $k=10.0$.

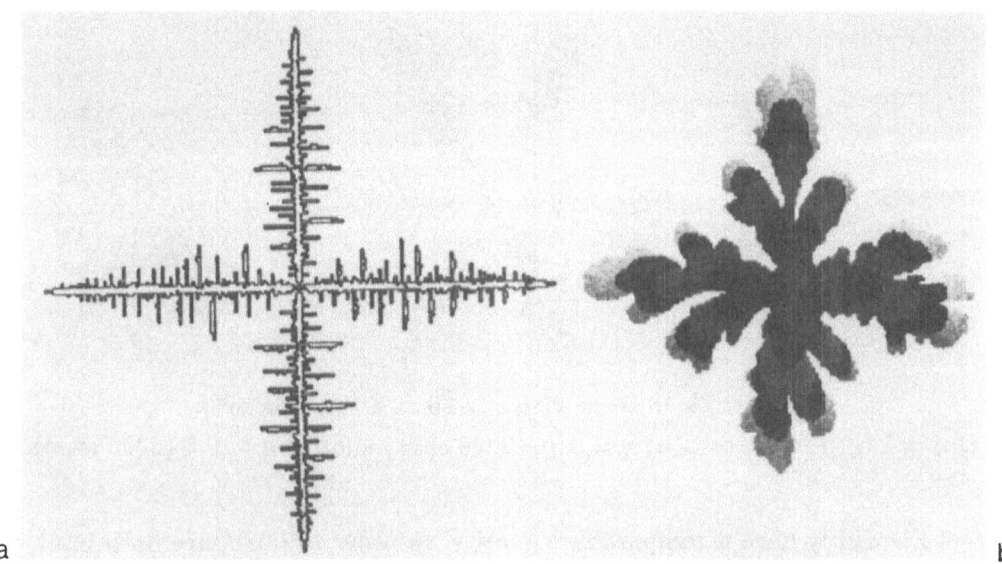

a

b

Fig.4: Comparison between DLA and DBM for $s = 200$,
(a) DLA (3000 sites), (b) DBM (6000 sites).

growth is evident which can however be largely reduced by allowing growth in the diagonal directions (8 neighbours).

Niemeyer et al. [10] have introduced a generalisation for the growth law equation (9)

$$u_i \propto (\nabla p_i)^\eta, \tag{13}$$

e.g. $\eta = 1$ for DLA and $\eta = 0$ for the Eden model. The exponent η controls the distribution of growth probabilities ∇p_i. For $\eta > 1$ the tips increase their probability of growth compared to the fjords. The growing structures are characterized by fewer but more elongated branches. In contrast, if $\eta < 1$ the fjords increase their probability of growth and more compact structures develop. In the case of noise reduction as mentioned above the lattice anisotropy manifests itself strongly. Using a triangular lattice we find for $\eta = 1.5$ (Fig.5a) a structure with three arms only. For $\eta = 1.0$ (Fig.5b) a structure with 6 arms of roughly equal length and random side branches develop. If $\eta = 0.5$ the growth on the tip is reduced and a more compact "snowflake" with long irregular side branches grows. Please notice that although a random growth process is at work a partially ordered structure has formed [16].

In snowflake formation the presence of interfacial tension in form of a Gibbs-Thomson relationship is important. In many fluid displacements the two fluids

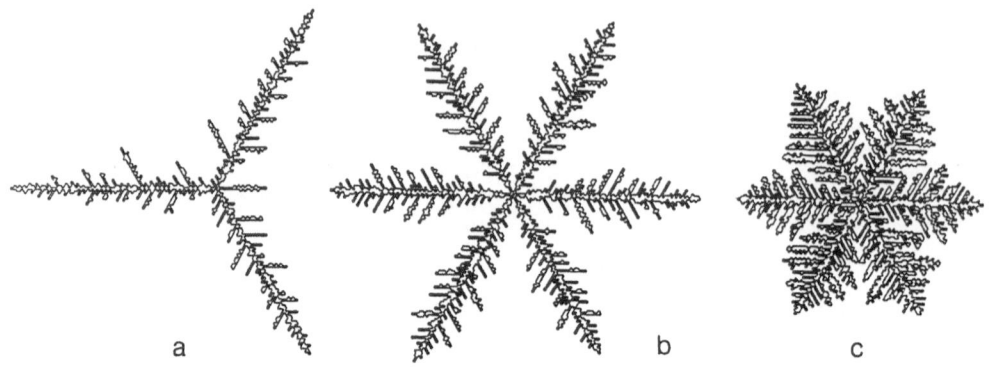

Fig.5: DLA clusters with s = 50 as a function of η:
(a) $\eta = 1.5$, (1200 sites), (b) $\eta = 1.0$, (2800 sites) and (c) $\eta = 0.5$ (3300 sites).

used frequently have a comparable viscosity. In order to introduce both inter-facial tension and a finite viscosity ratio we proceed as follows.

4. Growth Limiting Forces: Finite Viscosity Ratio and Surface Tension

So far we have assumed that the pressure p is constant in the displacing fluid. This corresponds to a zero viscosity ratio $m = \mu_2/\mu_1$. In order to study the role of a finite viscosity ratio we use modified equations (7) and (8) [17]:

$$Pr\{u_i\} = \frac{\nabla p_i}{\sum_l \nabla p_l} \tag{14}$$

$$p_i = \frac{\sum_n 1/(\mu_i + \mu_n)p_n}{\sum_n 1/(\mu_i + \mu_n)} \tag{15}$$

We can now also study the case where the low viscosity fluid is displaced by the high viscosity fluid. In Fig.6 a-c we show typical displacement patterns as a function of m.

We clearly see that a lowering of the viscosity difference does indeed diminish the tendency to grow predominantly on the tip. For the stable system $m \geq 1$, where a more viscous fluid displaces a less viscous fluid, we get a perfect radial flow.

A similar effect arising from a very different physical origin can be observed if we introduce surface tension [18]. The surface tension enters our problem as a boundary condition:

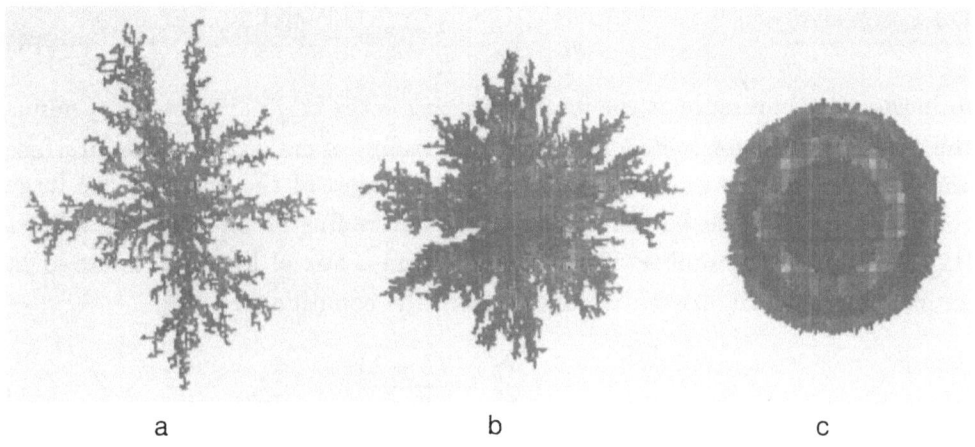

Fig.6: The role of a finite viscosity ratio m:
(a) $m = 0.01$ (5000 sites) (b) $m = 0.1$ (8000 sites) (c) $m = 10.0$ (8000 sites).

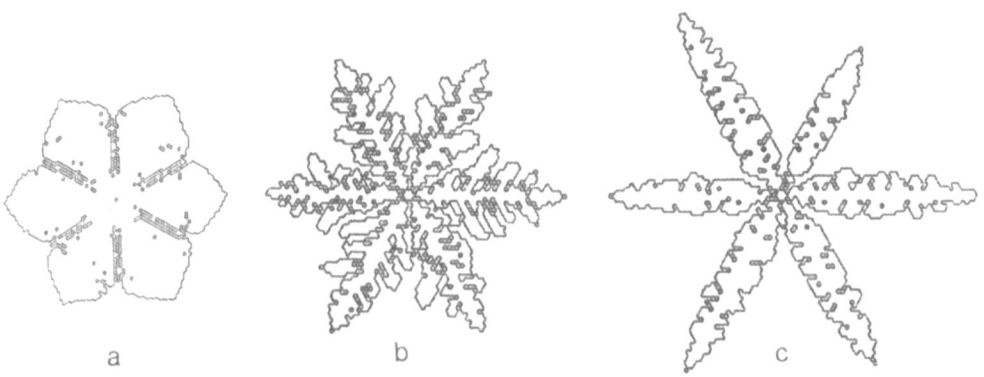

Fig.7: Snowflake pattern formation:
(a) $\eta = 0.5$, $\sigma = 0.7$ (3300 sites), (b) $\eta = 0.5$ (2000 sites), $\sigma = 0.1$,
(c) $\eta = 1.0$, $\sigma = 0.1$ (2000 sites)

Fig.8: Some examples of real snow crystals [20] for comparison with Fig.7

$$p_I = p_o - \frac{1}{R_c}\sigma; \tag{16}$$

p_I is the actual pressure at the interface which is the original pressure p_o minus the capillary pressure σ/R_c. R_c is the local radius of curvature of the interface and σ is the surface tension. If the spatial changes of the interface are large compared to a length L, one can estimate the radius of curvature as follows [19]: We count the number of sites (N_L) within a box of length L centered at an interface site. R_c can be calculated from the equation

$$\frac{1}{R_c} = \frac{1}{L}\left(\frac{N_L}{L^2} - \frac{(L-1)}{2L}\right). \tag{17}$$

For jagged snow crystal structures such as in Fig.5, we expect the local radius of curvature at a potential growth site to be of the order of a unit length of the lattice. Hence we take L equal to 3 and find R_c as follows:

$$\frac{1}{R_c} = \frac{((n/2+1) - n_L)}{n/2}, \tag{18}$$

where n is again the number of nearest neighbours and n_L is the number of occupied nearest neighbours. In Fig.7 we show how a finite surface tension modifies the snowflake pattern obtained in Fig.5. For comparison we have chosen, for Fig.8, some examples of real snow crystals photographed by Bentley and Humphreys [20].

5. Summary

We have given an outline of how a simple stochastic model can be designed to model physical mechanisms such as tip splitting and tip-stable growth. Based on the governing equation of a physical growth mechanism we explicitly introduce random fluctuations in the growth, which itself is the driving mechanism for pattern formation.

Acknowledgement

Much of this work was done in collaboration with H.E.Stanley. I would like to thank G.Daccord and C.Portella for Fig.1a and Fig.1b, respectively.

* Permanent Address: ÖMV Aktiengesellschaft, Technical and Scientific Systems, 1020 Vienna, Taborstrasse 1-3, Austria.

References:

1. Saffman,P.G. & Taylor,G.I., Proc.R.Soc. **A245**, 312, (1958), Chuoke,R.L.,Van Meurs,P. & Van der Poel,C.,J.Petrol.Tech. **11**, 64, (1959).

2. Langer,J.S., Rev.Mod.Phys. **52**, 1, (1980), Ben-Jacob,E., Goldenfeld, N.,Langer,J.S. & Schön,G., Phys.Rev.A **29**, 330 (1984), Brower,R.C., Kessler,D.A.,Koplik,J. & Levine,H., Phys.Rev.A **29**,1335 (1984).

3. Witten,T.A. & Sander,L.M., Phys.Rev.Lett. **47**, 1400, (1981).

4. Family,F. & Landau,D.P., eds.*Kinetics of Aggregation and Gelation*, North-Holland 1984, Pietronero,L. & Tosatti,E., eds.*Fractal in Physics*, North Holland, 1986, Stanley,H.E. & Ostrowsky,N., eds.*On Growth and Form: Fractal and Non-Fractal Patterns in Physics*, Nijhoff, 1985.

5. Paterson,L., J.Fluid.Mech. **113**, 513 (1981).

6. Mandelbrot,B.B., *The Fractal Geometry of Nature*, Freeman, 1982.

7. Nittmann,J., Daccord,G. & Stanley,H.E., Nature **314**, 141, (1985), Daccord,G., Nittmann,J. & Stanley,H.E., Phys.Rev.Lett. **56**, 336, (1986).

8. DeGregoria,A.J. & Schwarz, L.W., J.Fluid.Mech. **164**, 383, (1986), Shraiman, B.I. & Bensimon,D., Phys.Rev. A **30**, 2840 (1984).

9. Nittmann,J. & Stanley,H.E., Nature **321**, 663, (1986).

10. Niemeyer,L., Pietronero,L. & Wiesmann, H.J., Phys.Rev.Lett. **52**, 1033 (1984).

11. Paterson,L., Phys.Rev.Lett. **52**, 1621 (1984), Lenormand,R. & Zarcone,C., Phys.Chem.Hydrodyn. **6**, 497 (1985), Måløy,K.J., Feder, J. & Jøssang,T., Phys.Rev.Lett. **55**, 2688, (1985).

12. Tang,C., Phys.Rev. A **31**,1977 (1985).

13. Ben-Jacob,E., Goldbey,R., Goldenfeld,N.D., Koplik,J., Levine,H., Mueller,T. & Sander,L.M., Phys.Rev.Lett. **55**, 1315 (1985), Horwàrth,V., Vicsek,T. & Kertész,J., Phys Rev.A **35**, 2353 (1987).

14. Kertész,J. & Vicsek,T., J.Phys.A **19**, L257 (1986).

15. Ball,R.C., Brady,R.M., Samulski,E.T. & Thompson,B.R., preprint

16. Nittmann,J. & Stanley,H.E., submitted.

17. Sherwood,J.D. & Nittmann,J., J.Physique **47**, 15 (1986).

18. Kadanoff,L., J.Stat.Phys. **39**, 267 (1985), Liang,S., Phys.Rev.A **33**, 2663, (1986).

19. Vicsek,T., Phys.Rev.Lett. **53**, 2281 (1984).

20. Bentley,W.A. & Humphreys,W.J., *Snow Crystals*, Dover 1962.

Electrodeposition: Pattern Formation and Fractal Growth

L.M. Sander

Physics Department, University of Michigan,
Ann Arbor, MI 48109, USA

1. Introduction

In recent years it has become evident that a large number of different structures
and patterns can be formed by growth far from equilibrium [1]. A particular case
of such growth is diffusion-limited formation, i.e., a buildup of a structure
where the rate-limiting step is diffusion to or from the growing surface. For
example, ordinary dendritic crystallization [2] often is controlled by diffusion
of latent heat away from the crystal. Chemical precipitates and electrodeposits
grow when ions diffuse to their surfaces and attach.

An astonishing variety of patterns are known to arise from this single
process. Crystals growing near equilibrium have a shape governed by the well-
known Wulff construction [3]. Further from equilibrium the beautiful, complex
(but symmetric) patterns of dendritic crystals--most familiar as
snowflakes--occur. These are characterized by having stable growing tips

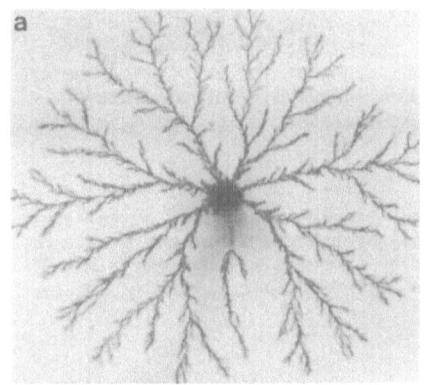

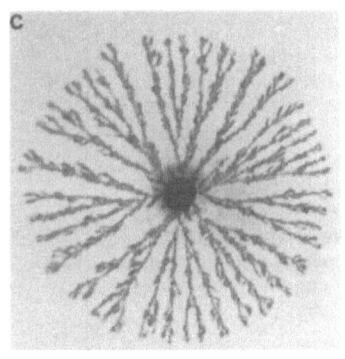

Fig. 1a) Fractal growth of zinc
electrodeposits

b) Dense Radial growth

c) dendritic growth

followed by complex sidebranching. In other cases a much more random pattern without stable tips [4,5] arises. Often it can be described by the diffusion-limited aggregation model (DLA) [6,7]. That is, as we will discuss in more detail below, diffusion-limited growth sometimes gives rise to fractals [8]. And, diffusion-limited growth, as we will see here, can make patterns which are neither dendritic nor are they fractal [9,10]. Figure 1 gives some examples of this variety for the case of electrodeposition. All of these patterns were produced in the same cell by varying voltage and concentration. The middle pattern is neither dendritic nor fractal.

It is clearly of great interest to try to understand how these different sorts of pattern arise; that is, to do a kind of morphological classification of the structures. This paper is devoted to that task. We will report a certain amount of progress, and a number of outstanding questions.

2. The Stefan Problem

The mathematical formulation of the problem of diffusion-limited growth has most often arisen in studies of dendritic crystallization [2]. In this conference, modern developments in the field, which has experienced an explosion of activity in recent years, are reviewed by CAROLI, and KESSLER and LEVINE [11, 12]. Here we will restate the usual formulation in a way which will make evident the various parameters which affect the general form of the growing structure.

In all the cases we are interested in, the structure grows as a result of the motion of some field outside the object, which we denote by $u(\vec{r},t)$. This could represent the temperature outside a growing crystal, or the flux of ions impinging on an electrodeposit. The fields in these cases are governed by the diffusion equation

$$\frac{\partial u}{\partial t} = B \nabla^2 u. \tag{1}$$

Here B is the appropriate diffusion coefficient. Note that for flow away or toward the surface, we get the identical equation. In what follows it will be convenient to slightly redefine u to allow a common language. For the case of chemical precipitation we take $n - n_{eq}$, where n is the ion density and n_{eq} is the equilibrium density near a flat interface. But, for heat diffusion, $u = T_m - T$, where T_m is the equilibrium melting temperature.

It is of interest to estimate the importance of the left-hand side of this equation, namely, the time derivative. We do this by assuming that the growth of the front is characterized by some average velocity, v, at a given time. Then, the equation can be replaced by

$$\frac{v}{B} \frac{\partial u}{\partial x} \sim \frac{\partial^2 u}{\partial x^2}. \tag{2}$$

The left-hand side involves a length, $\ell = B/v$, called the diffusion length. When ℓ is large, that is, when the growth is slow, the left-hand side can be neglected. More precisely, when ℓ exceeds all the relevant length scales of the growing structure, the diffusion equation (1) can be replaced by the Laplace equation

$$\nabla^2 u = 0. \tag{3}$$

In a practical situation, say, for crystallization, ℓ is not terribly large, and the full diffusion equation must be used. There are, however, two interesting cases when (3) is the appropriate equation governing u. One is the case of blowing a bubble in a viscous fluid whose flow obeys D'Arcy's law:

$$\vec{v} = K \nabla P. \tag{4}$$

258

Here, $\vec{v}$ is the fluid velocity, P the pressure, and K is a constant. This equation is appropriate [13] for fluid flow in a porous medium or in a Hele-Shaw cell (i.e., for flow between closely spaced parallel plates). If the fluid is incompressible then we can write

$$\nabla \cdot \vec{v} = 0 = \nabla^2 P .$$ (5)

That is, in this case, P obeys the Laplace equation, and P_0-P (with P_0 the pressure inside the bubble) plays a role of the field, u. The bubble, which has a complex shape because of the phenomenon of viscous fingering, is the growing structure.

Another case where the Laplace equation is appropriate to describe the field governing the growth is in a continuum description of the DLA model. In this model the growth is triggered by the arrival of random walkers at the growing structure. Now the probability density of random walkers obeys the diffusion equation. However, since the walkers are taken to arrive one at a time in DLA the growth is slow, indeed it is in the limit $v \to 0$. Thus the analysis of the average distribution of walkers in DLA must use (3) with u standing for the probability density.

In cases where the diffusion equation is relevant, the field, u, "heals" exponentially with characteristic distance ℓ away from the growing front, that is, u is undisturbed by the growth for distances larger than ℓ. The diffusion length is a kind of limit for the transmission of information about the growth. Within ℓ, or when (3) is appropriate, u has a power-law behavior.

The growth of the structure is governed by the flux of the field, u. Thus, the motion of the surface of a growing crystal is given by its ability to expel latent heat; an electrodeposit (or a DLA cluster) grows when particles arrive at the surface; a viscous fingering structure grows when fluid flows away. In all these cases, we can say that if v_n is the normal velocity of growth of the surface, then

$$v_n = C\hat{n} \cdot \nabla u$$ (6)

where C is a constant, and $\hat{n}$ is the normal to the surface. In the case of DLA, this equation only holds for the average growth. There are large fluctuations in the actual motion of the interface because for any given cluster particles arrive one at a time.

Finally, we must specify the boundary conditions on u. Far from the growing structure, u can be taken to be a constant, Δ, unless we are dealing with the Laplace equation in two dimensions. In that case, we fix $u = \Delta$ on a suitable stationary surface far from the growth. We refer to Δ as the undercooling.

The boundary condition on the growing surface represents the absorption of matter, and by our convention, it corresponds to $u = 0$. For example, for DLA, this is the well-known absorbing boundary condition for random walks. However, this is not the whole story. If the surface is curved, it effectively raises its temperature or pressure via the Gibbs-Thompson relation:

$$u_s = d_0 \kappa$$ (7)

where d_0 is a characteristic parameter related to the capillary length, and κ is the curvature of the surface. This innocent looking set of equations are a slightly generalized version of what mathematicians call the Stefan problem. They are extremely difficult to solve because of the non-linearity implicit in (6). However, a good deal is now known about various types of solutions [2, 11, 12] and general properties.

The key to the complex behavior generated is the fact that they imply a growth instability, the famous Mullins-Sekerka instability [14]. This is the tendency for a growing tip which has advanced ahead of the rest of the front to become ever sharper and to get even further ahead because it is closer to the source of diffusing material. The instability amplifies initially small features of the growing surface and is presumably the source of complexity of the system.

Two things should be noted here about the growth instability. One is that it occurs independent of ℓ and for all wavelengths, however long. The surface tension boundary condition (7) controls the instability at short wavelengths. In fact, in the absence of d_0, the theory is not well-defined in the sense that all but a small set of initial conditons develop cusps on the growing front in a finite time. It might be thought that DLA, which is usually formulated without surface tension is ill-defined in this sense. In fact, this is not true for two reasons. First, as we have noted, DLA is the Stefan problem plus noise. Probably, the noise prevents cusp formation on small scales. Even in the absence of noise [15] the finite particle size prevents cusp formation. The particle size plays the same qualitative role as d_0.

We note for later use that the control of the instability means that for a small enough total size, R_c, of the growing object the instability is completely suppressed. It is easy to show [14]

$$R_c \sim 20 \; d_0/\Delta . \tag{8}$$

The mathematical progress [11, 12] which has been made in this field has mainly been concerned with the existence of steady states, that is, growing objects with stable tips that move with fixed velocity. It is now clear that steady states can occur only in certain cases. For example, for growth with finite ℓ it seemed initially that a steady state tip was possible. In fact, the current view is that anisotropy (arising, say from an underlying lattice) is necessary for a steady state. Otherwise, the tip repeatedly splits and wanders. We must alter the boundary condition (7) to read

$$u_s = d_0(\theta) \; \kappa + \beta(\theta)v . \tag{9}$$

The first term is as above, but generalized to include a dependence on the angle, θ, between the crystal axis and the normal to the growing surface. The second term is called kinetic anisotropy [16]: it corresponds to a difference in growth kinetics arising, for example, from different densities of nucleation sites, for different crystal faces.

The anisotropic terms apparently stabilize tips even as $\ell \to \infty$, though no steady state is possible in this regime [17] because v depends on total size of the crystal, as a simple flux conservation argument shows.

Now we use these generalities to attempt to identify the regimes of dendritic (stable-tip) growth, and fractal growth. We tentatively identify fractal growth with the regime of repeated tip-splitting and $\ell \to \infty$.

This identification was first advanced by BEN-JACOB et al. [18]. Thus radial Hele-Shaw flow without any anisotropy should give rise to unstable patterns, and enough anisotropy should stabilize the tips. BEN-JACOB et al. showed this experimentally and SANDER et al. [19] did numerical solutions of the problem. Unfortunately, for various technical reasons, neither the experiment nor the simulations can be extended over a large dynamic range in R.

For this reason, my group in Ann Arbor [9] and others [10] have turned to another physical system, electrodeposition cells in a radial geometry, to attempt to cast light on these questions. This is the subject of the remainder of this paper.

260

3. Electrodeposition Experiments

The first workers to call attention to the relevance of DLA-type models for electrodeposition were BRADY and BALL [20]. They produced three-dimensional copper electrodeposits on a sharp tip and analyzed the scaling behavior of the resulting structures. Care was taken to ensure that the structures were always smaller than the relevant diffusion length for ions in solution.

The results were analyzed by putting

$$\rho = \rho_{cu}(a/R)^{D-3} , \tag{10}$$

where ρ is the average density of the deposit, ρ_{cu} that of bulk copper, R the mean radius, and a and D are fitting parameters. If we interpret the deposits as fractals, then D is the fractal dimension, and a the particle size. It was found that the experimental results were consistent with (10) for R's ranging from a few hundred angstroms (the particle size) to microns. However, a had to be taken to be a function of V, the voltage,

$$a(V) = c/(V - 0.23). \tag{11}$$

A possible explanation for this will be given below.

MATSUSHITA et al. [21] showed that two-dimensional patterns of zinc leaves attached to the interface between zinc sulphate and an oil also showed scaling symmetry. This work was further pursued by SAWADA, et al. [10] and by GRIER et al. [9] in Michigan using a somewhat different setup. The results of the Michigan work (which are consistent with those of SAWADA et al.) will be discussed here.

The experimental apparatus consisted of thin layer (0.1 mm) of $ZnSO_4$ confined between two layers of plexiglass. The electrolyte was surrounded by a ring-shaped anode of radius about 6 cm. Deposition took place on a small cathode in the center of the cell inserted through a hole in the plastic. Voltage was controlled over a range from 0.1 to 10 volts and concentrations of electrolyte from 0.01 to 2 molar were examined.

The morphology of the deposits fell into three basic types as illustrated in Fig. 1. At low concentrations and voltages DLA-like patterns are in fact produced in agreement with the earlier work mentioned above. At the highest voltages and concentration needle crystal structures with stable tips and side branches occurred. Thus the two major categories of morphology discussed above were reproduced. However, and unexpectedly, a third category of structure was evident for intermediate conditions. This dense radial pattern consists of many branches which tip-split repeatedly, but which fill space and have a stable overall outline. The three cases are illustrated in Fig. 1. It is of great interest to see how these three categories of structure fit into the general framework we have described above.

4. Phenomenology of Electrodeposition

The subject of electrodeposition of metals is an old one, and has a very large literature of its own [22]. However, a specific formulation in the terms which have become standard in the physics community for other kinds of crystallization has been lacking. We will give here a discussion based on work by LEVINE [23] which will allow us the use of the considerations of section 2, above.

In the electrodeposition of Zn from $ZnSO_4$, we have at least three relevant fields to take into account, rather than the u(r,t) of section 2. They are Zn^{++} concentration, which we call n_+, SO_4^{--} concentration, n_-, and the electrostatic potential ϕ, whose gradient is the electric field in the electrolyte. The

261

equation of motion of the n's is most easily given by writing, in the dilute solution approximation, an expression for the electrochemical potential:

$$\mu\pm = K_\pm(T) + k_BT \ln n_\pm \pm q\Phi. \tag{12}$$

Here K(T) is a function of temperature alone, and $q = 2|e|$ is the charge on the ion in solution. We simplify (12) by linearizing around the average density of ions:

$$n_+ = n_0 + m_\pm . \tag{13}$$

The number current of ions is, using the Einstein relation,

$$j_\pm = -\frac{n_0 B}{k_B T} \nabla \mu_\pm . \tag{14}$$

This current has two parts arising from the last two terms in (12), a diffusion and a drift current. The diffusion constant B will be taken, in this highly simplified model, to be the same for both species. Now the equation analogous to (1) is the continuity equation derived from (14):

$$\begin{aligned}
\frac{\partial m_\pm}{\partial t} &= \frac{n_0 B}{k_B T} \nabla^2 \mu_\pm \\
&= B \nabla^2 m_\pm \mp (n_0 B \, q/k_B T) \, \nabla^2 \Phi .
\end{aligned} \tag{15}$$

We also need an equation for ϕ: this, of course, is Poisson's equation:

$$\nabla^2 \Phi = -4\pi q \, (m_+ - m_-) . \tag{16}$$

The boundary conditions are

$$\hat{n} \cdot j{-}|_E = 0 , \tag{17a}$$

$$\Phi_2 - \Phi_1 = V , \tag{17b}$$

$$\mu_+ \, |_E = \mu_s - \gamma\kappa/n_s , \tag{17c}$$

$$\hat{n} \cdot j_+ \, |_E = v_n n_s . \tag{17d}$$

Here the subscript E means on either electrode. Equation (17a) expresses the fact that SO_4 has no chemistry in this system. In (17b) V is the battery voltage and 1, 2 mean the positive and negative electrode. In (17c) γ is the surface tension, μ_s the electrochemical potential in the bulk solid, n_s the density of solid zinc, and κ the curvature of the electrode. Equation (17d) is the growth condition analogous to (6).

Now in the quasi-static approximation we can write

$$\nabla^2 \mu_\pm = 0 . \tag{18}$$

The boundary condition (17a) means that $\mu_- = 0$. This corresponds to the drift and diffusion current cancelling for the negative species. We now have a problem of the Stefan form for μ_+ with qV as the "undercooling", Δ, and γ/n_s playing the role of d_0. This is a remarkable result. It shows that even though we are dealing with charged ions in electric fields, the same diffusion dynamics holds as for DLA if the diffusing field is identified with the electrochemical potential μ_+. The result was anticipated, in part, by BRADY and BALL [20].

262

Now the critical radius for instability, from (8), is

$$R_c \simeq 2\sigma\gamma/n_sqV \simeq 70\text{\AA}/V. \tag{19}$$

This is in agreement both in form and magnitude (except for the offset) with the results of BRADY and BALL [20]; compare (11).

Going beyond the quasi-static limit is more difficult in this system because of the presence of concentrated double layers at the electrodes where (12) and (15) fail to be valid approximations (Of course, double layers of some strength exist, in general, even in the quasi-static case). It is reasonable, however, that in these cases the importance of the left-hand terms in (15), and the range of non-locality will still be given by the diffusion length, $\ell = B/v$.

5. Morphology of Electrochemical Deposits of Zinc

We can now attempt to interpret the patterns in Fig. (1). We recall that in order to have a DLA pattern we need ℓ larger than the deposit itself. To this end we have measured the diffusion length for electrodeposition in our cells, Fig. (2).

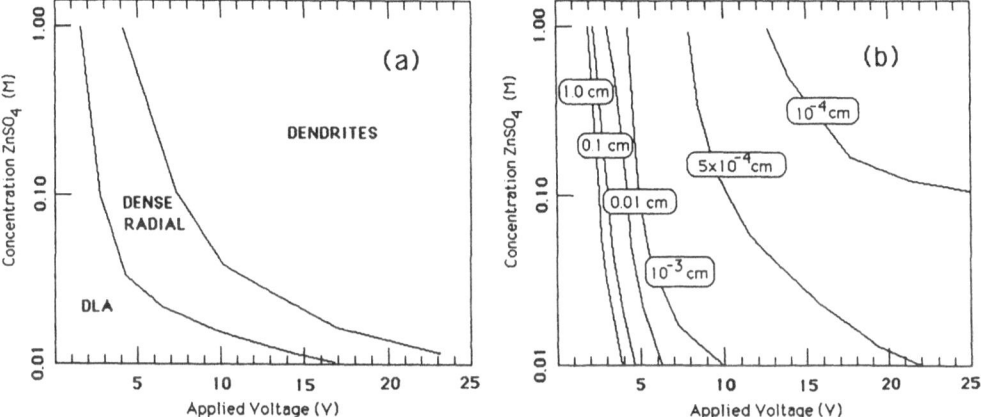

Fig. 2 a) The morphology diagram. b) Measured diffusion lengths, ℓ

Note that ℓ is larger than 1 cm, say, for low concentration and voltage, precisely in the regime where we find fractal patterns like those in Fig. (1a). In fact, the lines of constant diffusion length in Fig. (3) coincide, more or less, with the boundaries of the morphology diagram of GRIER et al. [9].

Thus as we pass out of the fractal deposition regime for larger concentration and voltage we can expect that ℓ will give a limit for the non-locality of the deformation of the diffusing field due to the growing deposit. That is, we should expect that the deposits will be fractal for lengths smaller than ℓ and compact when viewed on larger scales. This sort of crossover was discussed in the DLA literature some time ago [24]. We believe that in the dense radial regime (Fig. (1b)) we are in the situation where tip-splittings still dominate, but ℓ is small. On the scale of the experiment, the object is not fractal.

GOLLUB et al. [25] have, in fact, shown that a crossover from fractal to compact occurs in the dense radia regime and that the crossover length shifts as a function of voltage. We are in the process of remeasuring this crossover; we hope to show to that the crossover length is ℓ. In any case, the near coincidence of the line $\ell(n_+,V) \simeq 1.0$ cm with the dense radial-fractal boundary gives strong support to the conjecture.

Now we can easily understand why dense radial patterns are dense. The mean distance between branches cannot fall much below ℓ. However, we do not have even the vaguest notion of why they are round. The disc-like outline of the pattern in Fig. (1b) is not an illusion. For most of their history dense radial patterns are round. But, as we pointed out above, the Mullins-Sekerka instability occurs independent of ℓ and if we "smear out" the branches in Fig. (1b) it should operate just as well in this case, so that the outline in Fig. (1b) should be linearly unstable for scales larger than ℓ. The observation is that it is not unstable at all.

The mystery deepens when we consider the crossover from the dense radial regime to stable dendrites. Whether we are seeing a crossover driven by the elevation of a noise threshold as speed of growth increases, or whether the kinetic term in (9) plays an important role in this sytem is far from clear. Much more experimentation and analysis will be necessary to understand these points.

6. Morphology of Electrochemical Deposits of Copper

The Michigan group (N. Hecker and D. Grier) have done a brief survey of electrochemical deposition in two-dimensional cells for a variety of metals. This will be the subject of a forthcoming publication [26]. Most of the substances examined have similar behavior to zinc, with one exception, copper. The results in this case are so peculiar that we will briefly mention them here. More detail will be given in [26].

Copper in our 2-dimensional cells exhibits the dense radial pattern for a range of concentration and voltage. However, over a substantial range of these parameters something even more mysterious occurs. This is a phenomenon which we call the Hecker effect.

In brief, we find that copper electrodeposits exhibit an abrupt change in the density of their dense radial pattern, see Fig. (3). The change occurs at half

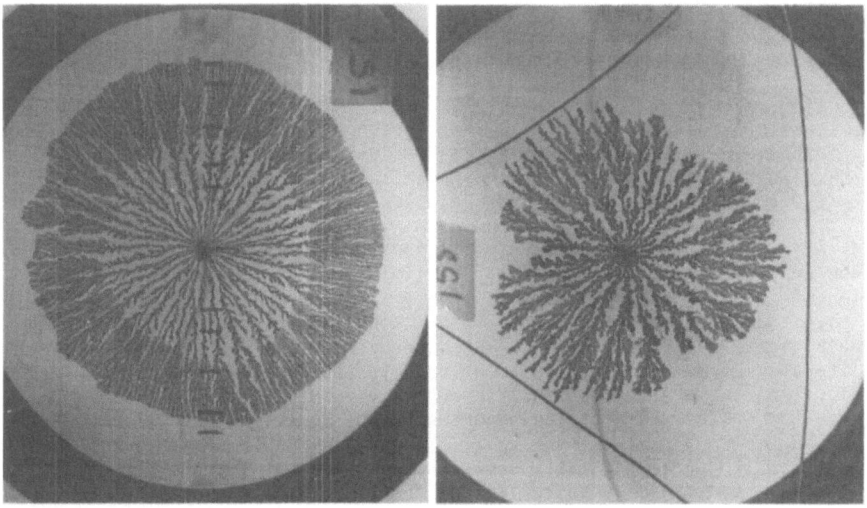

Fig. 3 Copper electrodeposition showing the Hecker effect

the radius of the cell independent of concentration and voltage over a suitable range. Further, it is clearly a size effect. Not only does the change always occur at half the cell size for the several cells we have used, it also could be deformed by changing the shape of the outer electrode, see Fig. (3). Note that the growth itself is always more or less round, but when the growing circle crosses the line halfway out, it changes density. Different parts of the deposit cross over at different times. When the deposit is quite near the boundary, it is, we expect, deformed by the larger fields there.

This mysterious effect may cast light on the dense radial pattern itself, or it may be unrelated. Its cause is a completely open question.

7. Discussion

In this paper, we have reported some progress towards understanding a particularly rich system which shows diffusion-limited growth.

We have been able to see that when the diffusion length, ℓ, is large, we should see DLA fractal patterns in our cells, and in fact, we do find them. For larger driving force ℓ decreases, and we find a crossover to compact aggregation.

Apparently, in this system ℓ becomes small before tip-splitting is suppressed. This seems to be the origin of the dense radial pattern. Then, at still larger velocities, and for reasons yet unclear, we supress tip splitting and produce orderly dendrites.

A number of very puzzling questions remain. For example, we do not know what stabilizes the outline of dense radial deposits. This may seem a small point, but it is sufficiently puzzling and apparently simple that we have tried various solutions. For example, we could guess that we are dealing with a finite size effect. This may be so in some sense. However, in my group RAMANLAL [27] has used the methods of [19] to test this possibility for large ℓ. In fact, the calculation shows that the instability is enhanced, not supressed, as the growth approaches the boundary, and there is no hint of the dense radial pattern for $\ell \to \infty$.

It is possible, of course, that there is something important about the physics or chemistry of electrodeposition that we have missed in our analysis. However, we think this not very likely. Recent work [28] has shown the dense radial pattern to be very widespread in nature, certainly not confined to electrodeposition. In fact, it is possible to make radial Hele-Shaw cells exhibit in some cases a pattern which looks dense. This is in direct contradiction with [19] and [27] since in this case, as we have pointed out, ℓ is very large. This, and the more exotic puzzle of the Hecker effect will have to be sorted out.

Acknowledgements

This work was supported in part by NSF grant DMR-8505474. My students, D. Grier, N. Hecker and P. Ramanlal, did much of the work on which I have based this account. I have learned a good deal in conversations with E. Ben-Jacob, J. Gollub, D. Kessler, and H. Levine.

References

1. L. Sander, Nature 322, 789 (1986).
2. J. Langer, Revs. Mod. Phys. 52, 1 (1980).
3. E. Lifshitz and L. Pitaevskiĭ, "Statistical Physics", Pergamon, 1980.
4. W.T. Elam, S.A. Wolf, J. Sprague, D. Gubser, D. vanVechten, G. Barz, and P. Meakin, Phys. Rev. Lett. 54, 701 (1985).
5. Gy. Radnoczi, T. Vicsek, L.M. Sander, and D. Grier, to be published.

6. T. Witten and L.M. Sander, Phys. Rev. Lett. 47, 1400 (1981).
7. T. Witten and L.M. Sander, Phys. Rev. B 27, 5686 (1983).
8. B. Mandelbrot, "The Fractal Geometry of Nature", Freeman, 1982.
9. D. Grier, E. Ben-Jacob, R. Clarke and L.M. Sander, Phys. Rev. Lett. 56, 1264 (1986).
10. Y. Sawada, A. Doughtery and J. Gollub, Phys. Rev. Lett. 56, 1260 (1985).
11. C. Caroli, these proceedings.
12. D. Kessler and H. Levine, these proceedings.
13. D. Bensimon, L. Kadanoff, S. Liang, B. Shraiman and C. Tang, Rev. Mod. Phys. 58, 977 (1986).
14. M.W. Mullins and R. Sekerka, J. Appl. Phys. 34, 323 (1963).
15. C. Tang, Phys. Rev. A 31, 1977 (1985).
16. E. Ben-Jacob, N. Goldenfeld, J. Langer and G. Schon, Phys. Rev. A 29, 330 (1984).
17. D. Kessler, J. Koplik and H. Levine, Phys. Rev. A 30, 2820 (1984).
18. E. Ben-Jacob, R. Godbey, N. Goldenfeld, J. Koplik, H. Levine, T. Mueller, and L.M. Sander, Phys. Rev. Lett. 55, 1315 (1985).
19. L. Sander, P. Ramanlal and E. Ben-Jacob, Phys. Rev. A 32, 3160 (1985).
20. R. Brady and R. Ball, Nature 309, 225 (1984).
21. M. Matsushita, M. Sano, Y. Hayakawa, H. Honjo and Y. Sawada, Phys. Rev. Lett. 53, 286 (1984).
22. J. Bockris and A. Reddy, "Modern Electrochemistry", Plenum, 1970.
23. H. Levine, unpublished.
24. M. Nauenberg, R. Richter and L.M. Sander, Phys. Rev. B 28, 1649 (1983).
25. J. Gollub and A. Dougherty, unpublished.
26. D. Grier, N. Hecker and L.M. Sander, to be published.
27. P. Ramanlal and L.M. Sander, to be published.
28. E. Ben-Jacob, G. Deutscher, P. Garik, N.D. Goldenfeld and Y. Lareah, Phys. Rev. Lett. 57, 1903 (1986).

Anisotropy and Fluctuation
in Diffusion Limited Aggregation

M. Kolb

des Solides, Bat. 510, Université de Paris-Sud,
F-91405 Orsay, France

Diffusion limited aggregation (DLA) models many different physical phenomena from dielectric breakdown to viscous finger instabilities. The resulting patterns can be analysed in terms of fractal geometries. They depend in a very sensitive way on the anisotropy and the noise of the growth rules. Here, several different variants of DLA are reviewed. In particular, simulations with enhanced growth in the direction of the axes of the underlying lattice are analysed to show that even a slight anisotropy distroys the spherical, self-similar structure in favor of a self-affine structure whose scaling properties depend on the symmetry of the lattice. Growth on a square lattice suggests that there is a crossover from off-lattice DLA with a fractal dimension of D=1.71 to a cross-like self-affine pattern with $D_{\parallel}=1.5$ along the axes and $D_{\perp}=2.0$ along the diagonals. Qualitative arguments are advanced to explain these results in terms of classical aggregation, i.e. neglecting fluctuations.

1. Introduction

Diffusion limited particle aggregation (DLA) is the first of a large number of irreversible aggregation processes which have been analysed in recent years to describe disordered growth [1]. For this model, it was found that the resulting structures are self-similar with distinct scaling or fractal properties. The model and numerous variants of it have been investigated extensively in order to fundamentally understand the mechanism and to isolate its essential features [2]. Despite all this effort, our understanding of DLA is far from complete and its structure appears to be much more complicated than anticipated initially. Most progress has come from numerical simulations though some promising analytic approaches have been proposed.

Early calculations [1,3] determined the density-density correlation function of aggregates of $\sim 10^3$ particles assuming that the cluster is an isotropic and homogeneous fractal. For clusters of this size this proves to be a correct assumption. Furthermore, the lattice appears to have no influence on the scaling indices - this is in agreement with ideas from equilibrium critical phenomena [4,5]. The fractal dimension D was obtained consistently either from the mass versus radius relation of the growing cluster

$$M = R^D , \qquad R \gg 1 , \qquad (1)$$

or from the spherically averaged density correlation function within the cluster

$$C(r) \sim r^{2-D} , \qquad 1 \ll r \ll R . \qquad (2)$$

Measured values for D range from 1.67 to 1.71 in two dimensions both for simulations on different types of lattices and for off-lattice calculations. Instead of growing aggregates spherically from a seed one can grow them on a surface. Their fractal properties are the same as long as the diffusion is isotropic. However, when the diffusion is anisotropic (the diffusion constant is larger in one direction than in the other) the value for D drops from 1.7 to 1.5. This is a first indication that anisotropies play an important role in determining the scaling properties of DLA clusters [6].

2. Large DLA aggregates

The direct implementation of the DLA process is numerically very time consuming, as the number of steps needed to add one particle to a cluster of radius R is of the order R^2. This restricts the maximum cluster size severely. In order to go to larger sizes, several methods have been proposed to improve the efficiency, both in terms of time and memory requirements [7-9]. The basic idea is to let the diffusing particles perform large jumps in the empty regions of space and only stick to the original growth rule (nearest neighbour jumps on the lattice) close to the aggregate. This reduces the time to grow a cluster considerably. The number of steps now is nearly constant (~log). Simultaneously, one maps out only the occupied regions of space - a gain in memory if the cluster is fractal. The lower the fractal dimension the higher the gain.

The first large scale calculation of DLA [7] revealed a most interesting and unexpected result on a square lattice: the envelope of clusters of ~10^5 particles is not spherical, but has the shape of a diamond, with the same symmetry as the lattice. Apparently the lattice favors growth along the axes over growth along the diagonals. Does this effect stabilize asymptotically or does the anisotropy become ever more pronounced? The growth of even larger aggregates is necessary to answer this question. Other interesting features appeared when large clusters were analysed more carefully. The aggregates grown from a seed are internally not isotropic: the center of the cluster can be identified when looking at a randomly chosen portion from the inside of the aggregate [10-12]. Put differently, the correlation function does not scale the same way radially and tangentially, both for lattice and off-lattice clusters.

3. Analytical approach, noise reduction

An elegant analytical description has been given based on classical aggregation where the fluctuations are neglected and assuming that the anisotropy stabilises asymptotically [13,14]. Its most interesting feature is an explicit expression for D which depends on the symmetry of the lattice

$$D=D(\Theta)=(3\pi-\Theta)/(2\pi-\Theta) \ , \qquad \Theta=2\pi/m \ , \qquad (3)$$

where Θ is the angle between the axes or, equivalently, m is the number of axes of the lattice (m=4 for the square lattice). This approach also provides a starting point for a more general multifractal analysis that was recently extended to the DLA problem [15]. High precision simulations [16,17] show that D does depend on the underlying lattice and the agreement with (3) is quite good. The assumption that the shape of the envelope stabilizes at a finite value is of crucial importance for the validity of (3). In contrast, exact solutions for classical aggregation suggest that a fluctuation becomes more and more singular during growth [18,19]. Even if the anisotropy does not stabilize, (3) suggests the correct asymptotic behaviour, as

D=1.5 in the limit when $\Theta=0$. Using scaling arguments and assuming that D for off-lattice growth is related to D for growth on a lattice, it was conjectured that lattice growth always becomes anisotropic in principle, but only for astronomically large cluster sizes when m is larger than five [20]. On the other hand analytical arguments based on a stability analysis and numerical calculations suggest that for small m the anisotropy remains finite whereas for large m (larger than about five) the lattice is irrelevant [21].

4. Variants of DLA growth

The slow change of the cluster envelope during the growth indicates that very large clusters are required to reach the asymptotic regime; beyond the reach of simulations even with improved algorithms. It is therefore desirable to control and vary the amount of anisotropy or noise in the growth artificially in order to cross over into the asymptotic region more rapidly. One way to reduce noise is as follows (analogous methods enhance the anisotropy) [22-25]: instead of adding a diffusing particle when and where it touches the cluster for the first time, a particle is added to a surface site only after this site has been hit by a given number h of diffusing particles. In this algorithm each diffusing particle simply increases the counter at its landing surface site by one without automatically adding a particle to the cluster. This new procedure has a tendency to render the growth more regular. On a lattice, the simulations show that the reduced noise increases the effect of the lattice anisotropy - for large values of h the growth is preferentially along the lattice axes - already for clusters of just a few hundred particles. However, noise reduction of this type is subtle: two similar but not identical schemes can lead to entirely different results [23,25].

5. Model with variable anisotropy: From self-similarity to self-affinity

Let me now present a different type of modification of the DLA model designed to study the asymptotic shape of the clusters on a lattice. As there is a natural tendency to grow preferentially along the axes, in this variant of DLA the growth is artificially enhanced along the axes such that the asymptotic regime is reached very rapidly. The results suggest that - provided the model is representative for lattice growth - DLA clusters are not self-similar but self-affine fractals [26] growing along the symmetry axes and scaling with two different exponents along or perpendicular to the primary growth directions. Very recently, clusters of › 10^6 particles seem to confirm this result [27].

The model of variable lattice anisotropy is defined as follows. A DLA cluster is grown on a square lattice by diffusing particles off-lattice one by one until they hit the cluster. They take long jumps far from the cluster (jumping the maximum distance which is possible without touching any part of the cluster) and they diffuse by jumps of one lattice spacing near the cluster. In Fig. 1 the way the particles are added to the cluster is illustrated. To have a minimum bias from the lattice, the particles diffuse until they actually hit the cluster (black squares) and then are placed on the site visited just prior to hitting the cluster (they back up by one step, shaded squares). The important difference with the standard rule is that a particle is placed either on a nearest neighbour or on a next to nearest neighbour site of the cluster, depending on its trajectory (trajectories A and B, respectively, in Fig. 1). The diffusion process

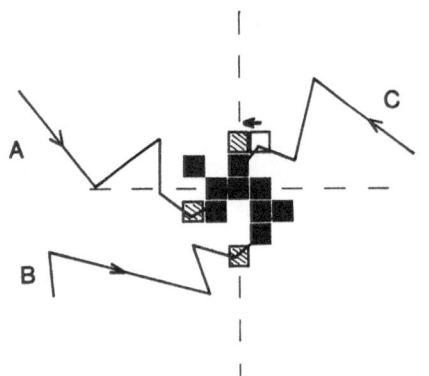

Fig. 1. Modified growth rule for DLA to enhance the lattice anisotropy

itself determines the relative frequency of the two choices - this diminishes the lattice effects considerably. If one wishes to encourage growth along the lattice axes, one imposes the additional rule (trajectory C in Fig. 1) that the particles landing on a second neighbour site are rotated by one step to the first neighbour site in the direction of the nearest lattice axis, with probability p. This is a global constraint tending to enhance the lattice anisotropy which is naturally present.The asymptotic region is reached faster by this procedure. The off-lattice diffusion makes the growth very rapid numerically. For reasonably large clusters (>10^5 particles) this new growth rule gives either isotropic or strongly anisotropic clusters, depending on the parameter p. For intermediate values of p one observes a crossover between the two types of behaviour.

6. Numerical results

The analysis of the scaling properties for the aggregates is performed separately for the different components of the coordinates of the particles: each particle coordinate (with the seed as the origin) is decomposed into a part along the nearest lattice axis and one along the nearest diagonal (for other lattices one determines the closest axis and the closest midline between the axes and decomposes each vector in components along these two directions). This allows one to determine different scaling exponents in the two directions. For comparison the standard radius of gyration is computed as well. These quantities provide a sensitive measure of the anisotropy. The quantities were measured separately for particles $N_{n-1}+1$ to N_n where $N_n \propto (1.4)^n$ (log mass scale). The calculation was done for 200 clusters of 20000 particles and separately for 20 clusters of 250000 particles, for several values of the anisotropy parameter p. The effective fractal dimension is estimated from

$$D(n) = \log(R_{n-1}/R_n) / \log(N_{n-1}/N_n) \qquad (4)$$

and tends towards D for large n. The exponents for the components are analysed the same way. Figure 2 shows the results for D, $D_\parallel$ and $D_\perp$ for p=0 (A), p=0.5 (B) and p=1 (C). For p=0 the three exponents coincide, indicating that the radius and the two components scale the same way - which is furthermore the exponent D for off-lattice DLA. Due to the new growth rule our lattice aggregation model has much weaker lattice effects for p=0 (visually, standard DLA of this size are distinctly diamond shaped whereas the clusters here appear perfectly spherical). For p=1, the maximum anisotropy that can be attained with this

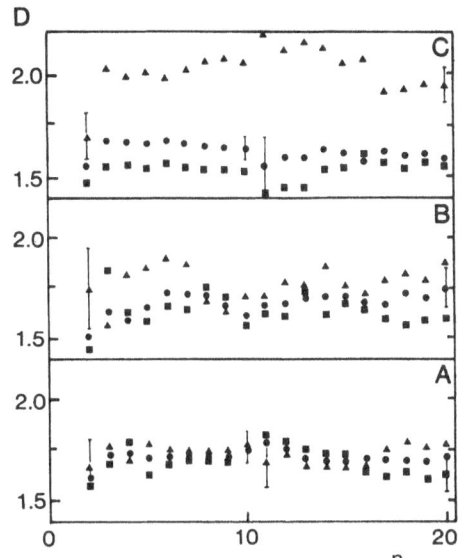

Fig. 2. Fractal dimension of the radius of gyration (D), of the axial component ($D_{||}$) and of the diagonal component ($D_\perp$) as a function of n (logarithmic mass scale) for p=0 (A), p=0.5 (B) and p=1 (C)

modified rule, clusters of a few hundred particles begin to show the effect of the anisotropy and yield distinctly different exponents along the axes and along the diagonals of the lattice. Part C of Fig. 2 shows the measured values for D, $D_{||}$ and $D_\perp$. One observes that the growth along the axis of the lattice and the usual radius of gyration are asymptotically governed by the same exponent. But the convergence of the radius of gyration is slower than the axial component. In Fig. 2, part B, the measured values for D, $D_{||}$ and $D_\perp$ for the intermediate value p=0.5 are comparable for small clusters (up to n ≅ 15) but they separate for larger n: $D_{||}$ and D tend to decrease towards their value for p=1 and $D_\perp$ increases towards its value in part C (the statistical precision for p=0.5 is not the same as for p=0 and p=1 as fewer clusters were sampled). This is the indication of a crossover between spherical, off-lattice growth (p=0) for R below a crossover length $\xi(p)$ and lattice dominated growth for R above ξ. The comparison of the measured values of the exponents with other calculations is as follows: for spherical aggregation (p≅0) all exponents are close to 1.71, the accepted value for off-lattice DLA. For p=1 the longitudinal exponent coincides with 1.5, a value already obtained for growth with anisotropic diffusion in strip geometry and also consistent with equation (3), for $\Theta=0$. The transverse and longitudinal exponents have also been calculated for isotropic diffusion in strip geometry, the result being consistent with the present calculation [28]. For the transverse exponent, $D_\perp=2.0$, a scaling argument will be given below.

To see the different growth rates in different directions more clearly, the ratio of the components is shown in Fig. 3 (log-log). The p=0 case (dots) shows a value for the log of the ratio r close to zero up to the maximum number of particles, indicating a perfect isotropy between axis and diagonal. For p=1, r follows a straight line with slope -1/6 already for the smallest values of n, indicating different scaling exponents, whereby $1/D_\perp - 1/D_{||} \cong -1/6$. Finally, for p=0.5 the r stays almost constant, but different from 0 for a large range of values of n. It finally seems to change over to the p=1 behaviour. The long, (almost) constant region could well be interpreted as a constant anisotropy region, as earlier simulations of standard DLA suggested.

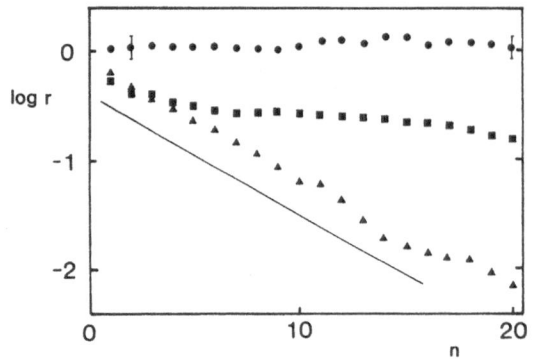

Fig. 3. Log of the ratio
(axial component) / (diagonal
component) as a function of n
(log of mass) for p=0 (dots),
p=0.5 (squares) and p=1 (triangles)

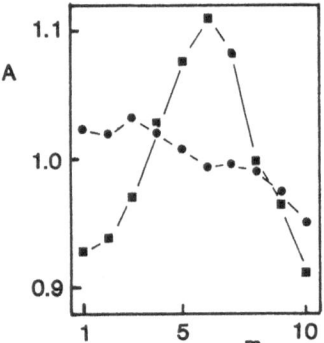

Fig. 4. Radius of gyration $R(\alpha)$ of a particle
added at an angle α (divided by the average
R of all previously added particles). These
ratios are binned and normalised such that
$\sum R(\alpha) = 1$. The dots are for clusters of
20000 particles, the square for clusters of
250000 particles, both for p=0.1

In a first approximation, the shape of DLA clusters can be
analysed by visual inspection. However, a much more sensitive measure
is to compute the radius of gyration as a function of the angle with
respect to the axes through the seed (normalised by the average radius
of gyration). Figure 4 shows this quantity (normalised) for p=0.1. For
this value of p the anisotropy between axis and diagonal is minimal
for clusters of 250000 particles. For a smaller cluster (20000
particles) - and presumably for larger ones as well - the anisotropy
is more pronounced. On the other hand the anisotropy at an angle $\alpha=\pi/8$
(eight symmetry axes) increases rapidly with increasing particle
number (from 20000 to 250000). The angular dependence is a sensitive
measure of the anisotropy; a difference of 1% can be readily
identified. These effects become visible to the eye only for much
larger clusters.

The growth process investigated here can be systematically
extended to study off-lattice DLA by a lattice model: for a given
cluster size one adjusts the parameter p such that the anisotropy
between $\alpha=0$ and $\alpha=\pi/4$ is zero, then one introduces a second parameter
rotating particles towards or away from the $\pi/8$ axis, analogous to p
in Fig. 1, and adjusts it to make the process isotropic at this level,
etc. This way a simpler lattice model can be used to investigate the
off-lattice model to any desired accuracy.

7. Scaling arguments to determine $D_\perp$

Let us now turn to the scaling of the anisotropic regime. The
asymptotic shape of the aggregate is sketched in Fig. 5. Globally, the

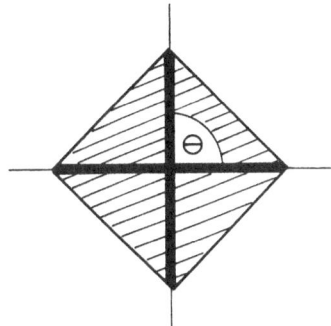

<u>Fig. 5.</u> Schematic diagram illustrating how the cross-like growth is globally determined by the lattice symmetry. Each branch is a self-affine fractal with different scaling powers in the different directions

cross-like appearence reflects the symmetry of the lattice whereas the individual branches internally have the self-affine structure determined in the above simulations. In order to relate the transverse exponent $D_\perp$ to the symmetry, one neglects the fluctuations. This means that the growth is assumed to be determined by uniform aggregation of a diamond shape as indicated in Fig. 5.

The growth rates can be estimated in two different ways: on one hand each branch grows at a rate $dN=R^D_\parallel{}^{-1}dR$ from the definition of $D_\parallel$; on the other hand from the cone angle argument one gets $dN=R_\perp^{D(\alpha)-1}dR=R^{(D(\alpha)-1)D_\parallel/D_\perp}dR$ where $D(\alpha)$ is taken from equation (3) and the width of the branch $R_\perp$ is expressed in terms of its length R. The difference with [13] is that here not the whole shaded area but only a small portion proportional to the width of the branches is absorbing. For $D_\parallel < D_\perp$ the tips become asymptotically sharp and $D_\parallel$ =1.5. From this one concludes that

$$D_\perp = (D(\alpha)-1)D_\parallel/(D_\parallel-1)=3\pi/(2\pi-\Theta) . \qquad (5)$$

For $\Theta=\pi/2$, this gives $D_\perp$=2.0, the value obtained in the simulations.For $\Theta=\pi$, one has $D_\perp$=3.0, which means that the clusters are compact, consistent with [14]. For $\Theta\to0$, $D_\perp\to D_\parallel$=1.5 which means that for very high symmetry the cluster becomes isotropic, but with a different exponent than off-lattice DLA.

<u>8. Conclusions</u>

Datailed investigations of the role of the anisotropy and the noise in DLA show that this growth mechanism is far more complex than initially anticipated. The resulting pattern is highly unstable in the sense that seemingly minor variations in how the algorithm is implemented change the resulting structure completely. Most importantly, the growth is not universal as the lattice on which the aggregate grows determines the transverse exponent.

The calculations were performed at the CIRCE in Orsay,France.

<u>References</u>

1   T.A.Witten and L.M.Sander, Phys.Rev.Lett. 47, 1400 (1981)
2   'Kinetics of Aggregation and Gelation', F.Family and D.P.Landau (eds.), North-Holland (1984)
3   P.Meakin, Phys.Rev. A27, 604,1495 (1983)

4   S.K.Ma, 'Modern Theory of Critical Phenomena', Benjamin (1976)
5   'Phase Transitions and Critical Phenomena', Vol. 6, C.Domb and
    M.S.Green (eds.), Academic Press (1976)
6   R.Jullien, M.Kolb and R.Botet, J.Physique 45, 395 (1984)
7   R.C.Ball and R.M.Brady, J.Phys. A18, L809 (1985)
8   P.Meakin, J.Phys. A18, L661 (1985)
9   M.Kolb, unpublished
10 P.Meakin and T.J.Vicsek, Phys.Rev. A32, 685 (1985)
11. M.Kolb, J. de Physique 46, L631 (1985)
12 R.Voss, unpublished
13 L.Turkevich and H.Sher, Phys.Rev.Lett. 55, 1026 (1985)
14 R.C.Ball, R.M.Brady, G.Rossi and B.R.Thompson, Phys.Rev.Lett.
    55,1406 (1985)
15 'On Growth and Form', H.E.Stanley and N.Ostrowski (eds.),
    Matinus Nijhoff (1985)
16 P.Meakin, Phys.Rev. A27, 1495 (1983)
17 P.Meakin, Phys.Rev. A33, 3371 (1986)
18 B.Shraiman and D.Bensimon, Phys.Rev. A30, 2840 (1984)
19 D.Bensimon, Phys.Rev. A33, 1302 (1986)
20 F.Family and H.G.E.Hentschel, preprint
21 R.Ball, talk at STATPHYS 16, Boston (1986)
22 M.Matsushita, H.Kondo, S.Ohnishi and Y.Sawada, J.Phys.Soc.Japan
    55, 61 (1986)
23 J.Kertesz and T.J.Vicsek, J.Phys. A19, L257 (1986)
24 M.Matsushita and H.Kondo, J.Phys.Soc.Japan 55, 2483 (1986)
25 J.Nittmann and H.E.Stanley, Nature 321, 663 (1986)
26 B.B.Mandelbrot, in 'Fractals in Physics', L.Pietronero and
    E.Tosatti (eds.), North Holland (1986)
27 P.Meakin,R.C.Ball, P.Ramanlal and L.M.Sander, preprint
28 P.Meakin and F.Family, Phys. Rev. A34, 2558 (1986)

The Classes of Fractals

B. Röhricht[1], *W. Metzler*[2], *J. Parisi*[1], *J. Peinke*[1], *W. Beau*[2], and *O.E. Rössler*[3]

[1]Physics Institute II, University of Tübingen,
  D-7400 Tübingen, Fed. Rep. of Germany
[2]Interdisciplinary Working, Group "Mathematization",
  University of Kassel, Gesamthochschule, D-3500 Kassel, Fed. Rep. of Germany
[3]Institute for Physical and Theoretical Chemistry,
  University of Tübingen, D-7400 Tübingen, Fed. Rep. of Germany

All fractals yield to a combined topological and measure-theoretical classification scheme proposed. The three criteria applied in succession are: (1) Reducibility, or not, to a product of a lower-dimensional fractal and a smooth manifold; (2) Connectedness, or not (line/web; islands); (3) Measure, within the fractal boundary, of the fractal points (positive; zero). To give an example, the surface of a fractal mountain is a trifractal (not reducible to less than 3 embedding dimensions) that is connected and strong. A Table of examples that is exhaustive up to dimension 4, with 34 entries, is provided.

1. Introduction

Fractals were introduced by Mandelbrot [1] as a generic term covering all kinds of Cantor-like objects. All these objects ideally have uncountably many substructures that are either identical ('self-similar') or alike, except for scaling.

Fractals are found everywhere in Nature. Examples range from coastlines, snowflakes and mountains over dendritic structures in physics and biology to the distribution of bubbles in cheese and cosmos [1].

So far, the only quantitative characterization available, due to Mandelbrot [1], consists in a single real number, the Hausdorff dimension of the object. (For elaborations, cf. [2].) The latter is strictly larger than the topological dimension, and smaller than or at most equal to the embedding dimension [1]. The Hausdorff dimension, which is easy to calculate in many realistic cases [2], obviously does not differentiate between qualitatively different fractal shapes. For example, the Hénon attractor [3], which is locally smooth, has a dimensionality of 1.26 for a certain combination of its two parameters [4]. In contrast, the Von Koch snowflake which is wrinkled like a coastline, also has a dimensionality of 1,26 (log4/log3) [1]. In spite of this, Mandelbrot himself still considers as "premature" any attempt at introducing a finer subdivision into his class (personal communication, January 1987).

In the following, a scheme is introduced that may allow for some reduction in complexity as a function of the embedding dimension. Only topological and measure-theoretic criteria are used, in a straightforward manner.

2. Proposed Procedure

Topologically, all objects - including fractals - can be divided up into three classes: Totally disconnected (dust); disconnected (islands beside islands, puppets within puppets), connected (boundaries and webs).

Specific to fractals is the existence of uncountably many components as mentioned. This of necessity introduces a measure-theoretic question: What is the Lebesgue measure of the smallest components, within the object ? This time it is two possibilities: Positive measure; zero measure.

Finally it is obvious that some fractals can be projected onto a lower dimension, without any attendant reduction in complexity, while many others cannot. This is the criterion of 'reducibility.' It naturally has to be applied first.

Let us briefly see how these three criteria come into play in the simplest case possible - the case in which the embedding space is one-dimensional.

As is well known, only one type of fractal is possible here (discounting the complement set as usual). It is the class of Cantor sets proper. A Cantor set certainly cannot be reduced to a lower dimension. It is totally disconnected topologically. As to the measure of the last-reached points in the set, it is positive [5]. Therefore, in 1 D, there is only one class of fractals, the $\underline{1}$ fractal. It is always totally disconnected and strong.

3. Proposed Classification

With the above example in mind, it is now possible to proceed more formally. If O is a fractal object embedded in R^n , let X be its boundary set. Only the latter set, X , will be considered in the following.

We call X FRACTAL, in a point x , if there exists a 1-D hyperplane h_x through x such that the intersection of h_x with X is homeomorphic to a Cantor set, in a neighborhood $\bar{x}$ to x .

Next, let us define reducibility. We call X REDUCIBLE, in a point x , if there exists a set of parallel hyperplanes $h_{\bar{x}}$ through x , such that the intersection I_x of $h_{\bar{x}}$ with X is smooth in $\bar{x}$. By selecting an appropriate $h_{\bar{x}}$, a maximal dimension r_x is obtained for I_x . We then call X r reducible, in x . (Note that any smooth - completely nonfractal - set X in R^n is $\underline{n}$ reducible.)

In consequence, any fractal point x of X in R^n becomes an n-dimensional fractal point (n FRACTAL for short) if it is completely nonreducible. More generally, it becomes an n-r fractal if it is $\underline{r}$ reducible.

For convenience, any $\underline{1}$ fractal may also be called a monofractal, any $\underline{2}$ fractal a bifractal, and so forth.

Note that the attractor of the Hénon map, mentioned previously, is a monfractal. The Von Koch snowflake, in contrast, is a bifractal.

Secondly, let us define connectedness. We call X CONNECTED, in a point x , if there exists in X a path from x to all points in a neighborhood, $\bar{x}$. (In the reduced case, we analogously choose $\bar{x}_r$.) We call X DISCONNECTED, in a

point x , if there exists a point in $\bar{x}_r$ to which there is <u>no</u> path from x in
X . And we call X <u>TOTALLY DISCONNECTED</u>, in a point x , if x is isolated in
$\bar{x}_r$.

Note that a coastline is connected, a Cantor cluster of circular islands is disconnected, and Cantor dust [1] is totally disconnected.

Thirdly and finally, let us define strength. Assume that all points x in X
have already been characterized in terms of their reducibility and fractality. Let
F^m be that subset of X which is <u>m</u> fractal, with <u>m</u> maximal after reduction. To
find out about the strength of F^m , it is first necessary to reduce X (which is
defined in R^n). Specifically, choose the dimension of X such that dim X^m
(the dimension of the reduced X) equals dim F^m , in all points of F^m .

We call X a <u>STRONG</u> <u>m</u> fractal, if F^m is dense in X^m . (Note that in this case
F^m has positive measure in X^m .) Conversely, we call X a <u>WEAK</u> <u>m</u> fractal, if F^m
has zero measure in X^m .

Note that the Mandelbrot set [6] is a strong (connected, bi-) fractal, including
dendrites. A coastline and the Von Koch snowflake also are strong connected bifractals.
The Sierpinski carpet (compare [1] for all these notions) is a weak connected bifractal. So is a locally smooth dendrite in 2 D.

4. A Synopsis

The maximum number of different types of fractals as a function of the embedding
dimensionality is given by the following formula:

$$N = 5[(n - 2)(n - 1)/2 + n - 1] + n . \qquad (1)$$

Here N is the maximum number of classes possible in the dimension in question, and
n is dimensionality. Eq.(1) results as an implication of the fact that with each
dimension, 5n - 4 new classes arise. This latter principle is a straightforward
consequence of the way the following infinite Table (with entries up to n = 4) is
built up. See Table 1, next page.

In every dimension, beginning with the first, one new class of monofractals
arises. In every dimension, beginning with the second, 5 new classes of bifractals
occur. In every dimension, beginning with the third, 5 new classes of trifractals
arise. And so forth through all n .

Most examples that the reader may think of can be fitted into Table 1, we tend
to hope. Unavoidably, some classes are a little bit crowded from the beginning.
This is because weak fractals (with fractal points of zero measure internally) can
of necessity assume a rather large variety of shapes, even after connectedness has
been taken into account. Here, 'line-like' and 'web-like' is a major subdistinction
that resurfaces each time. One could think of introducing it as a general further
subdivision. The reason we did not do so is that no similar need arises for the
other categories. So it appeared more convenient to put up with a (modest) complexity in a single subclass.

One of the messages of Table 1 is that in physical space-time (that is, R^4 ,
locally), no more than N = 34 types of classes of fractal shape can be expected

Table 1: The classes of fractals possible in n-space, demonstrated for n = up to 4

Embedding dimension n	Reduction by dims. r	Remaining dimensions n-r	Fractality	Connectedness	Strength	Examples	No. of types to add
1	0	1	Mono	Total disc.	Strong	Cantor set	1
2	1	1	Mono	Total disc.[a]	Strong	Cantor stripes; Hénon attractor	
	0	2	Bi	Total disc.	Strong	Cantor dust	
				Disconn.	Weak	Smooth Cantor islands	6
					Strong	Coastline-type Cantor islands	
				Connected	Weak	Sierpinski carpet; smooth dendrite; wild goose [7]	
					Strong	Coastline; Mandelbrot set; coastline dendrite	
3	2	1	Mono	Total disc.[a]	Strong	Cantor sheets; hyperchaotic attractor [8]	
	1	2	Bi	Total disc.[a]	Strong	Cantor needles; noodle map attractor [9]	
				Disconn.	Weak	Smooth Cantor tubes	
					Strong	Coastline-type Cantor tubes	
				Connected	Weak	Sierpinski fence	11
					Strong	Mandelbrot fence	
	0	3	Tri	Total disc.	Strong	Cantor dust; galaxies	
				Disconn.	Weak	Smooth Cantor spheres	
					Strong	Cantor rocks; Cantor caterpillars	
				Connected	Weak	Sierpinski sponge; smooth tree	
					Strong	Fractal mountain; barked tree; cloud	

n	r	n-r	Fractality	Connectedness	Strength	Examples	No. of types to add
4	3	1	Mono	Total disc.[a]	Strong	Cantor hyperplanes	
	2	2	Bi	Total disc.[a]	Strong	Nonparallel Cantor sheets	
				Disconn.	Weak	Smooth Cantor hypertubes	
					Strong	Coastline-type Cantor hypertubes	
				Connected	Weak	Sierpinski hyperfence	
					Strong	Mandelbrot hyperfence	
	1	3	Tri	Total disc.[a]	Strong	Nonparallel Cantor needles; cosmos	
				Disconn.	Weak	Smooth Cantor-sphere tubes	
					Strong	Cantor-rock tubes	16
				Connected	Weak	Sierpinski-sponge fence; smooth-tree fence	
					Strong	Fractal-mountain fence; barked-tree fence	
	0	4	Tetra	Total disc.	Strong	Cantor hyperdust	
				Disconn.	Weak	Smooth Cantor hyperspheres	
					Strong	Cantor hyper rocks	
				Connected	Weak	Sierpinski hyper sponge; smooth hyper tree	
					Strong	Fractal hyper mountain; barked hyper tree; hyper cloud	

[a]Footnote: A weak 'remnant' connectedness (of long-line [10] or Cantor-book [11] type) is admissible here

to occur. This is both an unexpectedly large and an unexpectedly small number, depending on what one is prepared to expect. It may (or may not) turn out to be a reasonable characteristic. The underlying scheme at least is fairly easy to remember.

It is perhaps worth pointing out that no need arose to introduce 'mixed' forms. All natural forms are taken care of directly, it appears. This is a feature any useful classification ought to provide. On the other hand, it also goes without saying that 'hybrid fractals' can be formed as a matter of course. All conceivable kinds of 'agglomerate' are possible. Hereby, all the crudely (or not so crudely) put together portions can still be classified individually according to the above scheme.

As a case in point, consider the 'extended' Mandelbrot set [12]. This set is, in the simplest case, defined in 3 D rather than in 2 D. The original 2-parameter plane of the complex-analytic logistic map [6] is augmented by one dimension as a real, analyticity-destroying parameter is introduced into the iteration. The original Mandelbrot set thereby survives as a cut through a 'tree trunk,' so to speak. At the time being it is known that this object in 3 D possesses regions that are bifractal (the Mandelbrot set proper), regions that are monofractal (stripe-like), and also regions that are non-fractal (smooth) [12]. The remaining question - of whether or not trifractal spots (and, with further parameters, even more complex regions) are also possible - is presently open.

5. Discussion

Fractals acquire more and more importance in applications that range from physics and chemistry and biology to dynamics proper. All chaotic attractors, for example, are fractals (cf. [1]). However, the types of fractals valid for attractors in differentiable dynamical systems, found so far, are very restricted. Only sheet-like (and thread-like) attractors are familiar at the time being, cf. [13]. This notwithstanding, it is a natural question to ask whether or not nondifferentiable (specifically, more than just monofractals and totally disconnected higher fractals of the type specified by Footnote a, in Table 1) fractals may be expected to exist in differentiable dynamical systems, too. The proposed answer is yes [14-16].

The occurrence of such 'cloud-like' objects may in fact be quite common. Nevertheless none has as yet been seen. Search for such attractors, both in explicit model systems and in realistic systems like dripping faucets [17] and low-temperature solid-state devices [18], can only begin once such structures have been defined in unambiguous terms. It is out of such application-bound considerations that the need for a synoptic classification first arose.

Eventually, a single measured time curve may suffice to characterize fractal objects of a complexity never before thought possible. Note that the state spaces of dynamical systems even of simple type are, unlike our own physical surroundings, in no way restricted to 4 dimensions locally. This means that numerical sentinels, of the type first introduced by Packard et al. [19], may acquire an even greater significance in the future.

Acknowledgments

We thank Werner Güttinger, Agnes Babloyantz and Rudolf P. Huebener for discussions.

References

1. B.B. Mandelbrot, The Fractal Geometry of Nature (Freeman, San Francisco 1982)

2. G. Mayer-Kress, ed., Dimensions and Entropies in Chaotic Systems (Springer-Verlag, New York 1986)

3. M. Hénon, A two-dimensional mapping with a strange attractor, Commun. Math. Phys. 50, 69-77 (1976)

4. H.G. Schuster, Deterministic Chaos (Physik-Verlag, Weinheim 1984)

5. As to the measure of the object, within the embedding space, there is also the possibility of 'fat' fractals (cf. D. Farmer, ref. [2], p. 54). This added distinction is independent of those considered in the present paper.

6. B.B. Mandelbrot, Fractal aspects of the iteration $z \rightarrow \lambda z(1-z)$ for complex λ and z, Ann. N.Y. Acad. Sci. 357, 249-259 (1980)

7. C. Kahlert and O.E. Rössler, Analogues to a Julia boundary away from analyticity, Z. Naturforsch. 42 a, 324-329 (1987)

8. O.E. Rössler, An equation for hyperchaos, Phys. Lett. 71 A, 155-157 (1979)

9. O.E. Rössler, Chaos and bijections across dimensions, In New Approaches to Nonlinear Problems in Dynamics, ed. by P.J. Holmes (SIAM, Philadelphia 1980) p.477-486

10. O.E. Rössler, Long Line attractors, In Iteration Theory and Its Functional Equations, ed. by R. Liedl, L. Reich and G. Targonski, Lect. Notes Math. 1163, 149-161 (1985)

11. R.F. Williams, The structure of Lorenz attractors, Lect. Notes Math. 615, 94-112 (1977)

12. J. Peinke, J. Parisi, B. Röhricht and O.E. Rössler, Instability of the Mandelbrot set, Z. Naturforsch. 42 a, 263-267 (1987)

13. O.E. Rössler, The chaotic hierarchy, Z. Naturforsch. 38 a, 788-801 (1983)

14. O.E. Rössler, C. Kahlert, J. Parisi, J. Peinke and B. Röhricht, Hyperchaos and Julia sets, Z. Naturforsch. 41 a, 819-824 (1986)

15. O.E. Rössler, J.L. Hudson and J.A. Yorke, Cloud attractors and noodle maps, Z. Naturforsch. 41 a, 979-980 (1986)

16. O.E. Rössler, C. Kahlert and J.L. Hudson, Bi-fractal basin boundaries in invertible systems, In Chaos in Biological Systems, ed. by A.V. Holden and L.F. Olsen (Manchester University Press, Manchester 1987) in press

17. R. Shaw, The Dripping Faucet as a Model Chaotic System (Aerial Press, Santa Cruz 1985)

18. J. Parisi, J. Peinke, B. Röhricht and K.M. Mayer, Self-organized formation of spatial and temporal dissipative structures in semiconductors, Z. Naturforsch. 42 a, 329-333 (1987)

19. N.H. Packard, J.P. Crutchfield, J.D. Farmer and R.S. Shaw, Geometry from a time series, Phys. Rev. Lett. 45, 712-716 (1980)

Structural Characterization of Aluminum Hydroxide Aggregates by S.A.X.S.: Comparison Between Simulations of Fractal Structures and Experiments

M.A.V. Axelos[1], D. Tchoubar[2], and R. Jullien[3]

[1]I.N.R.A.-L.P.C.M., rue de la Géraudière, F-44072 Nantes Cedex 3, France
[2]Laboratoire de Cristallographie, Université d'Orléans,
  B.P. 6759, F-45067 Orléans Cedex 2, France
[3]Laboratoire de Physique des Solides, Bat. 510,
  Université de Paris-Sud, F-91405 Orsay Cedex, France

Abstract

Colloidal particles of aluminum hydroxide $Al(OH)_x$ form different types of aggregates depending on the value of x. Small angle X-ray scattering (SAXS) functions of these aggregates are quantitatively compared with ones obtained from numerical simulations on cluster-cluster ($Cl - Cl$) aggregation models. While the $x = 2.6$ case is fitted by the standard $3d$ model, the $x = 2.5$ case can be fitted by a newly developed "tip-to-tip" $3d$ model ($T - T$), which considers cluster polarizability.

1. Introduction

The chemistry of aqueous solutions of $AlCl_3$ has recently received much attention because of the use of aluminum species as a new flocculating agent in water purification plants [1].

The association of ^{27}Al NMR spectroscopy and SAXS has been a determining factor for the knowledge of the nature and the structure of the ionic aluminum species [2],[3]. The predominant species is the polymer $Al_{13}O_4(OH)_x^{(31-x)+}$ which is named "Al_{13}". Depending on the hydrolysis ratio $x = (OH)/(Al)$, different types of macrostructure have been observed. For $x < 2.3$ the Al_{13} polymers are isolated and give a stable suspension of $12\mathring{A}$ radius spherical particles [4]. When x increases towards 2.6, the decrease of the electric charges induces aggregation phenomena: for $x = 2.5$ the resulting clusters exhibit a fractal dimension of $D = 1.45$; for $x = 2.6$, D is about 1.86 [5], [6].

These experimental results are compared with theoretical models of finite size clusters which have the same fractal dimension, the $3d$ $T - T$ model [7] is compared with $Al(OH)_{2.5}$ clusters and the $3d$ $Cl - Cl$ model [8] is compared with $Al(OH)_{2.6}$.

Quantitative fitting of SAXS functions allows us to determine the mean size of the aggregates.

2. Experiments

The colloidal samples of aluminum hydroxide were made by partial hydrolysis of aluminum chloride solution with sodium hydroxide [9]. The Al concentration remained constant: $(Al) = 0.1M$ and the samples were characterized by the hydrolysis ratio: $x = (OH)/(Al)$ with $x = 2.5$ and 2.6.

2.1 X-Ray Data

We used the synchroton radiation of the D.C.I. storage ring of L.U.R.E. (Université de Paris-Sud, Orsay) to benefit from a very intense X-ray beam associated with point collimation and a performance device which allows very small angles to be reached. Indeed the collected data cover the s-range from 0.001 to $0.06 Å^{-1}$ and from 0.0005 to $0.013 Å^{-1}$, where s is the scattering vector amplitude: $s = 2 \sin \theta / \lambda$, 2θ the scattering angle and λ the selected wavelength (here $\lambda = 1.6 Å$).

The experimental scattering function $I(s)$ of fractal aggregates is given by

$$I(s) = GI_0(s).(G(s) * F(s)), \qquad (2.1)$$

where $*$ denotes a convolution product, $I_0(s)$ is the scattering function of the subunit, and $G(s)$ is the interference function between these subunits. $G(s)$ scales as s^{-D} between the points $s \sim 1/2\xi$ and $s \sim 1/2r_0$ where r_0 is the subunit radius and ξ the characteristic length of the clusters. $F(s)$ is the natural cut-off function of the aggregate above ξ. From the exact interparticle distance distribution function $P(r)$, calculated from the simulated finite size clusters, the Fourier transform allows us to obtain directly the product $G(s) * F(s)$ without any assumption about the analytic expression of the $F(s)$ function. Then the theoretical expression of the scattering functions is given by [10]

$$I(s) = I_0(s) \left[1 + \frac{4\pi}{V_0} \int P(r) \frac{\sin 2\pi sr}{2\pi sr} dr \right], \qquad (2.2)$$

V_0 being the volume of the subunit. $I(s)$ is normalized such that $I(0) = V$, the volume of the cluster, which means that the distance distribution $P(r)$ is normalized such that

$$4\pi \frac{1}{V_0} \int P(r)dr = N - 1 \qquad (2.3)$$

with $P(0) = 0$, and N being the number of subunits in the cluster.

The comparison between the theoretical and experimental curves is performed by adjusting the intensity in the range $s > 1/2\pi r_0$, where the inter-particle interferences become negligible and the intensity is then only characteristic of the scattering by a single subunit.

2.2 Results

SAXS experiments carried out with $x = 2.5$ and $x = 2.6$ samples show that the aggregation of the Al_{13} subunits, represented in Fig. 1, yields fractal aggregates of dimension $D = 1.45$ and $D = 1.86$, respectively [5],[6]. The $I(s)$ scattering curves of the two samples are depicted in Fig. 2. From the bending of the curve for small s values one can estimate the characteristic length; in both cases $\xi \simeq 94\mathring{A}$ and $140\mathring{A}$ for $x = 2.5$ and 2.6.

Fig. 1: Schematic Keggin structure of the Al_{13} cation [4]

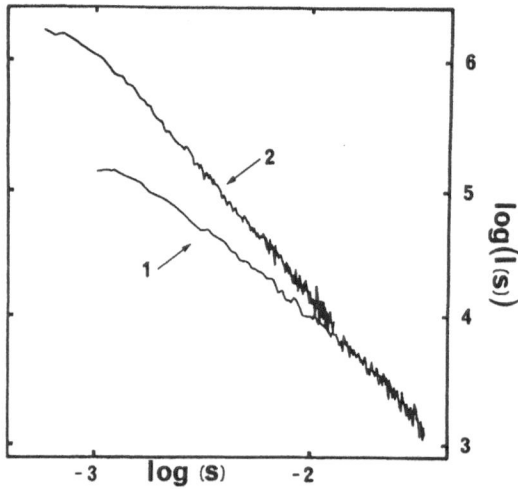

Fig. 2: Experimental scattering functions $I(s)$ in a $log_{10} - log_{10}$ plot. Curve 1: sample $x = 2.5$, curve 2: sample $x = 2.6$

3. Simulations

The "tip-to-tip" $(T - T)$ model is a recent extension of the $Cl - Cl$ model which considers the effect of cluster polarizability.The clusters systematically stick by their tips leading to a smaller fractal dimension 1.42 in $3d$, 1.26 in $2d$ [7].

The $Cl - Cl$ model is an off-lattice version with linear trajectories [11] and in the simulations proposed here we have systematically used the hierarchical version of these models [12].

In order to calculate the distance distribution function $\tilde{P}(x)$, $(x = r/d_0$, and $d_0 = 2r_0)$ we have generated 100 independent clusters of 64 to 1024 particles for the two models. For each size $\tilde{P}(x)$ is the histogram of inter-particle distances within the aggregate [6]. This distribution is such that

$$\int \tilde{P}(x)dx = N; \tilde{P}(0) = 1. \tag{3.1}$$

Figure 3 presents the $\tilde{P}(x)$ $3d$ curves in the case of the two models of 1024 particles whose fractal dimensions are 1.9 and 1.43. The effect of the fractal dimension on the size extension of the cluster is clearly seen.

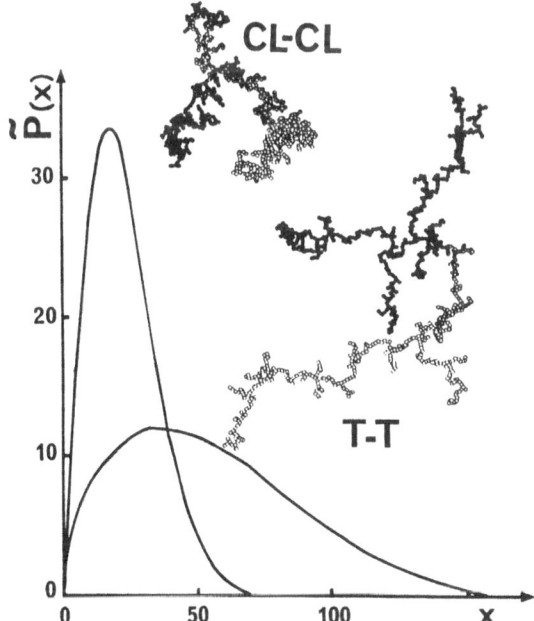

Fig. 3: Simulated distance distribution functions $\tilde{P}(x)$ for the $Cl - Cl$ model with $D = 1.90$ and $T - T$ model with $D = 1.43$ [7],[8]

From the above calculated $\tilde{P}(x)$ distribution functions, one can evaluate the theoretical $I(s)$ scattering functions through formula (2.1), by properly defining $P(r)$ by

$$P(r) = \frac{V_0}{4d_0}\tilde{P}(\frac{r}{d_0}).\qquad(3.2)$$

The quantitative fit with the $T-T$ $3d$ model in the case of $x = 2.5$ is shown in Fig. 4. The three theoretical scattering curves correspond to different numbers of particles in the aggregate. As expected, only the small s part of the curve is affected and by choosing the best fit with the experiment one can estimate the average number of particles to be $N = 64$.

The same kind of fit can be obtained between the $x = 2.6$ experimental curve and the simulated one using the $3d$ $Cl - Cl$ model as shown in Fig. 5. The best fit here is obtained for $N = 512$ particles.

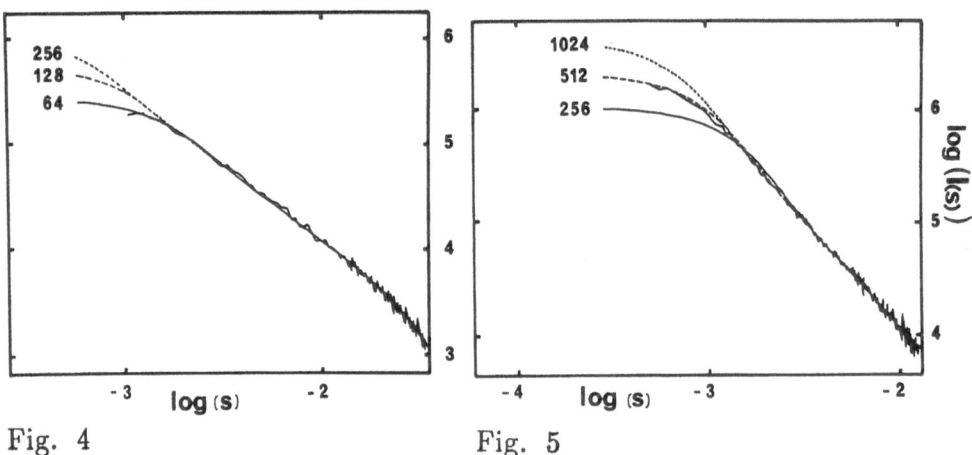

Fig. 4   Fig. 5

Fig. 4: Fitting of the sample $x = 2.5$ with the $T - T$ model

Fig. 5: Fitting of the sample $x = 2.6$ with the $Cl - Cl$ model

4. Discussion

One can develop some qualitative arguments to explain why the $T - T$ model is more adapted to the $x = 2.5$ case while the $Cl - Cl$ model describes the $x = 2.6$ case. The Al_{13} subunits are positively charged and water molecules are oriented near their surface, producing a surface field. The intensity of this field is directly linked with the importance of the charge on the subunits. For $x = 2.5$ the charge is strong and one can imagine that the polarization of the water molecules near the surface of the aggregates might induce a sticking mechanism like in the $T-T$ model. When x increases the charge on the subunits decreases, the water molecule surface charges decrease also and one can imagine that the standard diffusion-sticking mechanism could be recovered as in the regular $Cl-Cl$ model. Moreover, one can learn from the fits that, with the same initial concentration of Al_{13} and the same aging time for the aggregation process, the aggregates contain many more particles in the case $x=2.6$ than in the case $x=2.5$. We must remember that in the present simulations no size distribution has been considered for the clusters and the numbers of particles within the clusters must be considered more like average orders of magnitude than precise numbers. The increase in the number of particles is however clear when going from $x=2.5$ to $x=2.6$. This implies that the two models must have completely different kinetics.

To conclude, we were able to fit the entire experimental scattering function by using simulated clusters. This allowed a direct comparison between the different theoretical models and experiments. Further work is in progress to extend the present study.

Acknowledgements

We acknowledge support for the numerical computations from an ATP of the CNRS. Calculations were performed at CIRCE (Centre Interrégional de Calcul Electronique).

References

1 Bottero J.Y., Poirier J.E., Fiessinger F.:(1980), *Progress. Wat. Tech.*, **12**, 601

2 Akitt J.W., Farthing A.: (1981), *J. Chem. Soc. Dalton Trans.*, 1606

3 Bottero J.Y., Marchal J.P., Cases J.M., Poirier J.E., Fiessinger F.: (1982), *Bull. Soc. Chim. de France*, **I**, 439

4 Bottero J.Y.,Tchoubar D., Cases J.M., Fiessinger F.: (1982), *J. Phys. Chem.*, **86**, 3667

5 Axelos M.A.V., Tchoubar D., Bottero J.Y., Fiessinger F.: (1985), *J. Physique*, **46**, 1587

6 Axelos M.A.V., Tchoubar D., Jullien R.: (1986), *J. Physique*, **47**, 1843

7 Jullien R.: (1985), *Phys. Rev. Lett.*, **55**, 1697 and (1986),*J. Phys. A*, **19**, 2129

8 Meakin P.: (1983), *Phys. Rev. Lett.*, **51**, 1119 and Kolb M., Botet R., Jullien R.: (1983), *Phys. Rev. Lett.*, **51**, 1123

9 Bottero J.Y., Cases J.M., Fiessinger F., Poirier J.E.: (1980), *Phys. Chem.*, **84**, 2933

10 Guinier G., Fournet G.: (1955), In *Small-Angle Scattering of X-Rays* (John Wiley and Son)
11 Ball R., Jullien R.: (1984), *J. Physique*, **45**, 1031 and Meakin P.: (1984), *J. Colloid Interface Sci.*, **102**, 505
12 Botet R., Jullien R., Kolb M.: (1984), *J. Phys. A*, **17**, 275

Part V

Chaos and Turbulence

Spatial Disorder in Extended Systems

P. Coullet*, C. Elphick, and D. Repaux

Laboratoire de Physique Théorique, Université de Nice, Parc Valrose,
F-06034 Nice Cedex, France

1. INTRODUCTION

During the last few years much effort has been devoted to the study of temporal chaotic behavior of simple dynamical systems modeling turbulent-like phenomena in a great variety of macroscopic physical systems [1]. On the other hand the study of pattern formation occurring in extended systems has only recently become a very popular subject [2].

It is the aim of this work to present and study the basic mechanisms leading to spatial disorder in macroscopic systems [3]. As we will see in the next sections the phenomenon of spatial chaos in such systems is closely related to the presence of metastable disordered configurations of topologically stable defects and the mechanisms leading to such chaotic configurations are the analogs of those leading to temporal chaotic behavior. A first mechanism, related to Melnikov's theory of periodically driven Hamiltonian systems with one degree of freedom [4], is caused by the pinning of defects [5],[6]. A second mechanism , new to our knowledge, has to do with the intrinsic oscillatory nature of the defects and is related to Shilnikov's theory [4] of conservative systems with two degrees of freedom.

For a better understanding of the above ideas it is useful to recall the basic mechanisms leading to temporal chaos. Temporal complexity or chaotic behaviors arise in dynamical systems possessing the property of sensitive dependence on the initial conditions. This property explains the apparent incompatibility between the deterministic nature of the system and the unpredictability of its time evolution. A trivial system showing the above property is given by the map $x' = 2x$, $0 \leq x \leq 1/2$, $x' = 2(1 - x)$, $1/2 \leq x \leq 1$, which mimics the non-invertible map $x' = \mu x(1 - x)$, $0 \leq x \leq 1$, which is a simplified version of Smale's horseshoe map, which in turn is the simplest model for the Poincaré homoclinic transverse intersection property (a

---

* *Also Observatoire de Nice*

generic property leading to chaos in dynamical systems). Therefore temporal complexity is intimately related (through the Poincaré property) to horseshoe formation (h.f.) [4]. According to dynamical systems theory the basic mechanisms leading to (h.f.) are related to Melnikov's theory and Shilnikov's theory. Melnikov's theory deals with the existence of homoclinic orbits or heteroclinic loops [7] (separatrices) in periodically driven Hamiltonian systems. Shilnikov's theory states that chaotic behavior exists in the area of parameter space where a homoclinic or heteroclinic bifurcation occurs. This ends our review of chaos in spatially confined macroscopic systems.

We now consider extended macroscopic systems. These systems are of a much richer nature than constrained ones since they allow the possibility of complex behaviors such as temporal and/or spatial chaos, occurrence of pattern formation and the presence of topological defects in such patterns. A pattern is described by the stationary bounded solutions of a partial differential equation for a suitable order parameter describing the physical system. Among these solutions, localized ones represent defects of the pattern. These defects are an essential feature of spatially extended systems in a broken symmetry phase. The defect's core is the physical region where the order parameter vanishes, which means it represents the vestige of the disordered phase in an ordered one.

Physically interesting defects are those topologically stable, or in other words, elementary defects. For an n-vector order parameter describing a discrete symmetry phase a simple argument from homotopy theory shows that topologically stable defects have dimension $d' = d - 1$, where d is the real space dimensionality. The codimension of a defect is defined as $c = d - d'$. Then topologically stable defects in a broken discrete symmetry phase are those with $c = 1$ (codimension-one defects). We address the problem of spatial complexity in macroscopic systems and its connection with codimension-one topological defects (surfaces for $d = 3$, lines for $d = 2$, points for $d = 1$). Such defects are basically described by one-dimensional models, where the space variable (x) runs over the transverse dimension to the defect. In particular, the one-dimensional nature of these models will allow us, by replacing x by t, to make a parallel between the mechanisms leading to spatial disorder and those leading to temporal chaos. Defects can be classified according to their topological charge (the difference between the asymptotic values of the order parameter for codimension-one defects). A topologically charged defect is a kink-type solution connecting two equivalent equilibrium solutions [8] and corresponds to a heteroclinic solution (replacing x by t) for a dynamical system. A homoclinic orbit corresponds to an uncharged pulse-type solution which is shown to be topologically unstable. A chaotic ordering of such elementary defects represents a spatially chaotic pattern. A constraint on the observation of such patterns in nature concerns the possible instabilities leading to temporal behavior. These are ruled out in equilibrium systems because of the existence of a variational principle. In the case of non-equilibrium systems temporal instabilities can be avoided in large domains of parameter space, particularly near the onset of stationary instabilities. Such complex patterns have recently been observed by Lowe and Gollub [9] in an experiment on liquid crystal instabilities.

2. DYNAMICS OF CODIMENSION-ONE DEFECTS

Let us consider, for simplicity, an extended system with codimension-one topologically stable defects described by a scalar order parameter $A(x,t; y//)$, where x is the variable transverse to the defect and $y_{//}$ is a set of variables "parallel" to the defect. In many circumstances parallel dimensions play no crucial role in the defects' dynamics and their effect can be omitted in a first approach (they can be considered in a perturbative scheme by letting the defects' positions depend on them). We assume that A satisfies a dissipative-type partial differential equation (for simplicity we have not included propagative or non-variational terms) invariant under some discrete symmetry group:

$$A_t = L(\partial_x)A - V'(A) + P(x, A), \tag{1}$$

where L is a linear differential operator, P stands for a small perturbation and V is a potential function possessing a discrete set of minima A_α. Stationary solutions of (1) (with $P = 0$) connecting two equivalent minima are defect-type solutions (kinks or anti-kinks) $D(x - x_0)$ where x_0 is the defect's position. For a one-defect solution the position x_0 is constant since the defect is isolated (for a vanishing P). For a solution with more than one defect the defects' positions vary in time due to interactions between defects (and also due to the external perturbation). A reasonable ansatz [10] for a multi-defect solution near the i-th defect is given by

$$A(x,t) = D_i(x - x_i(t)) + D_{i+1}(x - x_{i+1}(t)) - D_{i+1}(-\infty)$$

$$+D_{i-1}(x - x_{i-1}(t)) - D_{i-1}(\infty) + R(x,t) \tag{2.a}$$

$$= D_i(x - x_i(t)) + w + R(x,t), \tag{2.b}$$

where we have assumed only nearest neighbor interactions and R stands for corrections (negligibly small in the dilute defect gas limit which we take as our basic hypothesis). By replacing ansatz (2) into equation (1) we obtain

$$(\partial_t + \mathcal{L})R = -\mathcal{L}w + N_v(D_i(x - x_i(t)), w)$$

$$+ \sum_j (-x_j \partial_x D_j(x - x_j(t)) + P(x, Dj(x - x_j(t)))) + \text{h.o.t.} \tag{3}$$

where j runs over $i-1, i, i+1$; h.o.t. stands for small corrections involving w, R, P. L is a linear operator given by

$$\mathcal{L} = L(\partial_x) - V''(D_i(x - x_i(t))) \tag{4}$$

and

$$N_v(X, Y) = V'(X + Y) - V'(X) - Y V''(X). \tag{5}$$

Since $\mathcal{L}\partial_x D_i(x - x_i(t)) = 0$, the defect equations of motion will be determined by the solvability condition

$$\int dx G_i(-\mathcal{L}w + N_v(D_i, w) + \sum_j (-x_j \partial_x D_j + P(x, D_j))) = 0, \tag{6}$$

where $\mathcal{L}^+ G_i(x - x_i(t)) = 0$. In the dilute defect gas hypothesis (and neglecting configurations with overlapping of kinks) condition (6) gives the defect equation of motion

$$\partial_T x_i = q_{i+1}(D_i(x_{i+1} - x_i) - D_i(\infty)) + q_{i-1}(D_i(x_{i-1} - x_i) - D_i(-\infty)) + p(x_i), \tag{7}$$

where q_j is the topological charge of the j-th defect, T is an appropriate scaled time and

$$p(x_i) = \int dx P(x - x_i, D_i(x)) G_i(x).$$

If L is a diffusion-type operator , perturbations along the defects (lines or walls) can be included by simply adding to the r.h.s. of (7) a diffusive term $\Delta_{//} x_i$.

3. SPATIAL CHAOS

3.1. Melnikov's mechanism

To understand the first basic mechanism leading to spatial complexity in macroscopic systems, we introduce the notion of pinning energy through the example of the Frankel-Kontorova (F-K) model. It is one of the most studied examples exhibiting spatial disorder due to the pinning of solitons [5], [6]. This model describes a one-dimensional chain of particles interconnected by springs and placed in a periodic potential $V(x)$ with a period b. Let us consider for example the case where $V(x) = V_0(1 - \cos(2\pi x/b))$. The potential energy of such a system is given by

$$U = \sum_i (\frac{l}{2}(x_{i+1} - x_i - a)^2 + V_0(1 - \cos(\frac{2\pi x_i}{b}))), \tag{8}$$

where a is the natural length of the spring and l its stiffness. Near the resonance $a = N(b + q)$ we can define a phase variable through the change of variable $x_i = Nb(i + \phi_i/2\pi)$ such that the potential reads

$$U = \sum_i (\frac{lN^2}{2}(\frac{b}{2\pi}(\phi_{i+1} - \phi_i) - q)^2 + V_0(1 - \cos N\phi_i)). \tag{9}$$

For q small enough, the phase variable is assumed to have a slow and smooth variation on i. This amounts to treating the phase as a continuous variable. Minimization of equation (9) leads to the pendulum equation

$$\phi_{nn} - \Gamma \sin N\phi = 0, \tag{10}$$

where $\Gamma = 8V_0\pi^2/lNb^2$. Hyperbolic points in the phase space are the homogeneous solutions, while solitons are the separatrices of the pendulum. The analytical expression of the separatrix is given by

$$\phi(n) = \frac{4}{N} \arctan(\exp(\sqrt{N\Gamma}n)). \tag{11}$$

A soliton-type solution corresponds to a local minimum of the energy. Translation of an elementary defect by one period b causes successive configurations in general not to be extrema of the energy. Consequently the energy of the configuration can just increase, be maximum, then decrease until it reaches its minimum value when the position of the soliton is equivalent to the initial position translated by b. The difference between the maximum and minimum values of the energy is by definition the pinning energy. This corresponds exactly to the Peierls-Nabarro [11] energy in solid state physics. The pinning effect is independent of the relative soliton positions, and is related to the breakdown of the translational invariance. For the F-K model this is due to its discrete nature and the pinning energy is simply the energy necessary to move a defect from one site to its nearest neighbor in the lattice.

In all the examples from equilibrium physics considered so far, the origin of spatial chaotic behavior is to be found in the very presence of a lattice. In the mean field approximation, the non-linear character of this theory generally displays localized structures of defect type. In this aproximation the translational invariance implies the arbitrariness of the defects' positions. When multi-defect solutions are considered, the effective interaction between defects has to be taken into account. For classical codimension-one defects (e.g. those found in the F-K model) the effective interaction force exponentially decreases with their separation. Thus, in the attractive case only, the static multi-defect solution, which consists of a periodic array of such localized structures, turns out to be dynamically unstable, while in the repulsive case it is stable. Chaotic configurations are possible when microscopic effects are introduced via the underlying lattice. This amounts to introducing a periodic forcing whose magnitude decreases as the validity of the continuous approximation becomes better. From the point of view of the localized solution this forcing is seen as a periodic potential. The position of the defects now becomes relevant. In particular, for sufficiently distant defects, because of the exponentially decreasing character of their interactions, the lattice plays a dominant role. In this limit, the defects behave as if they were free, and consequently can be found in any position which minimizes the periodic potential introduced via the lattice. This obviously leads to the existence of random distributions of such defects.

A quite similar situation arises in pattern formation. This problem is studied by means of an amplitude equation where the fast spatial scale is averaged out. This is the underlying lattice which breaks the translational invariance and allows chaotic behavior.

In order to illustrate this first mechanism we consider a simple model. Let A be a real field obeying a Ginzburg-Landau-type equation:

$$A_t = \mu A + A_{xx} - A^3 + \nu \sin(kx). \tag{12}$$

For $\nu = 0$, this model can be regarded as describing the relaxational dynamics associated with a ferromagnetic transition, in the mean field approximation, where A represents the magnetization. Topologically stable defects are associated with the localized solution whose analytic expression is given by

$$A(x) = \pm\sqrt{\mu}\tanh(\sqrt{\mu/2}x). \tag{13}$$

Equation (12) also admits a chain of such alternating defects and anti-defects. Using the technique presented in Sect. 2 we obtain that the dynamics of such a chain is described by (7) where its r.h.s. assumes the form

$$f(x_{i+1} - x_i) - f(x_i - x_{i-1}) - \gamma\nu\sin(kx_i) \qquad (14.a)$$

with γ a given constant and where the effective attractive interaction force reads

$$f(d) = 4\mu\exp(-2\sqrt{\mu}d), \qquad (14.b)$$

where d is the distance between two adjacent defects. The only possible static multi-defect solution is thus found to be periodic, with an arbitrary large period. These solutions are dynamically unstable, with a characteristic evolution time $(1/2\sqrt{2\mu})\exp(\sqrt{2\mu}d)$. This leads, in the case of a dilute gas approximation, to very long transients before the system reaches its final equilibrium. When one moves away from this approximation, (7) no longer describes the actual dynamics, and pairs of adjacent defects are annihilated. When ν is small enough, this model describes the effects of a macroscopic external periodic forcing on the ferromagnetic transition. Localized solutions are expected to persist. The dynamics of a chain of such solutions is then described by periodic solutions which no longer have arbitrary period. They are locked with the external forcing. This locking effect in a continuous model corresponds to the pinning effect in a lattice-type model . When the distance between defects is large enough, the third term in (14.a) dominates

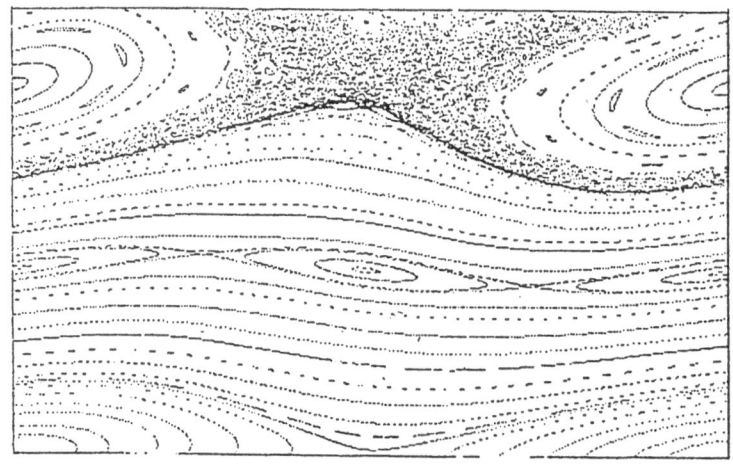

Figure 1
Successive iterations of the mapping defined by (15).

the dynamics. In this limit the positions of the defects are given by $x_i = 2\pi n_i/k$ where the n_i are arbitrary integers. This simple calculation thus demonstrates the existence of chaotic behavior.

The stability analysis is then straightforward: whenever an x_i corresponds to a minimum of the periodic potential, the corresponding state is stable; otherwise it is unstable. Stationary positions are described by the following mapping:

$$x_{i+1} = x_i + y_i - \frac{1}{\sqrt{2\mu}} \log(1 + \gamma\nu \sin(x_i) \exp(\sqrt{2\mu}y_i)), \qquad (15.a)$$

$$y_{i+1} = y_i - \frac{1}{\sqrt{2\mu}} \log(1 + \gamma\nu \sin(x_i) \exp(\sqrt{2\mu}y_i)). \qquad (15.b)$$

An alternative way to prove the existence of spatial chaos consists in looking at equation (12) without its left-hand side. For $\nu = 0$, the ordinary differential equation thus obtained is a one-degree-of-freedom Hamiltonian system. When $\nu \neq 0$, integrability is generically lost. More quantitatively Melnikov's analysis can be used to show explicitly the existence of horseshoes in this problem.

3.2. Shilnikov's mechanism

Another very important mechanism comes from Shilnikov's work on horseshoe formation. It is again related to the existence of a homoclinic or heteroclinic loop of a differential equation. Since we are particularly interested in defects with non-zero topological charge, we focus our attention only on the heteroclinic case. For example, the defect described by (13) is a heteroclinic solution (in the absence of an external field) of (12) without its left-hand side, which connects the two equivalent equilibrium solutions $\pm\sqrt{\mu}$. The linear behavior around these solutions dictates the asymptotic form of the heteroclinic solution. Linearization of (12) leads to opposite real eigenvalues $\alpha = \pm\sqrt{\mu}$. Such an equilibrium solution is termed (real) hyperbolic. Shilnikov's theory deals with homoclinic and heteroclinic solutions connecting complex hyperbolic equilibrium states. This very elegant theory allows one to demonstrate and code chaotic trajectories, under quite general conditions. In the same way that Melnikov's theory could be cast in the language of defects, we now perform the same analysis in the Shilnikov case.

We first give, without any particular model in mind, the equation to be satisfied by the Shilnikov defects, study it and then consider two models of physical interest which naturally display this kind of chaotic behavior. Because of the asymptotic oscillating character of the defects the interacting force now reads

$$f(d) = g \cos(\beta d) \exp(-\alpha d), \tag{16}$$

where α and β are respectively the real and imaginary parts of the eigenvalues associated with the complex hyperbolic equilibrium states. We consider the case where the only possible configurations consist of pairs of defects and anti-defects. This situation is quite analogous to what happens in the model previously discussed. There the dynamics was attractive, because of the opposite sign of the topological charge of two nearby defects. In the presence of spatial oscillations two defects of opposite topological charges can have locally repulsive interactions. The result of these oscillations is stabilization of a static configuration of defects and creation of chaotic states. Using the method of Sect. 2 we obtain that the overdamped defect dynamics obeys a gradient-type dynamical system whose potential reads

$$\mathcal{V}(x_j) = \frac{g}{(\alpha^2 + \beta^2)} \sum_j \exp(-\alpha(x_{j+1} - x_j))$$

$$- (\beta \sin(\beta(x_{j+1} - x_j)) - \alpha \cos(\alpha(x_{j+1} - x_j))). \tag{17}$$

A direct minimization of $\mathcal{V}$ shows that local minima exist. Actually this potential is unbounded from below, in such a way that the eventual evolution of the system cannot be captured in the defect dynamics picture. The natural tendency of such systems is again to eliminate defects by pair annihilation. The major difference between normal and Shilnikov defects is the possibility of constructing metastable states of periodic, quasi-periodic and even chaotic distributions of defects. Since our interest lies mainly in non-equilibrium systems, such metastable configurations, in practice, can be long lived , and thus play an important role in the long time behavior of the system. The nature of the chaotic behavior for Shilnikov type defects is quite different from the one associated with the presence of a "lattice". The periodic nature of the interaction leads to an infinite sequence of possible positions for the defects. Chaotic configurations arise by picking these positions at random.

We now discuss two models of physical interest displaying this kind of defect. First, an obvious generalization of (12) (without external field) reads

$$A_t = \mu A + \nu A_{xx} - A_{xxxx} - A^3. \tag{18}$$

Adding the fourth-order derivative to (12) only makes sense when the coefficient in front of the diffusion term is of the order $\sqrt{\mu}$. In this case the behavior of the

298

system has to be represented in the two-parameter space $\mu - \nu$. For ν positive and large enough, μ being positive, (18) can be reduced to (12). Thus, in this limit, (18) displays normal defects. Let us consider for a moment the differential equation obtained by dropping the left hand side of (18). It is straightforward to check that, by crossing the parabola $\mu = \nu^2/8$, the topological nature of the equilibrium solutions $\pm\sqrt{\mu}$ changes from real hyperbolic to complex hyperbolic. Thus the defects, inside the parabola, present spatial oscillations implying (through (7)) an interacting force given by (16).

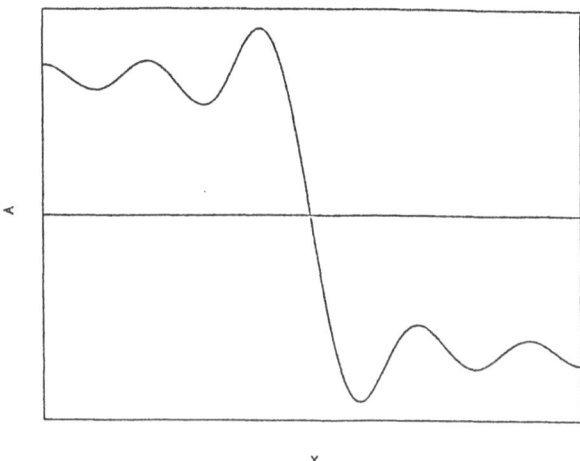

x

Figure 2
Defect-type solution of the Lifshitz equation (19) inside the oscillatory domain.

From a physical point of view (18) describes the dynamics associated with a so-called Lifshitz point [12] . We now turn our attention to a model which contains richer behavior. Let A be a complex field obeying the following partial differential equation:

$$A_t = (\mu - q^2)A + 2iqA_x + A_{xx} - |A|^2 A + \alpha \bar{A}^{n-1}. \tag{19}$$

This model has been used in the context of commensurate-incommensurate transitions in both equilibrium [12] and non-equilibrium [13 -15] situations. In the limit $\mu \ll q^2, \alpha^{2/4-n}$, (19) can be reduced to a phase-type dynamics described by the overdamped sine-Gordon equation

$$\phi_t = \phi_{xx} - \alpha \mu^{n-2/2} \sin(n\phi). \tag{20}$$

It displays kink and anti-kink solutions, periodic arrays of kink or anti-kink solutions and periodic arrays of kink anti-kink solutions. The analysis of these various solutions can be carried out in the same way as before. Away from the "phase approximation" the kinks present spatial oscillations, which results, in particular, in the appearence of chaotic behavior. This kind of chaotic behavior is likely to be observed in experiments of the type performed by Lowe and Gollub [9] for which model (19) has been devised.

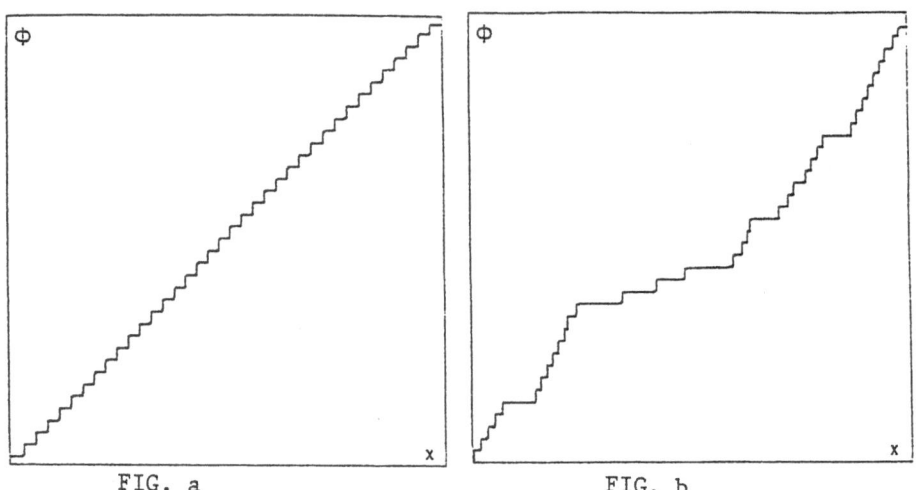

FIG. a   FIG. b

Figure 3

Sketch of a kink array solution of (19) reconstructed via a numerical simulation of the defect dynamics associated with the potential $\mathcal{V}$. a) Outside the oscillatory domain. b) Inside the oscillatory domain.

4. CONCLUSION

We have illustrated in this work the two basic mechanisms for spatial complexity in extended systems with codimension-one defects. They are closely related to the mechanisms of horseshoe formation in conservative dynamical systems with one and a half [17] and two degrees of freedom. The conservative nature of these dynamical systems is a direct consequence of the parity symmetry unbroken by a static pattern. Let us also remark that these two basic mechanisms can be combined. In particular, because of its translational invariance (19) does not exhibit pinning. One way to incorporate this phenomenon consists in adding a small forcing term, which mimics the effects of the periodic grid lost in the large scale approximation which lead to

(19). In the phase regime it led to the following model, which displays pinning, higher-order locked states, quasi-crystal effects and chaotic behavior:

$$\phi_t = \phi_{xx} - \alpha \mu^{n-2/2} \sin(n\phi) + R(x, \phi), \tag{21}$$

where R is a periodic function of x. This kind of model is likely to describe some of the transitions experimentally already observed by Lowe and Gollub. All the models considered in this paper have a variational nature. They can be cast into the form $A_t = -\delta \mathcal{F}/\delta A$ where the Lyapunov functional $\mathcal{F}$ plays the role of a free energy. Actually this is not a crucial fact. Similar equations for the dynamics of the defects can be derived for non-variational models. The overdamped character of this dynamics is just related to the dissipative character of the basic equations describing the physical system, and to the fact that the only continous symmetry is the translational one. Conservative systems, as for example the sine-Gordon equation, or Galilean invariant systems, would lead to inertial dynamics for the defects. For mixed systems, inertia and damping are expected. An interesting spatio-temporal complex dynamics should result from the interaction between Shilnikov kinks. The final question concerns fully developed spatio-temporal turbulence. We have described here strong spatial chaotic behavior and hope to consider, in the near future, their weak coupling with time-dependent effects. Fully developed spatio-temporal chaos involves more complex phenomena as for example creation and annihilation of pair defects, which cannot be taken into account with the methods used here.

Acknowledgments: This work has been partly supported by the CNRS through an ATP, the DRET and the CPAI.

REFERENCES

1. H. L. Swinney and J. P. Gollub (eds), Hydrodynamic Instabilities and the Transition to Turbulence, Topics in Appl. Phys., Vol. 45, 2nd ed. (Springer, Berlin, Heidelberg 1985).

2. P. C. Hohenberg and J. S. Langer, J. Stat. Phys. 28, 193 (1982); J. E. Wesfreid and S. Zaleski (eds), Cellular Structures in Instabilities, Lecture Notes in Phys., Vol. 210 (Springer, New York, 1984).

3. P. Coullet, C. Elphick and D. Repaux, Phys. Rev. Lett. 58, 431 (1987).

4. J. Guckenheimer and P. Holmes, <u>Nonlinear Oscillations, Dynamical Systems and Bifurcations of Vector Fields</u> (Springer, Berlin, Heidelberg 1983), and references quoted therein.

5. S. Aubry, in <u>Solitons and Condensed Matter Physics</u>, Springer Ser. Solid-State Phys., Vol. 8, edited by A. R. Bishop and T. Schneider (Springer, Berlin, 1979).

6. P. Bak, Rep. Prog. Phys. <u>45</u>, 587 (1982), and references quoted therein.

7. A homoclinic (heteroclinic) orbit is a solution which bi-asymptotically connects the same equilibrium solution (two different equilibrium solutions).

8. Two equilibrium solutions are equivalent if they belong to the same orbit of the discrete symmetry group which leaves the system invariant. In the cases considered here these are reflection symmetry and discrete translations. When two solutions are not equivalent in this sense, the defect connecting them is generally mobile and rather called a front.

9. M. Lowe and J. Gollub, Phys. Rev. <u>A31</u>, 3893 (1983).

10. K. Kawasaki and T. Ohta, Physica <u>116A</u>, 573 (1982).

11. F. Nabarro, <u>Theory of crystal dislocation</u> (Oxford University Press, Oxford 1967).

12. R. M. Hornreich, M. Luban and S. Shtrikman, Phys. Rev. Lett. <u>35</u>, 1678 (1975).

13. A. E. Jacobs and B. Walker, Phys. Rev. <u>21</u>, 4132 (1980).

14. P. Coullet, Phys. Rev. Lett. <u>56</u>, 724 (1986).

15. P. Coullet and D. Repaux, Europhys. Lett. <u>3</u>, 573 (1987).

16. E. Curado and C. Elphick," Effects of an almost resonant spatial thermal modulation in the Raleigh-Bénard problem: quasiperiodic behavior " , to appear in J. Phys. A.

17. The terminology used comes from dynamical systems theory. One and a half degrees of freedom Hamiltonian systems are systems with one degree of freedom periodically driven by an external force.

Nonlinear Gravity Waves and Resonant Forcing

K. Kirchgässner

Math. Institut A, Universität Stuttgart,
D-7000 Stuttgart 80, Fed. Rep. of Germany

1. Steady Water Waves and Resonant Forcing

It is well known that gravity supports nonlinear waves on the surface
of an inviscid fluid layer. Cnoidal (periodic) resp. solitary waves
are the most commonly observed. Here we discuss their behavior under
external pressure waves travelling with the same speed as the ground
wave (resonant forcing). This nonlinearly resonant phenomenon has at-
tracted much attention recently and has been analyzed for model equations by describing the far-field behavior in case of almost resonant
forcing [1], [2].

We describe a new mathematical approach which gives the totality
of possible solutions for the full Euler equations under the sole assumption of moderate amplitude flow. The steady problem is considered
"dynamically" in the unbounded space variable. Then, a wave is a
bounded orbit in a function space whose elements live on the cross-
section of the layer. Cnoidal waves correspond to closed orbits, solitary waves to homoclinic orbits.

Periodic external forces may break e.g. the homoclinic orbit and
yield transverse homoclinic points and thus lead to spacelike chaotic
behavior. Similarly one may ask how the fluid reacts to resonant
pressure waves having compact support, in particular, how far up-
stream the external pressure can be felt, and how this distance depends on the pressure amplitude.

The mathematical analysis is based on some recent progress on nonlinear elliptic systems in unbounded cylindrical domains which have
been developed by the author [3], and in particular by A. Mielke [4],
[5] who has given the strong results needed for the case under consideration. A general survey can be found in [6]. We treat steady
two-dimensional surface gravity-waves under two types of forcing:
local forcing - i.e. compact support - and global (periodic) forcing.

For the sake of simplicity we suppress obvious generalizations such
as the influence of capillarity or more general types of external
forcing. The interested reader is referred to [6].

2. The Mathematical Model and its Reduction

A layer of mean depth h filled with an inviscid fluid is considered
under the influence of gravity g and an external pressure distribu-
tion εp_0. In the unforced case ($\varepsilon = 0$) nonlinear waves exist on the
surface of the layer travelling with permanent form and constant
speed c from right to left. We assume that p_0 is a travelling wave
with speed c, and thus, the model is steady in a moving coordinate
system. Assuming irrotational motion for simplicity we obtain the
following system describing the physical situation

$$\operatorname{div}\underline{v} = \operatorname{curl}\underline{v} = 0 \qquad , \quad 0 < \eta < z(\xi)$$

$$v_2 = 0 \qquad\qquad , \quad \eta = 0$$

$$\tfrac{1}{2}|\underline{v}|^2 + p + \lambda z = \text{const}$$

$$\qquad\qquad\qquad , \quad \eta = z(\xi)$$

$$v_1 \partial_\xi z - v_2 = 0$$

(2.1)

where all quantities are in dimensionless form; $\underline{v} = (v_1, v_2)'$ is the
velocity, p the pressure, $z(\xi)$ the free surface, ρ the constant density and $\lambda = gh/c^2 = $ (Froude number)$^{-2}$. We seek solutions which are
bounded in the flow domain $\Omega = \{(\xi, \eta) \, / \, \xi \in \mathbb{R}, 0 < \eta < z(\xi)\}$. To obtain
waves of special forms, (2.1) has to be supplemented by boundary con-
ditions at $\xi = \infty$. For the forcing term we treat two cases:

1. Local forcing of gravity waves. Here, p_0 has compact support (ε
   measures the amplitude).

2. Global forcing of gravity waves; i.e. p_0 is periodic (the period
   being considered as a parameter).

According to our normalization, the flux Q equals one. Define the
stream-function Ψ by

$$\partial_\xi \Psi = -v_2 \quad , \qquad \partial_\eta \Psi = v_1 \quad , \qquad \Psi|_{\eta=0} = 0$$

(2.2)

then $\Psi(\xi, z(\xi)) = Q = 1$ and $\Psi = \eta + \psi$ where ψ can be considered small compared to η. To simplify (2.1), we apply two transformations which are
almost identical whenever ψ is C^1-small, thus defining global diffeomorphisms. Set

$$x = \xi \quad , \quad y = \Psi(\xi,\eta) = \eta + \psi(\xi,\eta)$$

$$W_1 = \frac{1}{2}((1+v_1)^2 + v_2^2 - 1) \quad , \quad W_2 = v_2 v_1^{-1} \tag{2.3}$$

then (2.1) is transformed into

$$\partial_x \underline{W} = K(\underline{W}) \partial_y \underline{W} \quad \text{in } D = \mathbb{R} \times (0,1)$$

$$W_1 + \lambda Z + \varepsilon p_0 = 0 \quad , \quad \partial_x Z - W_2 = 0 \quad , \quad \text{for } y = 1 \tag{2.4}$$

$$W_2 = 0 \quad , \quad \text{for } y = 0.$$

Here

$$Z(x) = z(\xi) - 1 = [\frac{1}{g}] - 1 \quad , \quad \text{where } [V] = \int_0^1 V \, dy \, ,$$

$$K(\underline{W}) = \begin{pmatrix} gW_2 & -g^3 \\ g^{-1} & gW_2 \end{pmatrix} \quad , \quad g(\underline{W}) = \left(\frac{1+2W_1}{1+W_2^2} \right)^{1/2} . \tag{2.4a}$$

It is easy to show that (2.4) represents a quasilinear elliptic system in the sense of Agmon [7]. We write it formally as a dynamical system by introducing α, the trace of W_1 on $y = 1$, as a new variable. Differentiating the first boundary condition at $y = 1$ in (2.4) yields

$$\partial_x \underline{w} = \widetilde{A}(\lambda) \underline{w} + \widetilde{F}(\varepsilon,x,\underline{w}) \quad , \quad \underline{w} = \begin{pmatrix} \alpha \\ W_1 \\ W_2 \end{pmatrix} \tag{2.5}$$

where

$$\widetilde{A}(\lambda) \underline{w} = \begin{pmatrix} -\lambda W_2(\cdot,1) \\ J \partial_y \underline{W} \end{pmatrix} \quad , \quad J = \begin{pmatrix} 0 & -1 \\ 1 & 0 \end{pmatrix}$$

$$\widetilde{F}(\varepsilon,x,\underline{w}) = \begin{pmatrix} -\varepsilon p_0(x) \\ (K(\underline{W})-J) \partial_y \underline{W} \end{pmatrix} . \tag{2.6}$$

$\widetilde{A}(\lambda)$ determines the linearisation of (2.4) about $\underline{W} = 0$. For $\varepsilon = 0$, system (2.5) has a trivial symmetry given by

$$\widetilde{A}(\lambda) R = -R\widetilde{A}(\lambda) \quad , \quad \widetilde{F}(0,\cdot,R\underline{w}) = -R\widetilde{F}(0,\cdot,\underline{w}) \tag{2.7}$$

where R is the reflection

$$R = \begin{pmatrix} 1 & 0 & 0 \\ 0 & 1 & 0 \\ 0 & 0 & -1 \end{pmatrix} .$$

In analogy to ordinary differential equations we call (2.5) *reversible* for $\varepsilon = 0$, i.e. if (2.7) holds. Reversibility will be broken by the external forcing ($\varepsilon \neq 0$), and it is this symmetry breaking which causes the interesting phenomena to occur.

A few words to the mathematics. We consider (2.5) in $C_b^1(\mathbb{R}, X) \cap C_b^0(\mathbb{R}, D(A))$ where "b" indicates boundedness, $X = \mathbb{R} \times (L_2(0,1))^2$, $D(A) = \mathbb{R} \times (H^1(0,1))^2 \cap \{W_1(1) = \alpha, W_2(0) = 0\}$. The spectrum of $\tilde{A}(\lambda)$ consists only of eigenvalues of finite multiplicities. According to (2.6), if σ is an eigenvalue then so is $-\sigma$. Thus, nonreal and non-imaginary eigenvalues appear in quadruples $\pm\sigma$, $\pm\bar{\sigma}$ (c.c.). The eigenvalue relation is

$$\lambda \sin \sigma = \sigma \cos \sigma . \tag{2.8}$$

Here, λ is real and positive, $\sigma \in \mathbb{C}$. It is readily seen that $\sigma = 0$ is a simple eigenvalue for all $\lambda \neq 1$ with eigenfunction $\underline{\varphi}_0 = (1,1,0)'$. For $\lambda < 1$ all other eigenvalues are real and simple. For $\lambda > 1$ two eigenvalues are not. They are imaginary, conjugate complex and simple. For $\lambda = 1$, the eigenvalue 0 has geometric multiplicity 1 and algebraic multiplicity 3, with the generalized eigenfunctions $\underline{\varphi}_1 = (0,0,-y)'$, $\underline{\varphi}_2 = (-\frac{1}{2}, -\frac{1}{2}y^2, 0)'$ complementing $\underline{\varphi}_0$.

For $0 < \lambda < 1$, $\underline{w} = 0$ is an isolated solution. Bifurcations from $\underline{w} = 0$ exist for each $\lambda \geq 1$. Pars pro toto, we study the case $\lambda = 1$, which is the most interesting one. The method we describe could be used to discuss the case $\lambda > 1$ as well. Write

$$\hat{A} = \tilde{A}(1) \quad , \quad F = \tilde{F} + \tilde{A}(\lambda) - \hat{A} \quad , \quad \lambda = 1 - \mu \tag{2.9}$$

and determine projections into the span of $\underline{\varphi}_0, \underline{\varphi}_1, \underline{\varphi}_2$ commuting with $\hat{A}$ in X. Such a projection is given by

$$S_0 \underline{w} = \sum_{i=0}^{2} a^j \underline{\varphi}_j \quad , \quad a^j = (\underline{w}, \psi^j) \quad ,$$

where $\underline{\psi}^0 = (-\frac{3}{10}, \frac{9}{5} - \frac{3}{2}y^2, 0)$, $\underline{\psi}^1 = (0,0,-3y)$, $\underline{\psi}^2 = (-3,3,0)$ are such that $(\underline{\varphi}_j, \psi^k) = \delta_j^k$. Set $\underline{w}_0 = S_0 \underline{w}$, $\underline{w}_1 = \underline{w} - \underline{w}_0$ then (2.5) reads - observing (2.9) -

$$\partial_x \underline{w}_0 = \hat{A}_0 \underline{w}_0 + F_0(\underline{\lambda}, \cdot, \underline{w}_0 + \underline{w}_1) \tag{2.10a}$$

$$\partial_x \underline{w}_1 = \hat{A}_1 \underline{w}_1 + F_1(\underline{\lambda}, \cdot, \underline{w}_0 + \underline{w}_1) \tag{2.10b}$$

where $\underline{\lambda} = (\lambda, \varepsilon)$, $\hat{A}_j = S_j \hat{A}$, $F_j = S_j F$. Since the spectrum of $\hat{A}_1$ is strictly bounded away from the imaginary axis, (2.10b) can be inverted yielding, for small $\underline{w}_0$, $\underline{w}_1$ in functional dependence of $\underline{w}_0$. However, one can prove more. Identify the span of $\varphi_0, \varphi_1, \varphi_2$ with $Z_0 \simeq \mathbb{R}^3$ and $Z_1 = S_1 \cap D(A)$. Restricting the analysis to solutions $(\underline{w}_0, \underline{w}_1)$ of (2.10) which are, for all $x \in \mathbb{R}$, in some suitably small neighborhood of O in $Z_0 \times Z_1$, then $\underline{w}_1$ is a $\underline{\text{pointwise}}$ function of $\underline{w}_0$; i.e. there exists a smooth function $h(\underline{\lambda}, x, \underline{w}_0)$ mapping $\Lambda \times \mathbb{R} \times Z_0$ into Z_1, Λ being a neighborhood of $\lambda = 1$, $\varepsilon = 0$, such that

$$\underline{w}_1(x) = h(\underline{\lambda}, x, \underline{w}_0(x)) \tag{2.11}$$

holds for all solutions $\underline{w}$ with $\underline{w}(x) \in U$, $x \in \mathbb{R}$. We call h the *reduction function*.

This by no means trivial result is obtained in analogy to the center-manifold for ordinary differential equations. The proofs can be found in Mielke [4]. We need subsequently only the following facts

$$h(\underline{\lambda}, \underline{w}_0) = h_0(\lambda, \underline{w}_0) + h_1(\varepsilon, \cdot, \underline{w}_0) \tag{2.12}$$

where $h_0 = O(|\lambda - 1| \|\underline{w}_0\| + \|\underline{w}_0\|^2)$, $\| \cdot \|$ any norm in Z_0, $h_1 = O(\varepsilon)$; h_0 is independent of x and satisfies

$$h_0(\lambda, R_0 \underline{w}_0) = R_1 h_0(\lambda, \underline{w}_0) \tag{2.13}$$

where $R_0 = R|_{Z_0}$, $R_1 = R - R_0$,

$$R_0 = \begin{pmatrix} 1 & 0 & 0 \\ 0 & -1 & 0 \\ 0 & 0 & 1 \end{pmatrix}. \tag{2.13a}$$

Moreover, if p_0 is periodic with period $d > 0$ then so is h_1. If p_0 has compact support, and if $\underline{\lambda} \in \Lambda$, $\underline{w}_0 \in U$ then, for every $\gamma > 0$, there exists a constant $c(\gamma, \Lambda, U)$ such that

$$\| h(\lambda, x, \underline{w}_0) \|_{D(A)} \leqslant c e^{-\gamma |x|}, \quad x \in \mathbb{R} . \tag{2.14}$$

3. The Reduced Equations and Results

From (2.10) and (2.11) we obtain the reduced system of ordinary differential equations

$$\partial_x \underline{w}_0 = \hat{A}_0 \underline{w}_0 + F_0(\underline{\lambda}, \cdot, \underline{w}_0 + h(\underline{\lambda}, \cdot, \underline{w}_0)) \tag{3.1}$$

which is of dimension 3. If we identify $\underline{w}_0$ with its Fourier-coefficients (a_0, a_1, a_2) we obtain explicitly, using (2.13),

$$a_0' = a_1 (1 + \frac{3\mu}{10} + \frac{39}{10} a_0 - \frac{3}{20} a_2 + r_{00}(\mu,\underline{a}))$$

$$+ \frac{3\varepsilon}{10} p_0' + r_{01}(\mu,\varepsilon,x,\underline{a}) \tag{3.2a}$$

$$a_1' = a_2 + \frac{3}{10} a_2^2 - a_0 a_2 - a_1^2 + r_{10}(\mu,\underline{a})$$

$$+ r_{11}(\mu,\varepsilon,x,\underline{a}) \tag{3.2b}$$

$$a_2' = a_1 (3\mu - \frac{1}{2} a_2 + 9a_0 + r_{20}(\mu,\underline{a}))$$

$$+ 3\varepsilon p_0' + r_{21}(\mu,\varepsilon,x,\underline{a}) \tag{3.2c}$$

where $\lambda = 1 - \mu$, $p_0' = \partial_x p_0$, and where

$$r_{k0} = O(|\mu||\underline{a}| + |\underline{a}|^2) \quad , \quad k = 0,2$$

$$r_{10} = O(|\mu||\underline{a}|^2 + |\underline{a}|^3) \quad \text{are even in } a_1 , \tag{3.3}$$

$$r_{k1} = O(\varepsilon|\underline{a}| + \varepsilon\mu) \quad , \quad k = 0,1,2 .$$

Due to the fact that we have used one of the basic equations in a differentiated form, the above equations are not independent. This can be seen by the validity of

$$W_1(1,\cdot) + \lambda([\frac{1}{g}] - 1) + \varepsilon p_0 = C \tag{3.4}$$

which follows from (2.4) and (2.4a). First we observe that

$$[\frac{1}{g}] - 1 = -[W_1] + \frac{1}{2}[W_2^2] + \frac{3}{2}[W_1^2] + O(|\underline{W}|^3) . \tag{3.5}$$

Moreover we have from the special form of the φ_j

$$W_1 = a_0 - \frac{1}{2} y^2 a_2 + h^1 ,$$

$$W_2 = -y a_1 + h^2 \tag{3.6}$$

where $h = (h^0, h^1, h^2)'$. An elementary calculation leads to

$$-\frac{1}{6}(2+\mu)a_2 + \mu a_0 + \frac{1}{6} a_1^2 + \frac{3}{2} a_0^2 - \frac{1}{2} a_0 a_2 + \frac{3}{40} a_2^2$$

$$+ h^1(1) - [h^1] + \varepsilon p_0 + O(|\underline{a}|^3 + \mu|\underline{a}|^2 + \varepsilon\mu + \varepsilon|\underline{a}|) = C . \tag{3.7}$$

Now, (3.2a,b,c) imply (3.7) for some C. On the other hand, near $\underline{a} = \underline{0}$, $\varepsilon = \mu = C = 0$, we can express a_2 as a unique function of a_0, a_1, μ, ε

308

and C. Thus we can replace (3.2c) by (3.7) and reduce (3.2) to a second order equation.

Another consideration concerns the term $h(1) - [h^1]$ in (3.7). It seems that one should calculate h up to order $\varepsilon + \mu|\underline{a}| + |\underline{a}|^2$. However, since h maps into Z_1, and Z_1 is the orthogonal complement of the ψ^k, we obtain, using $(\psi^2, h) = 0$, $h(1) - [h^1] = 0$. Concerning the free constant C, we observe that $C = 0$ for $\varepsilon = 0$, if $\underline{a}$ tends to 0 for $\xi \to -\infty$. This includes homoclinic solutions for $\varepsilon = 0$ as well as all solutions for $\varepsilon \neq 0$ if p_0 has compact support. In the global case, p_0 being periodic, C may be of order ε. But this constant can be incorporated into p_0. Therefore we set $C = 0$.

Using (3.7) for $C = 0$ and $h^1 - [h^1] = 0$ to eliminate a_2 in (3.2) we obtain

$$a_0' = a_1 (1 + \frac{3\mu}{10} + \frac{39}{10} a_0 + \tilde{r}_{00}) + \frac{3\varepsilon}{10} p_0' + r_{01}$$

$$a_1' = 3\mu a_0 - \frac{1}{2} a_1^2 + \frac{9}{2} a_0^2 + \tilde{r}_{10} + 3\varepsilon p_0 + r_{11}$$
(3.8)

where $\tilde{r}_{j0}$ is of the same order as r_{j0}, and r_{j1} is as in (3.3).

Scaling as follows

$$a_j(x) = |\mu|^{1+j/2} A_j(\xi) \quad , \quad \xi = |\mu|^{1/2} x \; , \quad j = 0,1,2$$
(3.9)

leads (3.8) to

$$A_0'' = 3 \mathrm{sign}(\mu) A_0 + \frac{9}{2} A_0^2 + O(\mu) + \frac{3\varepsilon}{\mu^2} P_0 (1 + O(\mu + \varepsilon)).$$
(3.10)

The bounded solutions of this simple equation determine all solutions of moderate amplitude of the original problem. In order to obtain an asymptotic formula for the free surface $z = 1 + Z$ we conclude from (3.7) and (3.9) $a_2 = O(\mu a_0 + a_1^2 + \varepsilon)$. Moreover (2.4a) and (3.6) yield

$$Z(x) = -|\mu| A_0(|\mu|^{1/2} x) + O(\mu^2).$$
(3.11)

Proposition 3.1:

Consider the unforced case $\varepsilon = 0$. Then, for $\mu > 0$ ($\lambda < 1$) sufficiently small, there exists a solitary wave of elevation, which is unique up to phase shifts in x.

For each $\mu < 0$ ($\lambda > 1$), $|\mu|$ sufficiently small, the trivial solution bifurcates into a two-parameter family of cnoidal waves.

The form of the free surface corresponding to these solutions is given - in lowest order - by (3.10) for $\mu = \varepsilon = 0$ and (3.11).

The proof is immediate, if one observes that in view of the reversibility of (3.10) for $\varepsilon = 0$, the term $O(\mu)$ in (3.10) depends on A_0, A_0'' and on $A_0'^2$. Thus, every solution for $\mu = 0$ can be continued to a solution of the full equation (c.f. [6]).

Proposition 3.2:

Consider the local forced case, i.e. $\varepsilon \neq 0$, p_0 periodic with period $d > 0$. Assume further that

$$<p_0> := \int_{-\infty}^{\infty} p_0(x)\, dx \neq 0 \; .$$

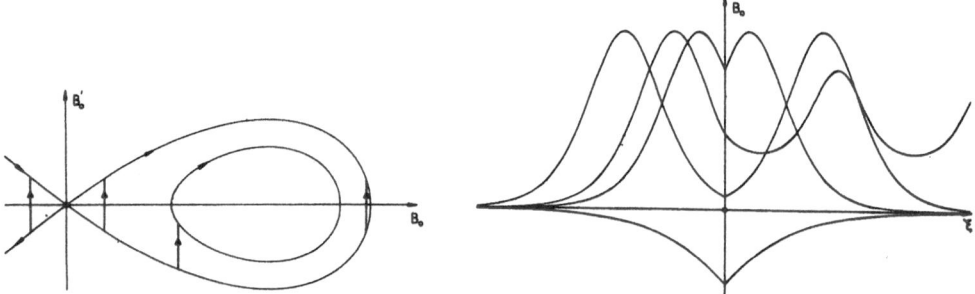

Fig. 1. Solutions of limiting equation (3.12) for $\mu > 0$, $0 < \eta < 4/3\sqrt{3}$

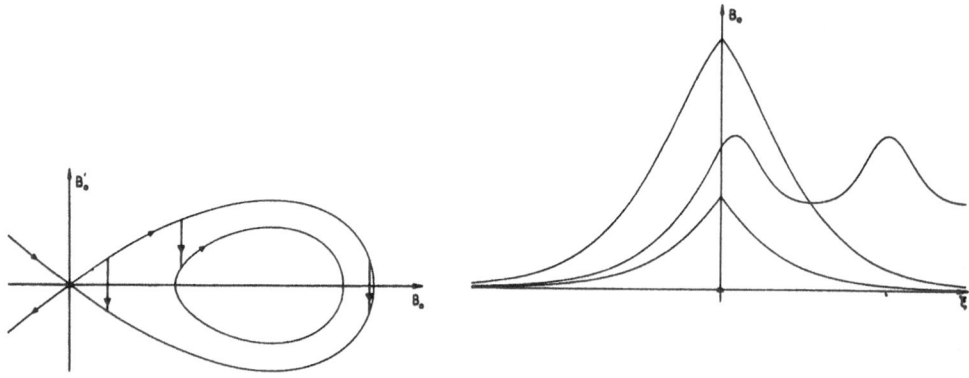

Fig. 2. Solutions of limiting equation (3.12) for $\mu > 0$, $-4/3\sqrt{3} < \eta < 0$

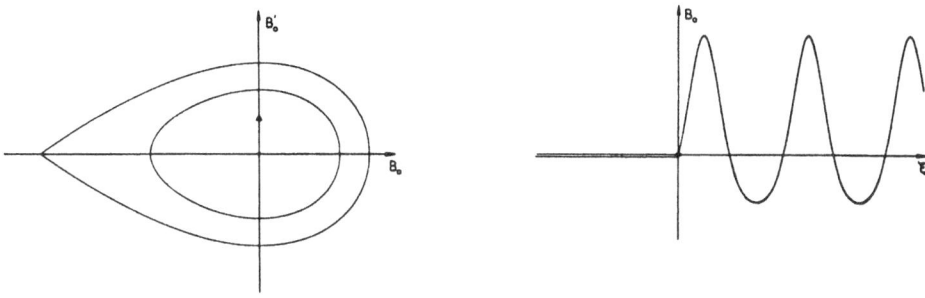

Fig. 3. Solutions of limiting equation (3.12) for $\mu < 0$, $\eta \in (0, 4/3\sqrt{3})$

Then, all solutions of moderate amplitude are determined, in lowest order in μ, by the limiting equation

$$A_O'' - 3 \operatorname{sign}(\mu) A_O - \frac{9}{2} A_O^2 - \eta \delta_O = 0 \qquad (3.12)$$

where $\eta = 3\varepsilon \langle p_O \rangle / |\mu|^{3/2}$; δ_O is the Dirac-distribution with support in 0.

The possible solutions of (3.12) are given in Figs. 1-3. They are smooth outside $x = 0$ and have a jump in the first derivative there.

The proof is by no means trivial. It was first given in [8], and a coarser version with simpler proof in [6]. The latter also contains a proof for the following proposition covering the global case of forcing.

<u>Proposition 3.3:</u>

For the global case, i.e. $\varepsilon \neq 0$, p_O periodic with period $d > 0$, assume $\mu > 0$ ($\lambda < 1$) and denote by Q_O the unique even homoclinic solution of (3.10) for $\mu = \varepsilon = 0$. Define

$$k_O(\beta) := \int_{-\infty}^{\infty} Q_O'(\xi) P_O(\xi - \beta)\, d\xi \quad , \quad \beta \in I\!R .$$

If, for some β_O, k_O has a simple zero, then there exists a transverse homoclinic point of (3.10) relative to A_O^*, the unique d-periodic perturbation in η of $A_O = 0$, where $\eta = 3\varepsilon/\mu^2$.

In particular, if $p_O = c \cos(2\pi x/d)$, then the above condition (Melnikov-condition) is satisfied for all sufficiently large d.

4. Literature

1. T.R. Akylas: J. Fluid Mech. 141 (1984) 455-466

2. R.H.J. Grimshaw, N. Smyth: Res. Rep. No. 14 (1985), Univ. Melbourne

3. K. Kirchgässner: J. Diff. Eqn. 45 (1982) 113-127

4. A. Mielke: to appear in Math. Meth. Appl. Sci. (1987)

5. A. Mielke: to appear in Math. Annalen (1987)

6. K. Kirchgässner: to appear in Adv. Appl. Mech. (1987)

7. T. Burak: Israel J. Math. 12 (1972) 79-93

8. A. Mielke: J. Diff. Eqn. 65 (1986) 89-116

Transitions to Chaos in a Finite Macroscopic System:
Direct Numerical Simulations
Versus Normal Form Predictions

A. Arneodo[1] *and O. Thual*[2]

[1]Laboratoire de Physique Théorique, Université de Nice,
 Parc Valrose, F-06034 Nice Cedex, France
[2]Centre National de Recherches Météorologiques,
 42 Av. Coriolis, F-35057 Toulouse Cedex, France

I INTRODUCTION

One of the major complications encountered in either experimental or numerical studies of the scenarios to chaos [1-10] in finite macroscopic systems is
the global character of these phenomena. As we move a constraint, the qualitative changes observed in the dynamical evolution of the system affect solutions
which are increasingly nonlinear and thus elude any theoretical analysis. For
years the temporal evolution of macroscopic continua has been investigated
through the study of artificially truncated models. For instance the Lorenz
equations [11] are a minimal truncation of the fluid equations which embody the
essential features of the transition to steady convective rolls. However, these
equations have been studied as the prototypic Ordinary Differential Equations
(O.D.E.'s) which display strange-attractor-like behavior [12]. The physical relevance of such a dynamical complexity is rather questionable since higher order
truncations lead to significantly different behaviors [13,14]. It has actually
been emphasized that for similar conditions, no strong chaos is observable in
the full Boussinesq system with sufficiently high resolution [15]. This raises
a serious problem in defining the domain of validity of the Galerkin truncation.
Such an arbitrary approach has also been used to study double convection with
more or less success. In the case of thermohaline convection, the numerical investigation of the lowest nontrivial truncated system displays chaotic behavior
very much like the ones found in the original Partial Differential Equations
(P.D.E.'s) [16-19]. For magnetoconvection, however, direct numerical simulations have failed to reproduce the chaotic dynamics predicted by the Galerkin
approximation [20-22].

A way out to this difficulty consists in reducing some aspects of the global dynamics to a local problem. The strategy we will adopt throughout this
paper relies on the fact that many transitions which are global as a function of
one control parameter become local when more than one such parameter is considered. Thus, in positioning the system near some polycritical surfaces in the
parameter space (by analogy with polycritical points in phase transitions [23])
on which several instabilities occur simultaneously [24,25], one can use the
center manifold theorem and the normal form theory [26-28] to reduce the
P.D.E.'s which model such continuous systems to a set of nonlinearly coupled
O.D.E.'s. If the number of competing instabilities is n, we usually speak about
a situation of criticality n. In order to cross the polycritical surface we
actually need to adjust n parameters. If no further conditions are required,
the codimension is equal to the criticality [28]. In principle one may reason-
nably expect from these methods only qualitative information on the nature of
the solutions in a neighborhood of such multidegenerate situations, the extent of
which is again difficult to determine theoretically. Generally one restricts
the reduction to an amplitude equation which is truncated to its lowest order
and one may wonder about the robustness of the predictions to the presence of

higher order terms. As far as codimension—one situations are concerned, the
Landau equation [29,30] for a stationary instability ζ (with special symmetry)
and the Hopf equation for the oscillatory instability ω [31], the effects of
higher order terms have been shown to be unsignificant. Bogdanov [32] has
proved that such a result generalizes to codimension—two problems (ζ^2 instability [33]) e.g. double convection [24,34,35]; the corresponding normal form is a
second-order O.D.E. leading to a two-dimensional (2-D) dynamics which can be
classified in the neighborhood of (bi)criticality. The asymptotic solutions are
steady states and periodic orbits, with the possibility that such solutions col-
lide to give rise to homoclinic or heteroclinic connections. The cascade of bifurcations leading to chaos observed in direct numerical simulations of thermohaline convection [18,19] does not enter this local picture. To apprehend such
a complicated dynamical evolution with perturbative technique one needs to consider higher degenerated situations [25,36].

    The aim of this paper is to illustrate the power of such a local approach
to study the transition to chaos in a multi-diffusive system. In section II, we
set up the P.D.E.'s which describe a fluid subject to triple convection and we
summarize the analytical derivation of the amplitude equations for the three
competing instabilities involved in this finite macroscopic system as originally
performed· in [37,38]. In section III, we emphasize that in the neighborhood of
the polycritical surface, the numerical solutions of these truncated equations
exhibit period-doublings and strange attractors, the nature and origin of which
can be understood in terms of the chaotic behavior which occurs on the way to
some homoclinic (heteroclinic) situation as shown by Shil'nikov [39]. We also
point out that this normal form locally accounts for different scenarios to weak
turbulence e.g. intermittency [40] and soft-mode instability (secondary Hopf
bifurcation [28,41]) leading to quasiperiodicity [42]. The delicate question
again is the influence of higher order perturbations on such structurally unstable systems. We answer this fundamental question in section IV through a comparison of the normal form predictions with the result of direct numerical investigations of the original P.D.E.'s. As already announced in a short note in
[43], there exists a remarkable agreement which attests to the robustness, at
least at a qualitative level, of these theoretical predictions in a wide range
of parameters at a distance from the polycritical situation. We conclude in
section VI.

II NORMAL FORM REDUCTION OF 2-D BOUSSINESQ THERMOHALINE CONVECTION IN A ROTATING LAYER

    Fluid dynamics is one of the branches of physics where a lot of effort has
been made to understand temporal chaos related to strange attractors. It has
long been suspected that convection with competing instabilities can manifest
chaotic dynamics [44]. Though the motion sometimes also has complicated spatial
structure, many numerical experiments are restricted to 2-D motion. Let us note
that recent laboratory experiments on convective chaos have been rendered nearly
two-dimensional by performing them on conducting fluids in strong horizontal magnetic field [45,46]. Here we will consider thermal stratification, saline
stratification and rotation as competing dynamical mechanisms. As already men-
tionned in [37], this study generalizes to semi-convection [47] and 2-D magnetoconvection with magnetic buoyancy under certain conditions [48].

II.1 Equations

    The equations of 2-D Boussinesq thermohaline convection in a plane-parallel
layer rotating about a vertical axis are [37,38]

$$\partial_t \Delta \Psi = \sigma \Delta^2 \Psi - \sigma R \partial_x \Theta + \sigma T S \partial_x \Sigma + \sigma^2 T \partial_z \Upsilon + J(\Psi, \Delta \Psi) \tag{1}$$

$$\partial_t \Theta = \Delta \Theta - \partial_x \Psi + J(\Psi, \Theta) \tag{2}$$

$$\partial_t \Sigma = \tau \Delta \Sigma + \partial_x \Psi + J(\Psi, \Sigma) \tag{3}$$

$$\partial_t Y = \sigma \Delta Y - \partial_z \Psi + J(\Psi, Y) \tag{4}$$

where $J(f,g) = \partial_x f . \partial_z g - \partial_z f . \partial_x g$, $\Delta = \partial_x^2 + \partial_z^2$. The motion is independent of y. The layer thickness d is the length unit while the time unit is d^2/κ_T where κ_T is the thermal diffusivity. Ψ is the stream function for the x and z velocities, Θ and Σ are the deviations of temperature and salinity from their static values, and Y is the y-velocity divided by $\sigma\sqrt{T}$. Let $(\Delta\rho/\rho)_T$ and $(\Delta\rho/\rho)_S$ be respectively the impressed density contrasts forced by the imposed temperature and salt vertical gradients. The parameter vector $\lambda=(R,S,T,\sigma,\tau,a)$ involves the Rayleigh numbers

$$R = (gd^3/\kappa_T \nu)(\Delta\rho/\rho)_T \qquad , \qquad S = (gd^3/\kappa_S \nu)(\Delta\rho/\rho)_S \tag{5}$$

while the Taylor number is defined from the rotation rate Ω about the z-axis:

$$T = (2\Omega d^2/\nu)^2 \ ; \tag{6}$$

$\sigma=\nu/\kappa_T$ and $\tau=\kappa_S/\kappa_T$ are the Prandtl and Lewis numbers respectively; a is the horizontal wave number of the box.

If we introduce the vectorial notation: $U(x,z,t) = \begin{Vmatrix} \Psi \\ \Theta \\ \Sigma \\ Y \end{Vmatrix}$ \qquad (7)

then the P.D.E.(1-4) can be reexpressed formally as

$$\partial_t U = M_\lambda U + N(U,U) \tag{8}$$

where M_λ and N are respectively a linear and a nonlinear operator independent of t. In order to make the reduction procedure to the normal form analytically tractable, the boundary conditions are conveniently chosen:

$$\Psi, \Sigma, \Theta, \partial_z Y, \partial_z^2 \Psi = 0 \text{ on } z=0,1$$

$$\Psi, \partial_x \Sigma, \partial_x \Theta, Y, \partial_x^2 \Psi = 0 \text{ on } x=0,\pi/a. \tag{9}$$

At this point let us note that these conditions will also make the direct numerical simulations of the P.D.E.'s (1)-(4) easily realisable.

II.2 Linear theory

The linear theory admits solutions of the form

$$U(x,z,t) = U_{mn}(t) * \Xi_{mn}(x,z) \tag{10}$$

where $U_{mn}(t)$ is a four-component vector and

$$\Xi_{mn}(x,z) = \begin{Vmatrix} \sin(max) \sin(n\pi z) \\ \cos(max) \sin(n\pi z) \\ \cos(max) \sin(n\pi z) \\ \sin(max) \cos(n\pi z) \end{Vmatrix} . \tag{11}$$

The operation A*B leads to a new four-component vector C, the components of which are the products of the respective components of A and B. The corresponding characteristic value equation reads

$$\prod_{m,n} \det(M_{mn} - s) = 0 \tag{12}$$

where M_{mn} is the matrix representation of the restriction of M_λ on Ξ_{mn}:

$$M_{mn}(\lambda) = \left\| \begin{array}{cccc} -\sigma q_{mn}^2 & ma\sigma R & -ma\sigma\tau S & -n\pi\sigma^2 T \\ -ma & -q_{mn}^2 & 0 & 0 \\ ma & 0 & -\tau q_{mn}^2 & 0 \\ -n\pi & 0 & 0 & -\sigma q_{mn}^2 \end{array} \right\| \qquad (13)$$

where

$$q_{mn}^2 = m^2 a^2 + n^2 \pi^2. \qquad (14)$$

As in Rayleigh-Bénard convection [49], the fundamental in z is the most unstable of the vertical modes, and we restrict our study to parameter values such that instability is encountered only for the mode $m=n=1$. For given m and n, the characteristic value equation (12) is a quartic. For $m\neq 1$ or $n\neq 1$, its roots have the property that $Re(s)<0$; they correspond to linearly damped modes. For $m=n=1$, this equation has one of its roots $s=-(1+2\sigma+\tau)q_{11}^2$ which is strictly real negative, therefore it can be reduced to a cubic:

$$s^3 + \mu_2 s^2 + \mu_1 s + \mu_0 = 0 \qquad (15)$$

where the parameters μ_i take the general form

$$\mu_i = \alpha_i(R-R_c) + \beta_i(S-S_c) + \gamma_i(T-T_c). \qquad (16)$$

We refer the reader to [38] for the detailed expressions of the coefficients α_i, β_i, γ_i, as functions of σ, τ and a. From (15) it is easy to deduce that (12) has zero as a triple root when $\mu=\mu_c=(\mu_0, \mu_1, \mu_2)=0$. These polycritical conditions define a codimension-three surface in the parameter space where the mode $m=n=1$ is triply degenerate. To position the system on this surface we need to adjust R, S and T to the value ($q^2 = q_{11}^2 = a^2 + \pi^2$)

$$R_c = \frac{q^6(\tau+2\sigma)}{a^2\sigma(1-\sigma)(1-\tau)} \qquad S_c = \frac{q^6\tau^2(1+2\sigma)}{a^2\sigma(\sigma-\tau)(1-\tau)} \qquad T_c = \frac{q^6\sigma(1+\sigma+\tau)}{\pi^2(\sigma-\tau)(1-\sigma)}. \qquad (17)$$

For typical values of a, when (17) holds, we find only one null vector of $M_{\lambda c}$, though its nullity is three. In the usual way [50], we complete the basis of the null space with generalized null vectors which satisfy

$$M_{\lambda c}\,\Phi_n = \sum_{m=1}^{3} J_{mn}\,\Phi_m \qquad \text{with} \qquad J = \begin{pmatrix} 0 & 1 & 0 \\ 0 & 0 & 1 \\ 0 & 0 & 0 \end{pmatrix}. \qquad (18)$$

We get

$$\|\Phi_n\|^T = q^{-2n}\|{-\delta_{n1}, \epsilon_n a, -\epsilon_n a/\tau^n, \epsilon_n \pi/\sigma^n}\| \qquad (19)$$

where the T means transpose, δ_{mn} is the Kronecker symbol and $\epsilon_n=(-1)^n$.

Near to polycriticality, there are an infinitude of rapidly damped modes, while the nearly marginal ones command our attention. We may split $U(x,z,t)$ into

$$U(x,z,t)=[A_1(t)\Phi_1(\lambda)+A_2(t)\Phi_2(\lambda)+A_3(t)\Phi_3(\lambda)]*\Xi_{11}(x,z) + S(x,z,t) \qquad (20)$$

where we require that $[\Phi_1(\lambda_c), \Phi_2(\lambda_c), \Phi_3(\lambda_c)]=[\Phi_1, \Phi_2, \Phi_3]$ and $S(x,z,t)$ is a linear superposition of stable modes. Then the amplitude vector, or order parameter, $A=(A_1, A_2, A_3)$ satisfies the linear amplitude equation

$$\dot{A} = J_{\mu}A \qquad \text{with} \qquad J_{\mu} = \begin{pmatrix} 0 & 1 & 0 \\ 0 & 0 & 1 \\ -\mu_0 & -\mu_1 & -\mu_2 \end{pmatrix} \qquad (21)$$

where J_{μ} is the Jordan-Arnold unfolding [26,28] of J (18) and μ_i are small parameters which account for a slight displacement off the tricritical surface.

II.3 Nonlinear analysis

To deal with the nonlinear terms in (1)-(4), we use the center manifold
theorem to eliminate the slaved modes [51] and the normal form technique to get
rid off the nonresonant nonlinearities in the amplitude equation [26-28]. Substantial amount of routine algebraic calculation is needed to find both the solution and the corresponding nonlinear amplitude equation. Such a calculation
can be performed using a general method [24] inspired from the ideas of the
asymptotic procedure of Krylov, Bogoliubov and Mitropolsky [52,53]. We refer
the reader to [38] for more details on such a perturbative approach. The solution is found to be of the form

$$U(x,z,t)=[A_1(t)\Phi_1(\lambda)+A_2(t)\Phi_2(\lambda)+A_3(t)\Phi_3(\lambda)]*\Xi_{11}(x,z)$$

$$\qquad\qquad\qquad (22)$$

$$+ \sum_{n=2}^{\infty} \sum_{k+l=0}^{n} u_{kl}^{(n)}(\lambda;x,z)A_1^k(t)A_2^l(t)A_3^{n-k-l}(t)$$

where $u_{kl}^{(n)}(\lambda;x,z)$ decomposes on both the marginal (Ξ_{11}) and stable $(\Xi_{mn}$,
m or $n\neq1)$ modes. The amplitude vector satisfies the nonlinear extension of
(21):

$$\dot{A} = J_{\mu}A + g(A) \qquad (23)$$

where g is a nonlinear vector-valued function which contains only odd order
terms in the A_i's because of the invariance of the P.D.E.'s (1)-(4) under the
transformation z ->1-z, x ->x, ψ ->-ψ, Θ ->-Θ, Σ ->-Σ, Y ->Y. Using the Arnold
gauge [26,28], g_3 is the only non-null component of g and the normal form (23)
transforms into the third-order O.D.E. (A_1=A, A_2=$\dot{A}$, A_3=$\ddot{A}$):

$$\dddot{A}+\mu_2\ddot{A}+\mu_1\dot{A}+\mu_0A = k_1A^3+k_2A^2\dot{A}+k_3A^2\ddot{A}+k_4A\dot{A}^2+k_5\dot{A}^3+k_6A\ddot{A}^2 \qquad (24)$$

where, as before, the small parameters μ_i of the linear terms account for a
small deviation from the polycritical conditions, while the coefficients k_i of
the six resonant cubic terms are computed on the tricritical surface. This amplitude equation has been truncated to its lowest order nonlinear terms (the
number of resonant terms grows linearly with the order of perturbation). The
detailed expressions of the coefficients k_i of the cubic terms as functions of
the physical parameters λ are already very complicated [38]. Let us specify the
simplest one:

$$k_1 = - \frac{a^4q^4\sigma\tau(1+\sigma+\tau)(a^2-\pi^2)}{8(1+2\sigma+\tau)(1-\sigma)(\sigma-\tau)}, \qquad (25)$$

which, for given σ and τ, can be positive or negative according to the value of a.
This property will be of main importance in the dynamical study of (24). The
actual reliability of this truncated normal form to describe the dynamics in the
neighborhood of the tricritical surface is the fundamental question we want to
address in this paper. In the following sections we will estimate the influence
of higher order terms which have been ignored in (24) through the comparison of
the normal form predictions with the results of direct simulations of the
P.D.E.'s (1)-(4).

II.4 Asymptotic normal form

One can still simplify the amplitude equation (24) when the parameters μ_i are made small in a suitable way [54]. Let ϵ be a small positive number and introduce the scaled variables

$$\tau = \epsilon t \quad , \quad X = \epsilon^{-3/2} A \quad , \quad \hat{\mu}_i = \epsilon^{i-3} \mu_i \quad (i=0,1,2) . \tag{26}$$

If we arrange matters so that the $\hat{\mu}_i$ are of order unity, the values of μ_i are near to the polycritical condition $\mu_i=0$ and the amplitude equation reduces to

$$\dddot{X} + \hat{\mu}_2 \ddot{X} + \hat{\mu}_1 \dot{X} + \hat{\mu}_0 X = k_1 X^3 + O(\epsilon) \tag{27}$$

where k_1 is defined in (25). Therefore, if we locate the system in that part of parameter space where for i=0,1,2,

$$\mu_i = O(\epsilon^{3-i}) \tag{28}$$

we may use the amplitude equation

$$\dddot{X} + \hat{\mu}_2 \ddot{X} + \hat{\mu}_1 \dot{X} + \hat{\mu}_0 X = \pm X^3 \tag{29}$$

(X has been rescaled in order to remove the constant k_1) as the asymptotic normal form for the instability ζ^3 with an error of order ϵ. Let us note that this equation contains only one nonlinear term $\pm X^3$, the sign of which depends on the sign of the coefficient k_1. This asymptotic result may be understood as the limit of very small dissipation $\mu_2=\det J_u>0$, taken in a way that preserves all of the terms of linear theory [25]. This asymptotic amplitude equation could have been obtained from the original P.D.E.'s (1)-(4) by application of the standard Poincaré-Lindsted approach as shaped for fluid dynamics by a series of authors [54,55]. The main distinction between this asymptotic procedure and the one which gives the truncated normal form (24) lies in the fact that, whereas to get the normal form we assume that the amplitudes and $|\mu|$ are small, in the presently illustrated asymptotic technique we additionally assume a relation (28) between their smallnesses. This somehow restricts the domain of applicability of the asymptotic normal form (29).

III THE SCENARIOS TO CHAOS IN THE NORMAL FORM OF 2-D BOUSSINESQ ROTATING THERMO-HALINE CONVECTION

In this section we use dynamical system theory [28] to find chaos in the previously derived truncated normal form of ζ^3. Our main goal is to show that well-known scenarios to chaos [1,2] occur as close as we want to the polycritical surface (17) ($\mu_0=\mu_1=\mu_2=0$) and thus can be captured in a perturbative analysis of the P.D.E.'s (1)-(4).

III.1 Homoclinic chaos near to tricriticality

* *Reduction of ζ^3 to ζ^2*: A way to study the truncated normal form (24) is to remark that when $|\mu_0|$, $|\mu_1| \ll \mu_2$, the characteristic value equation (15) factorizes into $[s+\mu_2][s^2+(\mu_1/\mu_2 - \mu_0/\mu_2^2)s+\mu_0/\mu_2]$, which means that one of the "three marginal modes" is more damped than both of the other ones. Then ζ^3 reduces to ζ^2 and the dynamics is locally described by a lower order amplitude equation:

$$\ddot{Y} + \nu \dot{Y} + \mu Y = \tilde{k}_1 Y^3 + \tilde{k}_2 Y^2 \dot{Y} \tag{30}$$

where

$$\mu = \mu_0/\mu_2 \qquad\qquad k_1 = k_1/\mu_2$$

$$\nu = \mu_1/\mu_2 - \mu_0/\mu_2^2 \qquad\qquad \tilde{k}_2 = k_2/\mu_2 - 3k_1/\mu_2^2 .$$

(31)

The dynamics resulting from the truncated normal form of the instability ζ^2 can be classified in two different diagrams in the (μ,ν) plane according to the sign of k_1 (Figure 1) [28]. (30) may display three steady states. In Case + ($k_1 > 0$), the two outer points are saddles when $-\mu < \nu^2/8$ and in Case − ($k_1 < 0$), the origin is a saddle when $\mu < \nu^2/4$. Of interest here are the orbits through the saddles as the ones shown in Figure 2. By standard methods [26,54] we find that the pair of heteroclinic orbits of Figure 2a and the pair of homoclinic orbits of Figure 2b occur close to the bicritical surface $\mu=\nu=0$ when the following conditions are fulfilled:

$$\nu = \mu\tilde{k}_2/5\tilde{k}_1 \quad \text{(Case +)} \quad ; \quad \nu = 4\mu\tilde{k}_2/5\tilde{k}_1 \quad \text{(Case −)}. \qquad (32)$$

* *Revolt of the slaved mode:* To find heteroclinic and homoclinic orbits in the normal form of ζ^3 one can start with parameter values for which ζ^3 is nearly ζ^2 ($|\mu_0|, |\mu_1| \ll \mu_2$). Then one can recover the orbits of Figure 2 numerically from the conditions (32) reexpressed in terms of the parameters μ_i:

$$\mu_1 = 2\mu_0/5\mu_2 \quad \text{(Case +)} \quad ; \quad \mu_1 = -7\mu_0/5\mu_2 \quad \text{(Case −)}. \qquad (33)$$

As we relax such a constraint and increase the departure from bidimensionality, one can follow the comitant deformations of the planar heteroclinic and homoclinic orbits of Figure 2a and b respectively. When the negative eigenvalue $-\mu_2$ associated with the slaved mode becomes comparable with the marginal eigen-

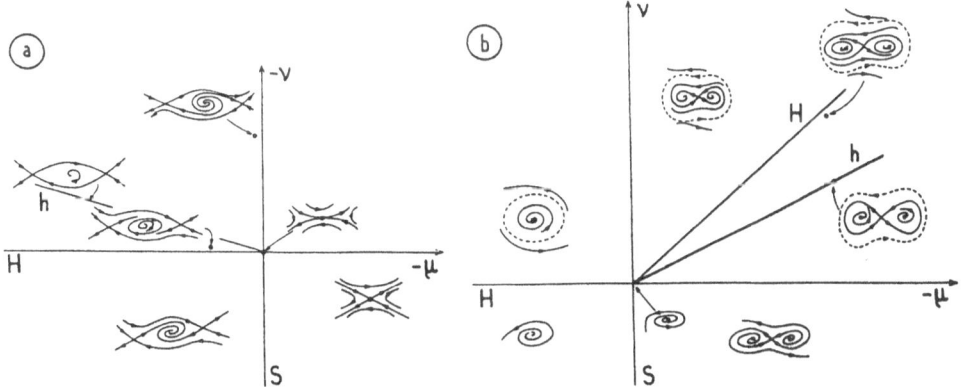

Figure 1: The parameter plane (μ,ν) for the normal form (30) of the instability ζ^2 with special symmetry: (a) $\tilde{k}_1 > 0$; (b) $\tilde{k}_1 < 0$. S=pitchfork bifurcation; H=Hopf bifurcation; h=heteroclinic (a) or homoclinic (b) bifurcation (dashed lines correspond to unstable periodic orbits).

(a)   (b)

Figure 2: (a) Heteroclinic orbits of (30) with $\tilde{k}_1 > 0$. (b) Homoclinic orbits of (30) with $\tilde{k}_1 < 0$.

values, then the saddles transform into spiral points and we get the tridimensional orbits displayed in Figure 3.

In Figure 4a and b we show the parameter plane $(\hat{\mu}_0, \hat{\mu}_1)$ and the curves Γ_1 and Γ_2 on which, for Case + and Case − we have respectively a pair of symmetric (with respect to the operation X →−X) heteroclinic and homoclinic orbits in the asymptotic normal form (29) of ζ^3. At this point let us note that Γ_1 and Γ_2 do not give the only points where heteroclinic or homoclinic conditions are fulfilled, but only those that grow out of the ones that are found from (33) (principal heteroclinic and homoclinic orbits [25]).

Another important qualitative change in the dynamics occurs when the curves Γ_1 and Γ_2 enter the regions on the right of the line labeled Σ in Figures 4a and b. Then the saddle foci which are involved in these heteroclinic and homoclinic connections are such that the unstable (positive) eigenvalue γ is greater than the absolute value of the real part of either of the complex conjugate eigenvalues $\rho \pm i\omega$. What gives significance to this seeming digression is a theorem by Shil'nikov [39] which ensures the existence of chaos (many infinite horseshoes [56] in an appropriate Poincaré map) in a neighborhood of these orbits. The Shil'nikov condition is

$$\gamma/|\rho| > 1 \qquad \text{i.e.} \qquad \begin{array}{ll} \hat{\mu}_1 < 2\hat{\mu}_0 - 2 & \hat{\mu}_0 > 1/2 \quad \text{(Case +)} \\[2mm] \hat{\mu}_1 < -\hat{\mu}_0 - 2 & \hat{\mu}_0 < -1 \quad \text{(Case −)} \end{array} \tag{34}$$

As experienced in [37,38,57,58], attracting chaos is expected to occur in a parameter neighborhood of heteroclinicity and homoclinicity when this condition

(a)   (b)

Figure 3: Sketches show the corresponding (as compared to Figure 2) three-dimensional (a) heteroclinic and (b) homoclinic orbits observed in the asymptotic normal form (29) of ζ^3 when we move away enough from the origin of Figure 4 along the curve Γ_1 and Γ_2 for Cases + and − respectively.

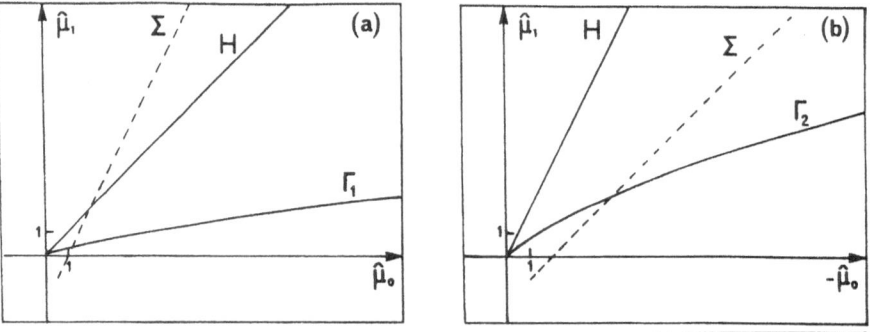

Figure 4: The parameter plane $(\hat{\mu}_0, \hat{\mu}_1)$ for the asymptotic normal form (29). (a) Case +: the line H corresponds to a Hopf bifurcation of the fixed point X=0; the curve Γ_1 is the location of the heteroclinic orbits which satisfy the Shil'nikov condition (34) on the right of the line labeled by Σ. (b) Case −: the line H corresponds to a Hopf bifurcation for the nontrivial fixed points X=±$(-\hat{\mu}_0)^{1/2}$; the curve Γ_2 is the location of the homoclinic orbits which satisfy the Shil'nikov condition (34) on the right of the line Σ.

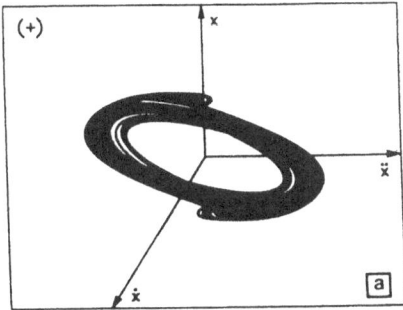
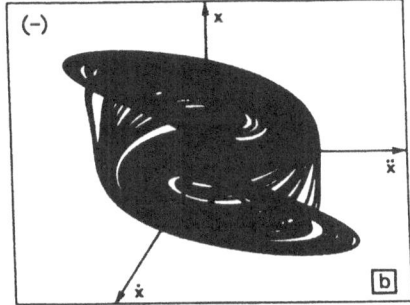

Figure 5: Numerical strange attractor obtained by integration of the asymptotic normal form (29). (a) Case +: $\hat{\mu}_2$=1, $\hat{\mu}_1$=3.5, $\hat{\mu}_0$=9.6; (b) Case -: $\hat{\mu}_2$=1, $\hat{\mu}_1$=3.5, $\hat{\mu}_0$=-5.5.

is fulfilled. In Figure 5, we show trajectories in the phase-space obtained by integrating the asymptotic normal form (29); they behave very much like strange attractors. Since the corresponding parameter values are such that $\hat{\mu}_0$, $\hat{\mu}_1$ and $\hat{\mu}_2$ are of order unity, this implies that letting $\epsilon \rightarrow 0$ in (28), chaos can occur as close as we want to the polycritical surface ($\mu_0=\mu_1=\mu_2=0$) in the normal form (24) of the instability ζ^3.

III.2 Period-doubling cascade and intermittency near to tricriticality.

In [38], Case + has been studied in great detail. Here we will mainly concentrate on Case -. In order to compare with direct numerical simulations of the P.D.E.'s (1)-(4), we investigate the truncated normal form (24) for values of the coefficients as obtained in our nonlinear analysis in section II.3 for σ=0.5, τ=0.2, $a=\pi\sqrt{2}$ namely:

$$k_1 = -557.17 \qquad k_2 = 211.07 \qquad k_3 = -45.75$$
$$k_4 = \cdot \qquad k_5 = -4.22 \qquad k_6 = -1.69 \ . \tag{35}$$

The results of ~ical numerical investigation of (24) when following a path to homoclinici _ : illustrated in Figure 6. We fix μ_2=1, μ_1=5 and we decrease μ_0 through negative values. As μ_0 decreases through zero, the rest A=0 loses stability to the steady convective solutions A=$\pm(\mu_0/k_1)^{1/2}$. As $-\mu_0$ is increased further, the steady convection solution becomes overstable at $-\mu_0$=2.5 and gives way to oscillatory solutions which preserve the exchange symmetry (A $\rightarrow$ -A) between them (Figure 6a). With even further increase of the parameter $-\mu_0$ we witness several period-doubling bifurcations as illustrated in Figure 6b and c. After the accumulation of the period-doubling cascade [59-61], the dynamics becomes chaotic. In Figure 6d we have given only one member of the pair of strange attractors that is found for $-\mu_0$=4.60. Beyond the reverse chaotic band-merging cascade [62], one observes an enrichment in the structure of the attractor and some intensification of the swirling motion in the vicinity of the origin. Finally, after a complicated alternating sequence of periodic and chaotic states, the two strange attractors join to form a large, symmetric attractor like the attractor in Figure 6e.

In order to have a better insight into the increasing complexity of the dynamics, we have defined in Figure 7 the corresponding Poincaré sections and 1-D maps. The swirl in the strange attractors is reflected in the spiralling shape of the Poincaré maps. When the symmetry is restored, the dynamics is described by a double spiral [57] as shown in Figure 7c. The 1-D map of Figure 7d looks

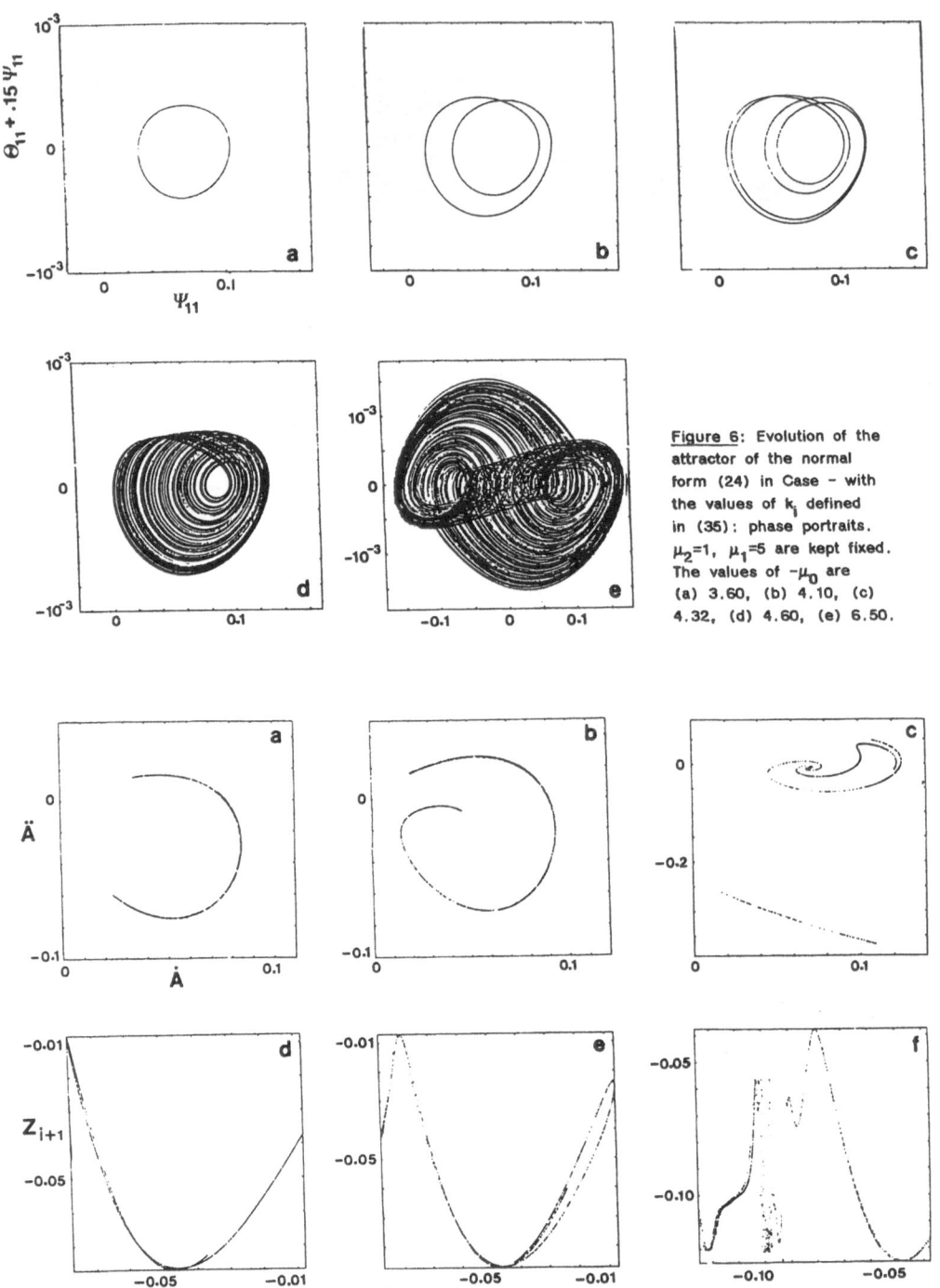

Figure 6: Evolution of the attractor of the normal form (24) in Case – with the values of k_i defined in (35): phase portraits. $\mu_2=1$, $\mu_1=5$ are kept fixed. The values of $-\mu_0$ are (a) 3.60, (b) 4.10, (c) 4.32, (d) 4.60, (e) 6.50.

Figure 7: (a), (b) and (c) represent Poincaré maps on the plane $A=(\mu_0/k_1)^{1/2}$ with $\dot{A}>0$, for the strange attractors observed in the normal form (24) of ζ^3 when following the numerical path described in Figure 6. The values of $-\mu_0$ are: (a) 4.60, (b) 4.90, (c) 6.50. (d), (e) and (f) are the corresponding 1-D maps $Z_{i+1}=f(Z_i)$ where $Z=\dot{A}\cos\theta+\ddot{A}\sin\theta$ with $\theta=3.41$.

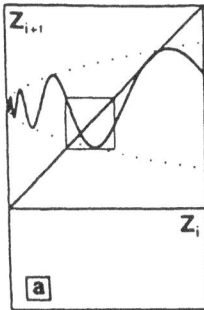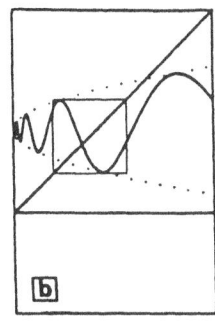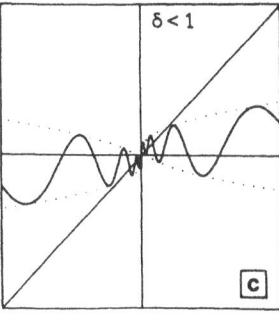

Figure 8: Graph of the unidimensional map (36) for parameter values satisfying the Shil'nikov condition $\delta < 1$: $\delta=.5$, $\xi=5$, $\eta=-2.7177$. The values of c are (a) 5.2, (b) 4.13, (c) 0.64. The squares isolate regions in the variable Z that are invariant under (36).

very much like the single-humped maps which have been the canonical model to study the universal properties of the cascade of period-doubling bifurcations leading to chaos [63]. As seen in Figure 7e, a second hump is involved in the dynamics when we approach homoclinicity. Like in [38] for Case +, a Poincaré map model can be constructed in the neighborhood of the homoclinic orbits. In the strong area contracting limit, this model reduces to the multi-humped 1-D map

$$
z_{i+1} = \begin{cases} c + z_i^{\delta} \sin(\xi \ln z_i + \eta) & z_i > 0 \\ -c - |z_i|^{\delta} \sin(\xi \ln|z_i| + \eta) & z_i < 0 \end{cases} \tag{36}
$$

where the two parameters $\delta=-\rho/\gamma$ and $\xi=\omega/\gamma$ characterize the linear dynamics near to the origin. When decreasing the parameter c to zero (c=0 corresponds to homoclinicity) this 1-D map reproduces much of the dynamics observed in the truncated normal form (24) of ζ^3. The comparison of the graph of this theoretical map presented in Figure 8 with the numerical 1-D maps in Figure 7 is very instructive about the reliability of (36) to account for the dynamics in nearby homoclinic conditions, even though the experimental 1-D maps are not strictly single-valued as the result of folding effects in the corresponding Poincaré maps.

Maps with double extrema such as the maps in Figures 7e and 8b have been extensively studied in the literature [64]. When one varies the control parameter one may encounter other phenomena such as transitions from periodic to chaotic regimes according to the Pomeau-Maneville theory of intermittency [40] and also discontinuous transitions between attractors with hysteresis [65,66]. These scenarios are also observed in the truncated normal form (24) of ζ^3 and are thus predicted to occur (like the subharmonic cascade) in the P.D.E.'s (1)-(4) in the neighborhood of the simultaneous onset of the competing instabilities.

III.3 Quasiperiodicity and fractal tori near to tricriticality.

* *Reduction of ζ^3 to $\zeta\omega$:* Now if we locate the normal form (24) of ζ^3 in the region of parameter space such that $|\mu_0|$, $\mu_2 \ll \mu_1$, then the characteristic value equation (15) factorizes into $[s+\mu_0/\mu_1].[s^2+(\mu_2-\mu_0/\mu_1)s+\mu_1]$. This means that ζ^3 reduces to $\zeta\omega$ whose truncated normal form reads [42,67,68]

$$
\begin{aligned}
\dot{X} &= -\mu X + kX^3 + k'X|Z|^2 \\
\dot{Z} &= (-\eta/2+i\sqrt{\mu_1})Z + \alpha \bar{Z}X^2 + \beta Z|Z|^2 \\
\dot{\bar{Z}} &= (-\eta/2-i\sqrt{\mu_1})\bar{Z} + \bar{\alpha}\bar{Z}X^2 + \bar{\beta}\bar{Z}|Z|^2
\end{aligned} \tag{37}
$$

323

where Z is a complex amplitude; $\mu=\mu_0/\mu_1$ and $\eta=\mu_2-\mu_0/\mu_1$ characterizes the distance to the bicritical surface where the stationary and oscillatory instabilities occur simultaneously ($\mu=\eta=0$). We have computed the coefficients of the nonlinear terms:

$$k = k_1/\mu_1 \qquad \beta = -(3k_1+\mu_1 k_3-3\mu_1 k_5+3\mu_1^2 k_6)/2\mu_1-i(k_2+3\mu_1 k_4)/2\sqrt{\mu_1}$$

$$\tag{38}$$

$$k' = (6k_1+2\mu_1 k_3-4\mu_1 k_5+2\mu_1^2 k_6)/\mu_1 \qquad \alpha = (\mu_1 k_5-3k_1)/2\mu_1- ik_2/2\sqrt{\mu_1}.$$

Setting $Z=\rho e^{i\Theta}$ and rescaling the variables in (37) we obtain

$$\dot{X} = X(\hat{\mu} + X^2 + B\rho^2) \tag{39a}$$

$$\dot{\rho} = \rho(\hat{\eta} + CX^2 + D\rho^2) \qquad D=\pm 1 \tag{39b}$$

$$\dot{\Theta} = \sqrt{\mu_1} + O(X^2,\rho^2) \tag{39c}$$

which attests that the normal form is equivariant to rotations about the X-axis. Thus the azimuthal component of (39) can be decoupled and we can work, at least initially, with the 2-D system (39a, 39b).

A case of interest arises when $D=-1$, $B>0$, $C<0$ and $D-BC>0$ [28,67,68]; then the central fixed point of the system (39a, 39b) may undergo a Hopf bifurcation as shown in Figure 9. This bifurcation, however, is degenerate and we must consider higher-order terms in the planar system to stabilize (in the case of a supercritical Hopf bifurcation) two attracting closed orbits of frequency $\omega\sim(\hat{\mu}^2, \hat{\eta}^2,\hat{\mu}\hat{\eta})^{1/2}$, which are exchanged by the symmetry $X\rightarrow-X$, $\rho\rightarrow\rho$. As $\hat{\mu}$ is varied beyond this bifurcation value, $\hat{\eta}$ being kept fixed (see Figure 10), the planar orbits grow from zero amplitude and approach the heteroclinic connections labelled Γ in Figure 9, while their frequencies go to zero. Note that when we restore the angular dependence described by (39c), a closed orbit of the planar system corresponds to an invariant torus for the three-dimensional normal form (37).

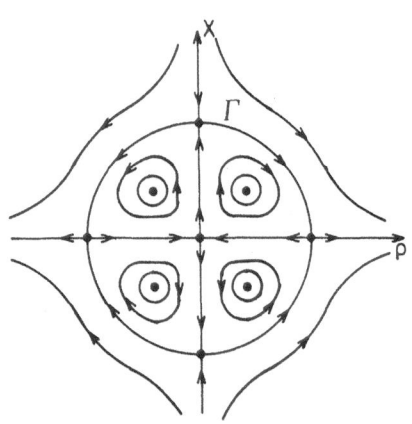

Figure 9: Phase portrait for the truncated normal form (39a, 39b) at $\hat{\eta} = \hat{\mu}(C-1)/(B+1)$: degenerate Hopf bifurcation ($D=-1$, $B>0$, $C<0$, $D-BC>0$).

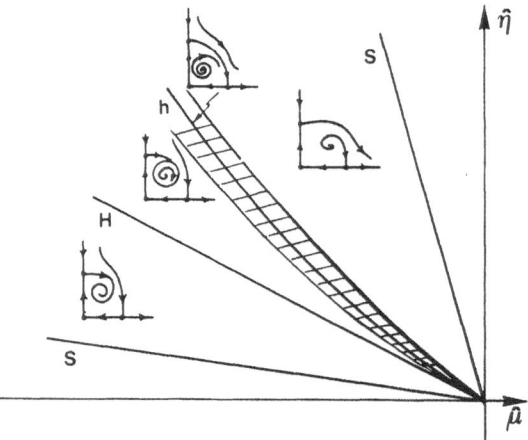

Figure 10: A possible completion of the unfolding of the truncated normal form (39a, 39b). S=stationary bifurcation, H=Hopf bifurcation, h=heteroclinic bifurcation. The dashed area corresponds to a fuzzy region where chaos is expected to occur when the angular dependence is restored in (39). ($D=-1$, $B>0$, $C<0$, $D-BC>0$). Only one quadrant is shown in the phase portraits.

324

Unfortunately the normal form calculation used to reduce ζ^3 to $\zeta\omega$ cannot be carried out in general to infinite order, since the series do not usually converge [26-28]. This results from the fact that the original P.D.E.'s (1)-(4) as well as the truncated normal form (24) of ζ^3 are nonaxisymmetric, in contrast to (37). Then the very exceptional situation Γ is expected to split yielding transverse intersections of the manifolds involved in these heteroclinic connections with creation of Smale horseshoes [56]. Consequently, on the way to heteroclinicity one observes after several frequency lockings, the breaking of the underlying torus which is associated with the appearance of chaotic dynamics [64,69-80]. Since transversal intersections of manifolds are preserved under higher order perturbations in (39), this suggests that the transition from quasiperiodicity to chaos predicted in the normal form of ζ^3 does exist in the full system (1)-(4) in nearly tricritical conditions. A characteristic of such a doubly-periodic flow is that one frequency associated to the angular variable Θ is finite ($\sim\sqrt{\mu_1}$) while the other frequency associated with the secondary Hopf bifurcation is smaller and smaller as we approach the bicritical surface where the instability $\zeta\omega$ takes place. One therefore expects to see the appearance of a slow modulation of a relative fast oscillation which is usually called "soft mode" instability [42].

In order to compare such an approach with the results of direct simulations of the P.D.E.'s (1)-(4), we investigate the truncated normal form (24) of ζ^3 for values of the coefficients as obtained in our nonlinear analysis in section II.3 for $\sigma=.15$, $\tau=.1$, $a=2\pi$ namely:

$$k_1 = -2840.24 \qquad k_2 = -474.48 \qquad k_3 = 1389.68$$

$$k_4 = -18.37 \qquad k_5 = -138.82 \qquad k_6 = -100.36 \ . \tag{40}$$

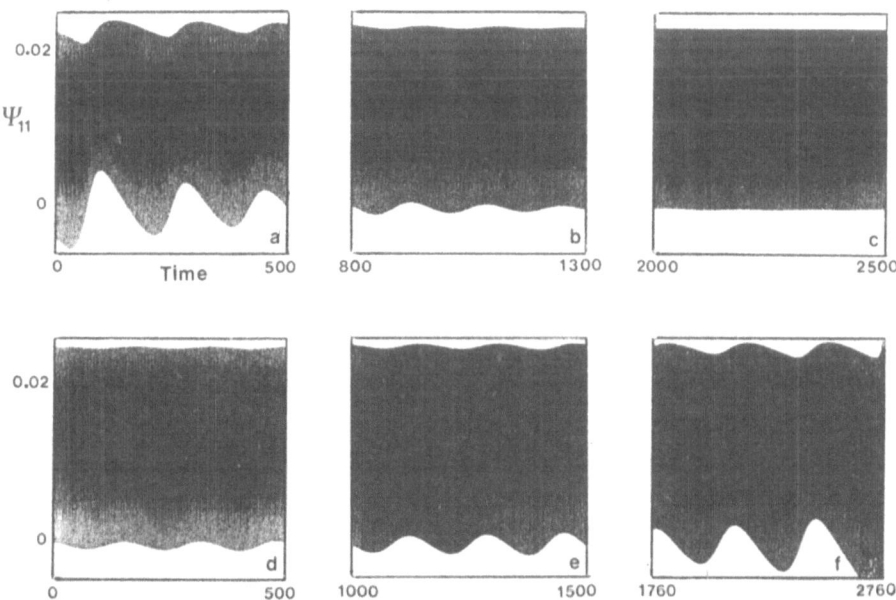

Figure 11: Integration of the normal form (24) of ζ^3 with the values of k_i defined in (40). $\mu_0 = -1$ $\mu_1 = 9$ are kept fixed. (a), (b) and (c) correspond to a value of $\mu_2 = .0370$ before the soft mode instability. (d), (e) and (f) correspond to a value of $\mu_2 = .0365$ beyond the soft mode instability.

The results of a typical numerical experiment are shown in Figure 11. We fix μ_o=-1, μ_1=9 and we decrease μ_2 through the secondary Hopf bifurcation value

$$\mu_2 = (kRe\beta + kk' - 2Re\alpha Re\beta)\mu_o \ / \ k(k' - Re\beta)\mu_1 . \qquad (41)$$

Before the soft mode instability the transients to the periodic solution display a slow modulation which is damped with a rate which goes to zero as we approach the transition value (Figure 11a, b, c). Beyond this value, such a modulation grows in size without manifestation of any nonlinear saturation of the soft mode instability (Figure 11d, e, f). This sugests that the secondary Hopf bifurcation is subcritical. Although we have succeeded in performing numerical investigations of the truncated normal form (24) of ζ^3 showing a supercritical bifurcation as predicted in Figure 10 (with adequate values of the coefficients k_i), all our attempts have failed when considering values of k_i as calculated from realistical values of the physical parameters σ, τ and a. This result has to be understood as an experimental finding and not as a proof.

IV THE SCENARIOS TO CHAOS IN 2-D BOUSSINESQ ROTATING THERMOHALINE CONVECTION: NORMAL FORM PREDICTIONS VERSUS DIRECT NUMERICAL SIMULATIONS

We solve the P.D.E.'s (1)-(4) using pseudo spectral methods [81,82]. The nonlinear terms are evaluated by fast Fourier transforms [83]. Time-stepping is done by a Crank-Nicholson scheme for diagonal linear terms and a leap-frog scheme for the remaining linear and nonlinear terms. Typical runs require from 10000 time steps in laminar regimes up to 100000 in chaotic regimes, consuming from 100 up to 1000s on the CRAY I computer for a 16x16 resolution. In the following simulations the values of the physical parameters σ, τ and a are conveniently chosen to ensure that Ξ_{11} is the only mode which destabilizes linearly along our experimental paths [43].

IV.1 The onset of convective rolls (test of the numerical code)

We perform our first simulation of (1)-(4) for the values σ=0.5, τ=0.2, a=$\pi\sqrt{2}$ which have been investigated in our normal form analysis in section III.2 (Case -). We adjust the parameters R, S and T such that using (16), μ_1=9 and μ_2=2 are kept fixed, while μ_o is decreased from positive values, for which the rest is stable, to negative values. In this subsection we mainly concentrate on the first bifurcation which occurs at μ_o=0 and corresponds to the onset of convective rolls (ζ instability). Near to the threshold of convection the dynamics is governed by the Landau equation [29,30]:

$$\dot{A} + (\mu_o/\mu_1)A = k_1(\mu_1,\mu_2)A^3 . \qquad (42)$$

From the analytic expression of $k_1(\mu_1,\mu_2)$ as a function of μ_1 and μ_2 (see section II.3), one deduces the following dependence of the coefficient k_1 in the truncated normal form (24) of ζ^3:

$$k_1(\mu_1,\mu_2) = k_1 + C_1(\lambda)\mu_1 + C_2(\lambda)\mu_2 \qquad (43)$$

with

$$C_1(\lambda) = \frac{-a^2\sigma^2[(1+\tau+\tau^2)(1+\sigma)+\tau^3-2\sigma^2(1+\tau)]+\pi^2\tau^2(1+\tau+\sigma)}{8\sigma\tau(1-\sigma)(\sigma-\tau)(1+2\sigma+\tau)}$$

$$\qquad (44)$$

$$C_2(\lambda) = \frac{q^2\tau(1+2\sigma+\tau)[a^2\sigma^2(1+\tau-\sigma)-\pi^2\sigma\tau]}{8\sigma\tau(1-\sigma)(\sigma-\tau)(1+2\sigma+\tau)} .$$

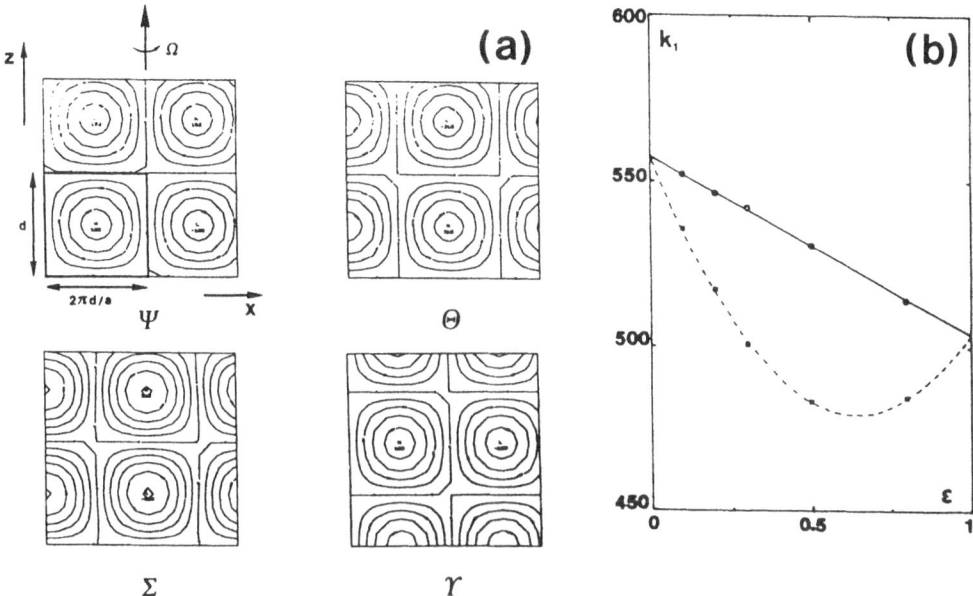

Figure 12: Steady two-dimensional convective rolls. (a) contours of constant fields. (b) Direct numerical estimate of the coefficient k_1 in the normal form (24) of ζ^3 when moving toward the tricritical surface with a linear (circles) and an asymptotic (squares) approach (see text). The continuous and dashed lines correspond to the analytical predictions obtained from (43) and (44).

From the measurement of ψ_{11} (in the limit $\mu_o \to 0-$) which, from (19) and (20), is equal to the amplitude A of the rolls, we derive an experimental estimate of the coefficient $k_1(\mu_1,\mu_2)$ which can be compared with the analytical expressions (43), (44). We can repeat this comparison along paths arbitrarily close to the tricritical surface ($\mu_o=\mu_1=\mu_2=0$). Two experimental sets of data obtained respectively for a linear ($\mu_1=\epsilon\hat{\mu}_1, \mu_2=\epsilon\hat{\mu}_2$) and an asymptotic (28) ($\mu_1=\epsilon^2\hat{\mu}_1, \mu_2=\epsilon\hat{\mu}_2$,) approach are presented in Figure 12. A remarkable agreement exists with the theoretical predictions. This may be seen as a nonlinear test of our numerical code. (43) can also be viewed as a first attempt to quantitatively apprehend the effect of higher order terms. We notice that along our main experimental path discussed in the next subsection, μ_1 and μ_2 are of order unity i.e. $\epsilon=1$, and $k_1(\mu_1,\mu_2)$ differs from the value obtained at tricriticality ($\epsilon=0$) by a factor ~10%.

IV.2 The cascade of period-doubling bifurcations leading to homoclinic chaos

To observe the cascade of period-doubling bifurcations in the P.D.E.'s (1)-(4), we position the system as suggested by our study of the normal form of ζ^3 in section III.2: $\sigma=0.5$, $\tau=0.2$, $a=\pi\sqrt{2}$, $\mu_2=1$, $\mu_1=5$. When increasing $-\mu_o$, the second bifurcation encountered after the stationary symmetry breaking bifurcation is a Hopf bifurcation from the convective state to periodic oscillations. When varying further μ_o one observes several period-doubling bifurcations as illustrated on the time series in Figure 13a-c. Beyond some critical value, several steps of the reverse chaotic cascade are observed before recording the chaotic state presented in Figure 13d. So far the asymptotic solution is not invariant under the symmetry (A $\to$ -A) which characterizes the original P.D.E.'s, but transforms into a symmetric solution under such an operation. When $-\mu_o$ is further increased, we witness a very complicated sequence of alternating periodic and chaotic regimes. Finally one reaches a value of $-\mu_o$ for which the sym-

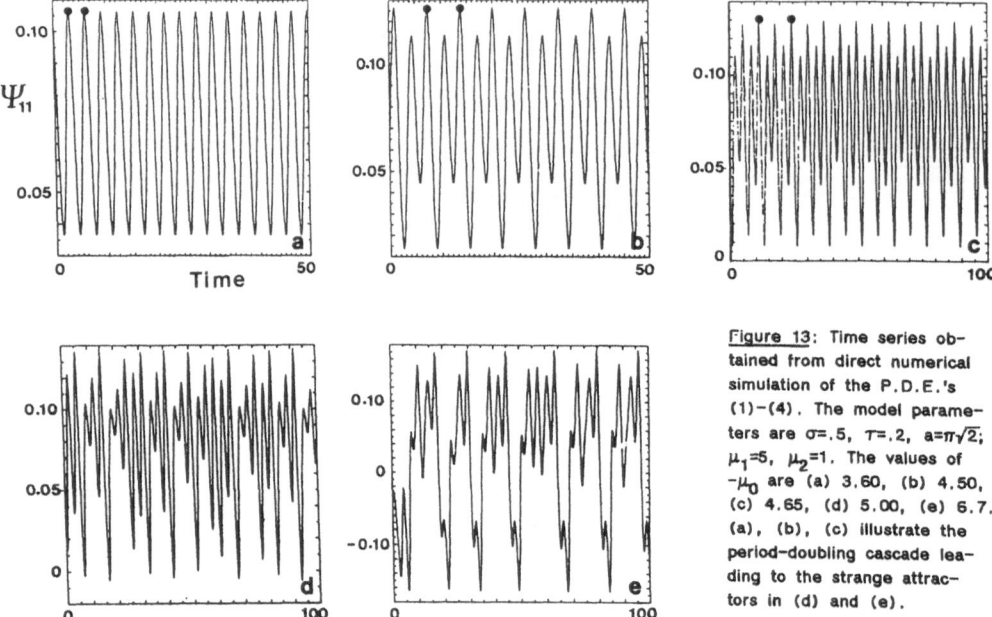

Figure 13: Time series obtained from direct numerical simulation of the P.D.E.'s (1)–(4). The model parameters are $\sigma=.5$, $T=.2$, $a=\pi\sqrt{2}$; $\mu_1=5$, $\mu_2=1$. The values of $-\mu_0$ are (a) 3.60, (b) 4.50, (c) 4.65, (d) 5.00, (e) 6.7. (a), (b), (c) illustrate the period-doubling cascade leading to the strange attractors in (d) and (e).

metry is restored as illustrated in the chaotic time series shown in Figure 13e. The corresponding power spectra are shown in Figure 14.

This scenario is very similar to the dynamical evolution computed with the truncated normal form (24). This can be checked when examining the phase portraits in Figure 15 comparatively to the predicted phase portraits in Figure 6. The parameter values where the successive bifurcations occur are slightly modified but by an amount of a few percents which is comparable to the relative size (when assuming no major trouble coming from the size of the corresponding coefficients k_i) of the neglected higher order terms as estimated from the computation of the average $\langle A^2(t)\rangle=\langle\psi_{11}(t)^2\rangle$ in the time-dependent regimes.

Therefore, one can conclude that the spiralling strange attractors observed in direct numerical simulations of the P.D.E.'s modeling thermohaline convection in a rotating layer (Figure 15) are very likely to be understood in terms of the chaotic behavior which exists in the neighborhood of homoclinic conditions as expected from Shil'nikov's work [39].

IV.3 Soft mode instability near to tricriticality

Using our study of the normal form (24) in section III.3 as a theoretical guide, we pursue our numerical investigation of the P.D.E.'s (1)–(4) for the physical parameter values: $\sigma=.15$, $T=.1$, $a=2\pi$, $\mu_0=-1$, $\mu_1=9$. When decreasing μ_2, we witness a soft mode instability which destroyes the periodic oscillations, very much like the scenario observed in the normal form calculation (Figure 11). In Figure 16 we show the transients to the periodic regime before the instability threshold. A modulation of the amplitude of these oscillations is observed to decay slower and slower as we approach the transition value. The frequency of this modulation is much smaller ($\sim 1/30$) than the fundamental frequency; this gives a hint of the proximity of our experimental path to the bicritical surface where the instability $\zeta\omega$ takes place. Beyond the transition value, this modulation becomes more and more pronounced when time evolves as expected for a secon-

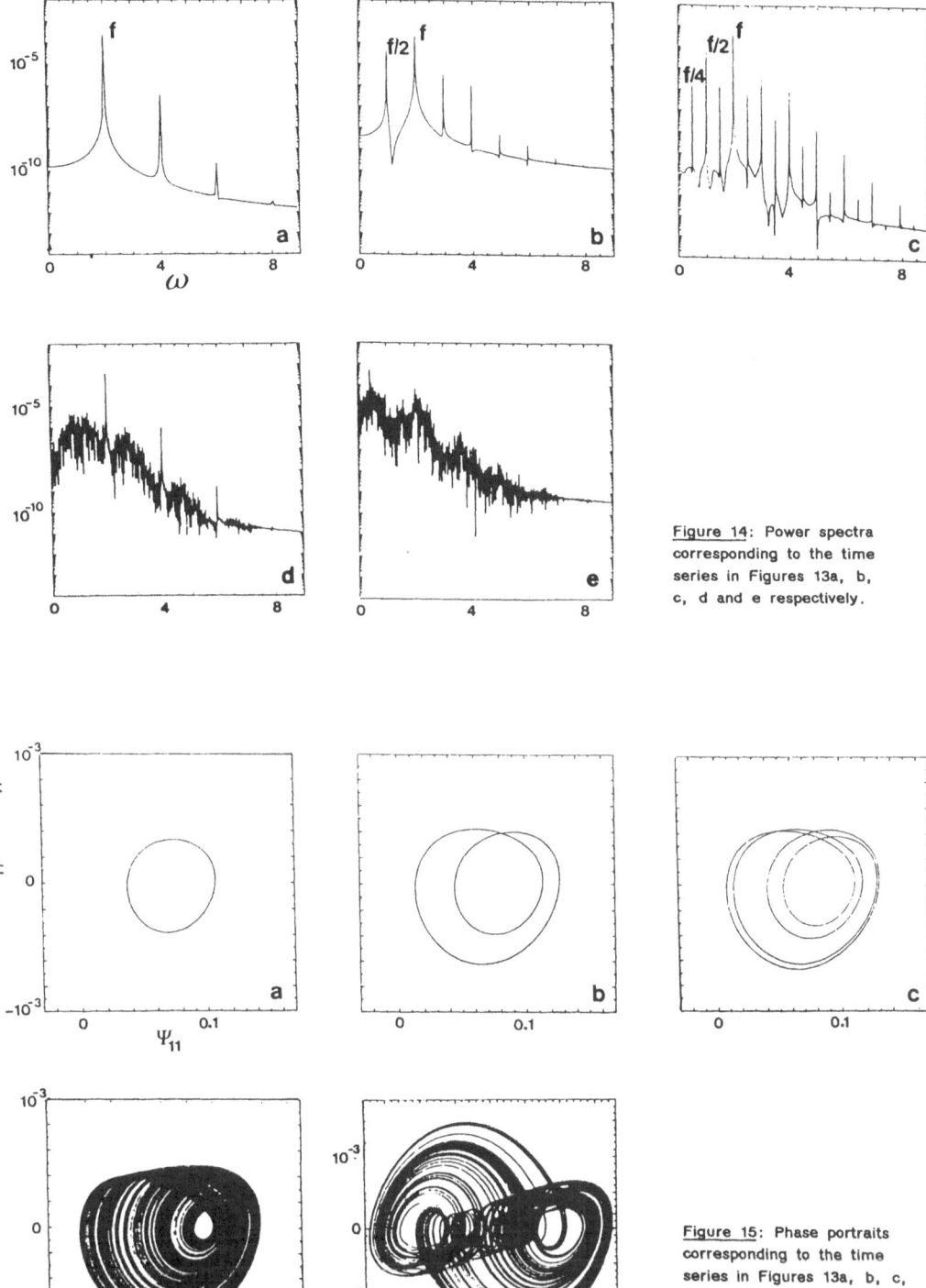

Figure 14: Power spectra corresponding to the time series in Figures 13a, b, c, d and e respectively.

Figure 15: Phase portraits corresponding to the time series in Figures 13a, b, c, d and e respectively. This evolution has to be compared with the normal form predictions in Figure 6.

329

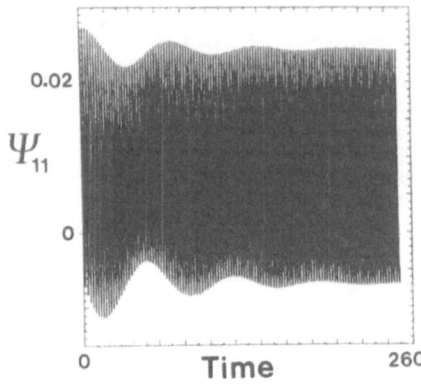

Ψ_{11}

0.02

0

0   Time   260

Figure 16: Time series obtained from direct numerical simulations of the P.D.E.'s (1)-(4). The model
parameters are σ=.15, T=.1, a=2π, μ_0=-1, μ_1=9,
μ_2=.001. The damped slow modulation of the amplitude of the periodic oscillation is a precursor to the
soft mode instability.

dary Hopf bifurcation. This growth of the modulation amplitude is not saturated
nonlinearly and the dynamics ends asymptotically on a nonlocal attractor. This
simulation confirms the predicted subcritical nature of this secondary bifurcation.

At this point, let us remark that we have performed several comparisons
between the normal form predictions and the results of direct simulations of the
P.D.E.'s (1)-(4). As far as the soft mode instability is concerned, the observed transition value can be significantly different from the theoretical prediction (41). This can be understood when looking at (38) which gives the expressions of the coefficients in the normal from of $\zeta\omega$ as functions of the coefficients in the normal form of ζ^3. These expressions are complicated combinations of the μ_i's and k_i's. Indeed a deviation of a few percent in these parameters, (in particular in the k_i's which have been computed on the tricritical
surface ζ^3) may induce a larger modification of the parameter k, k', α and β in
(37) and consequently a significant discrepancy in the secondary Hopf bifurcation value (41). However, thanks to the normal form analysis discussed in section II.3, we have always succeeded in any of our attempts to locate a soft mode
instability in the numerical investigation of the P.D.E.'s modeling thermohaline
convection in a rotating layer.

V CONCLUSION

We have presented a normal form approach of the transition to chaos which
points out that the lowest order nonlinear terms in the amplitude equations are
sufficient to account for the different scenarios to weak turbulence in finite
macroscopic systems. The extent of the domain of validity of the normal form
predictions is in fact determined by the relative importance of higher order
terms with respect to the leading nonlinear terms captured in the truncated normal form. The rotating thermohaline convection problem discussed in this paper
brings evidence that there exist systems such that normal form predictions and
direct simulations agree quantitatively in a rather close neighborhood of polycriticality, while a qualitative agreement spreads out in a wide domain of parameter space. Therefore for such systems the normal form approach looks very
promising from an experimental point of view. On the one hand, one can reasonably expect to learn much more from the recorded dynamics in real experiments; on
the other hand, one can hope to set up new experiments specially devoted to the
study of polycritical dynamics in finite macroscopic systems.

To conclude let us remark that the present study also enlightens the results of some recent experiments which have brought evidence of polycritical dy-

namics without any particular care in positioning the system in the constraint space. For example, it has been recently observed in the Belousov-Zhabotinskii reaction [84,85] quasiperiodic behavior resulting from the interaction (instability $\zeta\omega$) of two basic instabilities, namely the Hopf bifurcation at the origin of the oscillating nature of this reaction and the hysteresis bifurcation which accounts for the phenomena of bistability which is characteristic of such chemical systems. This scenario recorded when moving one control parameter looks very like the soft mode instability predicted by the normal form analysis. This strongly suggests that in the explored range of constraint values, the dynamics of this chemical reactions is governed by the lowest order nonlinearities which describe the interaction of both these elementary instabilities. Such a result has been corroborated by the observation of the breaking of the underlying torus into a fractal object and the appearance of chaos, as expected when locating the system in nearby bicritical conditions (section III.3).

We thank F.Argoul and P.Richetti for helpful discussions. The hospitality of the C.R.P.P. (Talence) is gratefully acknowledged. This work was supported in part by the N.S.F. under grant PHY 80-23721 and a CNRS ATP.

REFERENCES

[1] J.P.Eckmann, Rev. Mod. Phys. 53 (1981) 643.
[2] E.Ott, Rev. Mod. Phys. 53 (1981) 655.
[3] H.Haken, Chaos and Order in Nature (Springer, Berlin, 1981).
[4] H.L.Swinney and J.P.Gollub, Hydrodynamic Instabilities and the Transition to Turbulence (Springer, Berlin, 1981).
[5] C.Vidal and A.Pacault, Nonequilibrium Dynamics in Chemical Systems (Springer, Berlin, 1984).
[6] T.Tatsumi, Turbulence and Chaotic Phenomena in Fluids (North-Holland Amsterdam, 1984).
[7] B.L.Hao, Chaos (World Scientific Pub. Co, Singapore, 1984).
[8] P.Cvitanovic, Universality in Chaos (Adam Hilger, Bristol, 1984).
[9] P.Bergé, Y.Pomeau and C.Vidal, L'ordre dans le Chaos vers une Approche Déterministe de la Turbulence (Hermann, Paris, 1984).
[10] H.G.Schuster, Déterministe Chaos (Physik Verlag, Weinheim, 1984).
[11] E.N.Lorenz, J. Atmos. Sci. 20 (1963) 130.
[12] C.T.Sparrow, The Lorenz Equations: Bifurcations, Chaos and Strange Attractors (Springer, Berlin, 1982).
[13] J.H.Curry, Comm. Math. Phys. 60 (1978) 193; Phys. Rev. Lett. 43 (1979) 1013.
[14] V.Franceschini and C.Tebaldi, J. Stat. Phys. 21 (1979) 707.
[15] J.H.Curry, J.R.Herring, J.Loncaric and S.A.Orszag, J. Fluid. Mech. 147 (1984) 1.
[16] E.Knobloch and N.O.Weiss, Phys. Lett. 85A (1981) 127.
[17] L.N.DaCosta, E.Knobloch and N.O.Weiss, J. Fluid. Mech. 109 (1981) 25.
[18] D.R.Moore, J.Toomre, E.Knobloch and N.O.Weiss, Nature 303 (1983) 663.
[19] E.Knobloch, D.R.Moore, J.Toomre and N.O.Weiss, J. Fluid. Mech. 166 (1986) 409.
[20] N.O.Weiss, J. Fluid. Mech. 108 (1981) 247.
[21] E.Knobloch, N.O.Weiss and L.N.DaCosta, J. Fluid. Mech. 113 (1981) 153.
[22] E.Knobloch and N.O.Weiss, Physica 9D (1983) 379.
[23] S.K.Ma, Modern Theory of Critical Phenomena (Benjamin Reading Mass., 1976).
[24] P.H.Coullet and E.A.Spiegel, SIAM J. Appl. Math. 43 (1983) 774.
[25] A.Arneodo, P.H.Coullet, E.A.Spiegel and C.Tresser, Physica 14D (1985) 327.
[26] V.I.Arnold, Chapitres Supplémentaires de la Théorie des Equations Différentielles Ordinaires, Editions Mir, Moscow (1980).
[27] G.Iooss and D.D.Joseph, Elementary Stability and Bifurcation Theory (Springer, N.Y. 1980).
[28] J.Guckenheimer and P.Holmes, Nonlinear Oscillations, Dynamical Systems and Bifurcations of Vector Fields (Springer, N.Y. 1983).

[29] L.Landau, Dokl. Akad. Nauk. SSR 44 (1944) 339.
[30] L.Landau and E.M.Lifshitz, Fluid Mechanics (Pergamon Press, Oxford, 1959) 103.
[31] J.E.Mardsen and M.McCracken, Hopf Bifurcation and its Applications, Applied Math. Sci. 19 (Springer, N.Y. 1976).
[32] R.I.Bogdanov, Functional Analysis and its Applications, 9 (1975) 144.
[33] Throughout this paper we use the nomenclature defined in [25].
[34] E.Knobloch and M.R.E.Proctor, J. Fluid. Mech. 108 (1981) 291.
[35] J.Guckenheimer and E.Knobloch, Geophys. & Astrophys. Fluid. Dyn. 23 (1983) 247.
[36] J.C.Roux, P.Richetti, A.Arneodo and F.Argoul, J. Mec. Theo. & Appl. Numero special (1984) 77.
[37] A.Arneodo, P.H.Coullet and E.A.Spiegel, Phys. Lett. 92 (1982) 369.
[38] A.Arneodo, P.H.Coullet and E.A.Spiegel, Geophys. & Astrophys. Fluid. Dyn. 31 (1985) 1.
[39] L.P.Shil'nikov, Sov. Math. Dokl. 6 (1965) 163; Math. U.S.S.R. Sbornick 6 (1968) 427 and 10 (1970) 91.
[40] Y.Pomeau and P.Manneville, Commun. Math. Phys. 74 (1980) 189.
[41] G.Iooss, Bifurcation of Maps and Applications (North-Holland, Amsterdam, 1979).
[42] W.F.Langford, A.Arneodo, P.Coullet, C.Tresser and J.Coste, Phys. Lett. 78A (1980) 11.
[43] A.Arneodo and O.Thual, Phys. Lett. 109A (1985) 367.
[44] D.W.Moore and E.A.Spiegel, Astrophys. J. 143 (1966) 871.
[45] A.Libchaber, S.Fauve and C.Laroche, Physica 7D (1983) 73.
[46] S.Fauve and A.Libchaber, Bifurcation Theory, Mechanics and Physics C.P.Bruter et al. (eds) (Reidel Publishing Company, 1983) 257.
[47] E.A.Spiegel, Comments Astrophys. Space Phys. 1 (1969) 57.
[48] E.A.Spiegel and N.O.Weiss, Geophys. & Astrophys. Fluid. Dyn. 22 (1982) 219.
[49] S.Chandrasekhar, Hydrodynamic and Hydrodynamic Stability (Dover, 1981).
[50] B.Friedman, Principles and Techniques of Applied Mathematics (John Wiley and Sons, 1956).
[51] H.Haken, Synergetics (Springer, N.Y., 1978) 194.
[52] N.N.Bogoliubov and Y.A.Mitropolsky, Asymptotic Methods in the Theory of Nonlinear Oscillations (Gordon and Breach, N.Y., 1961).
[53] N.Minorsky, Nonlinear Oscillations (Van Nostrand, N.Y., 1962).
[54] A.H.Nayfeh, Perturbation Methods (Wiley, N.Y., 1973).
[55] S.Kogelman and J.B.Keller, SIAM J. Appl. Math. 20 (1967) 619.
[56] S.Smale, Bull. Amer. Math. Soc. 6 (1967) 803.
[57] A.Arneodo, P.Coullet and C. Tresser, Comm. Math. Phys. 79 (1981) 573.
[58] P.Glendenning, Phys. Lett. 103A (1984) 163.
[59] M.J.Feigenbaum, J. Stat. Phys. 19 (1978) 25; 21 (1979) 669.
[60] P.Coullet and C.Tresser, J. Phys. C5 (1978) 25.
[61] C.Tresser and P.Coullet, Compte Rendus Acad. Sci. 287A (1978) 577.
[62] S.Grossmann and S.Thomae, Z. Naturforsch 32A (1979) 669.
[63] P.Collet and J.P.Eckmann, Iterated Maps of the Interval as Dynamical Systems (Birkhauser, Boston-Basel-Stuttgart, 1980).
[64] R.S.Mackay and C.Tresser, Physica 19D (1986) 206.
[65] C.Tresser, P.Coullet and A.Arneodo, J. Phys. Lett. 41 (1980) 243.
[66] S.Fraser and R.Kapral, Phys. Rev. A25 (1982) 3223.
[67] G.Iooss and W.F.Langford, Ann. N. Y. Acad. Sci. 357 (1980) 489.
[68] W.F.Langford and G.Iooss, H.D.Mittelmann and H.Weber (eds) (Birkhauser Verlag, Basel, 1980) 103.
[69] D.G.Aronson, M.A.Chory, G.R.Hall and R.P.McGehee, Comm. Math. Phys. 83 (1982) 303.
[70] J.H.Curry and J.A.Yorke, Lect. Notes in Math. 668 (1978) 48.
[71] A.Ben-Mizrachi and I.Procaccia, Phys. Rev. A31 (1985) 3990.
[72] P.Coullet, C.Tresser and A.Arneodo, Phys. Lett. 77A (1980) 327.
[73] K.Kaneko, Prog. Theor. Phys. 68 (1982) 663; 69 (1983) 403.
[74] L.Glass and R.Perez, Phys. Rev. Lett. 48 (1982) 1772.
[75] M.Schell, S.Fraser and R.Kapral, Phys. Rev. Lett. A28 (1983) 373.

332

[76] S.Newhouse, J.Palis and F.Takens, Publ. Math. I.H.E.S. <u>57</u> (1983) 5.

[77] J.M.Gambaudo, P.Glendenning and C.Tresser, Phys. Lett. <u>105A</u> (1984) 97.

[78] P.Cvitanovic, B.Shraiman and B.Soderberg, Scaling laws for mode locking in circle maps, Nordita preprint 1985.

[79] M.H.Jensen and I.Procaccia, Phys. Rev. <u>A32</u> (1985) 1225.

[80] T.Bohr and G.Gunaratne, Phys. Lett. <u>113A</u> (1985) 55.

[81] S.A.Orszag, Stud. Appl. Maths <u>50</u> (1971) 293.

[82] O.Thual and P.L.Sulem, Note de travail de l'EERM, <u>81</u>, Météorologie Nationale, Toulouse (1984).

[83] E.O.Brigham, <u>The Fast Fourier Transform</u> (Prentice Hall, Inc. Englewood Cliffs, New-Jersy, 1974).

[84] F.Argoul, A.Arneodo, P.Richetti and J.C.Roux, "From quasiperiodicity to chaos in the Belousov-Zhabotinskii reaction: I Experiment", preprint (1986), to appear in J. Chem. Phys. (1987).

[85] P.Richetti, J.C.Roux, F.Argoul and A.Arneodo, "From quasiperiodicity to chaos in the Belousov-Zhabotinskii reaction: II Modeling and theory", preprint (1986), to appear in J. Chem. Phys. (1987).

Time-Dependent and Chaotic Behaviour in Systems with O(3)-Symmetry

R. Friedrich and H. Haken

Institut für theoretische Physik und Synergetik, Universität Stuttgart,
Pfaffenwaldring 57, D-7000 Stuttgart 80, Fed. Rep. of Germany

1. INTRODUCTION

It is by now well known that chaotic temporal behaviour may be generated by completely deterministic evolution laws. The main theoretical interest in chaotic phenomena has been devoted to the examination of the stochastic properties of low dimensional dynamical systems, although experimental confirmations of certain theoretical predictions have been mainly pursued in spatially extended systems, e.g. in lasers, hydrodynamic flows, and chemical reactions. In such systems low dimensional temporal chaotic behaviour evidently has to be strongly interrelated with coherent spatial patterns. The success of synergetics in the explanation of spontaneous pattern formation in selforganizing systems allows a straightforward treatment of time-dependent and chaotic behaviour in spatially extended systems of finite size which is caused by the nonlinear interaction of several order parameters, each connected with a coherent spatial pattern. In these cases a complete examination of the interrelations between temporal disorder and spatial coherence becomes possible. Spatial patterns are naturally connected to the symmetries of a system. Therefore, it can be expected that quite different systems, which, however, possess the same spatial symmetries, may show comparable behaviour. To demonstrate this we shall consider instabilities in systems with O(3)-symmetry. For the special case of mode interaction between two groups of modes belonging to two different irreducible representations of the O(3)-group with $\ell=1$ and $\ell=2$ we shall derive the order parameter equations. For the concrete case of thermal convection in spherical symmetries the occurrence of stationary, time-dependent, wavelike, and chaotic fluid flows is demonstrated. The examinations give sufficient evidence that disordered temporal behaviour and coherent spatial order are intimately related in this case. Therefore it is expected that similar behaviour can be observed in quite different systems exhibiting mode interaction with $\ell=1$ and $\ell=2$.

2. THE INSTABILITY OF A STATIONARY STATE IN A SYSTEM WITH O(3)-SYMMETRY

Near an instability the behaviour of a system with spherical symmetry is to a great extent independent of its very details. This remarkable property relies on two facts. On the one hand, as is by now well established, the behaviour of a system is entirely determined by few order parameters. On the other hand the underlying symmetry restricts the structure of the generalized Ginzburg-Landau equations (GGLE), the equations of motion of the order parameters, to a definite form. The structure of this dynamical system is solely given by the condition of O(3)-invariance. Special properties, specific for a concrete system, usually manifest only in relatively few coefficients in the nonlinearities of the GGLE. In this respect, a discussion of the instability of a spherically symmetric state becomes possible without a direct reference to a concrete system, and the obtained results thus may apply to quite different systems. This fact will transpire from the following brief outline of the mathematical description of an instability of a system with O(3)-symmetry.

We consider an evolution equation of the form

$$\partial_t \, \underline{q}(\underline{r},t) - L(\{\sigma\}) \, \underline{q}(\underline{r},t) = \underline{N}(\{\sigma\},\underline{q}(\underline{r},t)) \quad , \tag{1}$$

where $L(\{\sigma\})$ denotes a linear operator and $\underline{N}(\{\sigma\},\underline{q}(\underline{r},t))$ is a nonlinear operator which is at least quadratic in the state vector $\underline{q}(\underline{r},t)$, so that $\underline{q}^0(\underline{r},t) = 0$ is a solution of (1). This solution is supposed to become unstable at certain critical values $\{\sigma_u\}$ of the control parameters $\{\sigma\}$, which are determined simultaneously with the corresponding unstable modes by an investigation of the linear eigenvalue problem

$$[\ \lambda_k - L(\{\sigma\})\ ]\ \underline{\phi}_k(\underline{r}) = 0\quad . \tag{2}$$

If some eigenvalues λ_u as functions of the control parameters $\{\sigma\}$ cross the imaginary axis and obtain positive real parts, instability sets in (we restrict our attention to systems which have a discrete set of eigenvalues λ_k). Under certain conditions the emergence of a new stable state of the system can be interpreted as a process of selforganization and can be successfully handled by an application of the slaving principle of synergetics [1], [2], [3]: The newly evolving state is described by a superposition of the unstable modes $\underline{\phi}_u(\underline{r})$ and the stable modes $\underline{\phi}_s(\underline{r})$ of the linear eigenvalue problem (2):

$$\underline{q}(\underline{r},t) = \sum_u \xi_u(t)\ \underline{\phi}_u(\underline{r}) + \sum_s \xi_s[\xi_u(t)]\ \underline{\phi}_s(\underline{r})\quad . \tag{3}$$

As a result of the slaving principle the amplitudes ξ_s of the unstable modes are functions of the amplitudes $\xi_u(t)$ of the order parameters and do not depend explicitly on time. Therefore, the dynamics of the system near an instability point is exclusively determined by the order parameters $\xi_u(t)$. Their temporal behaviour is given by the GGLE

$$\xi_u(t) = \Lambda_{uu_1}[\{\sigma\}]\ \xi_{u_1} + f_u[\{\sigma\},\xi_u(t)]\quad . \tag{4}$$

In the general case the linear operator Λ_{uu_1} is equivalent to a Jordan normal form [2].

So far we have not yet taken into account that we are concerned with the instability of a spherically symmetric state. To this end we consider an arbitrary rotation of the coordinate system described by the matrix Q. This rotation transforms a solution $\underline{q}(\underline{r},t)$ to a new state $R\underline{q}(Q^{-1}\underline{r},t)$, which also has to be a solution of the equation (1), because the system under consideration is supposed to be spherically symmetric. If the state vector $\underline{q}(\underline{r},t)$ is a vector field the linear operator is just the Matrix Q, if $\underline{q}(\underline{r},t)$ is a scalar field then R is the identity. It is well known from group representation theory [4] that a rotation Q induces the existence of a linear operator D according to

$$R\ \underline{q}(Q^{-1}\underline{r},t) = D\ \underline{q}(\underline{r},t)\quad . \tag{5}$$

Since equation (1) defines an invariant decomposition into subspaces spanned by the unstable and stable modes, respectively, we can consider the representations induced by the following equation:

$$R \sum_u \xi_u\ \phi_u(Q^{-1}\underline{r}) = \sum_{u_1 u_2} D_{u_1 u_2}\ \xi_{u_1}\ \phi_{u_2}(\underline{r})\quad . \tag{6}$$

The transformation $D_{u_1 u_2}$ can be decomposed into its irreducible parts [4]. The index u now consists of three parts, $u = \{i, \ell, m\}$. The index ℓ classifies the transformation properties of the modes. There are $2\ell+1$ modes belonging to the ℓ-th irreducible representation which are distinguished by the index m. Furthermore, since there may be several groups of modes transforming according to this representation, these groups have to be distinguished by the index i. It follows from relation (6) that the order parameters $\xi_{i\ell m}$ transform in the following way:

$$\xi_{i\ell m} \rightarrow \sum_{m_1} D^{(\ell)}_{mm_1} \xi_{i\ell m_1} \ . \tag{7}$$

The GGLE (4) now take the form

$$d_t \, \xi_{i\ell m}(t) = \Lambda^{(\ell)}_{ii_1} \, \xi_{i_1\ell m}(t) + f_{i\ell m}(\{\xi_{i\ell m}(t)\}) \quad , \tag{8}$$

where the linear matrix $\Lambda^{(\ell)}_{ii_1}$ is, in the most general case, equivalent to a Jordan normal form. The O(3)-invariance of the evolution equation immediately turns over to the GGLE:

$$\sum_{m_1} D^{(\ell)}_{mm_1} \, f_{i\ell m_1}(\{u_{i\ell m}\}) = f_{i\ell m}(\{\sum_{m_1} D^{(\ell)}_{mm_1} u_{i\ell m_1}\}) \quad . \tag{9}$$

As usual, the nonlinearities $f_{i\ell m}(\{\xi_{i\ell m}\})$ have to be approximated by polynomial expressions of the form (we give an expression for the quadratic terms)

$$\alpha^{(2)}_{i\ell m; i_1\ell_1 m_1, i_2\ell_2 m_2} \, \xi_{i_1\ell_1 m_1} \xi_{i_2\ell_2 m_2} \ . \tag{10}$$

As a consequence of the invariance condition (9) there are various relations among the coefficients $\alpha^{(2)}_{i\ell m; i_1\ell_1 m_1, i_2\ell_2 m_2}$ reducing the effective number of unknown coefficients considerably. An algorithm for the determination of covariant polynomials of arbitrary order is described in references [5], [6], [9]. It turns out that the knowledge of only few coefficients is usually necessary to determine the behaviour of the system in the vicinity of an instability. Therefore, it becomes possible to discuss quite different systems and to classify their behaviour from a general point of view. In the following we want to exemplify this for the case where a spherically symmetric stationary state loses its stability against disturbances of two groups of modes belonging to the irreducible representations with $\ell=1$ and $\ell=2$.

A detailed analysis following the lines indicated above gives the structure of the GGLE for the case of mode interaction with $\ell=1$ and $\ell=2$ [7], [8]:

$$
\begin{aligned}
d_t \, \xi_m &= \lambda_1 \, \xi_m + (-1)^m \frac{\partial}{\partial \xi_{-m}} \, \{ \, \alpha \, Q^{(3)}(\underline{\xi}:\underline{\xi}:\underline{\eta}) + a/4 \, E_1^2 \\
&\quad + b/2 \, E_2 \, E_1 + c \, Q^{(4)}(\underline{\xi}:\underline{\xi}:\underline{\eta}:\underline{\eta}) \, \} \\
d_t \, \eta_m &= \lambda_2 \, \eta_m + (-1)^m \frac{\partial}{\partial \eta_{-m}} \, \{ \, - 2\beta \, Q^{(3)}(\underline{\xi}:\underline{\xi}:\underline{\eta}) + \delta \, Q^{(3)}(\underline{\eta}:\underline{\eta}:\underline{\eta}) \\
&\quad d/4 \, E_2^2 + (\, e/2 - f \,) \, E_2 \, E_1 - f \, Q^{(4)}(\underline{\xi}:\underline{\xi}:\underline{\eta}:\underline{\eta}) \, \} \quad .
\end{aligned} \tag{11}
$$

The amplitudes of the modes with $\ell=1$ and $\ell=2$ are denoted by ξ_m and η_m, respectively. E_1, E_2, $Q^{(3)}(\underline{\xi}:\underline{\xi}:\underline{\eta})$, $Q^{(3)}(\underline{\eta}:\underline{\eta}:\underline{\eta})$ and $Q^{(4)}(\underline{\xi}:\underline{\xi}:\underline{\eta}:\underline{\eta})$ are polynomial invariants of the O(3)-group. Explicit expressions can be easily derived from the covariant polynomials given in reference [8].

The form of the GGLE for an instability with only one group of modes with $\ell=1$ or $\ell=2$ can be obtained from equation (11) by putting either ξ or η equal to zero. In this case the dynamical system (11) reduces to a gradient system so that, besides transients, no time-dependent behaviour can occur. We observe that an instability with mode interaction with $\ell=1$ and $\ell=2$ is, up to degenerate cases, essentially determined only by the eigenvalues λ_1, λ_2, three coefficients α, β, δ of the quadratic terms and six coefficients a, b, c... of the cubic terms. Although this is certainly a difficult task, it is possible to consider these coefficients as parameters and to classify, if possible, the behaviour of the order parameter equations to obtain a complete overview of the temporal behaviour which may be induced by mode interaction with $\ell=1$ and $\ell=2$. This fact underlines the strong relations between temporal behaviour, governed by the order parameter equations, and the coherent structure of the emerging spatial patterns.

336

3. TIME — DEPENDENT AND CHAOTIC THERMAL CONVECTION IN SPHERICAL GEOMETRIES

In the following we shall be concerned with a concrete model rather than with a general investigation of the dynamical system (11). To this end we shall concentrate on the problem of thermal convection in spherical geometries [9], [10], [11], [12], [7], [8], which has turned out to be an appealing paradigm of a system exhibiting an instability of a spherically symmetric state and which simultaneously is of considerable interest for a more detailed understanding of fascinating geophysical as well as astrophysical fluid motions.

Let us consider a homogeneous viscous fluid contained between two concentric spherical shells in a radially symmetric gravity field. A spherically symmetric temperature gradient is maintained by a constant temperature difference between the two boundaries as well as homogeneously distributed heat sources inside the fluid. We are confronted with essentially the same situation as in the well-known Bénard experiment [13] except for the different geometry. For small temperature differences between the two boundaries we observe a stable purely heat conductive state. For sufficiently high temperature differences convection sets in. Using the Boussinesq-approximation [13], [14] the behaviour of the fluid is determined by the Navier-Stokes equations

$$P^{-1}\{ \frac{\partial}{\partial t}\underline{v}(\underline{r},t) + [\underline{v}(\underline{r},t)\cdot grad]\ \underline{v}(\underline{r},t) \} =$$

$$\Delta\ \underline{v}(\underline{r},t) + R\gamma(r)\ \underline{r}\ T(\underline{r},t)\text{-}grad\ p(\underline{r},t) \quad , \tag{12}$$

the incompressibility condition for the velocity field $\underline{v}(\underline{r},t)$

$$div\ \underline{v}(\underline{r},t) = 0 \quad , \tag{13}$$

and the equation of heat conduction

$$\frac{\partial}{\partial t}T(\underline{r},t) + [\underline{v}(\underline{r},t)\cdot grad]\ T(\underline{r},t) = \Delta\ T(\underline{r},t) + \tau(r)\ \underline{v}(\underline{r},t)\cdot\underline{r} \quad . \tag{14}$$

The scalar field $T(\underline{r},t)$ describes the contribution to the temperature due to convection. In the case of pure heat conduction $T(\underline{r},t)$ vanishes. The Prandtl number and the Rayleigh number are defined through

$$P = \nu\ /\ \kappa \qquad R = \tilde{\alpha}\ h^3 g_0 (T_i\text{-}T_a)\ /\ (\kappa\nu) \quad . \tag{15}$$

The functions $\gamma(r)$ and $\tau(r)$ are related to the gravitational field and the local temperature field of the purely heat-conductive state (A more detailed discussion can be found in reference [8]):

$$\gamma(r) = [1+\tilde{\gamma}(\frac{\rho}{r})^3]/[(1+\tilde{\gamma})\rho] \quad , \quad \tau(r) = [1+\tilde{\tau}(\frac{\rho}{r})^3]/\tau_0 \quad . \tag{16}$$

The two boundaries are assumed to be perfectly heat conducting and can be regarded to be either free or rigid. The Rayleigh number and the aspect ratio ρ, defined as the ratio of the radius of the inner sphere to the thickness of the fluid shell, will serve as the two most important control parameters.

An increase of the Rayleigh number R leads to an instability of the heat conductive state and eventually to the onset of macroscopic fluid motion. A detailed linear analysis of this instability can be found in references [14], [8]. A normal mode analysis of the form

$$\underline{v}(\underline{r},t) = curl\ curl\ \underline{r}\ S(\underline{r},t) + curl\ \underline{r}\ Z(\underline{r},t) \quad ,$$

$$T(\underline{r},t) = \sum_{i,\ell} \sum_{m=-\ell}^{m=+\ell} T_{i\ell}(r)\ Y_\ell^m(\theta,\phi)\ e^{\lambda_{i\ell}t} \quad , \tag{17}$$

and similar expressions for the scalar fields $S(\underline{r},t)$ and $Z(\underline{r},t)$, which define the poloidal and the toroidal velocity fields, respectively, lead to a linear

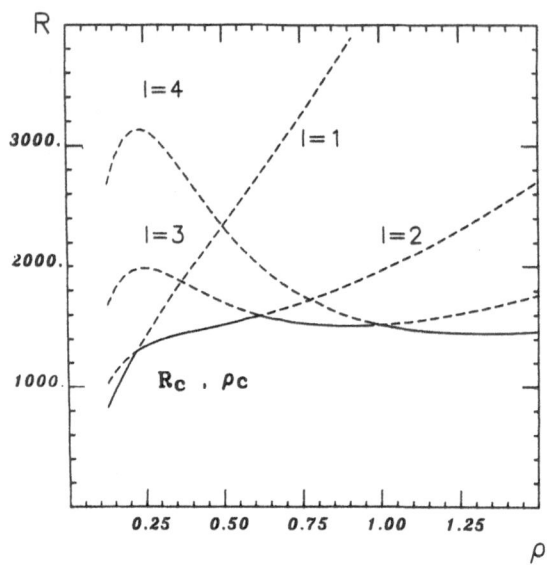

Fig. 1:
Neutral curves for disturbances
with $\ell=1,..,4$

$(\bar{\rho}=4,\ \tilde{\tau}=0,$
boundary conditions:
rigid / free)

eigenvalue problem which determines the critical Rayleigh number and the unstable
modes. The special form (17) of the scalar fields $T(\underline{r},t)$, $S(\underline{r},t)$, and $Z(\underline{r},t)$
is due to the spherical symmetry of the model and the eigenvalue $\lambda_{j\ell}$ is
typically $(2\ell+1)$-fold degenerate. The instability turns out to be
nonoscillatory. A typical result of the linear stability analysis is demonstrated
in Fig. 1, which shows the critical Rayleigh numbers for the disturbances of modes
with different values of ℓ as functions of the aspect ratio ρ. As a
consequence of the balance between the effectiveness of the convective heat
transport and the dissipative loss through viscous fluid flow the value ℓ of the
most unstable modes depends on the aspect ratio ρ. This value ranges from $\ell=1$
for small aspect ratios up to infinity for $\rho \to \infty$, where we obtain the limiting
case of the planar Bénard problem. We observe that there are special values of
the aspect ratio ρ, where two groups of modes classified by the values ℓ and
$\ell+1$ become unstable simultaneously. In the sequel we shall address our attention
to an instability involving the two groups of modes with $\ell=1$ and $\ell=2$.

To this end we introduce the following representation of the velocity and the
temperature fields

$$\underline{v}(\underline{r},t) = \sum_j x_j(t) \text{ curl curl } \underline{r}\ S_\ell^c(r)\ Y_\ell^m(\theta,\phi) + \underline{w}[\underline{x}(t),\underline{r}]\quad,$$
$$T(\underline{r},t) = \sum_j x_j(t)\ T_\ell^c(r)\ Y_\ell^m(\theta,\phi) + \tau[\underline{x}(t),\underline{r}]\quad,$$

(18)

where j is composed of the two indices (ℓ,m). Remember that this form is an
immediate consequence of the slaving principle of synergetics. Obviously, the
behaviour of the fluid is governed by the dynamics of the order parameters $\underline{x}(t)$.
Since we consider a situation with a simultaneous instability of the two groups of
modes with $\ell=1$ and $\ell=2$ the order parameter equations are given by the
equations (11), where now the coefficients α, β, ... have to be determined
starting from the basic equations. A detailed description of these calculations as
well as the numerical results can be found in reference [8].

The velocity field $\underline{w}[\underline{x}(t),\underline{r}]$ and the temperature field $\tau[\underline{x}(t),\underline{r}]$ is of
the order $\underline{x}(t)^2$ and thus small compared to the first terms. Therefore, the
velocity field $\underline{v}(\underline{r},t)$ and the temperature field $T(\underline{r},t)$ are mainly determined
by a time-dependent superposition of the unstable modes. The temperature fields of
the unstable modes are represented in Fig. 2a,b . There is essentially only one
convection pattern of the modes with $\ell=1$, which describes an axisymmetric

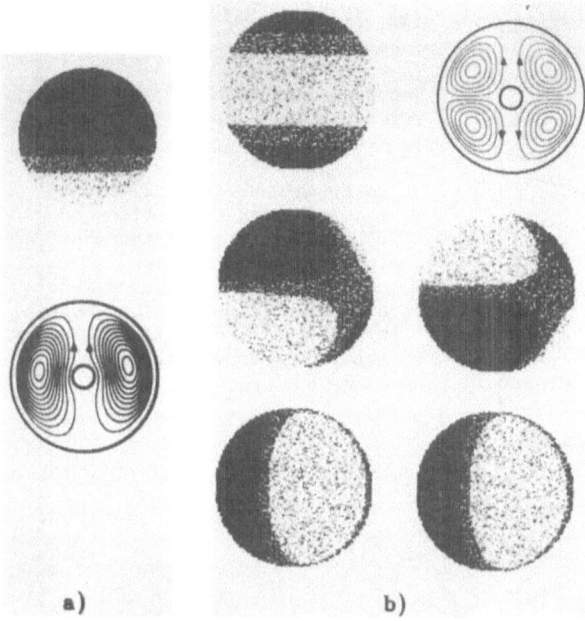

Fig. 2: a) Temperature pattern and streamlines in a meridional plane
for the mode with $\ell=1$, $m=0$
b) Temperature patterns of the modes with $\ell=2$
and streamlines of the mode with $\ell=2$, $m=0$

Hot regions with ascending fluid correspond to dark regions.
Cold and descending fluid is marked by the bright regions

convection cell. The two remaining modes differ only in the orientation of the
symmetry axis. The group of modes with $\ell=2$ contains two physically different
types of patterns, an axisymmetric and a nonaxisymmetric one, from which the
remaining ones again can be obtained by a pure rotation.

3.1 A Typical Phase Diagram

To demonstrate the richness of the spatio-temporal behaviour of a system
exhibiting mode interaction with $\ell=1$ and $\ell=2$ we describe the various
stationary and time-dependent convection flows which can be observed in the
vicinity of the critical point R_c, ρ_c (see Fig. 1). We consider the model
for the parameter values $\tilde{\gamma}=4$, $\tilde{\tau}=0$ and large Prandtl number P, where we
have supposed the inner boundary to be rigid and the outer boundary to be free.
The result of an examination of the GGLE, which can be done partly analytically,
partly numerically, is summarized in a phase diagram (see Fig. 3). At the onset of
convection the fluid flow is stationary and axisymmetric. For aspect ratios
greater than ρ_c the flow is described by the axisymmetric mode with $\ell=2$ so
that one observes two convection cells. The transition is slightly subcritical.
There are two different possible flow directions of the convection cells. In the
case of the α cells the fluid arises at the poles and descends at the equator.
The β cells possess the reversed flow direction (see Fig. 2b). Only one of the
two kinds of flow turns out to be stable. In the present case these are the β
cells. For aspect ratios smaller than the critical one the instability of the heat
conductive state gives rise to a flow pattern, which is, close to the boundary,
mainly described by the mode with $\ell=1$. The transition is supercritical. For

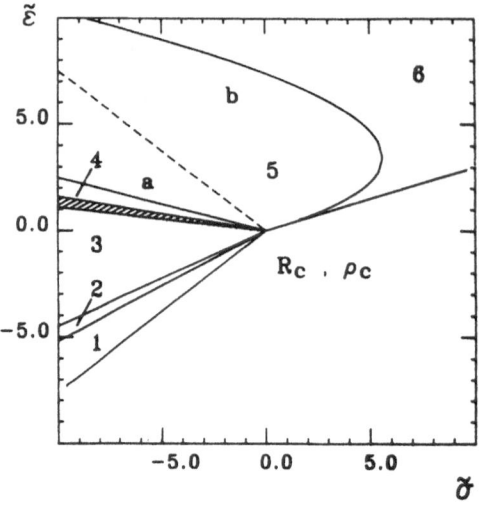

Fig. 3:
Phase diagram
($\tilde{\rho}=4$, $\tilde{\tau}=0$, large P,
boundary conditions:
rigid / free)
1, stationary convection ($\ell=1$)
2, rotating wave (type 1)
3, rotating wave (type 2)
4, quasiperiodic convection
   (shaded area: coexistence of
   rotating wave motion
   and quasiperiodic convection)
5, (a) aperiodic convection (type 1)
   (b) aperiodic convection (type 2)
6, stationary convection ($\ell=2$)

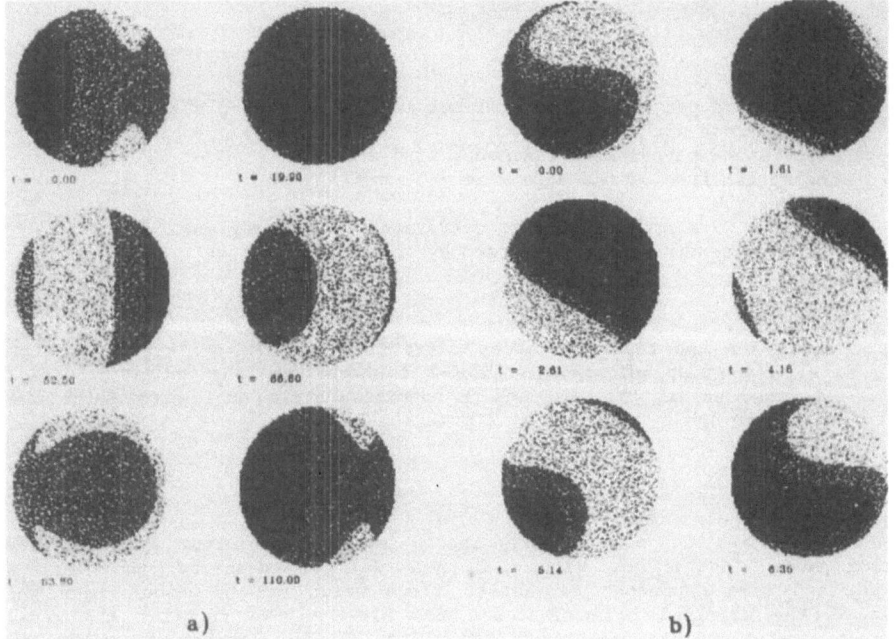

a)   b)

Fig. 4: a) Temperature patterns for rotating wave motion (type 1)
        b) Temperature patterns for rotating wave motion (type 2)

higher Rayleigh numbers the flow patterns are determined by a superposition of the
two axisymmetric modes with $\ell=1$ and $\ell=2$ so that the flow consists of two
axisymmetric convection cells which, however, have different sizes. In addition to
these two stationary convective states the fluid is able to undergo wavelike
motions which consist of two convection cells rotating around a certain axis. The
rotating wave, denoted type 1, occurs due to a supercritical instability of the
axisymmetric stationary state for ρ smaller than the critical one. The flow
pattern is mirror symmetric with respect to a plane perpendicular to the axis of
rotation. A secondary supercritical instability destroys this mirror symmetry but

340

the flow is still a wavelike motion (see Fig. 4a,b). A subcritical instability of this time periodic flow gives rise to a quasiperiodic fluid motion. In addition to a pure rotation of the flow patterns we now observe a time periodic deformation of the convection cells. These two kinds of motions could be decoupled by simply moving to a corotating coordinate system. Therefore, no frequency locking can occur and the fluid flow is quasiperiodic except for the cases where the ratio of the two frequencies as a function of the control parameters is rational. A further increase of the Rayleigh number leads to the occurrence of chaotic transients to the motion on the two-dimensional torus and eventually to an aperiodic fluid motion. We can distinguish between two slightly different kinds of chaotic fluid motions (see Fig.3) [8]. In the following we shall describe the main characteristics of the chaotic fluid motion denoted by type 2.

The main sequence of the temporal evolution of the type 2 chaotic fluid motion is demonstrated by the spatial patterns of the temperature field in Fig. 5. At $t=0$ the system is in a state approximately described by the axisymmetric mode with $\ell=2$ corresponding to the β cells. If the modes with $\ell=1$ were strongly damped, as is the case for values of ρ much larger than the critical one, this state would be stable. It corresponds to one of the two fixed points of the order parameter equations with $\xi=0$ describing α cells and β cells. In the vicinity of R_c, ρ_c this fixed point is unstable with respect to disturbances of the modes with $\ell=1$. As a consequence the temperature pattern changes to another one which is dominated by the mode with $\ell=1$ ($t=2.5$). The subsequent temporal evolution involves essentially only modes with $\ell=2$ and eventually leads to the same axisymmetric pattern as the one at $t=0$. But now the hot and cold regions have exchanged their positions ($t=4.$). The fluid flow is nearly axisymmetric so that it is possible to demonstrate this part of the sequence by an inspection of the streamlines of the flow in a plane containing the symmetry axis (see Fig. 6). One starts with the configuration of the β cells. One of the two convection cells begins to decrease, the other increases until there is only one cell left. Then a small cell appears at the opposite side. This cell grows in size until the flow reaches the configuration of the α cells. Thus, a reversal of the

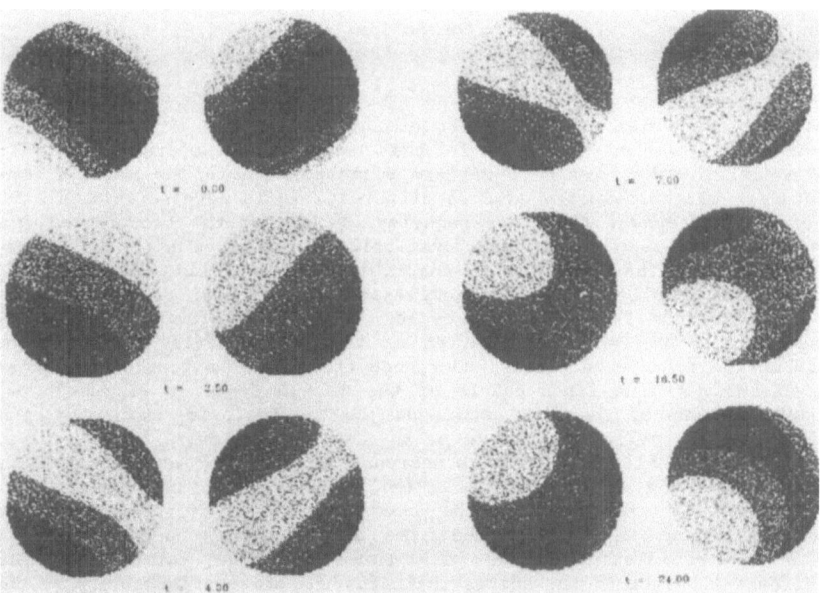

Fig. 5: Temporal evolution of the temperature patterns for chaotic convection (type 2)

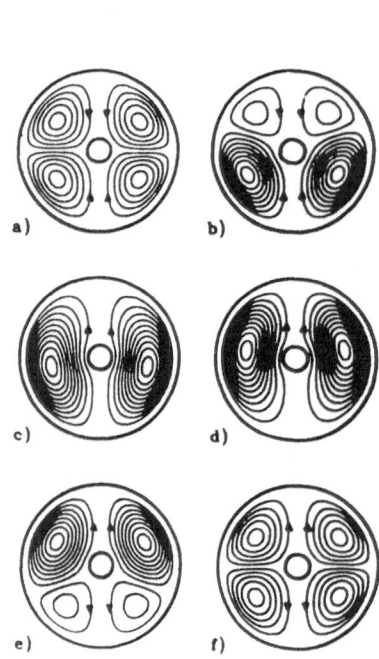

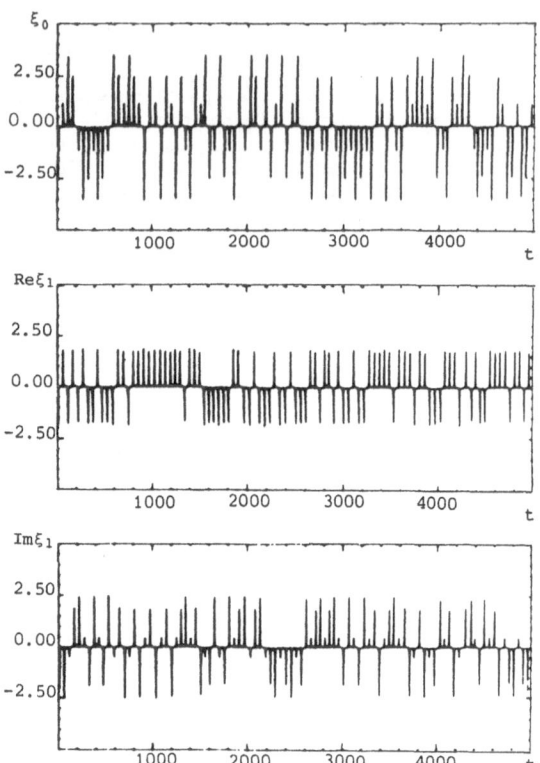

Fig. 6:
Streamlines in a meridional
plane during a reversal of the
flow direction (transition
from β cells to α cells)

Fig. 7:
Time evolution of the order
parameters $\xi(t)$ for chaotic
convection (type 2)

flow direction has taken place. There exist orbits joining the two fixed points $\xi_0=0$, $\eta_0\neq0$ in the ξ_0-η_0 phase plane. In the sequel the temperature pattern changes in a nonaxisymmetric way to a pattern corresponding to the β cells. But now the orientation of the symmetry axis is different, it is perpendicular to the former axis of the pattern at t=0. This behaviour is due to the fact that the α cells are unstable with respect to nonaxisymmetric disturbances with $\ell=2$: there exist orbits connecting the fixed point describing the α cells and the fixed point describing the β cells with a symmetry axis perpendicular to the symmetry axis of the α cells. Thus two additional cycles follow until the system reaches nearly the same state from where it started at t=0. The chaotic nature of the observed fluid motion can now be easily understood from the behaviour of the phase point in the vicinity of the fixed points of the α- and β- cells. It turns out that the chaotic element of the flow occurs due to the following mechanism: The transition from the β cells to the α cells can be achieved in two ways, because each of the two cells can begin to decrease so that the intermediate flow pattern with $\ell=1$ (see Fig. 6, pattern (c), (d)) can have two flow directions. This flow direction is indicated by the order parameters $\xi_k(t)$ whose time evolution is represented in Fig. 7. The time series of the order parameter $\xi_0(t)$ shows nearly equidistant spikes with three different extremal absolute values. The different height of the spikes indicates the three orientations of the elementary cycles described above. The sequence of the absolute value of the heights is regular, whereas the sign of the spikes apparently is stochastic. This means that it appears to be completely random which of the two cells disappears during the transition from the β- cell to the α- cell- configuration. In which way this reversal of the flow direction takes place depends on the very details of the spatial flow patterns. This fact clearly results in a sensitive dependence of the temporal evolution of the fluid flow on the initial conditions and is responsible for the chaotic nature of the fluid motion.

3.2 A further example of chaotic fluid motion

The above-described phase diagram and the observed fluid motions do not exhaust all possibilities of the spatio- temporal behaviour which can be induced by mode interaction with $\ell=1$ and $\ell=2$. So far no complete classification is available. To demonstrate a further behaviour we consider the model for large Prandtl number and for the parameter values $\bar{\tau}=4$, $\tilde{\gamma}=0$. That means that we consider a different gravitational field and a different local temperature gradient of the heat conductive state. For this case we find a chaotic fluid motion which is described in the phase space of the order parameters by a generalization of a Silnikov—type attractor [15], [16]. The usual Silnikov—type chaotic motion occurs due to the existence of a fixed point with a two—dimensional stable manifold, where the phase point spirals towards the fixed point, and a one dimensional unstable manifold, which brings the phase point back to the stable manifold. The corresponding chaotic attractor is shown in Fig. 8, where we have drawn the E_1-E_2- phase space. Remember that the quantities E_1, E_2 are proportional to the kinetic energies in the part of the fluid motion described by the unstable modes with $\ell=1$ and $\ell=2$. At the same time, they are the quadratic invariants and it is therefore impossible to distinguish between equal flow patterns with different orientations by merely looking at these quantities. A time series of the order parameters $\xi_k(t)$ (see Fig. 9) indicates the difference between the usual Silnikov-type chaos and the present aperiodic motion: whereas in the usual case the phase point returns to the same fixed point, in the present case the phase point approaches a different fixed point, which, however, describes the same flow pattern except for a different orientation. The time evolution of the corresponding temperature fields are represented in Fig. 10. At t=0 the temperature pattern is nearly described by the fixed point. It consists of a superposition of the modes with $\ell=1$ and $\ell=2$. This pattern is unstable, the contribution of the modes with $\ell=1$ first increases (t=18.) and then decreases again leading to a pattern similar to the one at t=0 but with a different orientation (t=24.). In the sequel one observes a pulsation of this pattern, which corresponds to the spiralling motion of the phase point towards the unstable fixed point, until a further cycle sets in.

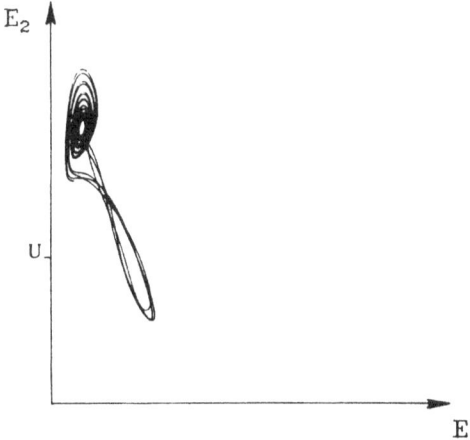

E_2

U

E_1

Fig. 8: Phase portrait of the Silnikov — type chaotic motion in the E_1-E_2 phase plane

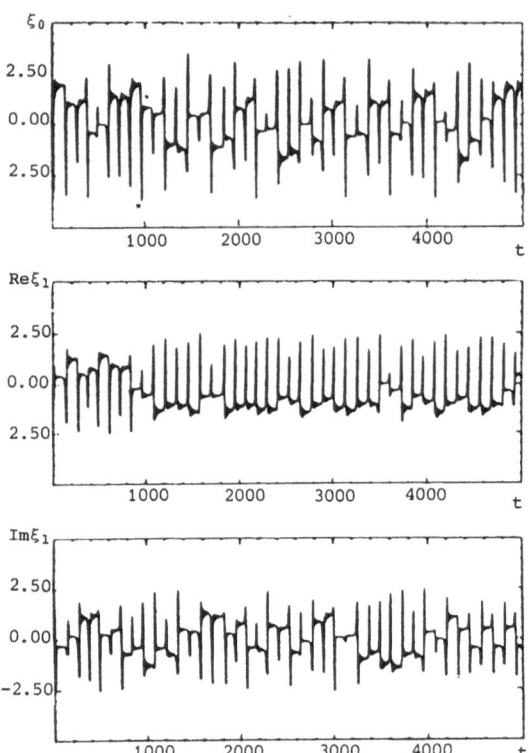

Fig. 9:
Temporal evolution of the order parameters $\xi(t)$ for the Silnikov-type chaotic motion

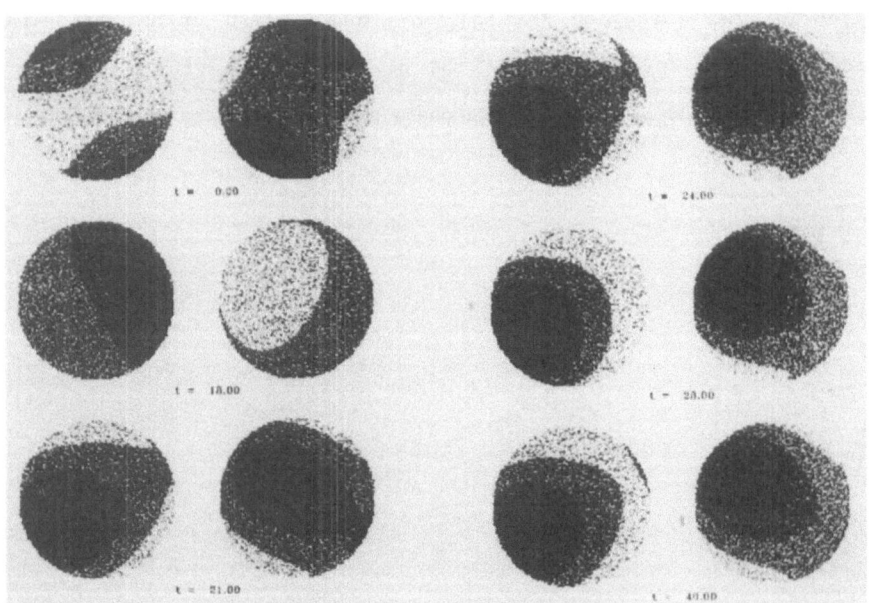

Fig. 10: Temporal evolution of the temperature patterns of the
Silnikov-type chaotic fluid motion

4. SUMMARY

Interesting and fascinating spatio- temporal behaviour can be observed in systems with O(3)-symmetry due to the interaction of modes belonging to different irreducible representation of the rotation group O(3). This has been exemplified for a situation where an instability occurs due to modes with $\ell=1$ and $\ell=2$ for the concrete case of thermal convection in spherical geometries. Various types of stationary, time-dependent, wavelike, and especially chaotic fluid motions can be observed and give considerable insight into the relations between coherent spatial flow patterns and a complex temporal evolution. Due to the generality of the order parameter equations it is expected that similar behaviour may be observed in quite different systems with O(3)-symmetry whenever mode interaction with $\ell=1$ and $\ell=2$ occurs.

ACKNOWLEDGEMENTS

We acknowledge the help of M. Bestehorn, H. Bunz, A. Fuchs, Th. Grauer, K. Marx, H. Ohno, and A. Wunderlin and wish to thank the Volkswagenwerk Foundation (Hannover) for financial support.

REFERENCES

[1] H. Haken, Synergetics. An Introduction, third edition
     (Springer-Verlag, Berlin, Heidelberg, New York 1983)
[2] H. Haken, Advanced Synergetics
          (Springer-Verlag, Berlin,Heidelberg, New York 1983)
[3] A. Wunderlin und H. Haken, Z. Physik. B 44, 135 (1981)
[4] M. Hammermesh, Group Theory and Its Application to
     Physical Problems (Addison- Wesley, Reading, Mass., 1962)
[5] D.H. Sattinger, J. Math. Phys. 19, 1720 (1978)
[6] D.H. Sattinger, Group Theoretic Methods in Bifurcation Theory
          Lecture Notes in Math., 762
          (Springer-Verlag, Berlin, Heidelberg, New York 1979)
[7] R. Friedrich und H. Haken, in Complex Systems-Operational Approaches,
     Proceedings of the International Symposium on
     Synergetics 1985, Schloß Elmau, edited by H. Haken
     (Springer-Verlag, Berlin, Heidelberg, New York 1985)
[8] R. Friedrich und H. Haken, Phys. Rev. A 34, 2100 (1986)
[9] F.H. Busse, J. Fluid Mech. 72, 65 (1975)
[10] F.H. Busse and N. Riahi, J. Fluid Mech. 123, 283 (1981)
[11] M. Golubitsky and D. Schaeffer, Commun. Pure a. Applied
          Math. 35, 81 (1982)
[12] P. Chossat : SIAM J. Appl. Math. 37, 624 (1979)
[13] F.H. Busse, Rep. Progr. Phys. 41. 1929 (1978)
[14] S.Chandrasekhar, Hydrodynamic and Hydromagnetic Stability
          (Clarendon, Oxford 1961)
[15] L. P. Silnikov, Sov. Math. Dokl. 6, 163 (1965);
          Sov. Math. Dokl. 8, 54 (1967); Math. USSR SB 10, 91 (1970)
[16] J. Guckenheimer und P. Holmes, Nonlinear Oscillations,
     Dynamical Systems and Bifurcation of Vector Fields
          (Springer-Verlag, Berlin, Heidelberg, New York 1983)

Selfsimilarity of Developed Turbulence

H. Effinger and S. Grossmann

Fachbereich Physik der Philipps-Universität, Renthof 6,
D-3550 Marburg, Fed. Rep. of Germany

One aspect of fluid flow with large Reynolds number and energy input is to understand its chaotic structure, the Lyapunov spectrum, the K-entropy, the invariant measure, etc., starting from the properties of the Navier-Stockes equations[1]. This would justify a statistical mechanics of turbulence. The question of which measurable statistical properties emanate from that have found even longer lasting interest[2]. A quantity of particular interest is the structure function

$$D(r) = \langle\!\langle |\vec{u}(\vec{x}+\vec{r})-\vec{u}(\vec{x})|^2 \rangle\!\rangle \tag{1}$$

in stationary, homogeneous, isotropic turbulence. $\vec{u}(\vec{x},t)$ is the Eulerian velocity field, $\langle\!\langle ... \rangle\!\rangle$ denotes the ensemble average (which then is independent of $\vec{x}$ and t). $\vec{u}$ is solenoidal and solves the Navier-Stockes equations (NS-E)

$$\partial_t u_i(\vec{x},t) = - u_j(\vec{x},t)\partial_{x_j} u_i(\vec{x},t) - \partial_{x_i} p(\vec{x},t) + \nu\Delta_x u_i(\vec{x},t) + f_i(\vec{x},t). \tag{2}$$

Here p denotes the kinematic pressure, ν the kinematic viscosity, and f_i the components of the external force. $D(r)$ describes the energy of an eddy of size r. There is general agreement and much support from measurements that in high Reynolds number flow the scaling behavior of $D(r)$ reveals a selfsimilar structure of the velocity field,

$$D(r) = b\varepsilon^{2/3}r^{2/3}, \qquad \eta \ll r \ll L. \tag{3}$$

This r-range is denoted as the inertial subrange (ISR), L being the scale at which energy is fed in, $\eta=(\nu^3/\varepsilon)^{1/4}$, ε the (mean) rate of energy dissipation, b=8.4 experimentally. Eq.(3) is believed to have corrections $\propto(r/L)^B$ due to intermittency.

This talk does not intend to contribute to the (small) intermittency corrections but, to show how the NS-E manage to lead to (3) in a kind of mean field approximation. Of course, the motivation is to provide a convenient and well understood point of reference for treating the fluctuations (intermittency), in a similar way as the Landau theory is a point of reference for dealing with critical fluctuations.

There were several attempts[3,4] to develop a classical field theory of the NS-E and to derive (3), together with a regular behavior for r in the viscous subrange (VSR)

$$D(r) = \frac{1}{3}\frac{\varepsilon}{\nu} r^2, \qquad 0 \le r \ll \eta, \tag{4}$$

including a proper transition from the ISR to the VSR. The effect of coupling random forces to the NS-E[5], moment closure methods[6], etc., have been studied. In particular R.Kraichnan's work[3] elucidated the most important role of the Galilean invariance of the NS-E, showed how to preserve it by introducing a Lagrangian description, and that the lowest order perturbation expansion (DIA) for the vertices (but with renormalized propagators) seem quite sufficient to get $D(r)$ or its Fourier transform, the spectrum $E(k)\propto k^{-5/3}$. His method[3], the introduction of a

mixed Eulerian-Lagrangian description with two-time fields, the various unusual
approximations made in order to derive the ALHDIA set of equations, unfortunately
make it hard to keep track of the physics, to derive systematic improvements, or
to include intermittency. We have offered an alternative approach[7]. Although on the
same general level of approximation, it is much simpler and therefore hopefully
will induce progress in the indicated direction.

In the following we outline our derivation of the structure function in a mean
field sense, with a DIA-like vertex approximation, propagator (D)-renormalized
closure, *starting from the NS-E*(2). By carefully preserving Galilean invariance
rather brute force methods seem quite well justified in deriving an equation for
$D(r)$, whose solution leads to (3), (4) *without* the necessity of any additional ansatz, scaling hypothesis, cut-off (since no infra-red singularity shows up), ad-
justable parameter, or the like. Our results are presented in Fig.1.

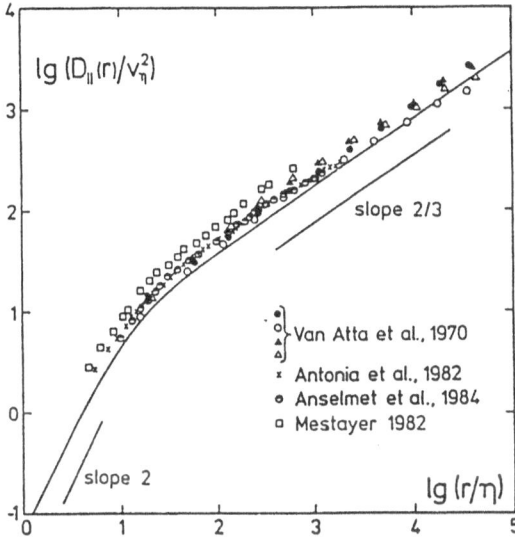

Van Atta et al., 1970

× Antonia et al., 1982
⊙ Anselmet et al., 1984
□ Mestayer 1982

Fig. 1. The longitudinal part $D_{\shortparallel}(r)$ of
the eddy energy distribution vs eddy
size r, present theory compared with
various experiments[8]

The basic idea is to look at the kinetic energy contained in the eddies of all
scales greater than some r and check how this energy *varies*[9] with r. Furthermore,
in contrast to most previous theories we work in position space, which is most
natural to formulate the Lagrangian motion of fluid elements. To define the separation of the sub-r- and the super-r-scales we introduce

$$u_i^{(r)}(\vec{x},t) = \frac{1}{V_r} \int d^3y\, u_i(\vec{x}+\vec{y},t) \equiv \left\langle u_i(\vec{x}+\vec{y},t)\right\rangle_y^{(r)} . \tag{5}$$

The integral extends over the sphere with radius r and volume V_r located at $\vec{x}$.
While $u_i^{(r)}$ contains all super-r-scales, the sub-r-scales are given by
$\tilde{u}_i^{(r)} = u_i - u_i^{(r)}$, which varies rapidly with $\vec{x},t$ and has a small magnitude, cf. Fig.2.
The scale boundary r is allowed to vary through all ranges. The energy in the
super-r-scales is intimately connected with the r-averaged structure function,

$$\left\langle\!\left\langle |\vec{u}^{(r)}|^2 \right\rangle\!\right\rangle = \left\langle\!\left\langle \vec{u}^2 \right\rangle\!\right\rangle - (1/2)\left\langle\!\left\langle \left\langle D(\vec{y}_1+\vec{y}_2)\right\rangle_{y_1}^{(r)} \right\rangle_{y_2}^{(r)} . \tag{6}$$

In observing that only derivatives will be needed, the super-scale energy is ex-
pressable in terms of the averaged Galilean invariant object D. The sub-scale
field $\tilde{u}_i^{(r)}$ is Galilean invariant. By r-averaging of the NS-E (2) one finds the

347

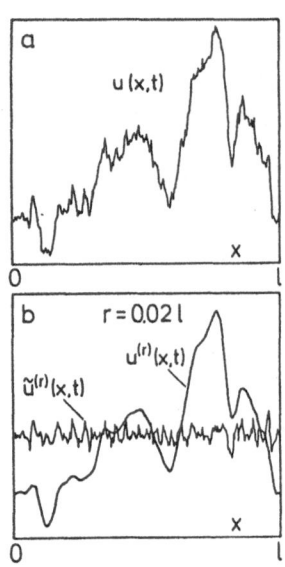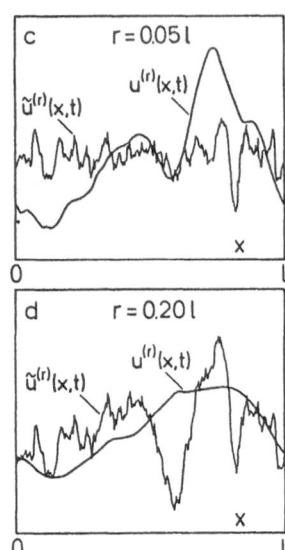

Fig. 2. Decomposition of the velocity into its sub- and super-r-scale parts (1-dimensional, about 100 Fourier coefficients)

rate of change of the super-r-scales, needed to calculate the power balance $\langle\!\langle u_i^{(r)} \partial_t u_i^{(r)} \rangle\!\rangle = 0$. The following terms contribute

$$ 0 = \langle\!\langle u_i^{(r)} f_i \rangle\!\rangle + \nu \langle\!\langle u_i^{(r)} \Delta_x u_i^{(r)} \rangle\!\rangle - \langle\!\langle u_i^{(r)}(\vec{x}) \langle \tilde{u}_j^{(r)}(\vec{x}+\vec{y}) \partial_{x_j} \tilde{u}_i^{(r)}(\vec{x}+\vec{y}) \rangle_y^{(r)} \rangle\!\rangle ,\quad (7) $$

The first term is the power input $E_{in}^{(r)}$ into *all scales above* r. It is assumed to be r-independent if r becomes smaller than L (L being the scale of the external forces f_i which induce turbulence) since all corresponding $u^{(r)}$ contain the L-scales. Then, from $r\to 0$ one concludes $E_{in}^{(r)} = \varepsilon$, $r<L$.

The second term in (7) is the direct dissipation, $-E_d^{(r)}$. With the basic relation (6)

$$ E_d^{(r)} = (\nu/2)\langle\!\langle \langle \Delta_{y_1} D(\vec{y}_1+\vec{y}_2) \rangle_{y_1}^{(r)} \rangle_{y_2}^{(r)}. \tag{8} $$

The third term, denoted by $-E_t^{(r)}$, describes the power transfer from the super-r-scales to sub-r-scales. This, of course, is the important but more difficult mechanism, which needs the typical approximation methods already mentioned. These will allow us to express also $E_t^{(r)}$ in terms of appropriate D-averages. A representative term will be given later in eq. (11). Therefore the power balance equation is cast into the form

$$ \varepsilon = E_d^{(r)}(\{D\}) + E_t^{(r)}(\{D\}). \tag{9} $$

This is a functional equation. Varying r allows one to calculate uniquely the turbulent structure function D(r). By the way, the sub-r-scale Lagrangian correlation function is obtained simultaneously in the one-pole approximation[10]. It is precisely its decay rate $\tilde{\gamma}(r)$, which is calculated from the NS-E (cf.[10]), whose ε-dependence enforces (3) in the ISR and whose limit $\tilde{\gamma}(r\to 0)$ implies (4) in the VSR. There are no singularities in (9), no cut-off is necessary, because our r-space formulation contains the range $r\approx 0$ where the finite viscosity guarantees regularity, while usually the k-space integrals contain the nonuniversal range at k=0. Furthermore, the consequent use of Lagrangian quantities implies a separation of the dynamical interaction from the advection, which has been one of the main problems[3]. The solution does not depend on the input power spectrum, if only $E(r)$ is constant for r below[11] the scale at which the fluid is stirred by the (non-random) forces f_i.

The arguments needed to derive[7] the functional $E_t^{(r)}$ consist of the following steps.

i) Write down the equation of motion for $\tilde{u}_i^{(r)}$, the velocity field on sub-r-scales, starting from the NS-E.

ii) Integrate this along the Lagrangian trajectory, terminating at time t in the relevant position. A representative contribution needed in $E_t^{(r)}$ is

$$\tilde{u}_i^{(r)}(\vec{x}+\vec{y},t) = \int_{-\infty}^{0} d\tau [-\tilde{u}_j^{(r)}(\tau|\vec{x}+\vec{y},t)u_{i|j}^{(r)}(\tau|\vec{x}+\vec{y},t)+...], \tag{10}$$

where $u(\tau|\vec{z},t)$ is the velocity of the fluid element at a time interval τ before it reaches position $\vec{z}$ at t.

iii) Introduce this Galilean invariant object into $E_t^{(r)}$ and decouple the arising (two time) fourth order moments of type $\langle\!\langle u^{(r)}u^{(r)}\tilde{u}^{(r)}\tilde{u}^{(r)}\rangle\!\rangle$ into the product $\langle\!\langle u^{(r)}u^{(r)}\rangle\!\rangle\langle\!\langle \tilde{u}^{(r)}\tilde{u}^{(r)}\rangle\!\rangle$. This factorization does *not* affect Galilean invariance, since $\tilde{u}^{(r)}$ does not contain advection on super-r-scales (by its definition as a difference). The factorization seems well justified since $u^{(r)}$ varies on large scales, whereas $\tilde{u}^{(r)}$ in small ones, cf. Fig.2.

iv) The temporal decay of the Lagrangian correlation is dominated by the small times characteristic of the sub-r-scales, during which the super-r-scales have not yet varied noticeably, i.e. $\langle\!\langle u^{(r)}u^{(r)}\rangle\!\rangle$ is approximated as a one-time object, even t-independent due to stationarity. Since it turns out that only spatial derivatives of it enter, it contributes only with its Galilean invariant part $\propto D$, cf.(6). The representative contribution to $E_t^{(r)}$ then reads

$$(N_{jk}^{(r)}/2)\left\langle\!\!\left\langle\left\langle \partial_{y_j}\partial_{y_k}D(\vec{y}+\vec{y}_1+\vec{y}_2)\right\rangle_y^{(r)}\right\rangle_{y_1}^{(r)}\right\rangle_{y_2}^{(r)}. \tag{11}$$

A comparison with eq. (8) suggests that one interpret $N^{(r)}$, denoting the Lagrangian correlation integral

$$N_{jk}^{(r)} = \int_{-\infty}^{0} d\tau \langle\!\langle \tilde{u}_j^{(r)}(\vec{x},t)\tilde{u}_k^{(r)}(\tau|\vec{x},t)\rangle\!\rangle, \tag{12}$$

as the "eddy viscosity" exerted by the sub-r-scales on the super-r-scales. $N^{(r)}$ is represented by a Kubo-formula, as it should be for a transport coefficient. As in its momentum space analogue, introduced in[9a], $N^{(r)}$ is not affected by advection.

v) Finally we trace back the eddy viscosity $N^{(r)}$ of the sub-r-scales to the structure function D(r). It is precisely here, where the details of the NS-E enter again, which in particular lead[10] to the explicit appearance of the energy dissipation rate ε in the expression for $N^{(r)}$. In the lowest order continued fraction approximation we obtained[7]

$$N^{(r)} = \tilde{C}^{(r)}\tilde{\gamma}^{(r)-1} = \tilde{C}^{(r)}\tilde{\Gamma}^{(r)-1}\tilde{C}^{(r)}. \tag{13}$$

The sub-r-scale correlation $\tilde{C}^{(r)}=\langle\!\langle \tilde{u}^{(r)}\tilde{u}^{(r)}\rangle\!\rangle$ can easily be expressed by sums of differently averaged D's. $\tilde{\Gamma}_{ij}^{(r)}(\vec{y})=-\langle\!\langle \tilde{u}_i^{(r)}L\tilde{u}_j^{(r)}(\vec{y})\rangle\!\rangle$, where L is the Liouvillian, can be evaluated to give

$$\tilde{\Gamma}^{(r)}(\vec{y}) =- 2\nu\langle\!\langle \tilde{u}_i^{(r)}\Delta\tilde{u}_i^{(r)}\rangle\!\rangle + \nu\Delta_y\langle\!\langle \tilde{u}_i^{(r)}\tilde{u}_i^{(r)}(\vec{y})\rangle\!\rangle. \tag{14}$$

The first term is up to small corrections just 2ε, the second can be reduced to D-averages. It is this $\tilde{\Gamma}^{(r)}$ which determines the scaling form of D by providing the relevant quantity ε from the NS-E. Thus $N^{(r)}$, describing the mean momentum drag of the sub-r-scale eddies on the super-r-scale motion, is expressed in terms of D, supporting the interpretation as a mean field theory.

It takes some work to elaborate all the details[7]: decompose the matrices into their longitudinal and transversal parts, perform the various y-averages (Eq.(5)), take proper care of the pressure contribution (we took the infinite space Green's function $\propto|\vec{x}-\vec{x'}|^{-1}$ to express p by $u_{i,j}u_{j,i}$), integrate over the wellknown[3] slit region. All this is contained in the final integro-differential equation (9), whose solution is presented in Fig.1, and which agrees rather well with measurements in highly turbulent atmospheric flows. The theoretical value for b is 6.3, the transition between the VSR and the ISR occurs at $r\approx9\eta$, in agreement with the data.

Some comments are now in order. a) Although we used a simple factorization scheme the r-dependence is dominated by the NS-E dynamics and not by advection. This is due to the fact that we derived an equation for a Galilean invariant quantity, D, in a Lagrangian approach in position space. Beyond the necessity to preserve this invariance, there are no peculiarities in the statistical theory of the NS-E: common arguments and approximations do well. No mixed Eulerian-Lagrangian two-time-fields are necessary.
b) Batchelor's[12] interpolation formula for D(r) agrees rather closely with the self-consistent numerical solution of (9). It also nearly coincides with the approximate analytical solution given below.
c) Since in the power law ranges (3), (4) the y-averages of D reproduce D(r) up to factors, one obtains a simple though approximate form of (9):

$$\varepsilon = [\nu + b^{-3}\varepsilon^{-1}D^2(r)]\,\frac{3}{2}\frac{1}{r}\frac{d}{dr}\,D(r) \ . \tag{15}$$

The second term in the brackets represents the eddy viscosity. One easily obtains the solution satisfying D(0) = 0,

$$D(r) = b(\varepsilon r)^{2/3}2^{-1/3}\{[1+\sqrt{1+\phi(r)}]^{+1/3} + [1- \sqrt{1+\phi(r)}]^{1/3}\}. \tag{16}$$

Here $\phi(r)\equiv4b^3(r/\eta)^{-4}$.
d) The scale-dependent eddy viscosity is the momentum drag of the sub-r-eddies on the super-r-eddies. The general expression (12) in the one-pole approximation (13) can be simplified further to

$$\nu_{eddy}(r) \cong b^{-3}\varepsilon^{-1}D^2 \equiv 2b^{-3}D(r)\gamma^{-1}(r). \tag{17}$$

It highly resembles the eddy diffusivity $K\cong2D(r)\gamma^{-1}(r)$ derived recently[13] and supports an ansatz by Kolmogorov[14]. One gets $\nu_{eddy}\propto r^{4/3}$ in the ISR and $\propto r^2$ in the VSR. It provides a NS-E derived expression for a generalized mixing length[15]

$$\ell^*(r) = \sqrt{D(r)}\ \gamma^{-1}(r) \cong v(r)\tau(r), \tag{18}$$

as already discussed in the literature[15].
e) To reiterate our point it might be useful to compare with the presently existing renormalization (RG) scheme (cf.ref.5, in particular 5d). This RG treatment studies how the NS-E *transforms* an additionally coupled noise spectrum into a velocity spectrum. For just one choice of the input spectrum the 5/3 velocity k-spectrum emerges (corresponding to the 2/3 r-power). Instead, we do not couple additional noise, but *derive*, why (not any exponent but just) the 2/3 is implied by the very structure of the NS-E. The present RG approach is restricted to the ISR while we derive its range of validity as well as the behaviour in the VSR. We cannot calculate the intermittency corrections since we restrict ourselves to the mean field approximation, but the existing RG approach[5] cannot do either (surprisingly enough), since it does not know in advance which proper input to choose. The RG method cannot find the prefactor (Kolmogorov constant) since this is proportional to the arbi-

trarily chosen noise strength (in ref.5d a guess is presented), we *calculate* b approximately but based on the properties of the NS-E unambiguously.

To conclude, we feel that the method outlined here and presented in detail in reference 7 to elaborate the statistical mechanics of a nonlinear field equation that contains no explicit length scale and thus provides selfsimilar solutions may serve as a guide for other cases and therefore not only is of interest for calculating the statistics of fully developed turbulent flow but also in other pyhsical fields. A systematic improvement using higher order approximations (e.g. an additional equation for the fourth order moment) seems to be possible. The variable scale separation presented here may also serve as a basis for a renormalization group treatment of fully developed turbulence which should give corrections to the classical exponents, still one of the unsolved problems.

1. D.Ruelle, Comm.Math.Phys.87,287(1982);93,285(1984)
2. A.N.Kolmogorov, A.M.Oboukhov, C.F.von Weizsäcker, W.Heisenberg, L.Onsager, in the 1940th, detailed references see ref.6
3. R.H.Kraichnan, J.Fluid.Mech.5,497(1959);83,349(1977); Phys.Fluids 7,1030; 1723(1964);8,575(1965)
4. P.C.Martin, E.D.Siggia, H.A.Rose, Phys.Rev.A8,423(1973); H.Horner, R.Lipowski, Z.Phys.B33,223(1979);C.Foias, O.P.Manley, R.Temam, Phys.Rev. Lett.51,617(1983)
5. D.Forster, D.R.Nelson, M.J.Stephen, Phys.Rev.Lett.36,867(1976) Phys.Rev. A16,732(1977); M.Lücke, Phys.Rev. A18,282(1978); M.Lücke and A.Zippelius, Z.Phys. B31,201(1978); V.Yakhot, S.A.Orszag, Phys.Rev.Lett.57,1722(1986)
6. S.A.Orzag, Les Houches Proc. 1973, Lectures on the Statistical Theory of Turbulence, Eds. R.Balian, J.L.Peube, Gordon & Breach, London, 1977
7. H.Effinger, S.Grossmann, Z.Phys.B66,289 (1987)
8. C.W.Van Atta, W.Y.Chen, J.Fluid Mech. 44,145(1970); R.A.Antonia, B.R.Satyaprakash, A.J.Chambers, Phys.Fluids 25,29(1982); F.Anselmet, Y.Gagne, E.J.Hopfinger, R.A.Antonia, J.Fluid Mech. 140,63(1984); P.Mestayer, J.Fluid Mech. 125,475(1982)
9. Such variable scale separation was introduced in momentum space: S.Grossmann, Z.Naturforsch. 26a,1782(1971); C.M.Tchen, Phys.Fluids 16,13 (1973)
10. S.Grossmann, S.Thomae, Z.Phys. B49,253(1982)
11. If the switch-over of $E_{in}(r)$ from ε to 0 in the range $r \cong L$ is given (assumed), D can be evaluated also in the non-universal input range
12. G.K.Batchelor, Proc.Camb.Phil.Soc. 47,359(1951)
13. H.Effinger, S.Grossmann, Phys.Rev.Lett. 53,442(1984) S.Grossmann, I.Procaccia, Phys.Rev. A29,1358(1984)
14. A.N.Kolmogorov, Izv.Acad.Nauk(SSSR), Ser VI, No 1-2,56(1942)
15. L.Prandtl, ZAMM 5,136(1925); B.E.Launder, D.B.Spalding, Math. Models of Turbulence, Academic Press, London, 1972; H.Tennekes, J.L.Lumley, A First Course in Turbulence, MIT-Press, Cambridge (Mass), 1972

Part VI

Bifurcations and Symmetry

Nonlinear Normal Modes
of Symmetric Hamiltonian Systems

J. Montaldi, M. Roberts, and I. Stewart

Mathematics Institute, University of Warwick,
Coventry CV47AL, United Kingdom

0.   Introduction

In the classical dynamics of anharmonic molecules and crystals an important role is played by *nonlinear normal modes* - families of periodic trajectories (parametrized by energy) with periods close to those of the normal modes of the linearization. These serve as *organizing centres*, a skeleton about which the rest of the dynamics can be arranged.   By the KAM Theorem, elliptic periodic trajectories are surrounded by invariant tori and so carry with them regions of regular motion in which the only instability is caused by Arnold diffusion.   Hyperbolic periodic orbits are typically connected by heteroclinic orbits, which in turn give rise to chaotic regions of phase space, CHURCHILL and ROD [3].   An excellent example of this organizing behaviour can be seen in the Hénon-Heiles system, see HÉNON [10], ROD and CHURCHILL [19], and §§1-2.

In the absence of resonance all normal modes persist as periodic trajectories of the nonlinear system. More precisely, we have the following result of LIAPUNOV [13]. Let P be the phase space, $\mathcal{H}$ the Hamiltonian function on P, and $X_{\mathcal{H}}$ the associated Hamiltonian vector field. Suppose p is a nondegenerate equilibrium point, so that $D\mathcal{H}(p)$ = 0 (equivalently $X_{\mathcal{H}}(p)$ = 0) and $D^2\mathcal{H}(p)$ is nondegenerate.   Suppose λ is a purely imaginary eigenvalue of the linearization L of $X_{\mathcal{H}}$ at p, which is *nonresonant* (i.e. it has multiplicity 1 and $n\lambda$ is not an eigenvalue for any integer n > 1). Liapunov's Theorem states that through p there passes a 2-dimensional submanifold of P, tangent to the space of normal modes with period $2\pi/|\lambda|$, consisting entirely of periodic trajectories of the Hamiltonian system with periods close to $2\pi/|\lambda|$, that is, of nonlinear normal modes. (By 'close to' we mean that on approaching p, the period tends to $2\pi/|\lambda|$.)

More recently, WEINSTEIN [21,22] and MOSER [17] have considered the resonant case. Again suppose that λ is a purely imaginary eigenvalue of L. Let V_λ be the *resonance space* for λ, that is, the real part of the sum of the (complex generalized) eigenspaces of L for λ and any integer multiples of λ. Suppose that the quadratic part of $\mathcal{H}$ is (positive) definite on V_λ. Then the linear system on V_λ is totally periodic, with every orbit having period $2\pi/|\lambda|$, though this need not be the *minimal* period of every orbit. The theorem of Weinstein and Moser states that on each (positive) energy surface near p there are at least $\frac{1}{2}$ dim V_λ periodic orbits with period close to $2\pi/|\lambda|$.   In the resonant case these orbits do not generally form a submanifold: instead they form surfaces with conelike singularities at p, DUISTERMAAT [5].

There are other results that do not require the assumption of definiteness. FADELL and RABINOWICZ [6] show that the number of families of periodic trajectories is bounded below in terms of the signature of the quadratic part of $\mathcal{H}$ on V_λ.

In general there will be more nonlinear normal modes than predicted by the Weinstein-Moser Theorem. In the 2:1 resonance there are three families, two of which have conical singularities at the origin. On detuning away from resonance, one Liapunov normal mode loses stability, at a small finite energy, to a 'mixed mode' with twice the period, DUISTERMAAT [5]. The quantum-mechanical consequences of this are apparent in the phenomena associated with Fermi resonance, for example in the CO_2 molecule, HELLER et al. [9], VOTH and MARCUS [20].

Symmetry in a Hamiltonian system typically forces 1:1 resonances, so we may expect to - and do - see extra nonlinear normal modes. In this article we describe some of our recent results, MONTALDI et al. [15], which refine the Weinstein-Moser Theorem when symmetry is present. The main idea is that the most symmetric normal modes of a linearization always persist to give nonlinear normal modes. This result, described in §1, is *model-independent*, that is, independent of the precise details of the anharmonicity. The symmetry of a normal mode is its spatio-temporal symmetry group, also defined in §1. In the Hénon-Heiles system all nonlinear normal modes can be obtained in this way, but in general there may be others.

The symmetry group of a periodic trajectory may also be used to study its stability. Indeed in certain cases it forces the trajectory to be spectrally stable, see §2. This occurs for two nonlinear normal modes in the Hénon-Heiles system.

In §2 we outline these general stability results. The remaining sections sketch various applications: lattice dynamics (§3), vibrating molecules (§4), and oscillations of a liquid drop (§5).

We employ the following notation for symmetry groups:

S^1 = the circle group.

Z_n = the finite cyclic group of order n (often written C_n).

D_n = the dihedral group of order 2n (symmetries of a regular n-gon).

$O(n)$ = the orthogonal group of rigid motions (rotation and reflection) in $\mathbb{R}^n$.

1. Existence of Symmetric Periodic Trajectories

We consider Hamiltonian systems that are invariant under the action of a discrete or continuous compact Lie group Γ. That is, Γ acts on phase space P by canonical transformations leaving the Hamiltonian $\mathcal{H}$ invariant.

Suppose that a group G acts on a space (or set) S. Then we define the *isotropy subgroup* of $x \in S$ to be

$$G_x = \{g \in G : gx = x\}.$$

If Σ is a subgroup of G then we define the *fixed-point space* (or set) of Σ to be

$$\text{Fix}(\Sigma;S) = \{x \in S : gx = x \text{ for all } g \in \Sigma\},$$

and write just $\text{Fix}(\Sigma)$ for this when S is unambiguous. If S has some special structure that is preserved by G, then Fix(Σ;S) usually inherits such a structure. Many results in equivariant dynamics depend on a simple but crucial fact:

Principle of Conservation of Symmetry Suppose that a group G acts on two sets S and T. Let f:S → T be a mapping that commutes with these actions (i.e. f(gx) = gf(x) for x ∈ S, g ∈ G). Then $G_{f(x)} \supset G_x$, or equivalently f(Fix(G;S)) ⊂ Fix(G;T).

Proof Suppose g ∈ G, x ∈ S, and gx = x. Then gf(x) = f(gx) = f(x).

For a simple application, suppose that an equilibrium point p ∈ P of the Hamiltonian system is fixed by Γ. Let K ⊂ Γ be an isotropy subgroup for the action of Γ on P. Then the submanifold Fix(K) is itself a Hamiltonian subsystem, with equilibrium p. Thus the theorems of Liapunov or Weinstein-Moser can be applied to the subsystem to get ½dim Fix(K) nonlinear normal modes, each with symmetry at least K. This argument easily yields *some* nonlinear normal modes, but in general the symmetry causes others, which can be obtained by looking at the action of a bigger group Γ×S¹ on periodic trajectories.

Let u(t) be a periodic solution with period T. Then so is χu(t) for all χ ∈ Γ. Let H be the subgroup of Γ that leaves u(t) invariant (as a set), and K the subgroup of H that fixes u(t) for each t ∈ ℝ. Then K is the *spatial symmetry group* of u. For full information we must know the internal effect of elements of H. Thus we introduce the 'internal symmetry group' of the periodic orbit, namely the circle group S¹ acting by phase shifts. The *spatio-temporal symmetry group* of u, for period T, is then
$$\Sigma_u = \{(\chi,\theta) \in \Gamma \times S^1 : \chi u(t+\theta T/2\pi) = u(t)\}.$$
Here S¹ is parametrized by [0,2π]. This group depends on how we specify the period T, which need not be minimal. In fact u has minimal period T/k if and only if $\Sigma \cap S^1 = \mathbb{Z}_k$, the finite cyclic group of order k.

In general Σ_u has the following form. Let H be a subgroup of Γ, and θ:Γ → S¹ a group homomorphism. Define H *twisted by* θ to be the group
$$H^\theta = \{(h,\theta(h)) : h \in H\} \subset \Gamma \times S^1.$$
Then Σ_u is always a twisted subgroup H^θ for suitable H, θ. If $\Sigma \cap S^1 = \mathbb{1}$, that is, we are considering the minimal period, we say that u is a
  standing wave if θ(H) = $\mathbb{1}$, $\mathbb{Z}_2$,
  travelling (or *rotating*) *wave* if θ(H) = S¹,
  discrete travelling wave otherwise.
If θ(H) ≠ $\mathbb{1}$ we often denote H^θ by $\tilde{H}$, with further information supplied by a superscript if necessary. For example $\widetilde{SO(2)}^k = \{(\theta,k\theta) \in SO(2) \times S^1\}$.

We can define Σ_u more abstractly. Let $C^r(T)$ be the set of T-periodic C^r maps u:ℝ → P. There is an action of Γ×S¹ on $C^r(T)$ defined by
$$(\chi,\theta).u(t) = \chi u(t+\theta T/2\pi).$$
The symmetry group of u is then the isotropy subgroup of u ∈ $C^r(T)$.

There is another action of $\Gamma \times S^1$, on (part of) the linearization L of $X_{\mathcal{H}}$ at p, and the existence theorem for nonlinear normal modes derives from the relationship between these two actions. Assume that L is nonsingular, and consider the differential equation

$$\dot{x} + Lx = 0. \qquad\qquad (*)$$

Let λ be an imaginary eigenvalue of L and define V_λ as in the introduction. Assume that the second variation of the Hamiltonian $\mathcal{H}$ is definite on V_λ. That is, assume:

(H1) $D^2\mathcal{H}_p$ is a nondegenerate quadratic form,

(H2) $D^2\mathcal{H}_p|V_\lambda$ is positive definite.

By (H2) $L|V_\lambda$ can be diagonalized over $\mathbb{C}$, so the solutions to $(*)$ are circular with common period $2\pi/|\lambda|$. The flow thus defines an action of the circle group S^1. Since V_λ is invariant under Γ we have a combined action of $\Gamma \times S^1$. The main existence theorem is stated in terms of this action.

Theorem 1 Suppose $\mathcal{H}$ satisfies (H1) and (H2). Then for every isotropy subgroup Σ of the action of $\Gamma \times S^1$ on V_λ, there are at least $\frac{1}{2}\dim \mathrm{Fix}(\Sigma)$ nonlinear normal modes with symmetry group containing Σ.

More pecisely, on each positive energy level near p there are at least $\frac{1}{2}\dim \mathrm{Fix}(\Sigma)$ periodic trajectories with symmetry group containing Σ. By 'positive energy level near p' we mean the surface on which $\mathcal{H}(z) = \mathcal{H}(p)+\varepsilon$ for small $\varepsilon > 0$. If further $\dim \mathrm{Fix}(\Sigma) = 2$, the periodic trajectories form a smooth 2-dimensional submanifold of P passing through p. For larger values of $\dim \mathrm{Fix}(\Sigma)$ we may expect conical singularities.

Example We consider the general system with two degrees of freedom and D_3 symmetry, e.g. the system of HÉNON and HEILES [11], see also ROD and CHURCHILL [19]. This has Hamiltonian

$$\mathcal{H}(q_1,q_2,p_1,p_2) = \tfrac{1}{2}(p_1{}^2+p_2{}^2)+\tfrac{1}{2}(q_1{}^2+q_2{}^2)+\tfrac{1}{3}q_1{}^3-q_1q_2{}^2.$$

Here (q_1,q_2) is position in configuration space $\mathbb{R}^2$, and (p_1,p_2) is momentum. The Hamiltonian is invariant under the standard action of D_3 on configuration space, where D_3 is the symmetry group of an equilateral triangle. This yields an action of D_3 on phase space, which we can – and do – view as the complexification of configuration space. Thus $P=\mathbb{R}^4\cong\mathbb{C}^2$.

Because the representation of D_3 on $\mathbb{R}^2$ is irreducible there is only one pair of eigenvalues, namely $\pm i$, with multiplicity 2. Thus we have an action of $D_3 \times S^1$ on $\mathbb{C}^2$, given by

$$(\gamma,\theta)(z_1,z_2) = e^{i\theta}(\gamma z_1, \gamma z_2).$$

The isotropy subgroups of this action are calculated in GOLUBITSKY and STEWART [7]. There are eight, each conjugate to one of the following:

$\bar{Z}_3 = \{(\theta,-\theta) \in \mathbf{D}_3 \times \mathbf{S}^1 : \theta = 0, 2\pi/3, 4\pi/3\}.$
$Z_2^{\kappa} = \{1,\kappa\}.$
$Z_2^{(\kappa,\pi)} = \{1,(\kappa,\pi)\}$

Here $\pi \in \mathbf{S}^1$, and κ is a reflection in $\mathbf{D}_3$.

Each subgroup has a 2-dimensional fixed-point space, so Theorem 1 yields two periodic trajectories with symmetry groups conjugate to $\bar{Z}_3$ and three periodic trajectories with symmetry groups conjugate to each of Z_2^{κ}, $Z_2^{(\kappa,\pi)}$. These are the only periodic trajectories of the Hénon-Heiles system near periodic trajectories of the linearization. We can identify these eight solutions with those shown in Figure 1.1, taken from CHURCHILL *et al*. [4], as follows:

Π_1, Π_2, Π_3 have symmetry groups conjugate to Z_2^{κ}.

Π_4, Π_5, Π_6 have symmetry groups conjugate to $Z_2^{(\kappa,\pi)}$.

Π_7, Π_8 have symmetry groups conjugate to $\bar{Z}_3$.

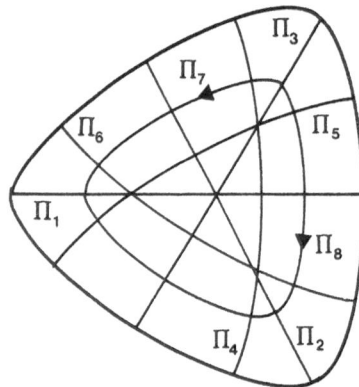

Fig. 1.1   Periodic trajectories in the Hénon–Heiles system, from CHURCHILL *et al*. [4]

2.   Stability

Let $u(t)$ be a periodic trajectory of minimal period T and let $z = u(0)$. Let $\sigma_T : P \to P$ be the time-T flow. Then z is a fixed point of σ_T. The *Floquet operator* is

$$M_u \equiv D(\sigma_T)_z : T_z P \to T_z P,$$

a linear map on the tangent space to P at z.   It is closely related to the Poincaré return map.   The eigenvalues of M_u are the *Floquet multipliers* for u.   We call u *spectrally stable* if all Floquet multipliers lie on the unit circle. (This idea may be more familiar for dissipative systems, where stability requires all Floquet multipliers to lie *inside* the unit circle.   In Hamiltonian systems the symplectic structure prevents this. Henceforth, when we say 'stable' we shall mean 'spectrally stable'.) We will see below that the symmetry group Σ_u places restrictions on M_u, and in some cases can force u to be stable.

In fact, for nonlinear normal modes, M_u commutes with an action of Σ_u on $T_z P$. Some caution is required here.   Because $\Gamma \times \mathbf{S}^1$ is *not* a symmetry of the vector field, there is no obvious action of Σ_u on T_zP, but only an action of K. But in fact the action of Σ on $V_\lambda \subset T_pP$ can be deformed to give an action of Σ_u on a subspace $\tilde{V}_\lambda \subset T_zP$, and this action is isomorphic to the action on V_λ.

Example The general system with two degrees of freedom and D_3 symmetry, e.g. the Hénon-Heiles system.

We have already seen that there are two nonlinear normal modes with $\tilde{Z}_3$ symmetry, namely Π_7 and Π_8 in Figure 1.1. Let u(t) be one of these, of minimal period T, and let z = u(0). The action of $\tilde{Z}_3$ at p decomposes into two inequivalent symplectic representations, $V_1 \oplus V_2$, where V_1 is 2-dimensional and trivial, and V_2 is the standard representation generated by rotation through $2\pi/3$. Thus the deformed action of $\tilde{Z}_3$ on T_zP is also of this form, and M_u commutes with it. Since Γ preserves the flow and the Hamiltonian, the trivial part is spanned by du/dt(0) and grad $\mathcal{H}(z)$. Thus with respect to the decomposition $V_1 \oplus V_2$ the Floquet operator has the form

$$M_u = \left[\begin{array}{cc|c} 1 & * & \\ 0 & 1 & 0 \\ \hline & 0 & N \end{array} \right]$$

where $*$ is an unknown real number, and N is a symplectic 2×2 matrix commuting with the standard action of $\tilde{Z}_3$. Thus N is a rotation, and has unit circle eigenvalues. Therefore the $\tilde{Z}_3$ normal modes of the general D_3-invariant system on $\mathbb{R}^4$, and in particular the Hénon-Heiles system, are linearly stable (at least for small energies).

If the group Γ is continuous, then further eigenvalues of M_u are forced to 1 by symmetry. How many, and how they relate to the action of Σ_u, may be determined from the momentum mapping.

We have shown that for systems with this D_3 symmetry, stability of the $\tilde{Z}_3$ solutions is purely a consequence of symmetry: it is *model-independent*. This fact does not involve calculations with higher order terms. Representations with this property – that every commuting linear map has unit circle eigenvalues – are called *cyclospectral*. They have a representation-theoretic characterization, described in MONTALDI *et al.* [15].

In contrast the other symmetry groups Z_2 and $\tilde{Z}_2$ do not have cyclospectral actions in this example. The stability of such nonlinear normal modes depends on the precise higher-order coefficients in the Hamiltonian. In general, it can be proved that generically exactly one of the Z_2 or $\tilde{Z}_2$ solutions is stable (for low energy).

When V_λ is not the entire space T_pP a separate argument is required to deal with eigenvalues that are not multiples of λ. This is provided by Krein Theory, see KREIN [12], ARNOLD [2], MOSER [16], MACKAY [14], which gives a simple criterion ensuring that under perturbation such eigenvalues do not leave the unit circle. This *Krein condition* is satisfied in particular if

(i) $D^2 \mathcal{H}$ is definite on each generalized eigenspace of L,

(ii)   There are no second order resonances $\mu_1 - \mu_2 = n\lambda$, where $n \in \mathbf{Z}$ and μ_1, μ_2 are eigenvalues of L, and

(iii) $\lambda/2$ is not an eigenvalue of L.

3. Crystal Lattices

One-dimensional Monatomic Lattice

Consider spatially periodic longitudinal vibrations of an infinite chain of identical atoms. This is equivalent to considering vibrations of a circular ring of N atoms, each constrained to lie on the circle.   We remove the possibility of translational motion of the lattice by restricting to configurations of atoms in which the sum of the displacements is zero. The configuration space is then an (N-1)-torus T^{N-1}, so the phase space is the (2N-2)-dimensional cotangent bundle $P = T^* T^{N-1}$.   We assume that the configuration of N equally spaced atoms is a nondegenerate minimum of the potential energy, and denote the corresponding equilibrium point in P by p.   The dihedral group D_N, the symmetry group of the regular N-gon, acts symplectically on P leaving p fixed.   The induced linear action of D_N on $T_p P$ decomposes as a direct sum of irreducible symplectic representations $\rho_1 + ... + \rho_{[N/2]}$ where ρ_r is defined as follows. Let $\varsigma \in D_N$ be rotation through $2\pi/N$, and κ a reflection.   If $0 \neq r \neq N/2$ then $\rho_r : D_N \rightarrow GL(2;\mathbb{C})$ is given by

$$\rho_r(\varsigma).(z_1, z_2) = (e^{2\pi i r/N} z_1, e^{-2\pi i r/N} z_2)$$
$$\rho_r(\kappa).(z_1, z_2) = (z_2, z_1).$$

If $r = N/2$ then $\rho_{N/2} : D_N \rightarrow GL(2;\mathbb{C})$ is given by

$$\rho_{N/2}(\varsigma).z = -z$$
$$\rho_{N/2}(\kappa).z = z.$$

This decomposition is the usual grouping of the normal modes of the linearization at p into symmetry types.   The normal modes transforming as ρ_r all have the same frequency ω_r. We assume there are no resonances between these frequencies.

We may apply the results of §2 to each ρ_r in turn, to yield a list of normal modes that must persist as periodic trajectories of the nonlinear system.   Note that for ρ_r the action of D_N factors through D_q where $q = N/h$, $h = \gcd(N,r)$, GOLUBITSKY and STEWART [8]. Thus the linear normal modes in ρ_r, and the corresponding nonlinear normal modes, are all fixed pointwise by the subgroup Z_h of D_N, and therefore correspond to lattice vibrations with spatial period h.   These could also be found by taking N = h. Thus for each N it is sufficient to consider ρ_r when N and r are coprime.

The Hénon-Heiles system of §2 corresponds to ρ_1 when N = 3. The results in the general case are similar.   For any r and $n \geq 3$ there are three types of maximally symmetric normal mode, that remain as periodic trajectories of the nonlinear system. If r = 1 these have the following symmetry groups:

$$Z_2^{\kappa} \qquad Z_2^{(\kappa,\pi)} \qquad \tilde{Z}_N \qquad \text{[N odd]}$$
$$Z_2^{\kappa} \oplus Z_2^{c} \qquad Z_2^{(\kappa,\pi)} \oplus Z_2^{c} \;\; \tilde{Z}_N \qquad \text{[N} \equiv \text{2 (mod 4)]}$$
$$Z_2^{\kappa} \oplus Z_2^{c} \qquad Z_2^{\kappa\zeta} \oplus Z_2^{c} \;\; \tilde{Z}_N \qquad \text{[N} \equiv \text{0 (mod 4)]}.$$

The notation used here is the same as in GOLUBITSKY and STEWART [8]. In terms of the 1-dimensional lattice:

Z_2^{κ} = reflection about a lattice site.

$Z_2^{\kappa\zeta}$ = reflection about a point midway between two lattice sites.

$Z_2^{(\kappa,\pi)}$ = reflection about a lattice point followed by a phase shift by π.

Z_2^{c} = translation by N/2 times the distance between lattice sites, followed by a phase shift by π.

$\tilde{Z}_N$ = translation by the distance between two lattice sites, followed by a phase shift by $2\pi/N$.

For other r the corresponding symmetry groups are isomorphic, but may have different realizations in $D_N \times S^1$. For N = 2 there is a single normal mode, with symmetry $Z_2^{(\kappa,\pi)} \oplus Z_2^{c}$. Figure 3.1 illustrates these normal modes for N = 2,3,4.

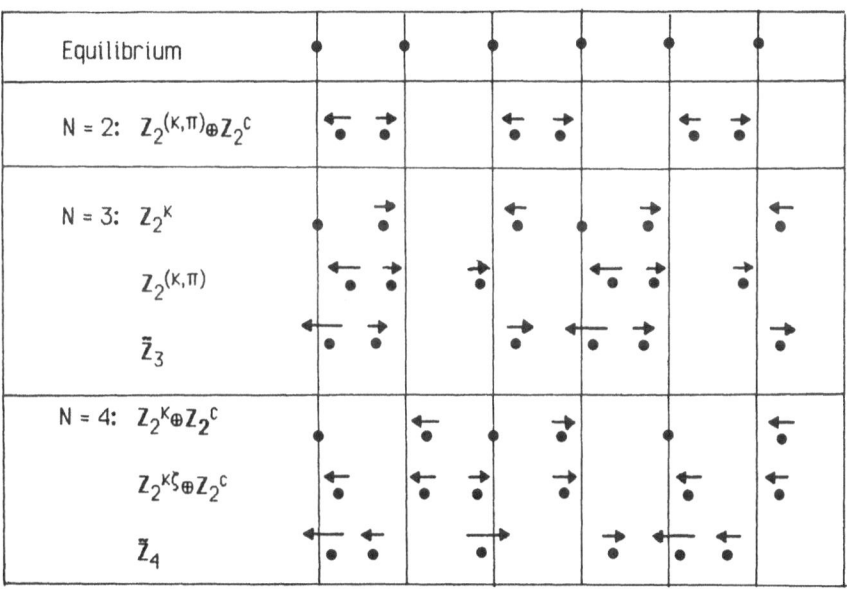

Fig. 3.1. Some nonlinear normal modes of a 1–dimensional monatomic crystal lattice

In MONTALDI *et al.* [15] we prove that the periodic orbits with symmetry group $\tilde{Z}_N$ are cyclospectral except when N = 4, and so lie at the centre of regions of phase space with 'regular' dynamics. The stability or otherwise of the other periodic orbits depends on the specific nonlinearities in the model.

For vibrations with a large spatial periodicity compared to the lattice site separation, we may consider a continuum approximation to the crystal, namely an elastic string. The D_N symmetry becomes an $O(2)$ symmetry, and the representations ρ_r are replaced by the actions of $O(2)$ on spaces consisting of linear combinations of a wave travelling to the left and its reflection travelling to the right. This is essentially the same as for a spherical pendulum near its stable equilibrium. In MONTALDI *et al.* [15] we show that the maximally symmetric normal modes are the pure travelling waves, with symmetry group $\widetilde{SO}(2)$, and standing waves with symmetry group $Z_2^K \oplus Z_2^c$. The travelling waves are the limits as $N \to \infty$ of the $\widetilde{Z}_N$ trajectories, and all other maximally symmetric normal modes in the discrete system converge to standing waves.

Planar Hexagonal Lattice

In principle the above techniques can be applied to higher dimensional lattices, though the calculations for normal modes with spatial periodicities with even 2 or 3 unit cells may prove complicated. Some indication of the complexity for normal modes with large spatial period may be gained by again passing to a continuum limit.

We consider transverse vibrations of a planar hexagonal monatomic lattice, i.e. vibrations perpendicular to the plane of the lattice. A typical normal mode is a plane wave travelling in the direction of a lattice vector. The lattice symmetry then relates five other plane waves to this one, the six waves being identical but travelling in directions separated by $\pi/3$. The resulting twelve-dimensional space of normal modes is an irreducible symplectic representation of the semidirect product $D_6 \ltimes T^2$, the D_6 being the point group of the lattice (corresponding to Z_2 in the linear lattice) and T^2 coming from the translational symmetry in two dimensions (corresponding to $SO(2)$ in the linear lattice). The maximally symmetric normal modes for this representation are calculated in ROBERTS *et al*. [18]. There are four types of standing wave (rolls, hexagons, triangles, and patchwork quilt, see Fig. 3.2); three types of travelling wave (rolls travelling perpendicular to their axes, and patchwork quilts travelling parallel to the sides of the patches); and a further four discrete travelling waves, one of which (the oscillating triangle) is shown in Fig. 3.3. Fig. 3.2. shows the amplitude of the displacement of each lattice point at some fixed time; Fig. 3.3 shows a series of 'snapshots'.

Further calculation shows that the three travelling waves and the oscillating triangle are cyclospectral, whereas the stability of the other solutions depends on the specific nonlinearities of the model.

4. Tetrahedral Molecules

In this section we consider the classical (i.e. not quantum- mechanical) theory of nonlinear vibrational modes of a molecule with tetrahedral symmetry (ignoring such effects as bond rotation), for example methane, CH_4. The equilibrium state has the carbon atom at the centre and four hydrogen atoms at the vertices of a regular tetrahedron. We consider the centre of mass to be fixed, obtaining a 12-dimensional

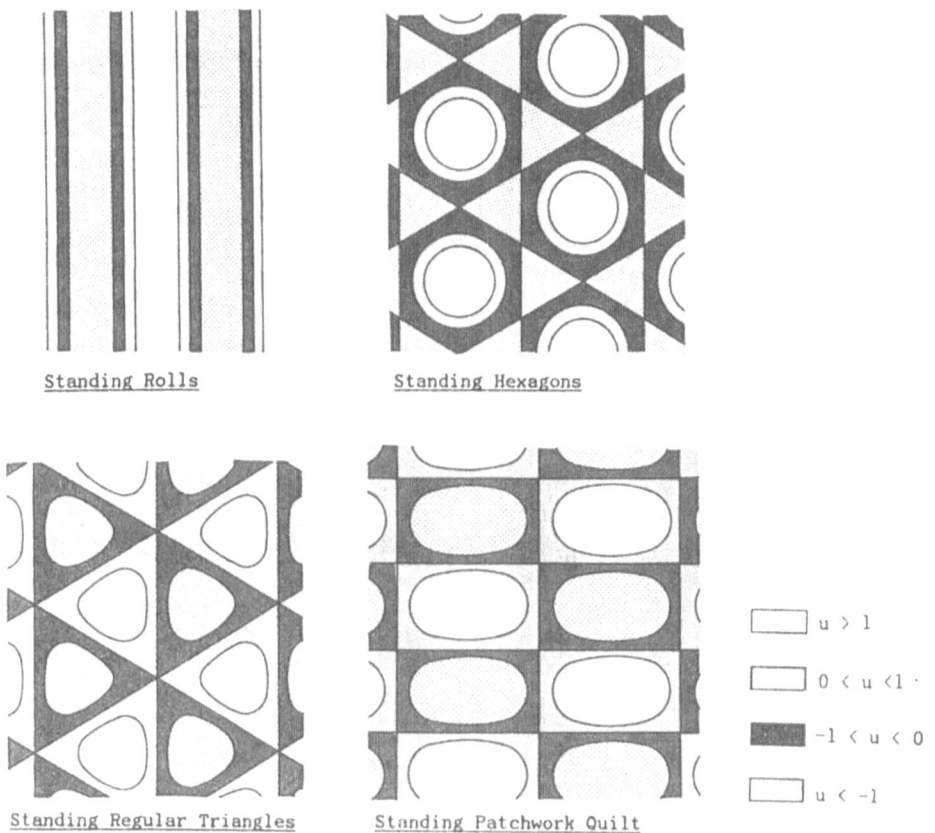

Standing Rolls

Standing Hexagons

Standing Regular Triangles

Standing Patchwork Quilt

$\square$ u > 1

$\square$ 0 < u < 1

$\blacksquare$ -1 < u < 0

$\square$ u < -1

Fig. 3.2. The four types of standing wave in a hexagonal lattice. From ROBERTS *et al.* [18]

configuration space. We also 'factor out' the rigid rotations. (In full rigour, it is not in fact clear whether there is a slice to the rotation orbit that is invariant under the dynamics, as there is for example in the linear crystal above, but we make the heuristic assumption that this reduction is possible.)   Thus we consider a 9-dimensional configuration space, with an action of the tetrahedral group S_4, which we think of as the permutations of the vertices 1234.

A standard result of linear theory is that the representation of S_4 on the 9-dimensional space decomposes as

$$\chi_0 \oplus \chi_2 \oplus 2\chi_3$$

where χ_0 is the trivial representation, χ_2 is the 2-dimensional representation (factoring through a representation of D_3, with kernel the Klein four-group), and χ_3 is the usual 3-dimensional representation. The 18-dimensional phase space has the same decomposition, but each irreducible should now be taken to be symplectic (so dim χ_0 = 2, dim χ_2 = 4, dim χ_3 = 6). Equivalently, if we think of phase space as the complexification of configuration space, we may take the representations to be unitary.

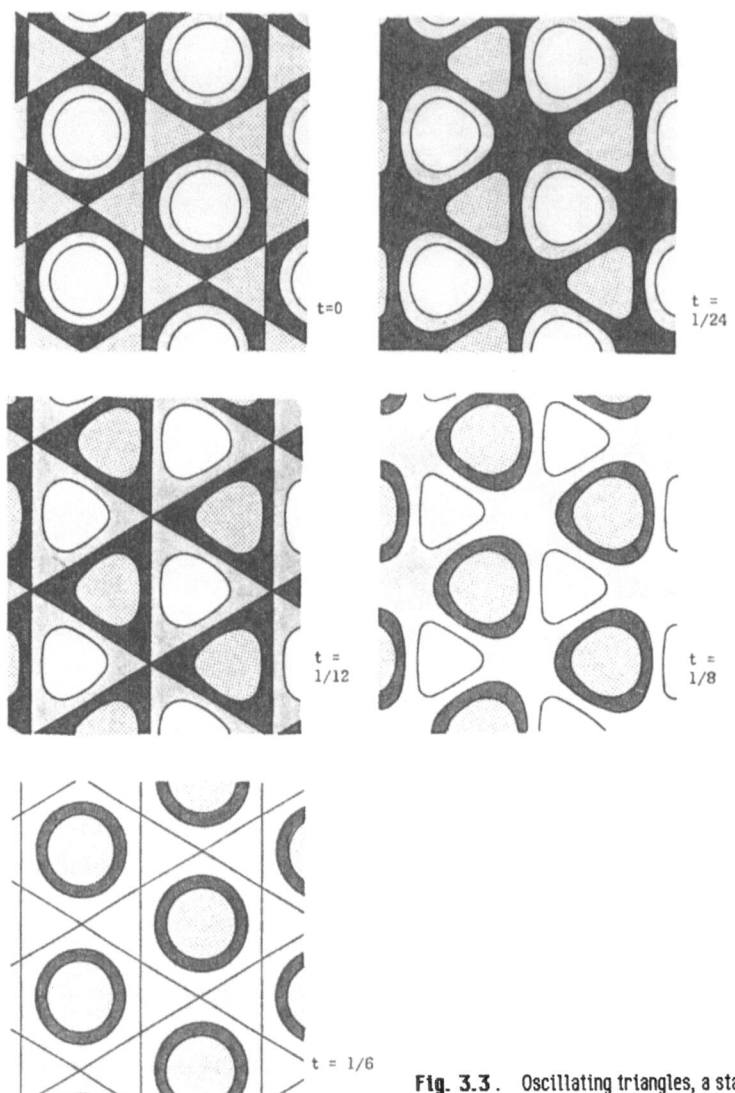

Fig. 3.3. Oscillating triangles, a stable discrete travelling wave. From ROBERTS *et al.* [18]

t=0

t = 1/24

t = 1/12

t = 1/8

t = 1/6

Generically there are four distinct eigenvalues of the linearization, corresponding to χ_0, χ_2, and some splitting of $2\chi_3 = \chi_3^1 \oplus \chi_3^2$. Generically there will also be no resonance (no eigenvalue an integer multiple of another). We analyse each in turn to find the nonlinear normal modes and their respective symmetries.

χ_0:   This is the subspace of phase space fixed by the S_4-action and hence, by the Principle of Conservation of Symmetry, is invariant under the nonlinear dynamics. It is thus a one-degree-of-freedom subsystem with a stable equilibrium point, so consists entirely of periodic orbits. By Krein theory these solutions are stable. Invariance of

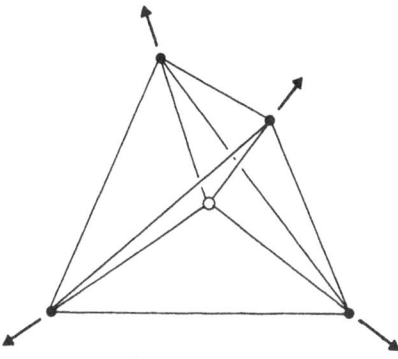

Fig. 4.1. The trivial 'breather' nonlinear normal mode of a tetrahedral molecule

the configuration under S_4 implies that this nonlinear normal mode has all four hydrogen atoms moving radially and in phase, Fig. 4.1.

χ_2:   Let $V \subset S_4$ be the Klein four-group, $V = \{1, (12)(34), (13)(24), (14)(23)\}$, which is isomorphic to $Z_2 \times Z_2$. Geometrically, these elements act on the tetrahedron by rotation through 180° about an axis joining midpoints of opposite edges. In the χ_2 representation, V acts trivially, leaving a representation of $S_4/V \cong D_3$.

Since $\chi_0 \oplus \chi_2$ is the fixed-point space of V, it is invariant under the nonlinear dynamics. However, χ_2 itself is invariant only under the linearized dynamics; but the nonlinear normal modes arising from χ_2 will be close to the subspace χ_2. The action of D_3 on χ_2 is equivalent to that in the Hénon-Heiles system (§2). Thus we get eight nonlinear normal modes. Three have symmetry Z_2, three $\tilde{Z}_2$, and two $\tilde{Z}_3$. These nonlinear normal modes are all fixed by V, hence the $\tilde{Z}_3$ ones resemble Fig. 4.2.

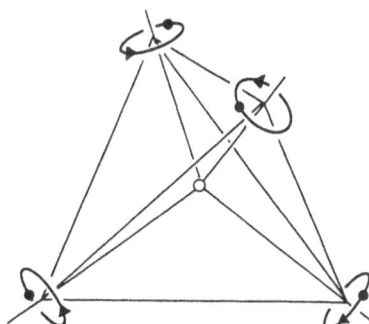

Figure 4.2. The stable $\tilde{Z}_3$ nonlinear normal mode fixed by the Klein four-group.

Without further analysis we can conclude that the $\tilde{Z}_3$ solutions are stable (at least, for low energy). To see this we use the same argument that applied to the Hénon-Heiles system to deal with the Floquet multipliers associated to χ_2, and Krein Theory to deal with the remainder. Generically, there is a 'choice of stability' between the other two types: depending on the coefficients of the Hamiltonian, either the Z_2 modes or the $\tilde{Z}_2$ modes are stable, but not both sets.

$2\chi_3$: On this space generically there are two distinct eigenvalues, but how the two eigenspaces sit inside $2\chi_3$ depends on the model (here the relative masses of the atoms and the interatomic forces). For example, if the H-H forces were vanishingly small compared to the C-H forces, then $2\chi_3$ would split into radial (stretching) and tangential (bending) modes. In general the splitting is determined by the linearization: we have

$$2\chi_3 \cong \chi_3^1 \oplus \chi_3^2$$

where the χ_3^j are the two eigenspaces of the linearization. Each χ_3^j carries a representation of $S_4 \times S^1$ having five types of isotropy subgroup. These are:

D_3	4 conjugates
$\tilde{D}_4$	3 conjugates
$Z_2 \times \tilde{Z}_2$	6 conjugates
$\tilde{Z}_4$	6 conjugates
$\tilde{Z}_3$	8 conjugates.

This yields a total of 27 nonlinear normal modes for each χ_3^j, none being cyclospectral. We have not analysed their stabilities. Thus in total we have found $1+8+27+27 = 63$ nonlinear normal modes of the tetrahedral molecule.

Briefly we describe the five isotropy subgroups arising from the representation χ_3, and the symmetry properties of the associated motions. For convenience we say that two hydrogen atoms are performing 'the same' motion if the motion is identical up to a suitable symmetry of the underlying tetrahedron.

D_3: This is a purely spatial symmetry group, which acts on the tetrahedron by fixing one vertex and permuting the other three, so acting on those three as the usual symmetry group of an equilaterial triangle. Motion fixed by D_3 must have one vertex moving radially and the other three each moving in a plane of reflection, with the radial motion of the first vertex having three times the frequency of the others. Thus symmetries can force harmonics, see also GOLUBITSKY and STEWART [8]. The space of such motions has one component in χ_0, and the other two are 'shared' between χ_3^1 and χ_3^2.

$\tilde{D}_4$: Abstractly this is the symmetry group of the square, but it sits inside $S_4 \times S^1$ as a twisted subgroup, namely, the two elements of order 4 in D_4 are combined with phase shifts by π. Writing $\rho = (1234)$, $\tau = (12) \in D_4 \subset S_4$, we have

$$\tilde{D}_4 = \{(1,0),(\rho,\pi),(\rho^2,0),(\rho^3,\pi),(\tau,0),(\rho\tau,\pi),(\rho^2\tau,0),(\rho^3\tau,\pi)\}.$$

Note that $\rho^2 = (13)(24)$ acts on $\mathbb{R}^3$ by rotation through π about an axis through the midpoints of opposite sides.

There are reasons to believe that for any generic Hamiltonian with this symmetry, the $\tilde{D}_4$ nonlinear normal modes are stable. This is suggested by Morse theory, and we sketch the line of reasoning. If the system is in Birkhoff normal form then the nonlinear normal modes can be viewed as critical points of a function on $\mathbb{C}P^{n-1}$ where $n = \dim V_\lambda$ ($= 3$ here). The Morse inequalities, combined with the numerology of this particular case (there are 3 $\tilde{D}_4$'s, 4 D_3's, etc.), show that the $\tilde{D}_4$ solutions have index 0 or 4, hence are stable. This calculation assumes that the 27 normal modes found above are the only ones: we expect this to be true for a generic Hamiltonian system with this symmetry but have not yet attempted to prove it.   A perturbation argument should then imply that the $\bar{\mathbf{D}}_4$ solutions for a system not in Birkhoff normal form remain stable.   The interesting feature of this argument is the way the group-theoretic and algebraic-topological structures interact to yield stability results. $\mathbf{Z}_2 \times \bar{\mathbf{Z}}_2$:Here $\mathbf{Z}_2$ acts by $((12),0)$ while $\bar{\mathbf{Z}}_2$ acts as $((34),\pi) \in \mathbf{S}_4 \times \mathbf{S}^1$. Thus atoms 1 and 2 vibrate in phase while 3 and 4 vibrate π out of phase.

$\bar{\mathbf{D}}_3$:   The spatial group $\mathbf{Z}_3$ is generated by rotation through $2\pi/3$ about the axis through a vertex. $\bar{\mathbf{Z}}_3$ acts in the same way but combined with a phase shift of one third of a period.   Thus the motions of vertices 2,3,4 are identical (when identified by a rotation of coordinates) but with a phase shift.   The motion of vertex 1 splits into radial and tangential components. The tangential frequency is the same as that of the other vertices, but the radial frequency is three times as great.

$\bar{\mathbf{Z}}_4$:   The spatial group $\mathbf{Z}_4$ is generated by $(1234) \in \mathbf{S}_4$, acting as rotation through $\frac{\pi}{2}$ followed by a reflection.   Under this identification of the vertices $1 \rightarrow 2 \rightarrow 3 \rightarrow 4 \rightarrow 1$, the motions are $\frac{\pi}{2}$ out of phase.

5.   Liquid Drops

Consider a 2-dimensional liquid drop in $\mathbb{R}^2$, having a circularly symmetric equilibrium state.   This is an infinite-dimensional system, but the results should remain applicable. Oscillations near equilibrium can be expected to occur for all 'modes', that is, irreducible symplectic representations of the symmetry group $\mathbf{O}(2)$ of the circle.   There is one such representation for each positive integer k, in which the rotations in $\mathbf{O}(2)$ act as k-fold rotations.   The symmetry groups in $\mathbf{O}(2) \times \mathbf{S}^1$ of periodic solutions are the dihedral group $\mathbf{D}_{2k}$ and the group

$$\widetilde{\mathbf{SO}(2)}^k = \{((\theta+2\pi l)/k, \theta) \mid l = 0,...,k-1\}.$$

Both are cyclospectral. The $\mathbf{D}_{2k}$ solutions are those in which a droplet shaped like a (rounded) regular k-gon oscillates as a standing wave.   The $\widetilde{\mathbf{SO}(2)}^k$ solutions correspond to a rotating structure with k-fold cyclic symmetry:like a k-armed spiral but close to circular form.   See Fig. 5.1.

Such oscillations are commonly observed in a droplet of water lying on a heated surface, AITTA [1].   To take this analysis further requires an explicit model, and a suitable reduction to finite dimensions, but the model-independent symmetry features should, broadly speaking, be as just outlined. Other kinds of dynamic behaviour are of course possible.

Similar remarks apply to the 3-dimensional liquid drop.   Now the symmetry group is $\mathbf{O}(3)$ and the vibrational modes correspond to irreducible symplectic representations of $\mathbf{O}(3)$.   These form an infinite family $V_l = U_l \oplus U_l$ where U_l is the space of spherical harmonics of degree $l \geq 0$. We have dim $U_l = 2l+1$, dim $V_l = 4l+2$. GOLUBITSKY and STEWART [7] classify a special class of maximally symmetric subgroups of $\mathbf{O}(3) \times \mathbf{S}^1$, namely, those having two-dimensional fixed-point spaces.   Table 5.1 shows

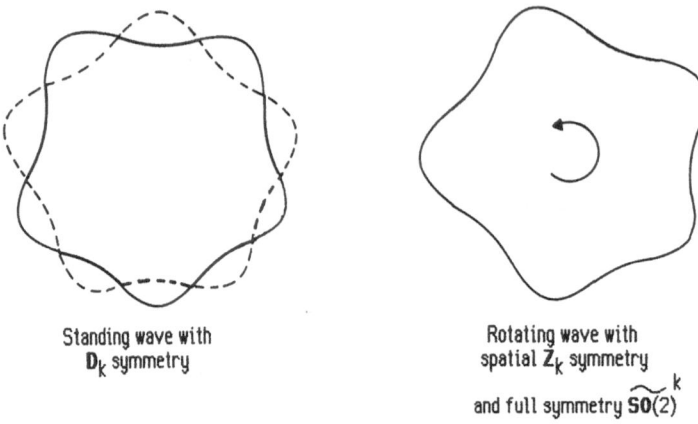

Standing wave with
D_k symmetry

Rotating wave with
spatial Z_k symmetry
and full symmetry $\widetilde{SO(2)}^k$

Fig. 5.1. Schematic illustration of symmetries of periodic solutions expected for the k–fold action of O(2).
Here k = 5

Table 5.1. Maximally symmetric periodic trajectories of the 3–dimensional liquid drop, $l \leq 3$

l	Σ	CYCLOSPECTRAL?	l	Σ	CYCLOSPECTRAL?
0	$O(3)$	YES	3	$\widetilde{O(2) \oplus Z_2^c}_1$	NO
1	$O(2) \oplus Z_2^c{}_1$	YES		$\widetilde{SO(2) \oplus Z_2^c}_2$	NO
	$SO(2) \oplus Z_2^c$	YES		$\widetilde{SO(2) \oplus Z_2^c}_3$	YES
2	$O(2)_1$	NO		$\widetilde{SO(2) \oplus Z_2^c}_2$	YES
	$SO(2)_2$	YES		$\mathbb{O} \oplus Z_2^c$	NO
	$SO(2)$	YES		$D_2 \oplus Z_2^c$	NO
	$\bar{D}_4$	NO		$D_3 \oplus Z_2^c$	NO
	$\mathbb{T}$	YES			

these symmetry groups when $l \leq 3$. For precise definitions see MONTALDI *et al.* [15].
The trajectories with symmetry groups $\widetilde{SO(2)}^k$ and $\widetilde{SO(2) \oplus Z_2^c}^k$ are rotating waves,
while the others are standing waves or discrete rotating waves.

We also analyse their Floquet operators when $l \leq 3$. In particular the table shows
which groups are cyclospectral, and hence give rise to stable vibrations.

The corresponding five types of vibration when $l = 2$ are shown in Fig. 5.2.
There are two standing waves, two cyclospectral rotating waves, and one discrete
rotating wave that takes the form of a 'pulsating cube' in which the three faces around a

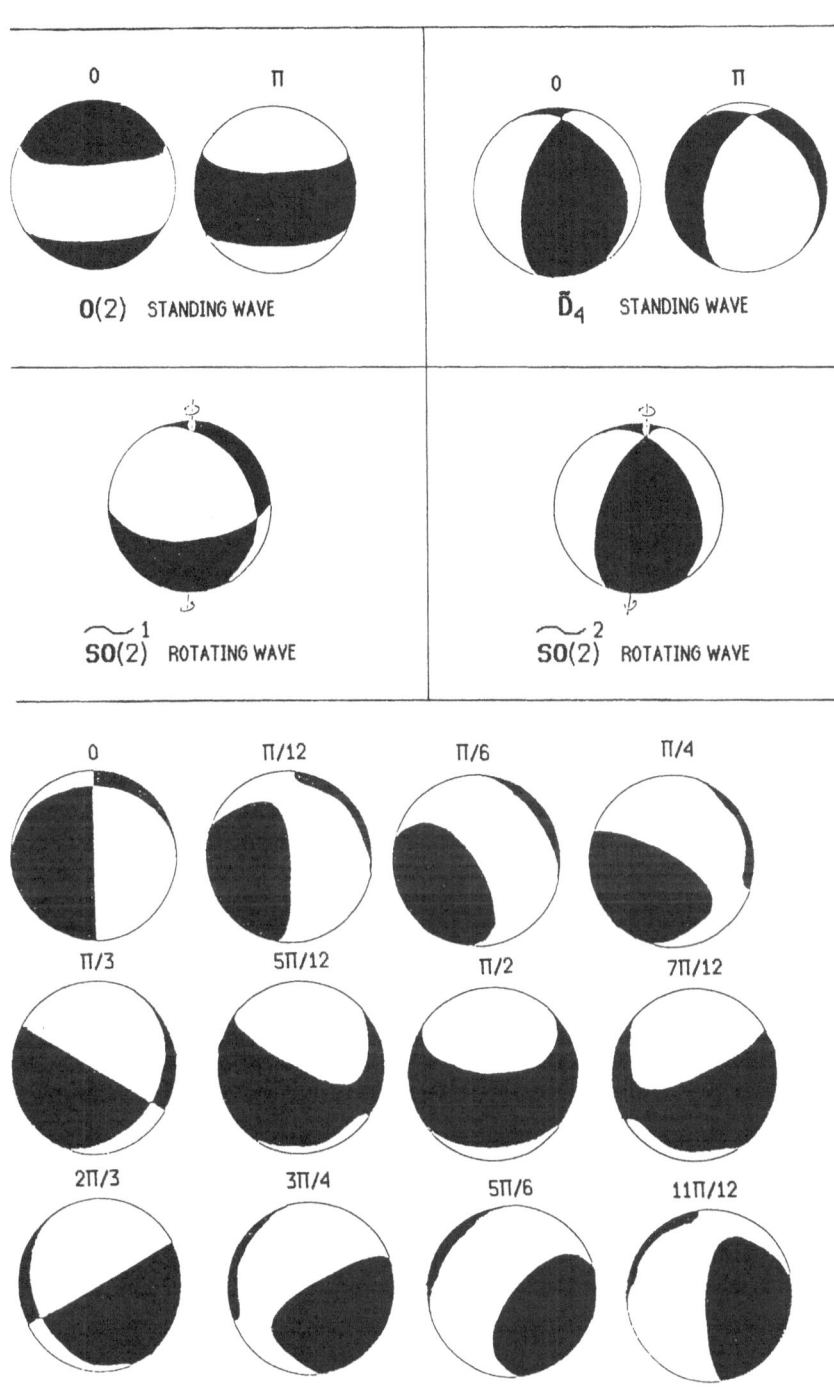

Fig. 5.2. Oscillations of the liquid drop, $l = 2$

vertex pulsate with phase lags of $2\pi/3$. This symmetry group is also cyclospectral, so the stability of the pulsating cube, like the rotating waves, is model-independent.

We have made a partial analysis of the cases $l > 3$. In particular, only *some* of the rotating wave solutions are then cyclospectral.

REFERENCES

1.   A.Aitta. Private communication, 1986.
2.   V.I.Arnold. *Mathematical Methods of Classical Mechanics* , Springer, New York, 1978.
3.   R.C.Churchill and D.L.Rod.   Pathology in dynamical systems III: analytic Hamiltonians, *J. Diff. Eq.* **37** (1980)23-38.
4.   R.C.Churchill, M.Kummer and D.L.Rod .   On averaging, reduction, and symmetry in Hamiltonian systems, *J. Diff. Eq.* **49** (1983) 359-414.
5.   J.J.Duistermaat.   Bifurcations of periodic solutions near equilibrium points of Hamiltonian systems, in *Bifurcation Theory and Applications - Montecatini 1983* (ed. L.Salvadori), Lecture Notes in Math. **1057**, Springer, Berlin, 1983.
6.   E.R.Fadell and P.H.Rabinowicz .   Generalized cohomological index theories for Lie group actions with an application to bifurcation questions for Hamiltonian systems, *Invent. Math.* **45** (1978) 139-174.
7.   M.Golubitsky and I.N.Stewart.   Hopf bifurcation in the presence of symmetry, *Arch. Rational Mech. Anal.* **87** (1985) 107-165.
8.   M.Golubitsky and I.N.Stewart.   Hopf bifurcation with dihedral group symmetry: coupled nonlinear oscillators, in *Multiparameter Bifurcation Theory* (ed. M.Golubitsky and J.Guckenheimer), *Contemporary Math .* **56** (1986) 131-173, Amer. Math. Soc., Providence R.I.
9.   E.J.Heller, E.B.Stechel, and M.J.Davis.   Molecular spectra, Fermi resonances and classical motion, *J. Chem. Phys.* **73** (1980) 4720-4735.
10.   M.Hénon.   Numerical exploration of Hamiltonian systems, in *Chaotic Behaviour of Deterministic Systems* (eds. G.Iooss, R.H.G.Helleman, and R.Stora), North-Holland, Amsterdam, 1983.
11.   M.Hénon and C.Heiles.   The applicability of the third integral of motion; some numerical experiments, *Astronom. J.* **69** (1964) 73-79.
12.   M.G.Krein.   The basic propositions of the theory of λ-zones of stability of a canonical system of linear differential equations with periodic coefficients,in *Topics in Differential and Integral Equations and Operator Theory* (by M.G.Krein), Operator Theory vol. 7, Birkhaüser, Basel 1983. (Translated from *Izdat. Akad. Nauk SSSR*, Moscow 1955, 413-498.)
13.   A.M.Liapunov. Problème générale de la Stabilité du Mouvement, *Ann. Fac. Sci. Toulouse* **9** (1907), reprinted by Princeton Univ. Press 1947. (Russian original 1895.)

14. R.S.MacKay. Stability of equilibria of Hamiltonian systems, in *Nonlinear Phenomena and Chaos*, (ed. S.Sarkar), Adam Hilger, Bristol 1986, 254-270.

15. J.A. Montaldi, R.M.Roberts, and I.N.Stewart. Periodic solutions near equilibria of symmetric Hamiltonian systems, preprint, Warwick 1986.

16. J.Moser. New aspects in the theory of stability of Hamiltonian systems, *Commun. Pure Appl. Math.* **11** (1958) 81-114.

17. J.Moser. Periodic orbits near equilibrium and a theorem by Alan Weinstein, *Commun. Pure Appl. Math.* **29** (1976) 727-747.

18. R.M.Roberts, J.W.Swift, and D.H.Wagner. The Hopf bifurcation on a hexagonal lattice, in *Multiparameter Bifurcation Theory* (ed. M.Golubitsky and J.Guckenheimer), *Contemporary Math*. **56** (1986) 283-318, Amer. Math. Soc., Providence R.I.

19. D.L.Rod and R.C.Churchill. A guide to the Hénon-Heiles Hamiltonian, in *Singularities and Dynamical Systems* (ed. S.Pneumatikos), Math. Studies **103**, North-Holland, Amsterdam, 1985.

20. G.A.Voth and R.A.Marcus. Semiclassical theory of Fermi resonance between stretching and bending modes in polyatomic molecules, *J.Chem. Phys*. **82** (1985) 4064-4072.

21. A.Weinstein. Normal modes for nonlinear Hamiltonian systems, *Invent. Math*. **20** (1973) 47-57.

22. A.Weinstein. Bifurcations and Hamilton's principle, *Math. Zeit*. **159** (1978) 235-248.

The Ubiquitous Astroid

D. Chillingworth

Department of Mathematics, University of Southampton,
Southampton SO95NH, United Kingdom

The aim of this lecture is to show how the form of the astroid arises naturally
in a variety of physical problems that in one way or another can be seen as
perturbations away from circular symmetry, and to give some mathematical insight
into why this should be the case.

The *astroid* is the 'star-like' curve (Fig. 1) whose equation in the (x,y)-
plane is

$$x^{2/3} + y^{2/3} = a^{2/3}$$

or, equivalently (exercise for the reader!):

$$(x^2+y^2-a^2)^3 + 27a^2x^2y^2 = 0 .$$

Parametric versions are

$$x = a \cos^3 t = \tfrac{1}{4}a(3\cos t + \cos 3t)$$

$$y = a \sin^3 t = \tfrac{1}{4}a(3\sin t - \sin 3t) .$$

The curve has a variety of standard geometric constructions, among which are the
following:

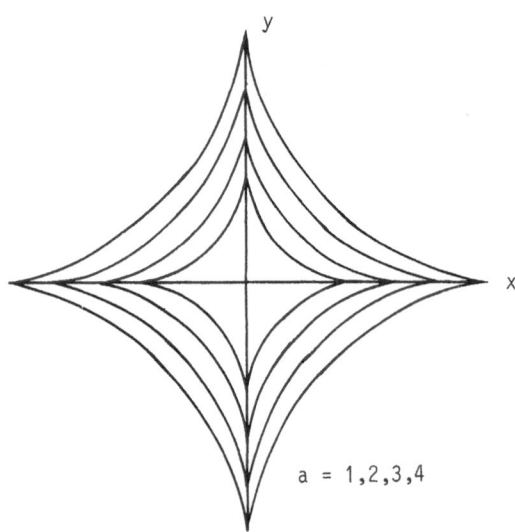

a = 1,2,3,4

Fig. 1: The astroid

1. Hypocycloid

The astroid is the locus of a point P on a circle of radius a when that circle is rolled inside a circle of radius 4a : see Fig. 2.

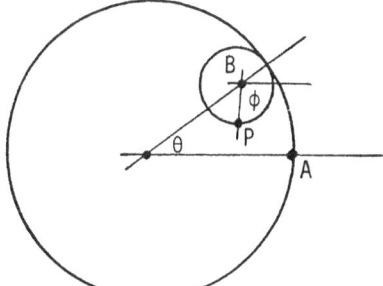

Fig. 2: The astroid as a
hypocycloid

The point B has coordinates $(3a\cos\theta,3a\sin\theta)$. The coordinates of P relative to B are $(a\cos\phi,-a\sin\phi)$. If P starts at A when $\theta = 0$ we see $\phi + \theta = 4\theta$; thus P is

$(3a\cos\theta+a\cos3\theta, 3a\sin\theta-a\sin3\theta)$.

2. The Envelope of a Ladder Sliding Against a Wall

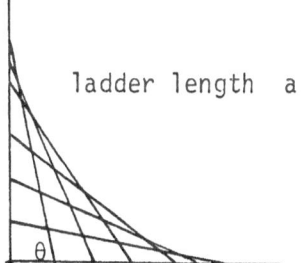

ladder length a

Fig. 3: Sliding ladder

The end-points of the ladder are $(a\cos\theta,0)$ and $(0,a\sin\theta)$. The equation of the line joining these is

$$x\sin\theta + y\cos\theta - a\cos\theta\sin\theta = 0 . \qquad (1)$$

For given (x,y) this equation has repeated root in θ (i.e. (x,y) lies on two 'consecutive' ladders) when the *derivative with respect to θ also vanishes*, i.e.

$$x\cos\theta - y\sin\theta - a(\cos^2\theta-\sin^2\theta) = 0 . \qquad (2)$$

When (1), (2) hold simultaneously we get

$$\begin{pmatrix}x\\y\end{pmatrix} = a\begin{pmatrix}s & c\\c & -s\end{pmatrix}^{-1}\begin{pmatrix}cs\\c^2-s^2\end{pmatrix} = \begin{pmatrix}ac^3\\as^3\end{pmatrix}$$

$(c = \cos\theta, s = \sin\theta)$.

3. The Envelope of the Family of Normals to an Ellipse

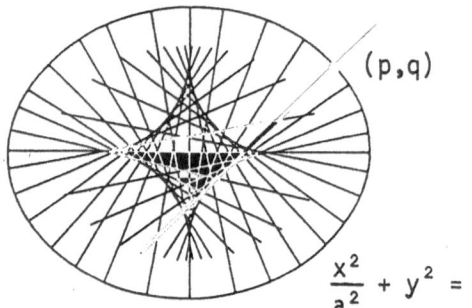

$$\frac{x^2}{a^2} + y^2 = 1$$

Fig. 4: Normals to an ellipse

Normal line through (p,q) has equation

$$(y-q) = \frac{2a^2q}{2p} (x-p)$$

i.e. $yp - a^2xq = (1-a^2)pq$.

Writing $(p,q) = (a\cos t, \sin t)$ this becomes

$$a^2 x \sin t - ay\cos t + a(1-a^2)\cos t \sin t = 0 .$$

Now (as for the sliding ladder) we find the envelope is

$$\left(\frac{-a^2}{1-a^2} x, \frac{a}{1-a^2} y\right) = (a\cos^3 t, a\sin^3 t)$$

which is the astroid rescaled.

4. The Envelope of a Family of Ellipses

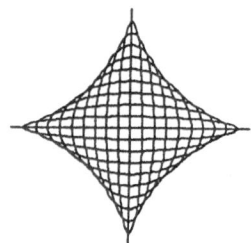

Fig. 5: Ellipses

Take the ellipses to be

$$\frac{x^2}{a^2} + \frac{y^2}{b^2} = 1$$

where $a + b = 1$ and a, b go from 0 to 1 . Rewrite this as

$$b^2x^2 + a^2y^2 - a^2b^2 = 0 .$$ (i)

For given (x,y) this equation has a repeated root in a (where b = 1-a) (i.e. (x,y) lies on two 'consecutive' ellipses) when the derivative with respect to a also vanishes, i.e.

$$bx^2 - ay^2 + ab(b-a) = 0 .$$ (ii)

When (i), (ii) hold simultaneously we see

$$\begin{pmatrix} x^2 \\ y^2 \end{pmatrix} = \begin{pmatrix} b^2 & a^2 \\ -b & a \end{pmatrix}^{-1} ab \begin{pmatrix} ab \\ b-a \end{pmatrix} = \begin{pmatrix} a^3 \\ b^3 \end{pmatrix}$$

so $x^{2/3} + y^{2/3} = 1 .$

After these mathematical manifestations of the astroid let us turn to some of its physical appearances. We shall use the same name now to refer to any four-pointed curve having the general appearance of a formal astroid.

5. Caustics by Refraction

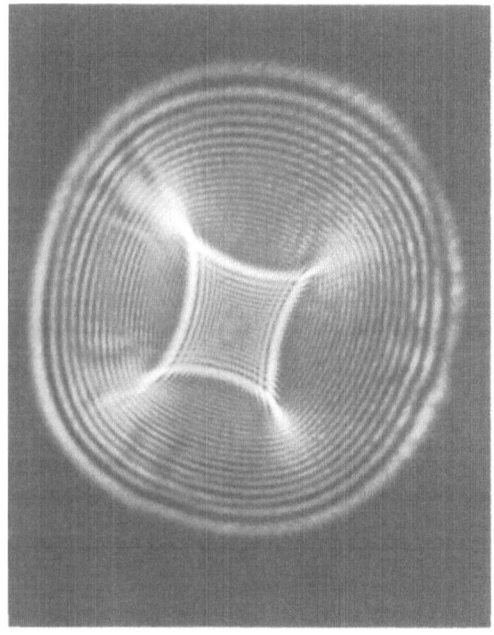

Fig. 6: Michael Berry's bathroom window

The photograph taken by BERRY [1] shows a caustic obtained when a point source was viewed through doubly-corrugated glass: the refracting medium has artificial 4-fold symmetry, but the caustic is not a square because the glass surface is not simply the superposition of two 1-dimensional corrugations at right angles. The astroid thus arises here as a natural perturbation of a degenerate 'idealized' form. See (12.) below for further discussion.

The sequence of pictures sketched by NYE [2] shows the caustic seen when a beam of light shines through a not-quite-spherical liquid drop onto a screen: as

375

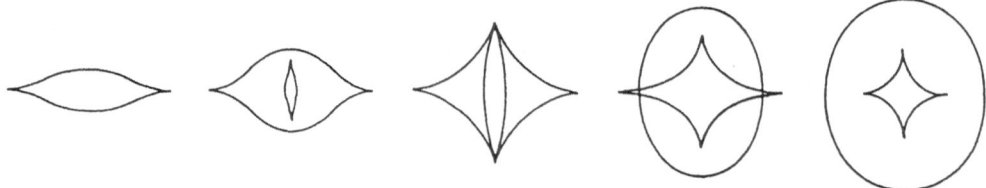

Fig. 7: John Nye's raindrop

the screen is moved further from the source (and the drop) the caustic typically
evolves through the sequence of forms as shown. See BERRY & UPSTILL [3], and
in particular the Appendix based on analysis due to HANNAY.

6. Caustics by Reflection

A signal-source at the focus of a parabolic reflector (in 3 dimensions) produces
a parallel beam. If the source is moved from the focus but remains on the axis
of the reflector then the resulting caustic has circular symmetry, but if the
source is slightly offset then the caustic geometry becomes more complicated.
Its structure can be explored by moving a screen as in the liquid drop lens
experiments above: not surprisingly the same geometric features are observed.
Figure 8 shows two pictures from a computer simulation involving many different
source points and screen positions as part of a general program on the ray-
tracing problem in the Geometric Theory of Diffraction: CHILLINGWORTH, DANESH-
NAROUIE & WESTCOTT [4].

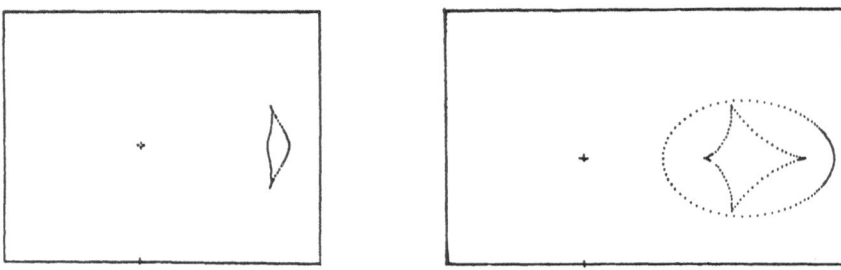

Fig. 8: Computer-simulated caustics by reflection

Now some mechanical examples.

7. The Zeeman Catastrophe Machine

This instructive device [5,6] shows how a simple mechanical system with more
than one stable equilibrium state can undergo sudden (= 'catastrophic') jumps
from one state to another when certain control parameters (in this case the
coordinates of the end-point of a piece of elastic) are slowly varied. The
bifurcation set is the curve in the parameter space across which some available
equilibrium states for the system coincide, and for this machine it has an
astroid-like 4-cusped shape: see Fig. 9.

376

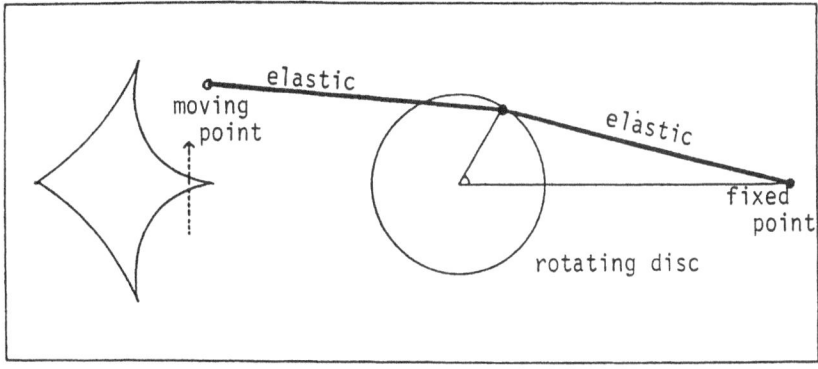

Fig. 9: Zeeman machine

8. Rolling an Elliptical Log

The log will rest in equilibrium on a horizontal plane when its centre of gravity G is directly above its point of contact P with the plane, i.e. G lies on the normal to the ellipse at P . Thus there are two, three or four equilibrium positions (choices of P) according to whether G lies outside, on or inside the astroid which is the envelope of the normals (see (3.) above).

When the plane is tilted by angle α the relevant line PG is no longer the normal but it makes (constant) angle α with the normal. As α varies the envelope of these 'equilibrium' lines changes as in Fig. 10. With G fixed the envelope can pass across G , meaning that the number of equilibrium lines through G goes down by two and so the log may be dislodged from one equilibrium and lurch to another (if available) or roll down the plane for ever.

In the next two examples, the astroid is more deeply buried in the mathematics but nevertheless still controls the bifurcation geometry.

9. The Traction Problem in Elastostatics

Consider an elastic body B in $\mathbb{R}^3$, subject to a force field or *load* ℓ consisting of body forces and surface tractions. Suppose that it is at rest in equilibrium at some initial configuration I_B with zero stress and zero load. The problem is to *find all equilibrium positions close to rigid rotations when small loads are applied*. Local bifurcation analyses for 1-parameter loads were given by SIGNORINI [7], STOPPELLI [8]. Here we adopt a more global geometric point of view: for details see CHILLINGWORTH, MARSDEN & WAN [9].

Under the assumption that equilibria are given by stationary points of a certain *stored energy function*, the problem can be shown to reduce to a search for critical points of a smooth function $f_\ell : SO(3) \to \mathbb{R}$. For most equilibrated loads ℓ (i.e. zero resultant and overall moment) this function f_ℓ has non-degenerate critical points, corresponding to equilibria of B close to the identity and to three rigid rotations of 180° about mutually perpendicular axes. However, if ℓ has a single *axis of equilibrium* (an axis μ in $\mathbb{R}^3$ such that rotations of the load field ℓ about μ continue to be equilibrated (see Fig. 11 for an illustration) then to first order f_ℓ has a whole circle S of critical points, and the problem then becomes one of analyzing critical points of certain second order terms on the circle. This in turn reduces simply to finding critical points of

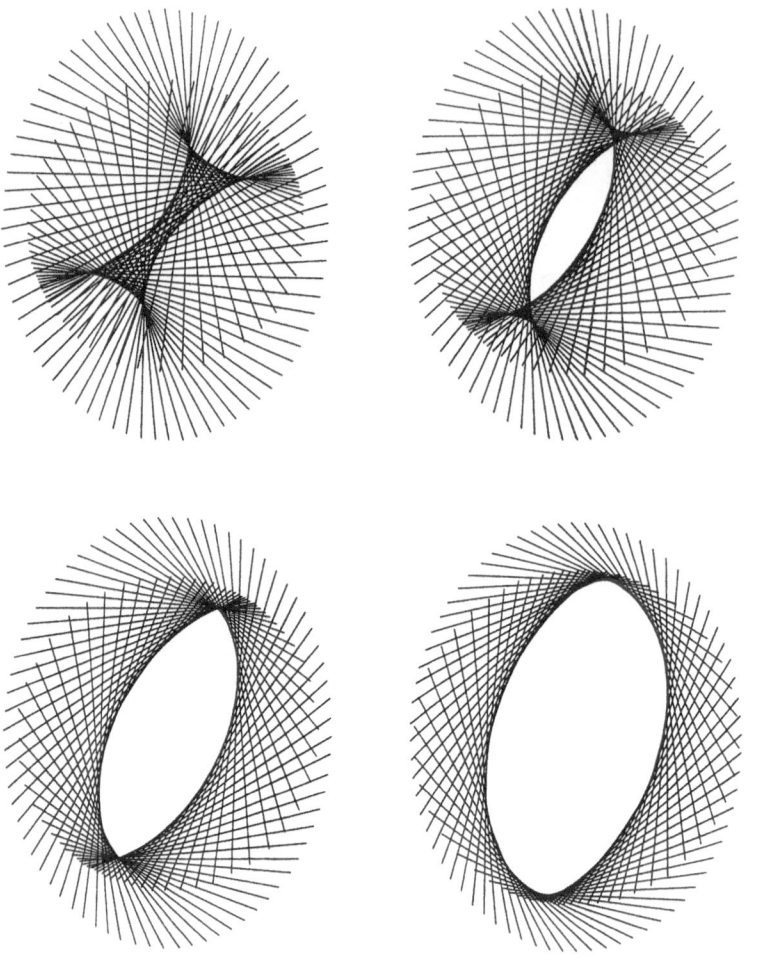

Fig. 10: Losing the astroid (hand-drawn pictures due to POSTON [10])

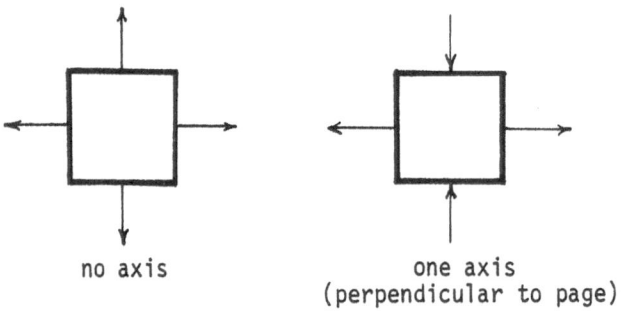

no axis

one axis
(perpendicular to page)

Fig. 11: Loads with zero or one axis of equilibrium

$$V = ax^2 + hxy + by^2 + cx + dy$$

on the circle $x^2 + y^2 = 1$. By rotating axes we first make $h = 0$, and then for fixed a, b the bifurcation set is the astroid

$$c^{2/3} + d^{2/3} = (2(a-b))^{2/3} .$$

Inside the astroid there are four critical points for V , while outside there are two. In terms of the original load parameters the bifurcation set looks like a cone as in Fig. 12, where the load ℓ is first scaled as $\lambda \ell_0$, and p_1, p_2 are certain integrals involving ℓ_0 over B and its boundary.

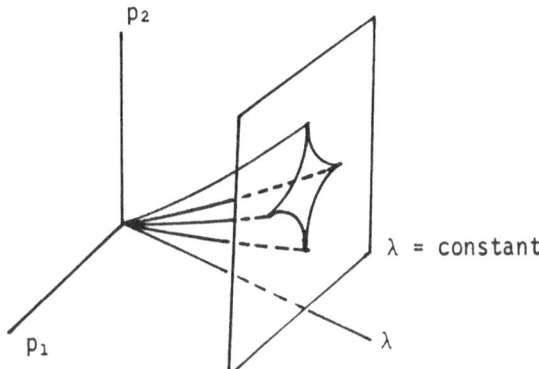

λ = constant

Fig. 12: Bifurcation set for the traction problem (load with one axis of equilibrium)

Note that the circular symmetry involved here is a subtle one, not merely a statement of symmetry in the body or the load field. Further axes of equilibrium may enter the problem, causing the circle S to be replaced by a copy of $\mathbb{RP}^2$ or even $SO(3)$ itself: see [9].

From the examples above we have seen how the astroid tends to appear whenever there is bifurcation involving breaking circular symmetry. Before looking at our final illustrative example, however, let us stand back and briefly consider the general question of bifurcation away from continuous symmetry in abstract terms. It reveals the astroid in yet another light.

10. Bifurcation from Continuous Symmetry

Consider a bifurcation problem

$$B(x,\mu) = 0 \qquad\qquad (P)$$

where x belongs to a smooth manifold M of positive dimension, $\mu \in \mathbb{R}^k$ and $B : M \times \mathbb{R}^k \to \mathbb{R}^p$ is a smooth map. The basic assumption for this problem is that B vanishes identically on the whole of M when $\mu = 0$:

$$B(x,0) = 0 \quad \text{for all} \quad x \in M . \qquad\qquad (H)$$

The aim then is to understand the structure of the set of solutions x on M when μ is perturbed away from zero.

Our strategy initially follows that in CHOW & HALE [11, Ch.11]. First write

$$B(x,\mu) \equiv (h(x) + O(\mu)).\mu$$

where $h : M \to \{p \times k \text{ matrices}\}$, this being possible in view of (H). Next, analyze the 'μ-linearized' problem

$$h(x).\mu = 0 \ , \quad \mu \in \mathbb{R}^k \tag{L}$$

and then try to make suitable generic hypotheses to ensure (using singularity theory) that the bifurcation picture obtained for (L) is structurally stable. This being the case, the bifurcation picture for (P) for small $|\mu|$ is then merely a small distortion (by diffeomorphism) of that for (L).

   The geometry is best understood by first replacing $\mu \in \mathbb{R}^k$ by the unit vector $u = \mu/|\mu|$ in S^{k-1} , since the bifurcation set for (L) is the *cone* (from $0 \in \mathbb{R}^k$) on the bifurcation set $C \subset S^{k-1}$ for the "spherical problem"

$$h(x).u = 0 \ , \quad u \in S^{k-1} \ . \tag{S}$$

   We shall not pursue this in complete generality here (a fuller account is in preparation [12]) but go immediately to the special case $p = 1$, $k = 3$ with $M = SO(2) \cong S^1$. Here we have

$$h : S^1 \to \{1 \times 3 \text{ matrices}\} \cong \mathbb{R}^3$$

and the problem (S) is: $h(x).u = 0, \ u \in S^2$.

   The bifurcation set C in S^2 is contained in

$$C = \{u \in S^2 : \exists x \in S^1 \text{ with } h(x).u = 0 \text{ and } h'(x).u = 0\} \ ,$$

which is the locus traced out on S^2 by the unit vectors $\pm u$ orthogonal to both $h(x)$ and $h'(x)$ as x traverses $M \cong S^1$. Provided $\{h(x), h'(x)\}$ are linearly independent for all $x \in M$ (a generic assumption) we see that u is simply parametrizing the family of tangent planes to the curve $\gamma = h(M)$ that pass through the origin in $\mathbb{R}^k$; in other words C is two antipodal copies of *the curve dual to the radial projection γ^* of γ into the unit sphere S^2 in* $\mathbb{R}^3$ - i.e. to the curve γ as 'observed' from $0 \in \mathbb{R}^3$. See Fig. 13. Observe that inflexions in the projected curve γ^* correspond to cusps in its dual; thus (again generically) C will consist of two antipodal copies of a curve in S^2 that is smooth everywhere except for a finite (even) number of $\frac{3}{2}$ - power cusps: moreover, this picture is structurally stable. Therefore the bifurcation set

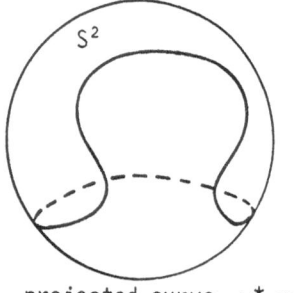

projected curve $\gamma^* \subset S^2$

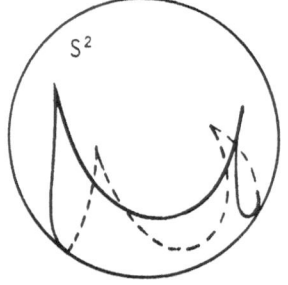

dual curve

Fig. 13: The curve γ^* and its dual

for (L) is the cone on this from $0 \in \mathbb{R}^3$, and for small $|\mu|$ the bifurcation set for the original problem (P) will be a 'curved' version of the cone. Compare Fig. 12 above.

This general description of the bifurcation set as a curved version of a cone on the dual to the radial apparent outline of h(M) goes over to higher-dimensional cases [12].

Now we turn to the last example in our list of appearances of the astroid, this time again in a physical problem.

11. Dynamics of a nearly-symmetric top

Here a *top* means a heavy rigid body Ω rotating in $\mathbb{R}^3$ under gravity about a fixed point (therefore perhaps better called a gyroscope). The configuration space is SO(3) , and the phase space (configuration and momentum) is the cotangent bundle T*SO(3) . There are two real-valued functions

energy $\qquad$ E : $T^*SO(3) \to \mathbb{R}$

angular momentum about vertical $\quad$ J : $T^*SO(3) \to \mathbb{R}$

which are of crucial importance in understanding the dynamics, because each of E, J remains constant along orbits of the motion in T*SO(3) . Thus in order to analyse how T*SO(3) is decomposed into all the orbits a first step is describe the set $(E \times J)^{-1}(h,p)$ for each $(h,p) \in \mathbb{R}^2$.

In 1970 SMALE [13] proposed a program for attacking this problem for general mechanical systems. The idea is to exploit the fact that $(E \times J)^{-1}(h,p)$ changes its topological structure only when (h,p) crosses the set of points Σ over which $E \times J$ is not locally trivial; in our case (any many others) this is precisely the set Σ of *singular values* of $E \times J$. The study of Σ is made easy by the following result, which is a general theorem of Smale specialised to our case.

Theorem $\quad$ The set Σ consists of those $(h,p) \in \mathbb{R}^2$ for which h is a critical value of a certain smooth function $\tilde{V}_p$: $S^2 \to \mathbb{R}$.

This $\tilde{V}_p$ is called the *reduced amended potential*, but its construction will not concern us here: it is enough to know that it can be calculated explicitly quite easily for this problem. The task then is to study how the set of critical values of $\tilde{V}_p$ changes as the parameter p is varied.

Some interesting results and appealing pictures have been worked out by IACOB [14], KATOK [15], TATARINOV [16]. The aim here is to view these in terms of some familiar (or less familiar) geometric objects found in singularity theory. Further details will appear in BRITT, CHILLINGWORTH & SWIFT [17].

Let the principal moments of inertia of Ω be I_1, I_2, I_3 with the initial symmetry assumptions that $I_1 = I_2$ and the axis for I_3 is the line joining the centre of mass C of Ω to the fixed point P . In this case $\tilde{V}_p$ is *rotationally symmetric* (about north-south axis NS, say), and the bifurcation picture in $(p^2, I_3/I_1)$ parameter space is easy to work out explicitly: see Fig. 14. The small pictures symbolize the critical points of $\tilde{V}_p$ on S^2 .

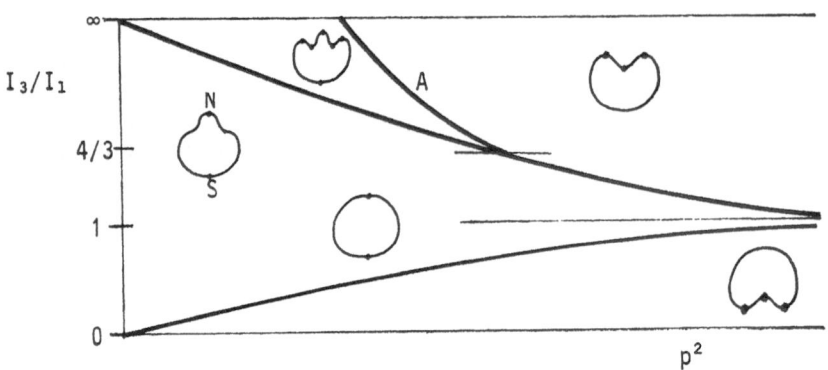

Fig. 14: Bifurcations of $\tilde{V}_p$

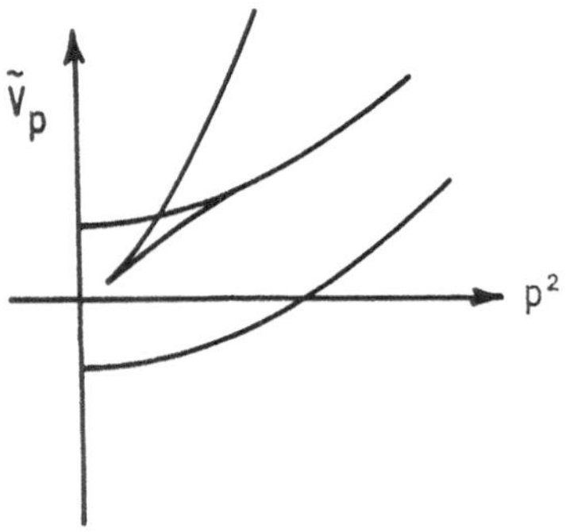

$\tilde{V}_p$

p^2

Fig. 15: Critical values of $\tilde{V}_p$ with
$I_1 = I_2$ and $CP = I_3$-axis

$4I_2 < 3I_3$

$4I_1 < 3I_3$
$3I_2 < 3I_3 < 4I_2$

$3I_2 < 3I_3 < 4I_1$

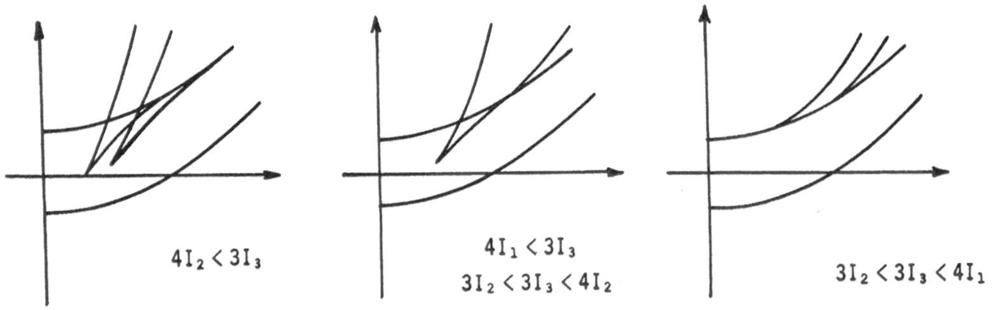

Fig. 16: Critical values of $\tilde{V}_p$ with $I_1 \neq I_2$ and $CP = I_3$-axis

We are interested in critical *values* of $\tilde{V}_p$. A sketch of these critical
values plotted against p^2 is as in Fig. 15 for choice of the parameter I_3/I_1
in the range $3I_3 > 4I_1$ (cf. IACOB [14]).

When $I_1 \neq I_2$ but $|I_1 - I_2|$ is small compared with I_3 then, provided CP
continues as the I_3-axis, the problem (and hence $\tilde{V}_p$) retains $\mathbb{Z}_2 \times \mathbb{Z}_2$
symmetry and the bifurcation diagram splits into two approximate copies of
Fig. 14 slightly displaced from each other. The plot of critical values like-
wise splits into two and we obtain pictures as in Fig. 16 (taken from KATOK [15]).

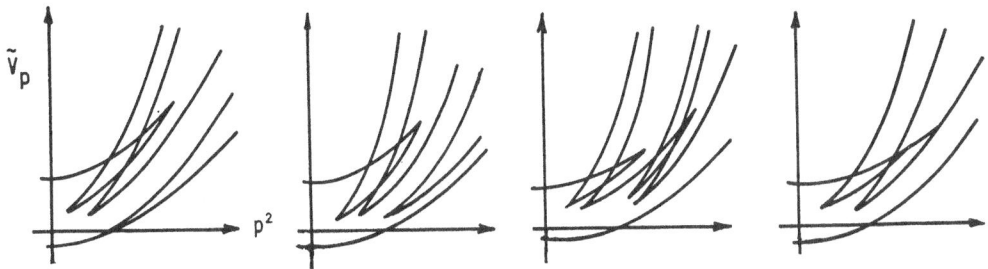

$\tilde{V}_p$

p^2

Fig. 17: Critical values of $\tilde{V}_p$ with $I_1 \neq I_2$ and $CP \neq I_3$-axis

When, finally, symmetry is totally removed by allowing the I_3-axis not to be CP , the Iacob/Katok pictures break up further, giving rise to the complicated configurations seen in the pictures produced by TATARINOV [16]: a sample is shown in Fig. 17.

How can all these families of pictures be unified? The approach indicated by singularity theory is to look for an organizing centre, i.e. a degenerate case from which the many nondegenerate configurations can be derived by *unfolding*. The most degenerate situation in this problem is when $I_1 = I_2 = I_3$: the next most degenerate one occurs when $I_1 : I_2 : I_3 = 3 : 3 : 4$ for a certain value of p^2 : see Fig. 14. This corresponds to a rotationally-symmetric 6th-degree degeneracy of $\tilde{V}_p$ at N on S^2 . We shall content ourselves here with some observations about degeneracy along the curve A in Fig. 14: here $\tilde{V}_p$ is quartic at the north pole N . In suitable local coordinates the leading term in the Taylor series at N is $(x^2+y^2)^2$, and so we see we have to do with the singularity X_9 in ARNOL'D's terminology [18]. Let us end, then, by looking more closely at X_9 .

12. Perturbing rotationally symmetric X_9

Because of the rotational symmetry there is no finite-dimensional versal unfolding (any number of new critical points could be introduced locally by suitable polynomial perturbations). However, we can attempt to grasp something of the response of $\tilde{V}_p$ to low-order perturbations by first considering the general X_9 singularity

$$V_t(x,y) \equiv x^4 + tx^2y^2 + y^4 , \quad t \neq 2$$

which does have a finite-dimensional versal unfolding, and then seeing what happens as $t \to 2$. The bifurcation geometry of X_9 has been studied in great detail by CALLAHAN[19,20]. Following some of his ideas, we focus attention on the family

$$V_t(x,y;\mu) = x^4 + tx^2y^2 + y^4 - \gamma(x^2+y^2) - \varepsilon(x^2-y^2) + \alpha x + \beta y ,$$

where $\mu = (\gamma,\varepsilon,\alpha,\beta)$, and look at the bifurcation diagram $B(\alpha,\beta)$ in the (α,β) plane for each fixed (t,γ,ε) . This turns out to depend only on the ratio γ/ε , so we take $(\gamma,\varepsilon) = (\cos\theta, \sin\theta)$ and sketch vignettes of $B(\alpha,\beta)$ in various regions of the (θ,t)-plane: see Fig. 18. The *butterflies* and *umbilics* will be familiar to readers acquainted with elementary catastrophe theory.

There is no room here to go into details of the relationship between these bifurcation pictures and the behaviour of $\tilde{V}_p$ in the rigid body problem: for this we need to say more about the geometry of the configuration of critical

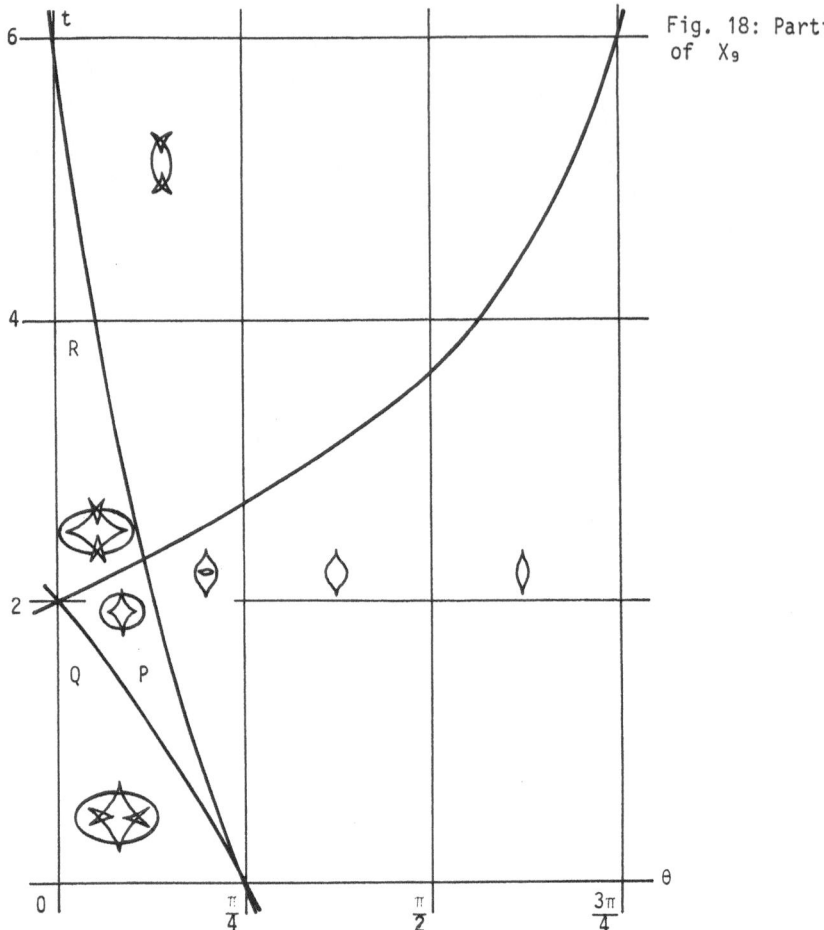

Fig. 18: Partial unfolding of X_9

values of unfoldings of X_9 . Corresponding to the astroid in Fig. 18 is a plot of critical values of $V_t(x,y;\mu)$ against (α,β) (for (t,γ,ε) fixed) as sketched in Fig. 19 (it is no coincidence that this Holy Grail also features in the classification of homogeneous binary quartics by STEWART & POSTON in [10, p.132]) which governs the critical value structure of $\tilde{V}_p$. For further details see [17].

Observe how $B(\alpha,\beta)$ undergoes the raindrop sequence (Fig. 7) as (θ,t) moves towards $(0,2)$ along the line $t = 2$: this corresponds to the approach towards circular symmetry in the bifurcation analysis of $V_t(x,y;\mu)$ as $\varepsilon \to 0$. Note particularly the astroid in the region P (where the bathroom window of Fig. 6 is located). We expect it not to be far away, since when $\varepsilon = 0$ and $\gamma \neq 0$ the function

$$x^4 + 2x^2y^2 + y^4 - \gamma(x^2+y^2)$$

has a circle of critical points and so further symmetry-breaking linear perturbations can be expected to yield an astroid as in (9.) above. Finally, however, note the more complicated cusped curves in regions Q, R : they seem to be just as accessible from the X_9 point as are the "simple" astroids.

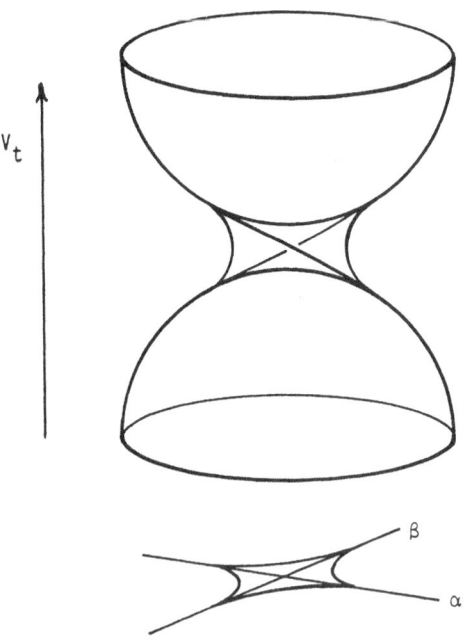

Fig. 19: The grail

v_t

β

α

It would be agreeable to end this lecture with a theorem stating (in precise mathematical terms) that the astroid is the canonical shape to be observed whenever a rotationally-symmetric X_9 is 'generically' perturbed. There probably is indeed such a theorem, but from Fig. 18 alone it is as yet unclear what it may be.

References

1. M.V. Berry: J. Phys. A8, 566-584 (1975)
2. J.F. Nye: Proc. R. Soc. Lond. A403, 1 (1986)
3. M.V. Berry, C. Upstill: In Progress in Optics, ed. by E. Wolf, Vol. XVIII (North-Holland, Amsterdam 1980) p.257
4. D.R.J. Chillingworth, G.R. Danesh-Narouie, B.S. Westcott: in preparation for IEEE Trans. AP
5. E.C. Zeeman: In Towards a Theoretical Biology, ed. by C.H. Waddington, Vol. 4 (Edinburgh Univ. Press 1972) p.276
6. E.C. Zeeman: In Catastrophe Theory: Selected Papers 1972-1977, (Addison-Wesley, Reading, Mass. 1977) p.409
7. A. Signorini: Proc. 3rd Int. Cong. Appl. Mech. 2, 80 (1930)
8. F. Stoppelli: Ricerche Mat. 6, 241 (1957); 7, 71, 138 (1958)
9. D.R.J. Chillingworth, J.E. Marsden, Y.-H. Wan: Arch. Rat. Mech. Anal. 80, 295 (1982); 83, 363 (1983)
10. T. Poston, I.N. Stewart: Taylor Expansions and Catastrophes, Res. Notes in Math, Vol. 7 (Pitman, London 1976)
11. S.-N. Chow, J.K. Hale: Methods of Bifurcation Theory (Springer, New York 1982)
12. D.R.J. Chillingworth: in preparation
13. S. Smale: Inventiones Math. 10, 305 (1970); 11 , 45 (1970)
14. A. Iacob: Rev. Roum. Math. Pures et Appl. XVI, 1497 (1971)
15. S.V. Katok: Uspekhi Mat. Nauk 27:2, 126 (1972)

16. Ja.V. Tatarinov: Vestnik Moskov, Univ. 28:5, 70 (1973); 29:6, 99 (1974)
17. J.P. Britt, D.R.J. Chillingworth, S.T. Swift: in preparation
18. V.I. Arnol'd: Uspekhi Mat. Nauk 30:5, 3 (1975) = Russian Math. Surveys 30:5. 1 (1975)
19. J.J. Callahan: Proc. London Math. Soc. (3) 45, 227 (1982)
20. J.J. Callahan: private communication

On the Hopf Bifurcation with Broken O(2) Symmetry

G. Dangelmayr[1] and E. Knobloch[2]

[1]Institute for Information Sciences, University of Tübingen,
 Köstlinstr. 6, D-7400 Tübingen, Fed. Rep. of Germany
[2]Department of Physics, University of California, Berkeley, CA 94720, USA

Abstract. Translation and reflection symmetries introduce the group O(2) into bifurcation problems with periodic boundary conditions. The effect on the Hopf bifurcation with O(2)-symmetry of small terms breaking the translation symmetry is investigated. Two primary branches of standing waves are found. Secondary and tertiary bifurcations involving two different types of modulated waves are analyzed in the neighborhood of secondary Takens-Bogdanov bifurcations. The effects of breaking the phaseshift (in time) and reflection symmetries are briefly considered.

1. Introduction

The theory of the Hopf bifurcation in the presence of O(2)-symmetry [6] has been remarkably successful, on a qualitative level, in accounting for a variety of phenomena [9,12] observed in recent experiments on binary fluid mixtures [1,8,13,14, 17,19]. In these experiments, characterized by a negative separation ratio S, the initial conduction state loses stability via a Hopf bifurcation as the layer is heated from below. To describe large aspect ratio phenomena it is both plausible and convenient to take the system to be translationally invariant, as a first approximation, and to assume periodic boundary conditions in the horizontal. Thus, if $\psi(x)$ is a solution for the stream function with an assumed period L, then $T_l \psi = \psi(x+l)$, obtained by translation through a distance l, is a new solution for all $0 < l < L$. If, in addition, the system is assumed to be symmetric under reflections in any vertical plane, then the translations and reflections together constitute the symmetry group O(2), the rotations and reflections of a circle. In the case of a Hopf bifurcation this O(2)-symmetry forces the number of eigenvalues on the imaginary axis to be doubled, leading to the appearance of two primary branches, the standing waves (SW) and the travelling (rotating) waves (TW). The normal form describing this bifurcation has O(2)xS(1)-symmetry [6,9]. The presence of the extra S(1) phase shift symmetry (in time) is a well-known feature of the Hopf bifurcation and leads to the possibility of spatio-temporal symmetries. In the present case the SW are invariant under the reflection Z(2), while the TW are invariant under the spatio-temporal symmetry group $\widetilde{SO}(2)$ [6], i.e. under spatial translations followed by appropriate (temporal) phase shifts [6,16].

In this paper we are interested in describing the effects of distant side walls. Unlike other investigations we do not employ a long wavelength expansion [2,9], and continue to study spatially periodic solutions. Instead, we note that one of the main effects of sidewalls, regardless of the detailed boundary conditions, is to break the translation symmetry that is present in the absence of sidewalls. We use this observation to motivate our study of the effects of breaking the SO(2) part of O(2) on the Hopf bifurcation with O(2) symmetry. We assume that the symmetry breaking is weak, and are thereby able to relate the solutions of the problem with broken symmetry (hereafter the "perturbed problem") to those of the fully symmetric problem (the "unperturbed problem"). The key to understanding the perturbed problem is the appearance of a subsidiary Takens-Bogdanov bifurcation whose local features are analyzed in the following section. A more global analysis is given elsewhere [4]. The results of section 2 predict several unexpected secondary bifurcations leading to modulated (i.e. two-frequency) waves, as well as homoclinic and heteroclinic bifurcations. In section 3 we note that the S(1) phase shift symmetry is only a symmetry of the normal form and not of the underlying physical system, and show that the expected breaking of the symmetry leads to the appearance of horse-shoes in the vicinity of the global bifurcation. Our results are discussed in relation to other work in section 4.

2. Symmetry-breaking

The unperturbed problem is described by the normal form equation [6,9]

$$\dot{v} = g(\lambda,|v|^2,|w|^2)v, \qquad \dot{w} = \bar{g}(\lambda,|w|^2,|v|^2)w \qquad (1)$$

for two complex amplitudes v,w, where $g : R \times R^2 \to C$ is a smooth function satisfying $g(0,0,0) = i\omega_0$ and $Reg_\lambda(0,0,0) \neq 0$. Here λ is the bifurcation parameter. The normal form (1) is equivariant (i.e. symmetric) under the three operations

$$SO(2) \quad : \quad (v,w) \to e^{ikl}(v,w) \qquad (2a)$$

$$Z(2) \quad : \quad (v,w) \to (\bar{w},\bar{v}) \qquad (2b)$$

$$S(1) \quad : \quad (v,w) \to (e^{i\omega\tau}v, \, e^{-i\omega\tau}w), \qquad (2c)$$

corresponding, respectively, to translations $x \to x+1$, reflections $x \to -x$ and phase shifts $t \to t+\tau$. In order to study the effects of breaking the $SO(2)$-symmetry we add terms to (1) which break the symmetry (2a), while respecting (2b,c). The dominant such terms are those that are linear in the amplitudes (v,w), and without loss of generality we study the system

$$\dot{v} = g(\lambda,|v|^2,|w|^2)v + e\bar{w}, \qquad \dot{w} = \bar{g}(\lambda,|w|^2,|v|^2)w + \bar{e}\bar{v}, \qquad (3)$$

where e is a complex parameter. We assume that $|e|$ and λ are of the same order. This approach to the problem of symmetry-breaking lacks the rigor of a mathematical theorem, and the possibility remains that other terms could be added to (3) resulting in phenomena not captured by the present analysis.

We proceed, as usual, by expanding g in a Taylor series:

$$\dot{v} = (\lambda + i\omega + a|v|^2 + bA^2)v + e\bar{w} \qquad (4)$$

$$\dot{w} = (\lambda - i\omega + \bar{a}|w|^2 + \bar{b}A^2)w + \bar{e}\bar{v},$$

where $\omega - \omega_0 = 0(\lambda)$. The expansion can be truncated at third order provided the complex coefficients a,b satisfy $a_r \neq 0$, $b_r \neq 0$, $a_r + 2b_r \neq 0$. Here the subscript r denotes the real part and A the total amplitude, $A^2 = |v|^2 + |w|^2$. In terms of the real variables

$$v = x_1 e^{i\theta_1}, \qquad w = x_2 e^{i\theta_2}, \qquad e = \varepsilon e^{i\alpha} \qquad (5)$$

one obtains

$$\dot{x}_1 = (\lambda + a_r x_2^2 + b_r A^2)x_1 + \varepsilon x_2 \cos(\theta - \alpha) \qquad (6a)$$

$$\dot{x}_2 = (\lambda + a_r x_1^2 + b_r A^2)x_2 + \varepsilon x_1 \cos(\theta + \alpha) \qquad (6b)$$

$$\dot{\theta} = a_i(x_2^2 - x_1^2) - \varepsilon \frac{x_2}{x_1}\sin(\theta + \alpha) - \varepsilon \frac{x_1}{x_2}\sin(\theta - \alpha) \qquad (6c)$$

where $\theta = \theta_1 + \theta_2$. The unperturbed problem ($\varepsilon = 0$) has three different types of stationary solutions (x_1,x_2): the trivial solution T, $(0,0)$, left- and right travelling waves $(x,0)$, $(0,x)$, and standing waves (x,x). There are six different generic cases for the unperturbed problem, distinguished by different regions in the (a_r,b_r)-plane. In this paper we confine ourselves to the case $a_r < 0$, $b_r < 0$. The resulting bifurcation diagram for this case is shown in Figure 1 [6,9].

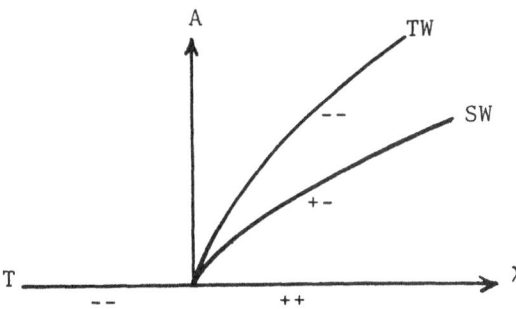

Fig. 1. Bifurcation diagram corresponding to the unperturbed system for $a_r<0$, $b_r<0$. The signs refer to the stability assignments.

To study the effect of the ε-perturbation we introduce the variables A,θ according to

$$x_1 = A \cos(\phi/2), \qquad x_2 = A \sin(\phi/2), \tag{7}$$

with the result

$$\dot{A} = A\{\lambda+(b_r +\tfrac{1}{2} a_r \sin^2\phi)A^2\}+\varepsilon A \sin\phi \cos\theta \cos\alpha \tag{8a}$$

$$\dot{\phi} = a_r A^2 \sin\phi \cos\phi+2\varepsilon(\cos\phi \cos\theta \cos\alpha-\sin\theta \sin\alpha) \tag{8b}$$

$$\dot{\theta} = -a_i A^2 \cos\phi - \frac{2\varepsilon}{\sin\phi} (\sin\theta \cos\alpha + \cos\phi \cos\theta \sin\alpha). \tag{8c}$$

The trivial solution $A=0$ loses stability to solutions in the form of standing waves ($\phi = \pi/2$) at $\lambda = \mp\varepsilon\cos\alpha$, the signs corresponding to $\theta = 0,\pi$, respectively. Thus there are two branches of standing waves, SW_0, SW_π, that bifurcate from $A=0$ and differ only in their total phase θ. Their amplitude is given by

$$\lambda + (\tfrac{1}{2} a_r+b_r)A^2 \pm \varepsilon\cos\alpha = 0. \tag{9}$$

There are no other bifurcations from the trivial solution, although there are secondary bifurcations from the SW-branches. To determine these we linearize (8) about $SW_{0,\pi}$. The amplitude eigenvalue is $s = (a_r+2b_r)A^2$ and vanishes only when $A=0$. The remaining two eigenvalues satisfy

$$s^2 + (a_r A^2 \pm 4\varepsilon\cos\alpha)s \pm 2\varepsilon|a|A^2\cos(\alpha-\gamma) + 4\varepsilon^2 = 0, \tag{10}$$

where we have set $a = |a|e^{i\gamma}$. We assume that $\pi/2<\gamma<\pi$ so that $a_r<0$. A steady state bifurcation occurs when

$$|a|A^2 \cos(\alpha-\gamma) \pm 2\varepsilon = 0 \tag{11}$$

and leads to a time-independent solution (A,ϕ,θ) satisfying equations (8). From (8b) it follows that as A (or λ) increases the angle ϕ approaches 0 or π and thus this solution looks like a TW away from the region influenced by the perturbation. We shall refer to it as TW although, since $\phi\neq0,\pi$ it is not a pure travelling wave, but a mixture of right and left travelling waves with slightly different frequencies, i.e., a wave modulated with a very low frequency.

A secondary Hopf bifurcation occurs on $SW_{0,\pi}$ when

$$A^2 = \mp(4\varepsilon/a_r) \cos\alpha > 0 \tag{12}$$

and gives rise to oscillations with frequency Ω given by

$$\Omega^2 = -4\varepsilon^2 \frac{\cos(2\alpha-\gamma)}{\cos\gamma} > 0. \tag{13}$$

389

This is an oscillation in both θ and ϕ, and hence in the amplitudes x_1, x_2. Such a wave is therefore a two-frequency wave, and we refer to it as MW. Some information about the connections between TW, MW and $SW_{0,\pi}$ can be obtained by choosing α close to a value where the Hopf and steady state bifurcations coincide in a codimension-two point on the $SW_{0,\pi}$-branch. At this point $\Omega^2=0$ which is satisfied on SW_π for $\alpha=\alpha_1,\alpha_2$ and on SW_0 for $\alpha=\alpha_3,\alpha_4$, where

$$\alpha_1 = \frac{\gamma}{2} + \frac{\pi}{4}, \qquad \alpha_2 = \frac{\gamma}{2} + \frac{3\pi}{4}, \qquad \alpha_3 = \frac{\gamma}{2} - \frac{\pi}{4}, \qquad \alpha_4 = \frac{\gamma}{2} + \frac{5\pi}{4}. \tag{14}$$

Near $\alpha=\alpha_i$ the flow on the center manifold for (8) is determined by a two-dimensional system which can be transformed to a Takens-Bogdanov normal form with $Z(2)$-symmetry. A lengthy calculation shows that the latter takes the form

$$\frac{1}{\epsilon}\dot{x} = y \tag{15a}$$

$$\frac{1}{\epsilon}\dot{y} = \{2\chi/r - 4\beta/\sin\alpha_i\}(x/\cos\alpha_i) \pm (2\chi/r)y + Mx^3 + Nx^2y, \tag{15b}$$

where

$$M = 2\cos(2\alpha_i) + \frac{1}{r}(3 - 4\cos^2\alpha_i) \tag{15c}$$

$$N = \pm (1/2r^2\cos\alpha_i)\{4r(r-1)\cos^2\alpha_i(1 - 6\sin^2\alpha_i) + 4\cos^2\alpha_i - 3\} \tag{15d}$$

$$\alpha = \alpha_i \pm \beta, \qquad \lambda = \pm (1 + 4b_r/a_r)\epsilon\cos\alpha_i \pm \epsilon\mu \tag{15e}$$

and

$$r = 1 + 2b_r/a_r. \tag{15f}$$

Here (x,y) provide a parametrization for the center manifold and (χ,β) describe (small) deviations from the codimension-two degeneracy. The dynamics of equations (15,a,b) are well understood [7, 10], enabling us to determine how the MW-branch interacts with TW and $SW_{0,\pi}$ near the codimension-two point. The resulting bifur-

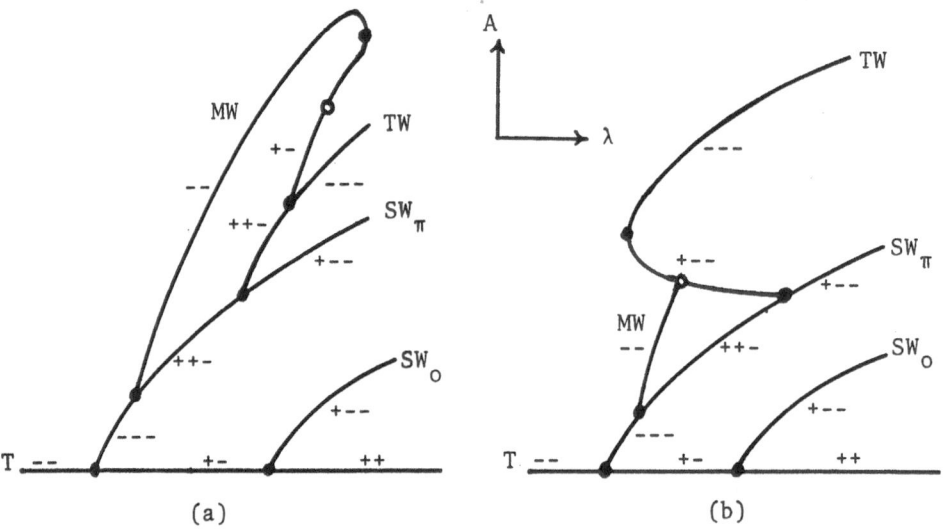

(a)   (b)

Fig. 2. Bifurcation diagrams for the perturbed system near (a) $\alpha=\alpha_1$ and (b)$\alpha=\alpha_2$ for $b_r<0$, $\pi/2<\gamma<\pi$, $\sin\gamma<2/3$ and r sufficiently large. The open circles indicate global (ambiclinic for (a), heteroclinic for (b)) bifurcations.

cation diagrams depend essentially only on the signs of the coefficients M and N. When r is sufficiently large and $\sin\gamma < 2/3$, N is always negative whereas $M<0$ for α_1, α_4 and $M>0$ for α_2, α_3. From this and from the results of [7,10] we find that the bifurcation diagrams near $\alpha=\alpha_1$ and $\alpha=\alpha_2$ are as shown in Figure 2(a) and (b), respectively. These diagrams should be compared with Figure 1.

3. Transition to Chaos

As pointed out earlier the $S(1)$ phase shift symmetry is not an exact symmetry of the problem, and one expects it to be broken at some level. According to normal form theory such $S(1)$ breaking terms can be put, by means of successive nonlinear near-identity coordinate changes, into the "tail" of equations (1), i.e. beyond any order in perturbation theory. In this paper we do not carry out such coordinate changes, and instead introduce $S(1)$ breaking terms already at linear order

$$\dot{v} = g(|v|^2, |w|^2, \lambda)v + \varepsilon e^{i\alpha}\bar{w} + \delta e^{-i\beta}w \tag{16a}$$

$$\dot{w} = \bar{g}(|w|^2, |v|^2, \lambda)w + \varepsilon e^{-i\alpha}\bar{v} + \delta e^{i\beta}v, \tag{16b}$$

where $\delta \ll \varepsilon$. The resulting normal form is symmetric under (2b) only. We are interested in determining the influence of the new terms in the limit $\delta \to 0$ on the dynamics near the codimension-two bifurcation, and specifically near the heteroclinic and ambiclinic orbits. The δ-perturbation introduces additional terms into equation (8) for $(\dot{A}, \dot{\phi}, \dot{\theta})$, given by

$$\delta(r\sin\phi\,\cos(\Delta-\beta),\ 2\cos\phi\,\cos(\Delta-\beta),\ -2\cot\phi\,\sin(\Delta-\beta)), \tag{17a}$$

where $\Delta=\theta_2-\theta_1$. Thus the phase difference Δ decouples no longer from the (A,ϕ,θ)-system. However, to leading order $\dot{\Delta} \approx -2\omega$ so that $\Delta \approx -2\omega t$ and (17a) becomes

$$\delta(r\sin\phi\,\cos(2\omega t+\beta),\ 2\cos\phi\,\cos(2\omega t+\beta),\ -2\cot\phi\,\sin(2\omega t+\beta)). \tag{17b}$$

Consequently, when adding (17b) to (8), we obtain a nonautonomous system for (A,ϕ,θ). We are interested in the effect of (17) on the codimension-two bifurcation. Although center manifold reductions are not possible in nonautonomous systems, the time-dependent terms here are very small and we can perform the same transformations as in the autonomous case of the preceding section. Then, to leading order, one variable decouples from the two others and vanishes for $t=o(1/\varepsilon)$. In this way we arrive again at (15a,b), but now with the terms

$$-\frac{\delta}{\varepsilon}(2x\mp y\cos\alpha_i)\ (\sin\alpha_i\sin(2\omega t+\beta_i),\ 2\cos(2\omega t+\beta_i)) \tag{18}$$

added to $(\dot{x}/\varepsilon, \dot{y}/\varepsilon)$, where $\beta_i=\beta+\alpha_i$. Next we rescale the variables according to $x=B\varepsilon X$, $y=C\varepsilon^2 Y$, $t=D\tau/\varepsilon^2$, with suitably chosen constants B,C,D to obtain

$$X' = Y + O(\delta/\varepsilon^2) \tag{19a}$$

$$Y' = \mu X + \sigma_1 X^3 + \varepsilon\{\nu Y + \sigma_2 X^2 Y - A(\delta/\varepsilon^4)XG(\tau/\varepsilon^2) + O(\delta/\varepsilon^3)\}, \tag{19b}$$

where

$$G(s) = \cos(\Omega s+\beta_i), \qquad (\sigma_1,\sigma_2) = (sgnM, sgnN), \tag{19c}$$

$$\Omega = 2\omega|M/N|, \qquad A = (2N/M)^2, \qquad ' = d/d\tau,$$

and (μ,ν) are linear combinations of $(\chi/\varepsilon^2, (\alpha-\alpha_i)/\varepsilon^2)$. Equations (19a,b) constitute a periodically forced and rescaled Takens-Bogdanov normal form with frequency of order $1/\varepsilon^2$. For $\varepsilon=0$ (19a,b) is a Hamilton system and possesses saddle loop trajectories (heteroclinic for $\sigma_1 = 1$, ambiclinic for $\sigma_1 = -1$) which are given by

$$(X(\tau),Y(\tau)) = \pm\sqrt{2}k(\sinh(k\tau)\operatorname{sech}(k\tau), \quad k\operatorname{sech}^2(k\tau)) \quad (\sigma_1=1) \tag{20a}$$

$$(X(\tau),Y(\tau)) = \pm\sqrt{2}k(\operatorname{sech}(k\tau), \quad -k\sinh(k\tau)\operatorname{sech}^2(k\tau)) \quad (\sigma_1=-1), \tag{20b}$$

where $k=|\mu/2|^{1/2}$ for $\sigma_1=1$ and $K=\sqrt{\mu}$ for $\sigma_1 = -1$. These saddle loops persist for $\delta=0$ and $\varepsilon\neq0$, provided (μ,ν) are on the straight half-line SL : $\nu=m(\varepsilon)\mu$, $\sigma_1\mu<0$, where the slope $m(\varepsilon)$ is a smooth function, given by $m(\varepsilon)= -1/5+0(\varepsilon)$ for $\sigma_1=1$ and $m(\varepsilon)= 4/5+0(\varepsilon)$ for $\sigma_1 = -1$ [2,7]. In order to elucidate the effect of the δ-perturbation near SL, we set

$$\gamma = \mu[m(\varepsilon) + (\delta/\varepsilon^4)c], \quad \sigma_1\mu<0 \tag{21}$$

i.e., we focus on a very small region around SL. Then we can write down a Melnikov-function [7], given by

$$Me(\tau_0,\varepsilon) = c\int_{-\infty}^{\infty} d\tau Y^2(\tau) - A\int_{-\infty}^{\infty} d\tau X(\tau-\tau_0)Y(\tau-\tau_0)G(\tau/\varepsilon^2). \tag{22}$$

The integrals in (22) are readily evaluated, giving to leading order

$$Me(\tau_0,\varepsilon) = ck_1 + 2\sigma_1 A\pi e^{-\pi\Omega/k\varepsilon^2}(\Omega/\varepsilon^2)^2\sin(\Omega\tau_0/\varepsilon^2+\beta_1), \tag{23}$$

where k_1 is some positive constant. In perturbation theory, εMe is considered as the leading term of the distance function $d_{\delta,\varepsilon}(\tau_0)$ between the unstable and stable manifolds of the saddle. Here, however, Me contains the exponentially small factor $\exp(-\pi\Omega/k\varepsilon^2)$ and estimates of the error have to be made. This difficulty has its origin in the different time scales of the periodic forcing and the autonomous part of (19a,b) and has, for general systems, been discovered by Sanders [18]. Sanders states that damping terms will provide $0(\varepsilon^k)$-contributions to d, and thus our Melnikov-function appears to be useless. We argue that the contributions from the damping terms to d have already been eliminated because we are perturbing about the ambiclinic or heteroclinic set $\nu=\mu m(\varepsilon)$. Thus any further contribution to d stems from the nonautonomous part and should also contain an exponentially small factor. We are aware that these arguments are not rigorous and will study the problem carefully on another occasion. Here we conjecture that the Melnikov-function (23) is a good approximation for the distance function. Consequently, if $Me(\tau_0,\varepsilon)$ has non-degenerate zeros, then (19a,b) encounter transversal intersections of the stable and unstable manifolds of the saddle and thus a horseshoe [7]. Obviously this is the case when

$$|c| < 2\pi(A/k_1)(\Omega/\varepsilon^2)^2 e^{-\pi\Omega/k\varepsilon^2} . \tag{24}$$

In terms of (μ,ν) this means that there is an exponentially small region around SL where chaotic behaviour occurs. Thus very close to the global bifurcation points of Figure 2 we can expect transient chaos.

4. Discussion and Conclusion

In this paper we have discussed the effects of breaking the translation symmetry SO(2) on the Hopf bifurcation with O(2)-symmetry. Our derivation of the amplitude equation (4) for this case was heuristic, but we expect it to describe the essential aspects of the breakup of the O(2)-symmetric bifurcation. Some of our results come as no surprise. In the O(2)-symmetric case the SW and TW branches bifurcate simultaneously. Without the translation symmetry pure travelling waves cannot exist, and indeed we find that such waves become a mixture of a pure TW together with a small contribution from an opposite TW. Such waves, therefore, exhibit a low frequency modulation in time. These waves cannot bifurcate from the trivial solution, and instead appear at a secondary bifurcation from a SW branch. There are two SW branches that differ by their overall phase. Such overall phase selection can be thought of as pinning made possible by the broken translation invariance. In addition, secondary Hopf bifurcations from the SW branches can result in a different kind of modulated wave, referred to here as MW.

There are many other ways in which the O(2) symmetry may be broken. When the translation invariance is preserved, but the Z(2) symmetry is broken, the standing waves disappear since these are equal amplitude superpositions of left- and right-travelling waves, and the symmetry between them is now broken [5,15]. Instead, there are two primary branches of travelling waves which bifurcate from the trivial solution at different values of the bifurcation parameter (i.e., one or other travelling wave is preferred), and modulated waves can appear in secondary bifurcations. This bears some resemblance to the problem studied here. In both cases only one type of solution (SW or TW) bifurcates from the trivial solution, but they come in pairs (SW$_0$, SW$_\pi$ or right- and left-travelling waves). In both cases secondary bifurcations produce modulated waves. As the imperfection parameter decreases, the modulated waves in the Z(2) symmetric case become travelling waves, while in the SO(2) symmetric case they become standing waves. In contrast, in the O(2) symmetric case modulated waves appear only when the nondegeneracy conditions ($a_r \neq 0, b_r \neq 0$, $a_r + b_r \neq 0$) are violated [11].

In the O(2) symmetric case the MW cannot frequency lock [3]. This is because one can go into a moving frame in which the TW is at rest, thereby eliminating one of the frequencies. However, when the translation symmetry is broken, frequency locking should be expected.

Acknowledgment. The work of GD was supported by Stiftung Volkswagenwerk and that of EK by the ACM program of DARPA.

References

1. G. Ahlers and I. Rehberg, Phys. Rev. Lett. 56, 1373-1376 (1986)
2. H.R. Brand, P.S. Lomdahl and A.C. Newell, Phys. Lett. A 118, 67-73 (1986)
3. P. Chossat, C.R. de l'Académie des Sciences, Paris (1985)
4. G. Dangelmayr and E. Knobloch, in preparation
5. S. van Gils, preprint
6. M. Golubitsky and I. Stewart, Arch. Rat. Mech. Anal. 27, 107-165 (1985)
7. J. Guckenheimer and P. Holmes, "Nonlinear Oscillations, Dynamical Systems and Bifurcations of Vector Fields", Springer 1983
8. R. Heinrichs, G. Ahlers and D.S. Cannell, preprint
9. E. Knobloch, Phys. Rev. A 34, 1538-1549 (1986)
10. E. Knobloch and M.R.E. Proctor, J. Fluid Mech. 108, 291-316 (1981)
11. E. Knobloch, in "Multiparameter Bifurcation Theory", M. Golubitsky and J. Guckenheimer (eds.), Contemporary Mathematics 56, Amer. Math. Soc., Providence R.I. 1986
12. E. Knobloch, A. Deane and J. Toomre, these proceedings.
13. P. Kolodner, A. Passner, C.M. Surko and R.W. Walden, Phys. Rev. Lett. 56, 2621-2624 (1986)
14. E. Moses and V. Steinberg, Phys. Rev. A 34, 693-696 (1986)
15. W. Nagata, in "Multiparameter Bifurcation Theory", M. Golubitsky and J. Guckenheimer (eds.), Contemporary Mathematics 56, Amer. Math. Soc., Providence R.I. 1986
16. D. Rand, Arch. Rat. Mech. Anal. 79, 1-38 (1982)
17. I. Rehberg and G. Ahlers, Phys. Rev. Lett. 55, 500-504 (1985)
18. J.A. Sanders, Celestial Mechanics 28, 171-181 (1982)
19. R.W. Walden, P. Kolodner, A. Passner and C.M. Surko, Phys. Rev. Lett. 55, 496-499 (1985)

Generic Spontaneous Symmetry Breaking in SU(n) – Equivariant Bifurcation Problems

C. Geiger and W. Güttinger

Institute for Information Sciences, University of Tübingen,
D-7400 Tübingen, Fed. Rep. of Germany

1. Introduction

An important feature of bifurcation problems with symmetry is the occurrence of
spontaneous symmetry breaking. A basic state of a bifurcation equation loses
stability when a distinguished bifurcation parameter passes through a critical
value. Typically, then, new solutions branch off that basic state which possess a
lower symmetry. One of the fundamental problems of equivariant (or covariant)
bifurcation theory is to determine the symmetry of the bifurcating solution branches,
i. e., their isotropy subgroups. The approach to this problem (e. g. [5,6]) was so
far mainly based on the fixed-point subspaces corresponding to the maximal isotropy subgroups. In this paper we focus attention on a complementary approach
which rests upon the geometry of the orbit strata in the space of the basic invariants. This concept has been introduced in [1,10,11] into field theory, but till
now is disregarded in bifurcation theory. Our objective here is to demonstrate in
terms of bifurcation problems with SU(n)-symmetry that the dimensions of the strata
corresponding to maximal isotropy subgroups are closely related to the number k of
basic nonquadratic invariants of lowest order. Specifically for k = 1 generic spontaneous symmetry breaking leads to 1-dimensional strata whereas for k > 1 strata of
higher dimensions may occur. The basic group theory is summarized in Section 2.
Bifurcation problems with SU(n)-symmetry are analysed in Section 3 and a discussion
including consequences for bifurcation theory and particle physics is postponed
to Section 4.

2. Equivariant Bifurcation Problems

Let Γ be a compact group acting orthogonally and absolutely irreducible in $\mathbb{R}^n$,
endowed with the standard scalar product $(.,.)$. A smooth function $f: \mathbb{R}^n \to \mathbb{R}$
is called invariant if $f(\gamma x) = f(x)$ for all $\gamma \in \Gamma$. We assume that the ring of smooth
invariant functions is generated by an integrity basis $\{\Theta_0,\ldots,\Theta_p\}$, where $\Theta_0 = (x,x)$
and each $\Theta_j(x)$ $(j \geq 1)$ is a homogeneous polynomial of degree n_j with $2 < n_1 \leq n_2 \leq \ldots$
$\leq n_p$. This means that any invariant smooth function can be represented in the form
$g(\Theta)$ where $\Theta = (\Theta_0,\ldots,\Theta_p)$ and $g: \mathbb{R}^{p+1} \to \mathbb{R}$. A mapping $X: \mathbb{R}^n \to \mathbb{R}^n$ is called
equivariant (or covariant) if $X(\gamma x) = \gamma X(x)$ for any $\gamma \in \Gamma$. We assume that the gra-

dients of the basic invariants generate the module of smooth invariant mappings so that an equivariant $X(x)$ is given by

$$X(x) = a_0(\Theta)x + \sum_{j=1}^{p} a_j(\Theta)X_j(x), \quad X_j(x) := \nabla\Theta_j(x).$$

The fact that there is only one linear equivariant, $X_0(x) = x$, follows from Schur's lemma.

For a fixed $x \in \mathbb{R}^n$ the subgroup of Γ defined by $\Sigma_x = \{\sigma \in \Gamma | \sigma x = x\}$ is called the isotropy subgroup of x. The fixed-point subspace of an isotropy subgroup Σ of Γ is $V(\Sigma) = \{x \in \mathbb{R}^n | \sigma x = x$ for all $\sigma \in \Sigma\}$, and consists of all points in $\mathbb{R}^n$ whose isotropy subgroups include Σ. We denote the Γ-orbit of $x \in \mathbb{R}^n$ by $\Omega(x)$, i. e., $\Omega(x) = \{y \in \mathbb{R}^n | y = \gamma x$ for some $\gamma \in \Gamma\}$. The isotropy subgroups of different points $y' = \gamma y$ of a Γ-orbit are conjugate: $\Sigma_{y'} = \gamma\Sigma_y\gamma^{-1}$. This allows one to associate with each orbit an isotropy subgroup. All orbits corresponding to the same isotropy subgroup (up to conjugacy) are collocated in a stratum S of $\mathbb{R}^n$. Since the invariants Θ_j are constant along an orbit, orbits may be considered as points in the space $\mathbb{R}^{p+1}$ of the basic invariants, and to each stratum S in $\mathbb{R}^n$ there corresponds a stratum S' in $\mathbb{R}^{p+1}$. Consequently the image in $\mathbb{R}^{p+1}$ of $\mathbb{R}^n$ under the mapping $x \to \Theta(x)$ forms a semi-algebraic variety consisting of strata S' which correspond to the isotropy subgroups of Γ. The results of [14] imply that $\dim S' \leq \dim V(\Sigma)$ where Σ is the isotropy subgroup of the stratum S' in $\mathbb{R}^{p+1}$, i. e., isotropy subgroups Σ with $\dim V(\Sigma) = 1$ correspond to 1-dimensional strata S'.

Consider a bifurcation problem of the form

$$G(x,\lambda) = g_0(x,\lambda)x + \sum_{j=1}^{p} g_j(x,\lambda)X_j(x) = 0, \tag{1}$$

where λ is a distinguished bifurcation parameter and $g(0,0) = 0$, $g_\lambda(0,0) = 1$. For all λ the equation (1) has the trivial solution $T: x = 0$. Because 0 is fixed by any $\gamma \in \Gamma$, the isotropy subgroup of T is the full group Γ. However, in virtue of $d_x G(0,0) = 0$, one expects that nontrivial solutions branch off T at $\lambda = 0$ so that (1) gives rise to spontaneous symmetry breaking. The nontrivial solutions are certainly not fixed by the full group Γ and thus possess a lower symmetry, determined by some proper subgroup. This raises three questions: (i) What is the symmetry of a bifurcating solution? Specifically one would like to know whether generically the bifurcating solutions correspond to maximal isotropy subgroups. This question is related to (ii): Which stratum S' corresponds to the solution? Let $\Sigma_2 \subset \Sigma_1$ be successors in the lattice of isotropy subgroups and let S_1', S_2' be their strata. Then $\dim S_2' > \dim S_1'$ and S_1' is an open subset of the boundary of S_2'. Thus, going down along a path in the lattice is equivalent to moving in $\mathbb{R}^{p+1}$ from one stratum to another in a specific way which provides a geometrical method for determining the symmetry of a solution. In particular, the strata corresponding to maximal isotropy subgroups have only the origin as boundary. (iii) What is the dimension of the fixed-point subspace corresponding to the solution? Even if its isotropy subgroup is maximal it may happen that the fixed-point subspace has a dimension > 1. Golubitsky [5] and Ihrig and Golubitsky [7] conjectured that any bifurcating solution has generically a maximal symmetry if the fixed-point subspaces associated with maximal isotropy subgroups are all 1-dimensional. The counter examples discussed in [2,8] show, however, that this conjecture is not true in full generality.

Our approach to spontaneous symmetry breaking is essentially based on the structure of the strata in $\mathbb{R}^{p+1}$, specifically of those corresponding to the maximal isotropy subgroups. We wish to point out, in terms of two examples with SU(n)-symmetry, that the number of the basic invariants of lowest order plays an important role for the strata dimensions and for generic spontaneous symmetry breaking. The examples indicate that in the case $n_1 < n_2$ the strata of the maximal isotropy subgroups are of dimension 1 whereas for $n_1 = n_2$ higher-dimensional strata may occur. In both examples only solutions with maximal symmetry exist and the dimensions of the fixed-point subspaces and the strata coincide.

3. Examples with SU(n)-Symmetry

We consider first the adjoint representation of SU(n) (the group of unitary n × n-matrices γ with det $\gamma = 1$) for general n, but specify to n = 5 in more detailed calculations below. Our objective is to show that for this representation generic bifurcations lead to maximal isotropy subgroups. Here SU(n) acts on $\mathbb{R}^{n^2-1}$ which is identified with the space E_n of hermitian n × n -matrices x satisfying Trx = 0. The action is given by $\gamma x \equiv \gamma x \gamma^*$ where $\gamma \in SU(n)$. The general algebraic features of the adjoint representation can be described by two independent algebra-laws [11], defined by

$$x_\vee y \equiv (\sqrt{n}/2)(xy+yx) - (2/\sqrt{n})(x,y)I$$
$$x_\wedge y \equiv -\frac{i}{2}(xy - yx),$$

where I denotes the n × n -unit-matrix. The ring of invariant smooth functions is generated by $\{\Theta_0(x),\ldots, \Theta_{n-2}(x)\}$, where

$$\Theta_0 = (x,x), \quad \Theta_1 = (x,x_\vee x),\ldots, \quad \Theta_{n-2} = (x,(x_\vee(\ldots(x_\vee x)\ldots))) \ .$$

We denote by $X_j(x) \equiv \nabla\Theta_j(x)$ $(0 \le j \le n-2)$ the generators for the equivariant mappings. Observe that there exists only one quadratic equivariant $X_1(x)$. This follows from the fact that we have only two basic algebra laws, where the asymmetric algebra law defines the Lie - algebra. A general bifurcation problem can be written in the form

$$G(x,\lambda) = \sum_{k=0}^{n-2} g_k(\Theta,\lambda)X_k(x) = 0,$$ (2)

where $\Theta = (\Theta_0,\ldots, \Theta_{n-2})$ and $g(0,0) = 0$. We suppose that the non-degeneracy conditions $\partial g_0(0,0)/\partial\lambda > 0$, $\partial g_0(0,0)/\partial\Theta_0 > 0$ are satisfied. Then one can construct a contact transformation in the sense of Golubitsky ans Schaeffer [6] which transforms (2) into the normal form [3, 4]

$$2(\lambda + \Theta_0)x + 3x_v x = 0.$$ (3)

This equation can be rewritten as

$$P(x)\xi = 0$$ (4)

where $P(x)$ is the $2\times(n-2)$ -matrix with elements $P_{ik}(x) = (X_i(x), X_k(x))$ $(i = 0,1; k = 0,\ldots, n-2)$ and $\xi = (\Theta_0+\lambda, 1)^T$. Since $P_{ik}(x)$ is an equivariant homogeneous polynomial of degree $i + k + 2$ we can represent it in the form $P_{ik}(x) = P'_{ik}(\Theta)$ where P'_{ik} is a polynomial in Θ. Nontrivial solutions of (4) can only exist if rank $P'(\Theta)=1$ which is exactly the condition for $\Theta(x)$ being located on a 1-dimensional stratum. Any 1-dimensional stratum of the adjoint representation of $SU(n)$ corresponds to one of the maximal isotropy subgroups $S(U(p)\times U(q))$ where $p + q = n$. Consequently, spontaneous symmetry breaking for the adjoint representation of $SU(n)$ leads generically to a maximal isotropy subgroup.

We exploit these considerations in more detail for the case of $SU(5)$, where $x \in \mathbb{R}^{24}$ is identified with a hermitian traceless 5×5 -matrix. It is convenient to choose a modified set of generators for the invariant functions, namely, $\Theta_k(x) - \text{Tr}x^{k+2}$ $(0 \leq k \leq 3)$. In this basis the equations for the 1-dimensional strata have the forms

$$S(U(4)\times U(1)): \Theta_1^2 = \frac{9}{20}\Theta_0^3, \quad \Theta_2 = \frac{13}{20}\Theta_0^2, \quad \Theta_3 = \frac{17}{20}\Theta_0\Theta_1$$ (5a)

$$S(U(3)\times U(2)): \Theta_1^2 = \frac{1}{30}\Theta_0^3, \quad \Theta_2 = \frac{7}{30}\Theta_0^2, \quad \Theta_3 = \frac{13}{30}\Theta_0\Theta_1$$ (5b)

and x and $x_v x$ are linearly related according to $\Theta_1 x_v x = \alpha\Theta_0^2 x$ where $\alpha = (9/20)\sqrt5$ for (5a) and $\alpha = (1/30)\sqrt5$ for (5b). This allows us to reduce (3) to scalar equations along the strata (5a,b). Setting $\Theta_0 = \epsilon^2$ and solving these equations yields the following representation for the solutions of (3),

$$S(U(4)\times U(1)): \quad x = \epsilon y_1, \quad \lambda = -\frac{9}{4}\epsilon - \epsilon^2$$ (6a)

$$S(U(3)\times U(2)): \quad x = \epsilon y_2, \quad \lambda = -\frac{\sqrt6}{4}\epsilon - \epsilon^2$$ (6b)

where $\varepsilon \in \mathbb{R}$ and y_1, y_2 are unit vectors spanning the corresponding fixed-point subspaces,

$$y_1 = (1/\sqrt{20})\text{diag}(1,1,1,1,-4), \quad y_2 = (1/\sqrt{30})\text{diag}(2,2,2,-3,-3).$$

Obviously both bifurcations are transcritical in contrast to SU(4) [3, 4].

As a second example we discuss the group $\Gamma \equiv SU(3) \times SU(3)$ acting on $\mathbb{R}^{18}$ which is identified with the space of complex 3×3 -matrices, F_3 [12]. The action can be written in the form $\gamma x \equiv \gamma_+ x \gamma_-^*$, where $(\gamma_+ , \gamma_-) \in SU(3) \times SU(3)$. This representation of Γ is absolutely irreducible on the reals but not irreducible when complexified. There are four basic invariant functions, given by

$$\Theta_0 = (x,x), \quad \Theta_1 = (x,x_\top x)$$

$$\Theta_2 = (x,ix_\top x), \quad \Theta_3 = (x,x_\top(x_\top x)),$$

where $_\top$ denotes the following algebra law on F_3,

$$x_\top y \equiv \frac{1}{2} \{(\text{Tr}x^*)(\text{Tr}y^*) - \text{Tr}(x^*y^*)\}I$$

$$- \frac{1}{2} (\text{Tr}y^*)x^* - \frac{1}{2} (\text{Tr}x^*)y^* + \frac{1}{2} (x^*y^* + y^*x^*).$$

A general bifurcation problem can also be represented in the form (2) with $n = 5$, $\Theta = (\Theta_0, \ldots, \Theta_3)$ and $X_0 = x$, $X_1 = x_\top x$, $X_2 = ix_\top x$, $X_3 = (x_\top x)_\top x$. Similarly as in the SU(n)-case it is possible to construct a contact-transformation which transforms the general bifurcation problem to the normal form

$$(\lambda + \Theta_0)x + \sqrt{3}(x_\top x \cos \alpha + ix_\top x \sin \alpha) = 0, \tag{7}$$

provided the non-degeneracy conditions

$$\partial g_0(0,0)/\partial \lambda > 0, \; \partial g_0(0,0)/\partial \Theta_0 > 0 \text{ and}$$

$$(g_1(0,0), g_2(0,0)) \neq (0,0),$$

are satisfied. The equation (7) can be rewritten as

$$P'(\Theta)\xi = 0 \tag{8}$$

where P' is now a 3×4 -matrix with elements $P'_{ik} = (X_i, X_k) \; (0 \leq i \leq 2; \; 0 \leq k \leq 3)$, and $\xi = (\lambda + \Theta_0, \sqrt{3} \cos \alpha, \sqrt{3} \sin \alpha)^\top$. In order that a nontrivial solution of (8) exists we must have $1 \leq \text{rank } P' \leq 2$. This condition is satisfied on the strata S'_1, S'_2, given by

$$S'_1: \Theta_0 > 0, \; \Theta_1 = \Theta_2 = \Theta_3 = 0 \qquad (\dim S'_1 = 1) \tag{9a}$$

$$S'_2: \Theta_1^2 + \Theta_2^2 = \frac{1}{3} \Theta_0^3, \; \Theta_3 = \frac{1}{3} \Theta_0^2 \qquad (\dim S'_2 = 2) . \tag{9b}$$

The stratum S_1' corresponds to the isotropy subgroup $\Sigma_1 = SU(2) \times SU(2) \times U(1)^d$ whereas S_2' is associated with $\Sigma_2 = SU(3)^d$, the diagonal action of $SU(3)$: $x \to \gamma \times \gamma^*$. Both isotropy subgroups are maximal, but dim $V(\Sigma_1) = 1$ whereas dim $V(\Sigma_2) = 2$.

Along the 1-dimensional stratum S_1' all nonlinear basic equivariants vanish. Consequently the solution of (7) is of pitchfork-type and can be represented parametrically in the form $\lambda = -\varepsilon^2$, $x = \varepsilon y$ where $\varepsilon \in \mathbb{R}$ and the unit-vector $y = (1/\sqrt{6})\text{diag}(1,1,-2)$ spans $V(\Sigma_1)$. Along S_2' we find the following relation among the generators,

$$\Theta_1 x_T x + \Theta_2 i x_T x = \frac{1}{3} \Theta_0^2 x, \quad x_T(x_T x) = \frac{1}{3} \Theta_0 x. \tag{10}$$

The fixed-point subspace $V(\Sigma_1)$ is most conveniently parametrized by polar coordinates,

$$x = (\varepsilon/\sqrt{3})e^{i\phi}I; \quad \varepsilon \in \mathbb{R}, \ 0 \le \phi < \pi, \tag{11}$$

which yields the following form for the basic invariants and equivariants,

$$\Theta_0 = \varepsilon^2, \quad \Theta_1 + i\Theta_2 = (\varepsilon^3/\sqrt{3})e^{3i\phi}, \quad x_T x = (\varepsilon^2/3)Ie^{-2i\phi}. \tag{12}$$

By using (9b), (10), (11) we find that the solution of (7) corresponding to Σ_2 can be parametrized according to

$$\lambda = -\varepsilon - \varepsilon^2, \quad x = (\varepsilon/\sqrt{3})e^{i\alpha/3}I; \quad \varepsilon \in \mathbb{R}, \tag{13}$$

which gives rise to a transcritical bifurcation. A section $\Theta_0 = 1$, $\Theta_2 = 0$ through the strata is shown in Figure 1. The full section $\Theta_0 = 1$ in $(\Theta_1, \Theta_2, \Theta_3)$-space is obtained via rotation about the Θ_3-axis.

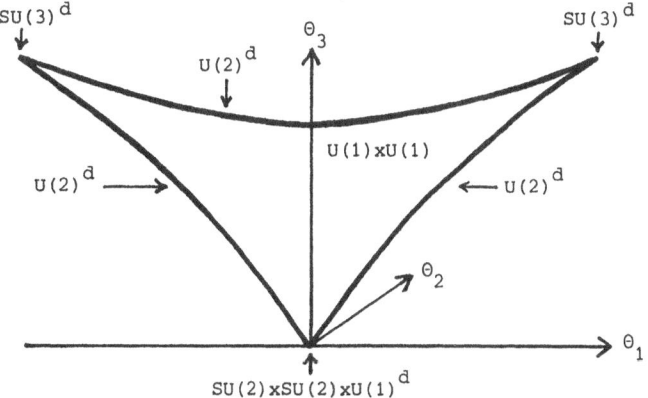

Figure 1 Section $\Theta_0 = 1$, $\Theta_3 = 0$ through the strata in $\{\Theta_j\}$-space corresponding to the representation of $SU(3) \times SU(3)$.

4. Conclusions

The main difference between the two examples presented in the previous section consists in the presence of 1 and 2 basic cubic invariants for case I (SU(n)) and case II (SU(3)×SU(3)), respectively. In the first case generic symmetry breaking always led to 1-dimensional strata whereas for case II also a solution in a 2-dimensional stratum appeared. We conjecture that this is a special case of a general feature of generic spontaneous symmetry breaking: The number of nonquadratic basic invariants of lowest order is equal to the maximal dimension of the strata in which the bifurcating branches can live. Although for example II the 2-dimensional stratum corresponded to a maximal isotropy subgroup this need not be the case in general. Thus we expect that in the presence of several ($k > 1$) nonquadratic basic invariants of lowest order spontaneous symmetry breaking to submaximal isotropy subgroups may be generic. On the other hand, for $k = 1$ only 1-dimensional strata will appear. The relation between the strata and the fixed-point subspaces follows from the results of [1, 14]: Let $x \in \mathbb{R}^n$, $\Theta(x) \in S'$ and let N_x be the normal space of the orbit $\Omega(x)$ at x. Denote by N_x^0 the subspace of N_x fixed by the isotropy subgroup Σ_x, i. e., $N_x^0 = V(\Sigma_x) \cap N_x$. Then N_x^0 is isomorphic to the tangent space of the stratum S', i. e., $\dim N_x^0 = \dim S'$. In virtue of this result and the examples discussed in the preceding section we argue that the dimensions of the invariant normal spaces play a more important role for generic spontaneous symmetry breaking than the dimensions of the fixed-point subspaces themselves. A more detailed and general discussion of these points will be presented elsewhere.

In this paper we have concentrated on generic spontaneous symmetry breaking with unitary symmetries because these occur in various forms in field theories, e. g. as gauge theories or as dynamical symmetries in particle physics. The groups SU(n) and SU(n)xSU(n) are relevant for the physics of strong interactions and occur also as geometrical symmetries in unified field theories. Finally the combination of SU(n) with local supersymmetry is important for supergravity and the Kaluza-Klein theory. It is generally believed that spontaneous symmetry breaking is the basic mechanism for structure formation during the evolution of the universe with the temperature (or another cosmological quantity) playing the role of the bifurcation parameter. In order to analyze the low energy limit of a fundamental field theory one frequently deals with effective models. For example in the context of supergravity or supersymmetric Kaluza-Klein theories local ($N = 1$)-supersymmetry seems to describe physics at an energy scale below the Planck mass [13]. Then physical constraints very often require fine tuning of dimensionless parameters of the effective model to quite absurd values. For example, in inflationary models fine tuning seems to be a major problem [9]. Spontaneous symmetry breaking through generic bifurcations does not allow any fine-tuned constraints in the equations. The question whether it is possible to arrive generically at higher dimensional strata with submaximal isotropy subgroups is therefore of special importance. Based on the discussion presented here we conjecture that this is only possible for group actions which possess more than one basic nonquadratic invariant of lowest order.

References

1.   M. Abud and G. Sartori, Phys. Lett. B 104 (1981), 147

2.   P. Chossat, C.R.Acad.Sc. Paris, Série I, t.297 (1983), 639

3.   C. Geiger, Thesis (1985)

4.   C. Geiger, W. Güttinger and P. Haug, in: H. Haken (ed.), "Complex Systems - Operational Approaches", Springer 1985

5.   M. Golubitsky, in: C. P. Borter et al. (eds.), "Bifurcation Theory, Mechanics and Physics", Reidel 1983

6.   M. Golubitsky and D. Schaeffer, "Singularities and Groups in Bifurcation Theory, Vol.I", Springer 1985

7.   E. Ihrig and M. Golubitsky, Physica 13D (1984), 1

8.   R. Lauterbach, Contemp. Math. 56 (1986), 217

9.   A. Linde, Phys. Lett. 108B (1982), 389

10.   L. Michel, CERN-preprint TH 2716 (1979)

11.   L. Michel and L. A. Radicati, Ann. Inst. Henri Poincaré 18 (1973), 185

12.   L. Michel and L. A. Radicati, in: "Evolution of Particle Physics", Academic Press 1970

13.   D. V. Nanopoulos, in: C. Konnas et al. (eds.), "Grand Unification with and without Supersymmetry and Cosmological Implications", World Scientific 1984

14.   G. Sartori, J. Math. Phys. 24 (1983), 765

Computer Algebraic Tools for Applications of Catastrophe Theory

F.J. Wright and R.G. Cowell

School of Mathematical Sciences, Queen Mary College,
University of London, Mile End Road, London E1 4NS, United Kingdom

Abstract

We propose a "model-mapping" approach to the understanding of smooth local structure. Describing such structure using catastrophe theoretic models of gradient systems leads to algebraic problems, which we sketch and indicate how computers can help to solve.

1. Motivation and Background

1.1. How Can a Structure Be Described?

Fig. 1: What is this structure?

(a) *A heart shape.* But what does that mean? In mathematical language it means that there exists an equivalence class of shapes whose members are all equivalent to one that is conventionally called a *heart*. The equivalence relation has to preserve the essential "heartiness" of the heart, which only fairly smooth transformations will do. An allowed transformation must therefore be differentiable up to some given order, i.e. it must be a *diffeomorphism*. Having fixed the equivalence relation, the equivalence class of hearts can be fixed completely by defining one member of it. This *canonical heart* should be in some sense simple, perfect perhaps, and should probably have as much symmetry as possible. A mathematical discussion of the aesthetics of the heart shape is given by Donald Knuth in his book [1] on Metafont, a computer language for generating typefaces. A good candidate for the canonical heart would be that defined by Knuth.

The description given so far is very superficial (topological, the mathematician would say) and in no way distinguishes the heart shape in Fig. 1 from other hearts. So what more might be said about it?

(b) (i) It is a certain width (perhaps too wide to be elegant);
(ii) it points down and a little sideways;
(iii) it is distorted (asymmetrical) – it is pushed up at the left and down at the right.

In mathematical terms, (b) adds *geometrical* information to the previous *topological* information. Point (iii) only makes sense relative to some preconceived notion of a heart shape, which we take to be the canonical heart. Points (i) and (ii) can also usefully be interpreted this way. If we consider the mapping

M : actual heart → canonical heart,

then the linear part L of this map represents mathematically the information contained in (b) above. Roughly, (i) relates to the scale, which is contained in the diagonal part of L; (ii) relates to the orientation, which is contained in the anti-symmetric part of L; and (iii) relates to the shear, which is contained in the rest of L.

More might be said about the heart in Fig. 1, in addition to the most basic geometrical information given so far. For example:-

(c) It is too bulbous – the curvature is unaesthetic (Knuth would not like it!).

In mathematical terms, (c) relates to the *quadratic* part of the mapping M, although not very precisely. One could continue *ad infinitum* in this fashion, progressively refining the information available.

Summary. A structure may be described (at least locally) by giving:

(1) a canonical model (as a representative of an equivalence class of such structures);

(2) the mapping: actual → canonical to *sufficient* degree.

This approach is only intended for fairly simple, mainly smooth structures – in fact, those to be discussed in the rest of the paper.

1.2. Structure Described by Elementary Catastrophe Theory (ECT)

ECT [2] describes the structure of the *critical* or *stationary* points of typical families of functions of the form

$$\phi : \mathbf{R}^n \times \mathbf{R}^K \to \mathbf{R}.$$

Each member of the family is a function of *n state variables* s_i, and the members of the family are parametrised by *K control variables* c_j. A critical point s_c is a solution of the equations $\partial\phi/\partial s_i = 0$, $i = 1$ to n, and depends on c. ϕ is called an *unfolding* of a *singularity*, which is a function in the family for which the maximum possible number of critical points has coalesced. It is convenient to make a translation in $(R^n \times R^K, R)$ so that the singular point is at the origin; then the singularity is the function $\phi(s,0)$, which has a critical point at $s = 0$ such that $\phi(0,0) = 0$.

ECT provides suitable mathematics for analysing systems governed by equilibria, optima or extrema. Equilibria correspond to minima of some real-valued function, usually called a potential function, as in quasi-static mechanics and thermodynamics. Optima correspond to maxima of a payoff function, or minima of a cost function, as in economics, sociology, behavioural modelling and game theory. Extrema correspond to maxima, minima and stationary points in general of a function such as the distance function used to describe evolutes in geometry and caustics in geometrical optics. ECT describes *typical* systems, in which no special conditions such as symmetries or other constraints apply. In contrast, more conventional applied mathematics tends to deal with precisely these specialised systems, about which it is easier to make (geometrically) precise statements. (However, there are variants of ECT [2] that apply to typical systems subject to specified symmetries or constraints.)

The *observable* structure described by ECT is usually in the control space R^K, which represents the space of parameters through which a system interacts with its environment, whereas the state space represents behaviour internal to the system. For example, in applications to geometrical optics, the control space corresponds to the physical space in which the light waves propagate, and in which any caustics occur, whilst the state space is a space of labels for individual rays.

The principal structure observed in the control space is the *bifurcation set* where the critical point structure of ϕ changes. A point on the bifurcation set corresponds to a *degenerate* critical point. In the simplest case this is a coalescence of a maximum and a minimum, so that as the bifurcation set is crossed a maximum and a minimum either appear or disappear, depending on the direction of the crossing, leading to a discontinuous change in the state of the system.

If the state of the system is determined by either minima only, or maxima only, the change will be *observable* only if the bifurcation set is crossed in the direction in which extrema coalesce and disappear, which will force the system to *jump* to a new stable equilibrium or optimal state. Such jumps are the origin of the term *catastrophe*. If the state of the system is determined by extrema (i.e. both maxima and minima), then this change will always be observable. For example, in optics the bifurcation set is directly observable as the bright caustic, and the number of rays passing through nearby points on either side of the caustic differs by two.

1.3. Elementary Catastrophe Theory: Equivalence and Models

The method we proposed earlier for describing a structure is to place it in an equivalence class of related structures, and to specify in sufficient detail the mapping that relates the given structure to the canonical representative of its class. The equivalence relation used in ECT is called *right equivalence*. Two families of real-valued functions ϕ and ϕ' are right equivalent, denoted $\phi \sim \phi'$, if

$$\phi(s,c) = \phi'(s'(s,c),u(c)) + \lambda(c) \quad \text{where} \quad s'(0,0) = 0, \quad u(0) = 0, \quad \lambda(0) = 0. \qquad (*)$$

All functions and mappings used in ECT are assumed to be infinitely differentiable, and are strictly defined only in the neighbourhood of the main singular point being considered, which is assumed to have been translated to the origin. The mapping $u(c)$ is precisely the diffeomorphism that we proposed in §1.1; $s'(s,c)$ is slightly more complex and plays a different rôle – it is a differentiable family of diffeomorphisms parametrised by the control variables c. The literature of ECT [2] is concerned only with the existence of right equivalences, and not with their practical construction.

The canonical representative of the right-equivalence class of ϕ should be as simple as possible; it is chosen to be polynomial, of lowest possible degree in s and linear in c. It is a non-trivial theorem that such a *normal form* exists. However, it is not unique, which allows the most appropriate normal form to be chosen for each application. For example, sometimes it may be desirable to choose a normal form that either does or does not manifest a particular symmetry. Some of the simpler normal forms are listed in [2] pp121-2 and in §1.4 below.

Assuming that the interesting structure is in the control space the appropriate *structure map* is $u(c)$, which maps the physical control space into the canonical one. In general, this map cannot be constructed completely, but we have already argued that only *some* information is required from it, rather than *all possible* information. In the local context of ECT this means that the map need be constructed only up to some given degree in c, which is determined by the problem in hand. The method that we will describe for finding $u(c)$ is to construct the map $s'(s,c)$ that *reduces ϕ to normal form to the chosen degree in c*.

We shall need the following nomenclature later. Let $\phi'(s',u)$ be the normal form for a catastrophe (an unfolding of a singularity). The dimension of s' is the *corank* of the catastrophe (strictly of the singularity unfolded), and is not necessarily equal to the dimension of s. If it is not, then the dimensions missing from s' are made up by adding to the normal form a diagonal quadratic (or Morse) form $\pm f_1^2 \pm \cdots \pm f_r^2$ where r is the difference in dimensions. This Morse form is simply the normal form for a function of r variables in the neighbourhood of a non-degenerate critical point. The dimension of u in the normal form is the *codimension* of the catastrophe (again strictly of the singularity unfolded), and is not necessarily equal to the dimension of c. If it is larger, then some of the control variables c are "inessential" in the sense that they are not necessary to explore the whole structure of the system; if it is smaller then there are not sufficient control variables c to explore the whole of the structure. If the dimensions of s and s', or of c and u, do not match up then (*) above is not a right equivalence, because it cannot be in any sense invertible, but is a generalization of a right equivalence.

1.4. Example of the Model-Mapping Approach

The model-mapping philosophy may be likened to the method of solving differential equations by using Laplace transforms to map a difficult analytical problem into an algebraic problem that is hoped to be easier, and back again. It is illustrated in the context of ECT by Fig. 2.

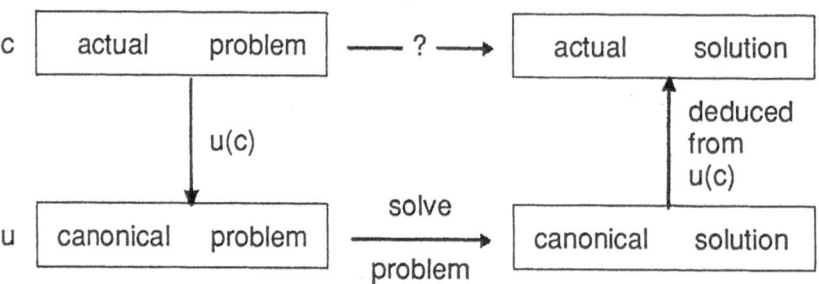

Fig. 2: The model-mapping approach

The following example is taken from DANGELMAYR & WRIGHT [3]. Suppose that a twisted space curve S is being examined remotely by high frequency radar, or that S is a source of radiation of uniform phase – the geometry in the two cases is almost identical. In practice, S might represent a wire or pipe, or the edge of a curved surface. If it is to be regarded as a source of radiation, then it would have to be an explosive source in order to have both uniform phase and high frequency; unfortunately this would not be a good model for a radio antenna. The problem is to find the local nature of the caustic generated by S at any point (which geometrically is just its evolute) and also to describe the nature of the diffraction around the caustic (which has no geometrical analogue), in the limit that the wavelength of the radiation is much smaller than any other significant distance.

Since this is a local analysis, a convenient way to specify the shape of the curve S is in terms of its *Frenet parameters*: the unit vectors **t**, **n**, **b** and the principal curvature κ, torsion τ and their derivatives, all as functions of the arc length s as a convenient parameter. The required information about the system is contained in the function $\phi(s;X,Y,Z)$, which is the distance from the point on S with parameter s to an observation point (X,Y,Z). Since it is not possible to describe the *complete* structure of the caustic and its diffraction in any meaningful way, we settle for describing their *dominant* structure, which reduces to calculating ϕ to sufficient degree in the control variables (X,Y,Z) to adequately determine the structure map $u(X,Y,Z)$.

According to ECT only the three local caustic types listed below typically occur. For each we regard the *dominant* structure as its existence, orientation and the following additional information:

(a) fold – the principal curvature of the caustic surface, and the fringe spacing of the diffraction pattern;

(b) cusp – the principal curvature of the rib (cusp line) of the caustic, and the scale and degree of shear of the diffraction pattern;

(c) swallowtail – the scale and degree of shear of the diffraction pattern.

DANGELMAYR & WRIGHT [3] performed the reduction of φ to normal form by hand (since no computerised methods had then been developed) working to at most quadratic degree in the controls. In fact, the precise degrees retained in various terms were determined by whether they affected the information required. The exercise of such careful judgement was necessary when working by hand in order to keep the number of terms involved to the absolute minimum. The results are summarized below. They give the required information relative to the canonical diffraction catastrophes [4], and the caustics are sketched in Fig. 3.

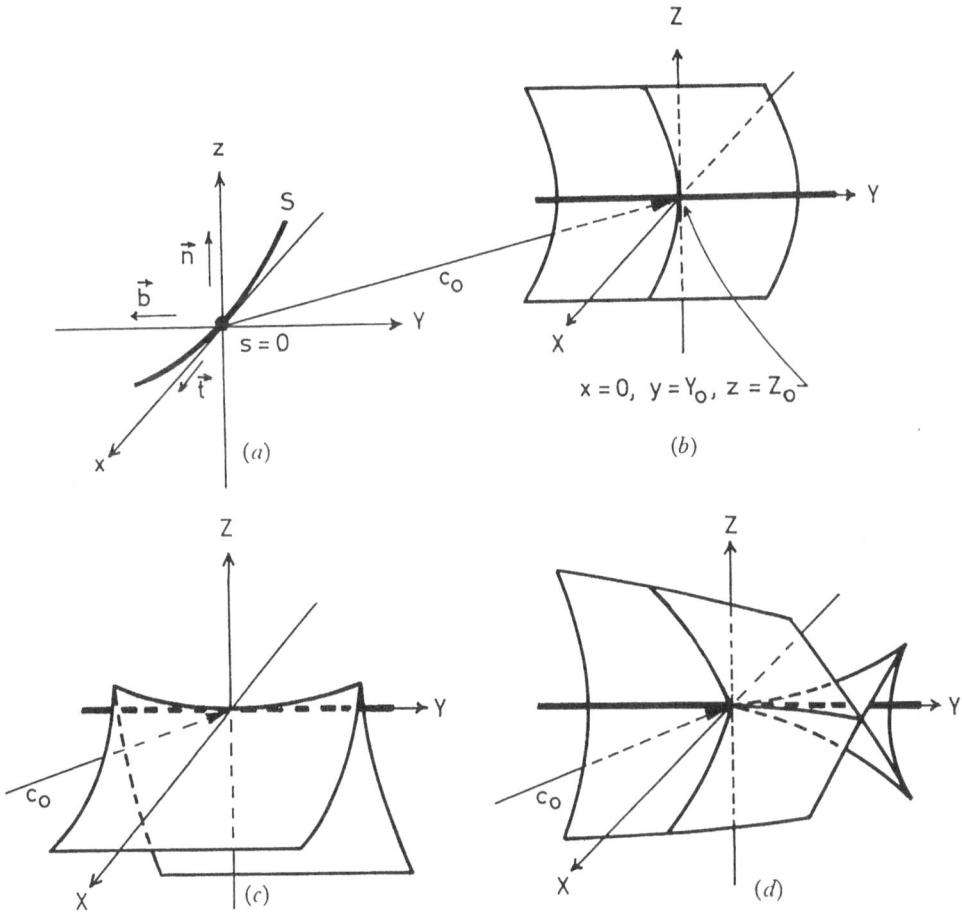

Fig. 3: Basic local geometry of a line source (a) and its three typical caustic types: fold (b), cusp (c) and swallowtail (d)

A coordinate system oriented as shown in Fig. 3 is chosen so that the point of S under consideration is at the origin and the caustic point generated is at $x = 0$, $y = Y_0$, $z = Z_0$. To simplify the formulae, we will leave them in terms of parameters c_i, which together with Y_0 and Z_0 are easily found [3] in terms of the Frenet parameters of S. The most important of these parameters is c_0, which is simply the length of the principal ray as shown on Fig. 3.

For a fold point of the caustic

$$\phi \sim \phi'(s',u) \equiv s'^3 + u_1 s', \quad \text{where} \quad u_1 = \frac{-1}{c_0 c_3^{1/3}} \left[X + \frac{\kappa^2 Z^2}{12 c_0 c_3} \right] + \cdots .$$

The caustic in the canonical model has the equation $u_1 = 0$, which clearly maps back to the real problem as $X = -(\kappa^2/12 c_0 c_3) Z^2 + \cdots$. Thus the caustic is locally cylindrical, with principal curvatures 0 and $\kappa^2/6|c_3|c_0 \equiv \kappa^3/|\kappa^2 \tau Y_0 - \dot{\kappa}|$. The ratio between the canonical fringe spacing and the physical fringe spacing is given by the factor $-1/c_0 c_3^{1/3}$ in the expression for $u_1(X,Y,Z)$ above.

At a cusp point of the caustic

$$\phi \sim \phi'(s',u) \equiv \text{sgn}(c_4) s'^4 + u_2 s'^2 + u_1 s', \quad \text{where}$$

$$u_1 = \frac{-1}{c_0 |c_4|^{1/4}} X + \cdots \quad \text{and} \quad u_2 = \frac{-1}{c_0 |c_4|^{1/2}} \left[\frac{\kappa Z}{2} - \frac{c_5 X}{4 c_4} + \frac{\kappa^2 \tau^2 Y^2}{96 c_4 c_0} \right] + \cdots .$$

The rib of the caustic in the canonical model has the equation $u_1 = u_2 = 0$, which maps back to the real problem as $X = 0$, $Z = -(\kappa \tau^2/48 c_4 c_0) Y^2$ to quadratic degree, giving the principal curvature of the rib to be $\kappa \tau^2/24|c_4|c_0 \equiv \kappa^3 \tau^3/|\kappa \tau[\kappa \tau^2 - \ddot{\kappa}] + \dot{\kappa}[2\dot{\kappa}\tau + \kappa \dot{\tau}]|$. It is the relative coefficients of Z and Y^2 in u_2 that give the principal curvature. The coefficient of X in u_2 gives the degree of shear of the caustic and diffraction pattern around the rib. The overall factors multiplying X in u_1 and Z in u_2 give the scales of the diffraction pattern in two directions relative to the canonical pattern. The term $\text{sgn}(c_4)$ in the normal form determines which way up the cusp points, and has no analogue in the fold case.

The results for the swallowtail are similar, but more complicated [3]. They show that a swallowtail generated in this way is typically sheared both along the Z axis and in the YZ plane; it is impossible to have no shear at all.

2. Algebraic Problems and Computer Algebra

One of the major theoretical components of ECT is the reduction of various analytical problems to algebraic ones [2]. We are writing a system called CATFACT (Computer Algebraic Tools For Applications of Catastrophe Theory) in the algebraic-computing language REDUCE 3, and as we describe the algebraic problems we will indicate how CATFACT attempts to solve them. We assume that ϕ is given algebraically, and that its main singularity occurs at $s = 0$, $c = 0$.

2.1. The Recognition Problem

The singularity $f(s) \equiv \phi(s,0)$ can be classified by topological invariants, the most important of which are *corank, determinacy* and *codimension*. The corank of $f : \mathbf{R}^n \to \mathbf{R}$ at $0 \in \mathbf{R}^n$ is defined to be the corank of the Hessian matrix of $f(s)$ evaluated at $s = 0$, i.e. $n - \text{rank}(\partial^2 f / \partial s_i \partial s_j)_{s=0}$. The Hessian relates to the quadratic part of f, whose normal form is the Morse form referred to earlier; the latter has only simple critical points and is hence non-degenerate. There are standard techniques for determining the rank of a matrix, which can be implemented by hand or computer. In CATFACT the corank of f is determined as a by-product of diagonalising its Hessian by Gaussian elimination.

Determinacy and codimension require a little more mathematical machinery. All calculations are performed in the ring $E(n)$ of functions on $\mathbf{R}^n$. Let $<a, \cdots, b>$ denote the *ideal* in $E(n)$ generated by $\{a, \cdots, b\}$; then $<s> \equiv <s_1, \cdots, s_n>$ is the *maximal ideal* of all functions that vanish at the origin and $<\partial f / \partial s> \equiv <\partial f / \partial s_1, \cdots, \partial f / \partial s_n>$ is the *Jacobian* ideal. Roughly, the Jacobian ideal represents the orbit of f under right-equivalence. Then

codimension of $f \equiv \dim_\mathbf{R} <s>/<\partial f / \partial s>$

where the ideals are regarded as real vector spaces. This is the codimension of the orbit of f in $<s>$, i.e. the number of missing directions that make f singular. A universal unfolding of f puts these missing directions back in the simplest possible way – linearly.

The determinacy k of f is the lowest degree of all polynomials that are right-equivalent to f. Finite determinacy implies finite codimension, which implies a finite unfolding and the existence of a normal form. This definition of (ordinary) determinacy is not easy to test algorithmically, whereas the restricted definitions of *strong* and *local* determinacy are. Let $<s>^p$ denote the ideal of all functions in $E(n)$ that are of order p, i.e. both they and their partial derivatives up to order $p-1$ vanish at the origin. Then the singularity f is

strongly–k–determined if and only if $<s>^{k+1} \subseteq <s>^2 <\partial f / \partial s>$,

which implies that f is ordinarily k-determined, and

locally–k–determined if and only if $<s>^{k+1} \subseteq <s><\partial f / \partial s>$,

which implies that f is ordinarily $(k+1)$-determined. Ordinary k-determinacy implies local k-determinacy. A lemma due to Nakayama allows elements of $<s>^{k+2}$ to be ignored so that the above conditions reduce to simple *finite* vector space problems. Probably the most straightforward way to solve them is to turn them into matrix problems as follows.

2.1.1. The OCRM Program

The first computer program to find the determinacy, codimension and unfolding of a singularity was written by Olsen, Carter and Rockwood in a dialect of Algol60, and is published as Appendix 1 of [2]. This program was corrected and enhanced by Karl Millington [5], and will be referred to as the OCRM program. It uses a dense representation of polynomials, in which the coefficients are stored as elements of an array indexed by the monomials. The various vector spaces referred to above are spanned by sets of polynomials, and the arrays representing these comprise the rows of a matrix that is the main data structure used by the OCRM program. (In fact, there are two matrices, each with integer elements representing respectively the numerators and denominators of the coefficients of the polynomials.)

With this representation it is necessary to decide in advance the maximum degree of polynomials because this determines the size of the matrix. Since by Nakayama's lemma it suffices to work with polynomials of degree $k+1$ when using the above determinacy tests, this sets an upper limit on the determinacy that can be found. Assuming for illustration a 2-dimensional state space, the matrix looks like this:

$$
\begin{array}{l}
\text{column labels:} \\
n(=2) \text{ rows:} \\
n^2(=4) \text{ rows:} \\
\text{the rest:}
\end{array}
\left[
\begin{array}{l}
s_1 \; s_2 \; s_1^2 \; s_1 s_2 \; s_2^2 \; \cdots \; s_2^{k+1} \\
\text{span} (<\partial f/\partial s> - <s><\partial f/\partial s>) \\
\text{span} (<s><\partial f/\partial s> - <s>^2<\partial f/\partial s>) \\
\text{span} (<s>^2<\partial f/\partial s> - <s>^{k+1})
\end{array}
\right]
$$

The matrix is then reduced to echelon form by elementary row operations (the main error in the OCR program was to use column operations), when bases for $<\partial f/\partial s>$, $<s><\partial f/\partial s>$ and $<s>^2<\partial f/\partial s>$ become obvious.

As a concrete example, consider the singularity $f(x,y) = ax^2y + by^2$ which we wish to test for 3-determinacy. By Nakayama's lemma only polynomials of degree up to 4 are required, so that only $<s>^p<\partial f/\partial s>$ with $p = 0, 1, 2, 3$ need be considered. This gives the matrix of non-zero ideal generators shown in Fig. 4. From this matrix it is clear that the singularity is locally 3-determined, but not strongly 3-determined (and hence it is strongly 4-determined), because x^4 cannot be obtained from the bottom two bands alone.

The symbolic constants a and b would not be allowed in the OCRM program itself, but an algebraic computing language like REDUCE allows the same algorithm to be implemented in a considerably more general, useful and friendly way (for reasons described in more detail in [5]), which we have done.

This example illustrates clearly the inefficiency of the data structure, since almost all elements of the matrix are 0. This will frequently be the case, and the lower triangle must *always* be zero. We have also tried an implementation in REDUCE using its standard internal representation for polynomials (a sparse representation) rather than the OCR array representation. Neither version is consistently faster than the other, and we suspect that both are slower than the Algol version (although since they have to be run on different machines we have not made a careful comparison).

The current version of CATFACT does not use this algorithm at all, but rather uses an algorithm based on Siersma's trick, which is well described in

x	y	x^2	xy	y^2	x^3	x^2y	xy^2	y^3	x^4	x^3y	x^2y^2	xy^3	y^4
0	0	0	2a	0	0	0	0	0	0	0	0	0	0
0	2b	a	0	0	0	0	0	0	0	0	0	0	0
0	0	0	0	0	0	2a	0	0	0	0	0	0	0
0	0	0	0	0	0	0	2a	0	0	0	0	0	0
0	0	0	2b	0	a	0	0	0	0	0	0	0	0
0	0	0	0	2b	0	0	a	0	0	0	0	0	0
0	0	0	0	0	0	0	0	0	0	2a	0	0	0
0	0	0	0	0	0	0	0	0	0	0	2a	0	0
0	0	0	0	0	0	0	0	0	0	0	0	2a	0
0	0	0	0	0	0	2b	0	0	a	0	0	0	0
0	0	0	0	0	0	0	2b	0	0	a	0	0	0
0	0	0	0	0	0	0	0	2b	0	0	a	0	0
0	0	0	0	0	0	0	0	0	0	2b	0	0	0
0	0	0	0	0	0	0	0	0	0	0	2b	0	0
0	0	0	0	0	0	0	0	0	0	0	0	2b	0
0	0	0	0	0	0	0	0	0	0	0	0	0	2b

Fig. 4: Matrix of ideal generators

[2]. It amounts to finding a monomial basis for as much of $<\partial f/\partial s>$ as possible, and adding in any remaining polynomials that do not fall into the space spanned by monomials. Since typically only a few basis monomials plus a few additional polynomials need to be stored, this gives a very efficient data structure, although the algorithm required to manipulate it is not quite as straightforward as Gaussian reduction of the matrix. The canonical method for this problem is probably to use a Gröbner basis [6]. CATFACT also attempts to find ordinary determinacy (for which the OCR code contains a hook on which nothing was ever hung) using a conjecture relating nilpotent vector fields to diffeomorphisms.

2.2. The Mapping Problem for Unfoldings

As an introductory example taken from a recent optical application of ECT [7] consider the reduction to normal form of the unfolding

$$\phi(x,y;c_1,c_2) = ax^2y + by^2 + c_1x + c_2y$$

of the singularity considered above, where x, y are state variables, c_1, c_2 are control variables and $a, b \neq 0$ are arbitrary constants. Completing the square in y gives immediately

$$\phi = by'^2 - \frac{a^2}{4b}x^4 - \frac{ac_2}{2b}x^2 + c_1x - \frac{c_2^2}{4b} \quad \text{where} \quad y' = y + \frac{ax^2 + c_2}{2b}.$$

In the reduced form of ϕ the first term is Morse (non-singular) and the remaining terms correspond to the normal form $\pm s^4 + u_2(c) + u_1(c) + \lambda(c)$ for this (cusp) catastrophe, where x still has to be rescaled to map it into s. The

411

"structure map" $u(c)$ shows how the structure described by ϕ maps into the canonical cusp structure. Incidentally, the normal form shows that the singularity has ordinary determinacy 4.

2.2.1. Truncation Theorems

The general approach works with Taylor series, truncated at degrees determined by various *truncation theorems* [8], which can be thought of as generalisations of determinacy. The degree M in the controls c to which the structure map is sought implies the degrees in s and c to which the unfolding $\phi(s,c)$ is required and the degree to which the state-space mapping $s'(s,c)$ must be constructed. From this work has emerged the importance of the *maximum possible degree of canonical unfolding monomials*, or equivalently the maximum degree of terms not in the Jacobian ideal of a singularity, which we denote by κ. This has to lie between $k-2$ and k inclusive [8], where k is the (ordinary) determinacy of the singularity.

The truncation theorem for unfoldings of singularities with vanishing Hessian states that terms in ϕ are required only with homogeneous degree j in s and l in c satisfying

$$j \geq 0, \ M > 0, \ M \geq l \geq 0, \ M(k-1)+\kappa \geq j+l(k-1),$$

where $j+l(k-1)$ constitutes a "generalised degree" with the state and control variables having weightings of 1 and $k-1$ respectively. This theorem generalises one proved by WRIGHT & DANGELMAYR [9] in the special case of catastrophes involving only one state variable, for which $\kappa = k-2$. The theorem and the related reduction algorithm generalise further [8, II] to singularities with non-vanishing Hessian to which the splitting lemma [2] has to be applied.

2.2.2. Reduction Algorithm

As an example, we sketch the reduction algorithm for an unfolding of a singularity with vanishing Hessian. It is assumed that the determinacy k of the unfolding $\phi(s,c)$ has been determined as indicated in §2.1, and that at every stage of the reduction ϕ will be truncated as permitted by the above theorem.

(1) Expand $\phi(s,c)$ as a multivariate Taylor polynomial (the REDUCE code to do this takes no more than half a page including liberal comments).
(2) Reduce the singularity $\phi(s,0)$ to an appropriate normal form $\psi(s) + O(s^{k+2})$ – the method and the reason why the tail should be $O(s^{k+2})$ rather than $O(s^{k+1})$ are explained in §2.2.3 below.
(3) Organise the truncated unfolding into the form

$$\phi(s,c) = \psi(s) + \sum_{j=k+2}^{J_0} \phi_{j,0}(s,c) + \sum_{l=1}^{M} \sum_{j=0}^{J_l} \phi_{j,l}(s,c)$$

where $J_l \equiv (M-l)(k-1) + \kappa$. The k-determinacy of $\psi(s)$ implies that $\langle s \rangle^{k+2} \subseteq \langle s \rangle^2 \langle \partial\psi/\partial s \rangle$, and hence there exist functions $q_\alpha(s)$ of order s^2 such that any function of order s^{k+2}, and in particular $\phi_{k+r,0}$ with $r \geq 2$, can be expressed in the form

$$\phi_{k+r,0}(s,c) = \sum_{\alpha=1}^{n} q_\alpha(s)\frac{\partial\psi}{\partial s_\alpha}(s) + O(s^{k+r+1}).$$

Then the transformation

$$T_{k+r,0} : s = \exp\left(-\sum_{\alpha=1}^{n} q_\alpha(s')\partial/\partial s_\alpha'\right)s'$$

removes the term $\phi_{k+r,0}(s,c)$ from ϕ.

(4) Because the unfolding terms in ϕ belong to $\langle s \rangle / \langle \partial\psi/\partial s \rangle$ there exist functions $q_\alpha(s,c)$ such that all terms $\phi_{j,l}, \ j,l \geq 1$, which are linearly independent of the unfolding monomials $\{\chi_i(s), \ i = 1 \text{ to } K'\}$ where $K' \equiv$ codimension ϕ, can be expressed in the form

$$\phi_{j,l}(s,c) = \sum_{\alpha=1}^{n} q_\alpha(s,c)\frac{\partial\psi}{\partial s_\alpha}(s) + \sum_{i=1}^{K'} u_i(c)\chi_i(s) + O(s^{j+1},c^l).$$

Then the transformation

$$T_{j,l} : s = s' - q(s',c)$$

removes the term $\phi_{j,l}(s,c)$ from ϕ.

The transformations T have to be applied in a suitable order to remove undesired terms from ϕ without introducing new terms with degrees that have already been removed. There is considerable flexibility in this sequence, and various sensible choices are given in [8, 9, 10].

2.2.3. Strong Versus Ordinary Determinacy

Strong equivalence of singularities is equivalence under a diffeomorphism whose differential at the singular point is the identity, so that it has the form $s = s' - q(s')$ where $q(s') = O(s'^2)$. The transformation $T_{k+r,0}$ used in step 3 of the reduction algorithm described above is a strong diffeomorphism. A singularity that has ordinary determinacy k is strongly $(k+1)$-determined, so that an $O(s^{k+2})$ tail can be removed by a strong diffeomorphism. Only if the singularity is also strongly k-determined can an $O(s^{k+1})$ tail be removed by a strong diffeomorphism, and there are singularities, such as E_7, for which the strong determinacy is not the same as the ordinary determinacy. The weak diffeomorphism that must be used in general to remove the terms of degree $k+1$ has a linear part that is not the identity, and cannot be constructed as simply as in step 3 above – it involves *several* applications of that transformation. Alternatively, some *ad hoc* approach could be used, such as substituting a formal power series for the transformation and equating coefficients. (In fact, a complete reduction algorithm can be based on this technique [5].)

2.2.4. An Example

We conclude with a simple example of the output produced by CATFACT for the unfolding

$$ax^5 + xy^3 + x^3 + c_1x + c_2y,$$

where x, y are the state variables, c_1, c_2 are control variables and a is an arbitrary symbolic constant. The *recognition* stage gives:

singularity is E7
corank = 2
determinacy = +4
strong determinacy > determinacy
codimension, excluding constant term = +6
modality = 0

which shows that it is a *partial* unfolding of the singularity E_7 mentioned above. Relative to the normal form

$$xy^3 + x^3 + u_6y^4 + u_5y^3 + u_4y^2 + u_3xy + u_2y + u_1x + u_0$$

the control space (structure) map to quadratic degree, shown essentially in the form in which REDUCE displays its output by default, is:

$$u0 = -829/81 {}^*a^5 {}^*c2^2 - 55/27 {}^*a^4 {}^*c1 {}^*c2 + 1/9 {}^*a^3 {}^*c1^2$$

$$u1 = 513016/729 {}^*a^8 {}^*c2^2 + 60181/243 {}^*a^7 {}^*c1 {}^*c2 + 20/9 {}^*a^6 {}^*c1^2 + a {}^*c2 + c1$$

$$u2 = 9517/81 {}^*a^7 {}^*c2^2 + 10312/243 {}^*a^6 {}^*c1 {}^*c2 + 10/3 {}^*a^5 {}^*c1^2 + c2$$

$$u3 = -60217517/6561 {}^*a^{10} {}^*c2^2 - 28975865/6561 {}^*a^9 {}^*c1 {}^*c2 - 13217/27 {}^*a^8$$

$${}^*c1^2 - 38/9 {}^*a^3 {}^*c2 - a^2 {}^*c1$$

$$u4 = -40175581/19683 {}^*a^9 {}^*c2^2 - 1045742/729 {}^*a^8 {}^*c1 {}^*c2 - 22615/81 {}^*a^7 {}^*c1^2$$

$$- 4/3 {}^*a^2 {}^*c2 - a {}^*c1$$

$$u5 = 1601199340/59049 {}^*a^{11} {}^*c2^2 + 352142545/19683 {}^*a^{10} {}^*c1 {}^*c2$$

$$+ 768997/243 {}^*a^9 {}^*c1^2 + 76/9 {}^*a^4 {}^*c2 + 34/9 {}^*a^3 {}^*c1$$

$$u6 = -3412688773943/11337408 {}^*a^{13} {}^*c2^2 - 35623557988/177147 {}^*a^{12} {}^*c1 {}^*c2$$

$$- 74328349/2187 {}^*a^{11} {}^*c1^2 - 17024/243 {}^*a^6 {}^*c2 - 716/27 {}^*a^5 {}^*c1$$

The actual REDUCE code used to produce this result, which is only about a page long, and further details of the implementation of CATFACT are given in [11].

References

1. D. E. Knuth: *The METAFONT Book* (Addison Wesley, Reading, Massachusetts 1986) p134

2. T. Poston, I. N. Stewart: *Catastrophe Theory and its Applications* (Pitman, London 1978)

3. G. Dangelmayr, F. J. Wright: *Caustics and diffraction from a line source*, Optica Acta 32, *441-62 (1985)*

4. M. V. Berry, C. Upstill: *Catastrophe optics: morphologies of caustics and their diffraction patterns* in *Progress in Optics* Vol. 18, ed. by E. Wolf (North-Holland, Amsterdam 1980) pp257-346

5. K. Millington, F. J. Wright: *Algebraic computations in elementary catastrophe theory*, in *Eurocal'85*, Lecture Notes in Computer Science 204 (Springer, Berlin, Heidelberg 1985) pp116-25

6. D. Armbruster: *Bifurcation theory and computer algebra: an initial approach*, in *Eurocal'85*, Lecture Notes in Computer Science 204 (Springer, Berlin, Heidelberg 1985) pp126-37

7. J. F. Nye, D. R Haws, R. A. Smith: *Use of diffraction gratings with curved lines to study optical catastrophes*, Optica Acta, to appear (1987)

8. R. G. Cowell, F. J. Wright: *Truncation criteria and algorithm for the reduction to normal form of catastrophe unfoldings - I: Singularities with vanishing Hessian; II: Singularities with non-vanishing Hessian*, in preparation

9. F. J. Wright, G. Dangelmayr: *On the exact reduction of a univariate catastrophe to normal form*, J. Phys. A18, *749-64 (1985)*

10. F. J. Wright, G. Dangelmayr: *Explicit iterative algorithms to reduce a univariate catastrophe to normal form*, Computing 35, *73-83 (1985)*

11. R. G. Cowell, F. J. Wright: *CATFACT: Computer Algebraic Tools For Applications of Catastrophe Theory*, submitted for EUROCAL'87

Computer Algebra Programs
for Dynamical Systems Theory

D. Armbruster

Mathematical Sciences Institute, Cornell University,
Ithaca, NY 14853, USA and
Institut für Informationsverarbeitung, Universität Tübingen,
D-7400 Tübingen, Fed. Rep. of Germany

1. Introduction

This is a description of the current state of that part of the program outlined in [1], that is concerned with the development of Computer Algebra tools for the analysis of dynamical systems. At the moment there is only one program generally accessible (i.e. published) which fits into this context. It deals with nonlinear near identity transformations [4] and can thus be used to determine center manifolds and normal forms. We present here two additional programs in MACSYMA which do averaging for nonautonomous o.d.e's and which do a Lyapunov - Schmidt reduction for a steady state bifurcation, respectively. Further programs implementing various perturbation methods, Lie-series and a more general Lyapunov - Schmidt reduction which allows to treat some classes of p.d.e's will be published in [3].

2. Averaging

For an introduction to this subject consult [5]. We assume that we have a n-dimensional nonautonomous system of ordinary differential equations of the form

$$\dot{x} = \varepsilon f(x,t) \qquad \varepsilon << 1 \tag{1}$$

where the time t is scaled in such a way that f is 2π periodic in time. The averaging theorem then states that there exists a near identity change of coordinates $x = y + \varepsilon w(y,t)$ such that

$$\dot{y} = \varepsilon \overline{f}_1(y) + \varepsilon^2 f_1(y,t,\varepsilon) \tag{2}$$

where $\overline{f}_1$ is the mean $\frac{1}{2\pi}\int_0^{2\pi} f(y,t,0)$ dt. Then solutions of (1) and solutions of (2) stay close, of $0(\varepsilon)$, to each other for a time of $0(\frac{1}{\varepsilon})$. Basically one can prove that the qualitative features of (2) are the 2π periodic qualitative features of (1). Hence averaging is a method to transform a nonautonomous system of o.d.e.'s into an autonomous system. Our program allows to average n-dimensional o.d.e.'s to arbitrary order. I.e. for mth order averaging we calculate the transformation

$$x = y + \varepsilon w_1(y,t) + \varepsilon^2 w_2(y,t) + \ldots + \varepsilon^{m-1} w_{m-1}(y,t)$$

such that (1) becomes

$$\dot{y} = \varepsilon \overline{f}_1(y) + \varepsilon^2 \overline{f}_2(y) + \ldots + \varepsilon^m \overline{f}_m(y) \, . \tag{3}$$

Let us demonstrate the program with the van der Pol oscillator in polar coordinates (x_1, x_2) in the form given in [5] .

```
(C13)  TT:1-X1^2*COS(T+X2)^2;
Time= 2350.0 msecs.
                                    2      2
(D13)                          1 - X1   COS (X2 + T)

(C14)  EQ1:TT*X1*SIN(T+X2)^2;
Time= 150.0 msecs.
                            2      2              2
(D14)              X1 (1 - X1   COS (X2 + T)) SIN (X2 + T)

(C15)  EQ2:TT*SIN(T+X2)*COS(T+X2);
Time= 117.0 msecs.
                                    2      2
(D15)          COS(X2 + T) (1 - X1   COS (X2 + T)) SIN(X2 + T)

(C16)  SETUP();
number of differential equations
2;

enter the ordinary differential equations, use x1..xn as variables
use t as the time variable
scale time such that you average over 2* %pi
rhs( 1 )=eps*...
EQ1;
rhs( 2 )=eps*...
EQ2;
Time= 12000.0 msecs.
                                    [ dW1    dW1 ]
                                    [ ---    --- ]
                                    [ dY1    dY2 ]
(D16)                               [            ]
                                    [ dW2    dW2 ]
                                    [ ---    --- ]
                                    [ dY1    dY2 ]

(C17)  AVERAGE();
to which order
1;
Time= 16900.0 msecs.
                                            3
                            EPS X1    EPS X1
(D17)                      [------ - -------, 0]
                              2         8

(C18)  AVERAGE();
to which order
2;
```

```
            3
    EPS Y1  SIN(4 Y2 + 4 T) - 8 EPS Y1 SIN(2 Y2 + 2 T) + 32 Y1
  [----------------------------------------------------------,
                                32


                2                               2
        EPS Y1  COS(4 Y2 + 4 T) + (4 EPS Y1  - 8 EPS) COS(2 Y2 + 2 T) + 32 Y2
        ------------------------------------------------------------------]
                                        32
Time= 80500.0 msecs.
                        3        2    4         2   2        2
             EPS X1  EPS X1    11 EPS   X1   3 EPS   X1    EPS
  (D18)      [------ - -------, - ----------- + ---------- - ----]
                2        8          256            16        8
```

The differential equations that we consider are $\dot{x}_1 = EQ1$ *and* $\dot{x}_2 = EQ2$. The average to first order (line D17) gives us the familiar limit cycle of the van der Pol oscillator at $x_1 = 2$. Averaging to second order (line D18) gives information about the behaviour of the phase x_2. The intermediate output on line C18 is the near identity transformation $(x_1, x_2) \rightarrow (y_1, y_2) + \varepsilon(w_{11}, w_{12})$. The following is a MACSYMA listing of the functions necessary to perform mth order averaging:

```
/*this is a set of programs to perform mth. order averaging on an n-dimensional
system of nonautonomous o.d.e's. To  operate it type SETUP() and enter the
input information, then enter AVERAGE() and average to the desired order*/
SETUP():=(
        N:READ("NUMBER OF DIFFERENTIAL EQUATIONS"),
        PRINT("ENTER THE ORDINARY DIFFERENTIAL EQUATIONS, USE X1..XN AS VARIABLES"),
        PRINT("USE T AS THE TIME VARIABLE"),
        PRINT("SCALE TIME SUCH THAT YOU AVERAGE OVER 2* %PI"),
        INPUT:MAKELIST(READ("RHS(",I,")=EPS*...")*EPS,I,1,N),
        F:EXPAND(EXPONENTIALIZE(INPUT)),
        XOLD:VECMAKE(X,N),
        XNEW:VECMAKE(Y,N),
        WT:VECMAKE(W,N),
        DEPENDS(WT,CONS(T,XNEW)),
        TRAFO:XNEW,
        XSUB:MAPLIST("=",XNEW,XOLD),
        WNULL:VFUN(WT,0),
        JACOB[I,J] := DIFF(WT[I],XNEW[J]),
        JAC:GENMATRIX(JACOB,N,N))$
AVERAGE():=(
        M:READ("TO WHICH ORDER"),
        IF M > 1 THEN (
                FOR I :1 THRU M-1 DO (
                        TEMP1:MAPLIST("=",XOLD,XNEW+EPS^I*WT),
                        TEMP2:TAYLOR(SUM((-EPS)^(I*J)*JAC^^J,J,0,M-1).
                                (EV(F,TEMP1) - DIFF(WT,Z)*EPS^I),EPS,0,M),
                        TEMP3:MATTOLIST(TEMP2,N),
                        TEMP4:MAP(EXPAND,TAYLOR(EV(TEMP3,WNULL,DIFF),EPS,0,I)),
                        MEAN:APPLY1(TEMP4,EXPO),/*the ith order mean*/
                        TEMP6:EXPAND(TEMP4-MEAN),
                        TEMP7:EV(INTEGRATE(TEMP6,T),EPS=1),
                        TEMP8:MAPLIST("=",WT,TEMP7),
                        TEMP9:RATSIMP(TEMP8),
                        F:EXPAND(EV(TEMP3,TEMP9,DIFF,XSUB,INFEVAL)),/*the transformed d.e.*/
                        TRAFO:EXPAND(TRAFO+EV(EPS^I*WT,TEMP9))),
                PRINT(RATSIMP(DEMOIVRE(TRAFO))),
                APPLY1(F,EXPO))/*this is the final averaging*/
        ELSE (/*this is first order averaging*/
                TM1:TAYLOR(F,EPS,0,1),
                TM2:MAP(EXPAND,TM1),
                DEMOIVRE(APPLY1(TM2,EXPO)))))$
```

418

```
VFUN(LIST,VALUE):=MAP(LAMBDA([U],U=VALUE),LIST)$
VECMAKE(NAME,ORD):=MAKELIST(CONCAT(NAME,I),I,1,ORD)$
MATTOLIST(MAT,DIM):=IF DIM>1 THEN MAKELIST(MAT[I,1],I,1,DIM) ELSE [MAT]$

IN_T(EXP):= NOT FREEOF(T,EXP)$
MATCHDECLARE(Q,IN_T)$
DEFRULE(EXPO,EXP(Q),0)$
```

SETUP organizes the input and transforms any trigonometric function into exponential form. First order averaging is then nearly trivial: We perform a Taylor expansion to order ε, expand the whole expression and set every exponential term to zero. The remaining is the first order mean. Higher order averaging is a bit more complicated: Let us consider second order averaging. We write $x = y + \varepsilon wt$ in TEMP1 and insert this into the differential equation $\dot{x} = \varepsilon f(x,t)$. This gives

$$\dot{y} = \varepsilon[I + \varepsilon D_y wt]^{-1}[f(y + \varepsilon wt) - \frac{\partial wt}{\partial t}] \tag{4a}$$

$$= \varepsilon[I - \varepsilon D_y wt][f(y + \varepsilon wt) - \frac{\partial wt}{\partial t}] \tag{4b}$$

$$= \varepsilon[f(y + \varepsilon wt) - \frac{\partial wt}{\partial t}] + 0(\varepsilon^2). \tag{4c}$$

This calculation is performed in TEMP2. Then we determine the mean as before and find $w = \int f(y + \varepsilon wt)$ - mean dt in TEMP7. Inserting the so found transformation into (4b) and calculating the mean again gives us the second order averaged equations.

3. The Lyapunov - Schmidt Reduction for o.d.e.'s:

For a comprehensive introduction see [2]. Let

$$\dot{y} = f(y, \lambda) \tag{5}$$

be a n-dimensional system of o.d.e.'s with a bifurcation parameter λ. Assume that at $\lambda = \lambda_c := 0, f(0,0)$ is zero and hence $y = 0$ is a stationary solution of (5). Let $L = D_y f(0,0)$ be the Jacobian of f with respect to y at $y = 0, \lambda = \lambda_c$. Our program deals only with nondegenerate steady state bifurcations hence we require that rank $L = n - 1$. Then one can reduce the n-dimensional algebraic system

$$f(y, \lambda) = 0 \tag{6}$$

to a one-dimensional equation $g(A, \lambda) = 0$ whose solutions are locally in one to one correspondence with the steady state solutions of (5). Let us call

$CFUN$ the zero eigenvector for the Jacobian L, i.e. $L\ CFUN = 0$. Since we are dealing with a system of o.d.e.'s we can identify cokernel and range of L and split $\mathbf{R}$ into $CFUN + $ range L. We project onto the kernel of L via the inner product with the adjoint eigenvector $AFUN$ (i.e. $L^*AFUN = 0$) and thus have an equivalent system for (6):

$$< AFUN, f(y, \lambda) > = 0 \qquad (7a)$$

$$f(y, \lambda) - \frac{< AFUN, f(y, \lambda) >}{< AFUN, CFUN >} CFUN = 0 \ . \qquad (7b)$$

Eq (7b) is now, locally near $y = 0$ and $\lambda = \lambda_c$, a nonsingular algebraic system of $(n-1)$ variables which can be solved. Its solution is then inserted into (7a) which becomes the bifurcation equation, i.e. an algebraic equation depending only on the amplitude A of the critical eigenvector and the bifurcation parameter λ .

Let us write $y = A\ CFUN + w(A, \lambda)$ where $Lw = w$ such that $< AFUN, w > = 0$ and hence w comprises the $n - 1$ components of range L. We split $f(y, \lambda)$ into $Ly + \tilde{f}(y, \lambda)$ and write (7) as

$$< AFUN, \tilde{f}(A\ AFUN + w, \lambda) > =: g(A, \lambda) \qquad (8a)$$

$$w + \tilde{f}(A\ AFUN + w, \lambda) - \frac{< AFUN, \tilde{f}(A\ AFUN + w, \lambda) >}{< AFUN, CFUN >} CFUN = 0. \qquad (8b)$$

Since $\tilde{f}$ as a function of w and λ is at least quadratic, w can recursively be calculated as a polynomial in A and λ up to any desired order. If we assume w of $O(A^n)$ and insert it into (8a) then we can determine the bifurcation equation $g(A, \lambda)$ up to $O(A^{n+1})$. Now let us demonstrate the program with a nontrivial example:

```
(C14)  R1DOT;
                                       2         2
(D14)            R1 (R2 COS(THETA) + DD1 R2  + CC1 R1  + A1)

(C15)  R2DOT;
                           2         2         2
(D15)            R2 (DD2 R2  + CC2 R1  + A2) - R1  COS(THETA)

(C16)  THETADOT;
                                2
                              R1
(D16)                     - (--- - 2 R2) SIN(THETA)
                              R2

(C17)  RSUB;
```

420

$$\text{(D17)} \quad [R10 = -\;\frac{SQRT(2)}{DD2 + 2\;CC1}, \quad R20 = -\;\frac{1}{DD2 + 2\;CC1}, \quad A1 = \frac{DD2 - DD1}{(DD2 + 2\;CC1)^2},$$

$$A2 = -\;\frac{3\;DD2 + 2\;CC2 + 4\;CC1}{(DD2 + 2\;CC1)^2}]$$

(C18) RHO1DOT:RATSIMP(TAYLOR(EV(R1DOT,R1 = RHO1+R10,R2 = RHO2+R20),[RHO1,RHO2,THETA],0,3));

$$\text{(D18)} \quad - ((R10\;RHO2 + R20\;RHO1 + R10\;R20)\;THETA^2 + (- 2\;DD1\;RHO1 - 2\;DD1\;R10)\;RHO2^2$$

$$+ ((- 4\;DD1\;R20 - 2)\;RHO1 - 4\;DD1\;R10\;R20 - 2\;R10)\;RHO2 - 2\;CC1\;RHO1^3 - 6\;CC1\;R10\;RHO1^2$$

$$+ (- 2\;DD1\;R20^2 - 2\;R20^2 - 6\;CC1\;R10^2 - 2\;A1)\;RHO1 - 2\;DD1\;R10\;R20^2 - 2\;R10\;R20^2$$

$$- 2\;CC1\;R10^3 - 2\;A1\;R10)/2$$

(C19) RHO2DOT:RATSIMP(TAYLOR(EV(R2DOT,R1 = RHO1+R10,R2 = RHO2+R20),[RHO1,RHO2,THETA],0,3));

$$\text{(D19)} \quad ((2\;R10\;RHO1 + R10^2)\;THETA^2 + 2\;DD2\;RHO2^3 + 6\;DD2\;R20\;RHO2^2$$

$$+ (2\;CC2\;RHO1^2 + 4\;CC2\;R10\;RHO1 + 6\;DD2\;R20^2 + 2\;CC2\;R10^2 + 2\;A2)\;RHO2^2$$

$$+ (2\;CC2\;R20 - 2)\;RHO1^2 + (4\;CC2\;R10\;R20 - 4\;R10)\;RHO1 + 2\;DD2\;R20^3$$

$$+ (2\;CC2\;R10^2 + 2\;A2)\;R20 - 2\;R10^2)/2$$

(C20) THETANEWDOT:RATSIMP(TAYLOR(EV(THETADOT,R1 = RHO1+R10,R2 = RHO2+R20),[RHO1,RHO2,THETA],0,3));

$$\text{(D20)} \quad - ((2\;R20^4 - R10^2\;R20^2)\;THETA^2 + (6\;R10^3\;RHO2$$

$$+ (- 12\;R10\;R20\;RHO1 - 12\;R20^2 - 6\;R10^2\;R20)\;RHO2 + 6\;R20^2\;RHO1^2 + 12\;R10\;R20^2\;RHO1$$

$$- 12\;R20^4 + 6\;R10^2\;R20^2)\;THETA)/(6\;R20^3)$$

```
(C23) SETUP();
NUMBER OF EQUATIONS
3;

ENTER VARIABLE NUMBER 1
RHO1;
ENTER VARIABLE NUMBER 2
RHO2;
ENTER VARIABLE NUMBER 3
THETA;
ENTER THE CRITICAL EIGENVECTOR AS A LIST
[0,0,1];
ENTER THE ADJOINT CRITICAL EIGENVECTOR
CFUN;

ENTER THE DIFFERENTIAL EQUATION
DIFF( RHO1 ,T)=
RHO1DOT;
DIFF( RHO2 ,T)=
RHO2DOT;
DIFF( THETA ,T)=
THETANEWDOT;
Time= 35300.0 msecs.
```

$$\text{(D23)} \qquad\qquad\qquad\qquad\qquad 1$$

```
(C24) LS();
WHICH COEFFICIENT IN THE BIFURCATION EQ DO YOU WANT TO DETERMINE?
AMP^I*LAM^J,ENTER I
1;
ENTER J
0;
Time= 4870.0 msecs.
```

$$\text{(D24)} \qquad\qquad\qquad\qquad \frac{2\;R20^2 - R10^2}{R20}$$

421

```
(C25)  EV(%,RSUB);
Time= 133.0 msecs.
(D25)                                              0


(C27)  WINV()$
WHICH COEFFICIENT IN W(AMP,LAM) DO YOU WANT TO DETERMINE?
AMP^I*LAM^J,ENTER I
2;
ENTER J
0;
Time= 15100.0 msecs.

(C28)  EV(%,RSUB,RATSIMP);
Time= 7000.0 msecs.
                                          2
         dW1       dW2       dW3       d W1
(D28)  [---- = 0,  ---- = 0,  ---- = 0,  ----- =
         dAMP      dAMP      dAMP         2
                                       dAMP


                  SQRT(2) DD2 - 2 SQRT(2) DD1 + 4 SQRT(2) CC1
 - -----------------------------------------------------------------------,

      2
   2 DD2  + (- 4 DD1 + 2 CC2 + 8 CC1) DD2 + (- 4 CC2 - 8 CC1) DD1 + 4 CC1 CC2

   2
  d W2                           DD2 + CC2 + 4 CC1
 ----- = - -----------------------------------------------------------------,
     2        2
 dAMP       DD2  + (- 2 DD1 + CC2 + 4 CC1) DD2 + (- 2 CC2 - 4 CC1) DD1 + 2 CC1 CC2

    2
   d W3
 ----- = 0]
    2
 dAMP


(C29)  LS()$
WHICH COEFFICIENT IN THE BIFURCATION EQ DO YOU WANT TO DETERMINE?
AMP^I*LAM^J,ENTER I
3;
ENTER J
0;
Time= 31300.0 msecs.

(C30)  EV(%,RSUB,RATSIMP);
Time= 10200.0 msecs.
                         6 DD2 + 12 DD1 + 12 CC2 + 24 CC1
(D30)    - -------------------------------------------------------------------
             2
           DD2  + (- 2 DD1 + CC2 + 4 CC1) DD2 + (- 2 CC2 - 4 CC1) DD1 + 2 CC1 CC2
```

We consider a bifurcation from the three equations R1DOT, R2DOT, THETADOT given in line D14-D16 at the point fixed by RSUB (line D17). Here we are bifurcating from a nontrivial solution branch. In fact a thorough discussion of this bifurcation would reveal that it is a degenerate bifurcation involving the interaction of a Hopf and a steady state bifurcation. We could choose A_1 or A_2 as bifurcation parameter λ and get the steady state component of the bifurcation equation in the form $g(A, \lambda) = c_1 A^3 + c_2 A\lambda$. For this demonstration we only calculated the coefficient c_1 . The output line D25 shows that we are indeed at a bifurcation point since $\frac{dg}{dA}$ evaluated at the point given by RSUB is zero. In order to determine the cubic coefficient in g we need the corresponding quadratic term in w. This calculation is performed by calling WINV in D28 and the coefficient c_1 is given in D30.

The following is a listing of the functions necessary to perform the reduction. SETUP reads the input to create a list EQS of equations and sets the unknowns $Y = AMP\ CFUN + W(AMP, LAM)$. LS inserts Y into EQS and asks which Taylor coefficient of the bifurcation equation $g(AMP, LAM) = 0$ it should calculate. This is then done by projecting onto the critical eigenvector $CFUN$ and exploiting the fact that w is in range L. If there is not yet enough known about the corresponding Taylor expansion of $W(AMP, LAM)$, which can be calculated via the function WINV and is stored in the list DATABASE, then LS will print an error message.

```
/*these functions allow to determine any particluar Taylor coefficient of the bifurcation
equation G(AMP,LAM) = 0 */

SETUP():=(/*the function SETUP needs the dimension N of the system of o.d.e., the
            n-dimensional list of variables Y, the system of o.d.e's called EQS, the
            critical eigenvector CFUN and the adjoint critical eigenvector AFUN.*/
          N:READ("NUMBER OF EQUATIONS"),
          Y:MAKELIST(READ("ENTER VARIABLE NUMBER",I),I,1,N),
          CFUN:READ("ENTER THE CRITICAL EIGENVECTOR AS A LIST"),
          AFUN:READ("ENTER THE ADJOINT CRITICAL EIGENVECTOR"),
          PRINT("ENTER THE DIFFERENTIAL EQUATION"),
          EQS:MAKELIST(READ("DIFF(",Y[I],",T)="),I,1,N),
          W:MAKELIST(CONCAT(W,I),I,1,N),
          ZWLIST:MAKELIST(CONCAT(ZW,I),I,1,N),
          DEPENDS(APPEND(ZWLIST,W),[AMP,LAM]),
          WNULL:VFUN(W,0),
          SUB:MAPLIST("=",Y,AMP*CFUN+W),
          DATABASE:MAP(LAMBDA([U],DIFF(U,AMP)=0),W),
          ZWNULL:VFUN(ZWLIST,0),
          NORM:INNER_PRODUCT(AFUN,CFUN));


LS():=BLOCK(/*Provided all  necessary knowledge about the Taylor coefficients of W(AMP,LAM)
            is stored in the list DATABASE, LS calculates any specific Taylor coefficient
            of the bifurcation equation G(AMP,LAM)= 0 via the projection onto CFUN.*/
          PRINT("WHICH COEFFICIENT IN THE BIFURCATION EQ DO YOU WANT TO DETERMINE?"),
          I:READ("AMP^I*LAM^J, ENTER I"),J:READ("ENTER J"),
          WMAX_DIFF:MAP(LAMBDA([U],'DIFF(U,AMP,I,LAM,J) = 0),W),
          TEMP1:EV(EQS,SUB,DIFF),
          TEMP2:DIFF(TEMP1,AMP,I,LAM,J),
          TEMP2:SUBST(WMAX_DIFF,TEMP2),
          TEMP3:SUBST(DATABASE,TEMP2),
          IF FREEOF(PART(DIFF(QQ(XY),XY),0),TEMP3) THEN
                           TEMP4:EXPAND(EV(TEMP3,AMP=0,LAM=CLAM,WNULL))
                     ELSE
                           (PRINT(TEMP3),
                           RETURN(PRINT("GET LOWER ORDERS FIRST AND CALL WINV()"))),
          BIFEQ:RATSIMP(TEMP5:INNER_PRODUCT(TEMP4,AFUN)));


WINV():=BLOCK(/*The function WINV allows to calculate iteratively, i.e. starting with the
            lowest order, all Taylor coefficients of W(AMP,LAM) by projecting onto
            the complement of CFUN*/
          PRINT("WHICH COEFFICIENT IN W(AMP,LAM) DO YOU WANT TO DETERMINE?"),
          I:READ("AMP^I*LAM^J, ENTER I"),J:READ("ENTER J"),
          WMAX_DIFF:MAP(LAMBDA([U],'DIFF(U,AMP,I,LAM,J) = CONCAT(Z,U)),W),
          TEMP2:DIFF(TEMP1,AMP,I,LAM,J),
          TEMP2:SUBST(WMAX_DIFF,TEMP2),
          TEMP3:SUBST(DATABASE,TEMP2),
```

```
IF ERRCATCH(TEMP4:EV(TEMP3,AMP=0,LAM=CLAM,WNULL))=[] THEN
                    (PRINT(TEMP3),
                    RETURN(PRINT("GET LOWER ORDERS FIRST")))
            ELSE
                    TEMP5:INNER_PRODUCT(EV(TEMP4,ZWNULL),AFUN),
    TEMP6:TEMP4-TEMP5/NORM*CFUN,
    TEMP7:SOLVE(TEMP6,ZWLIST),
    TEMP8:EV(ZWLIST,TEMP7),
    TEMP9:INNER_PRODUCT(TEMP8,AFUN),
    IF RATSIMP(TEMP9)=0 THEN
                    ADDBASE:MAP("=",MAKELIST('DIFF(W[U],AMP,I,LAM,J),U,1,N),TEMP8)
            ELSE
                    (TEMP10:RATSIMP(TEMP8- TEMP9/NORM*CFUN),
                    ADDBASE:MAP("=",MAKELIST('DIFF(W[U],AMP,I,LAM,J),U,1,N),TEMP10)),
    DATABASE:APPEND(DATABASE,ADDBASE));


INNER_PRODUCT(L1,L2):=SUM(L1[I]*L2[I],I,1,N);

VFUN(LIST,VALUE):=MAP(LAMBDA([U],U=VALUE),LIST);
```

The following theoretical results have gone into the program: (i) It can be shown that $\frac{dw}{dA}(0,0) \equiv 0$. Therefore we make this the first entry into the list DATABASE. (ii) If we are interested in the Taylor coefficient of g of the form $A^i\lambda^j$ then the term $\frac{d^{i+j}w}{dA^i d\lambda^j}$ is annihilated by the projection onto the kernel of L and we therefore set this term to zero at the earliest possibility, here at TEMP2 of LS. (iii) The same term defines the unknowns in WINV when one wants to determine the Taylor coefficients of w of the form $A^i\lambda^j$. Hence they are set to ZW1...ZWN at TEMP2 of WINV.

Our experience with this program is that it is usually much faster to calculate the bifurcation equation via the Lyapunov - Schmidt reduction than to perform a center manifold reduction. This is obvious if one recalls that the center manifold is found via a near identity transformation of a n-dimensional problem. However since we finally are interested only in one dimension we perform much more calculations than really necessary leading to untolerable response times. Let us mention that the computation times of the demonstration runs apply for a Symbolics 3670.

Acknowledgments: This work was supported by the NSF (Grant 85-09481),the U.S. Army Research Office and by a travel grant from DFG.

References:

1. Armbruster, D.: (1985) *Bifurcation theory and computer algebra: an initial approach.* In: (Caviness, B.F., ed.) Proceedings of EUROCAL 85 Vol **II**, Springer Lec. Notes Comp. Sci. 204, 126-137.
2. Golubitsky, M., Schaeffer, D.: (1985) *Singularities and groups in bifurcation theory* **I**, Springer

3. Rand, R.H., Armbruster, D.: *Perturbation Methods, bifurcation theory, and computer algebra*, Springer to appear 1987

4. Rand, R.H., Keith, W.L.: 1985 *Normal form and center manifold calculations on MACSYMA*. In: (Pavelle,R. ed.) Applications of computer algebra, Kluwer, 309-328.

5. Sanders, J.A., Verhulst,F.: (1985) *Averaging methods in nonlinear dynamical systems*, Springer

Index of Contributors

L. Rensing, U. an der Heiden, M. C. Mackey (Eds.)

Temporal Disorder in Human Oscillatory Systems

Proceedings of an International Symposium, University of Bremen, 8–13 September 1986

1987. 133 figures. Approx. 270 pages. (Springer Series in Synergetics, Volume 36). ISBN 3-540-17765-5

Contents: Temporal Disorder. – Introductory Remarks. – Theoretical Concepts. – Neural and Neuromotor Systems. – Heart and Respiration. – Circadian Clocks. – Ovarian Cycles. – Index of Contributors.

F. T. Arecchi, R. G. Harrison (Eds.)

Instabilities and Chaos in Quantum Optics

1987. 135 figures. XIII, 253 pages. (Springer Series in Synergetics, Volume 34). ISBN 3-540-17282-3

Of the variety of nonlinear dynamical systems that exhibit deterministic chaos optical systems both lasers and passive devices provide nearly ideal systems for quantitative investigation due to their simplicity both in construction and in the mathematics that describes them. In view of their growing technical application the understanding, control and possible exploitation of sources of instability in these systems has considerable practical importance.

The aim of this volume is to provide a comprehensive coverage of the current understanding of optical instabilities through a series of reviews by leading researchers in the field. The book comprises nine chapters, five on active (laser) systems and four on passive optically bistable systems.

Instabilities and chaos in single- (and multi-) mode lasers with homogeneously and broadened gain media are presented and the influence of an injected signal, loss modulation and also feedback of laser output on this behaviour is treated. Both electrically excited and optically pumped gas lasers are considered, and an analysis of dynamical instabilities in the emission from free electron lasers are presented.

Instabilities in passive optically bistable systems include a detailed analysis of the global bifurcations and chaos in which transverse effects are accounted for. Experimental verification of degenerative pulsations and chaos in intrinsic bistable systems is described for various optical feedback systems in which atomic and molecular gases and semiconductors are used as the nonlinear media. Results for a hybrid bistable optical system are significant in providing an important test of current understanding of the dynamical behaviour of passive bistable systems.

Springer-Verlag
Berlin Heidelberg New York
London Paris Tokyo

E. Schöll

Nonequilibrium Phase Transitions in Semiconductors

Self-Organization Induced by Generation and Recombination Processes

1987. 165 figures. Approx. 330 pages. (Springer Series in Synergetics, Volume 35). ISBN 3-540-17582-2

Semiconductors can exhibit electrical instabilities like current runaway, threshold switching, current filamentation, or oscillations, when they are driven far from thermodynamic equilibrium. This book presents a coherent theoretical description of such cooperative phenomena induced by generation and recombination processes of charge carriers in semiconductors. It goes beyond older work by taking into account the nonlinearities of these generation-recombination processes and the dynamic degrees of freedom of free and trapped carriers, and by developing and applying theoretical concepts and methods from nonequilibrium thermodynamics. The theory is illustrated by recent experimental findings, in particular in connection with current filamentation and chaotic oscillations. The book might be useful both to applied scientists and to theoretical physicists.

Springer-Verlag
Berlin Heidelberg New York
London Paris Tokyo

Springer